AF841373

Concrete Permeability and Durability Performance

Modern Concrete Technology Series

A series of books presenting the state-of-the-art in concrete technology.

Series Editors

Arnon Bentur
National Building Research Institute
Faculty of Civil and Environmental Engineering
Technion-Israel Institute of Technology
Technion City, Haifa 32000
Israel

Sidney Mindess
Department of Civil Engineering
University of British Columbia
6250 Applied Science Lane
Vancouver, B.C. V6T 1Z4
Canada

17. Sustainability of Concrete
P. C. Aïtcin and S. Mindess

18. Concrete Surface Engineering
B Bissonnette, L Courard and A Garbacz

19. Textile Reinforced Concrete
A. Peled, A Bentur and B Mobasher

20. Durability of Concrete: Design and Construction
M.G. Alexander, A. Bentur and S Mindess

21. Ultra High Performance FRCs
E. Denarié

22. Shotcrete: Materials, Performance and Use
M. Jolin and D.R. Morgan

23. Concrete Permeability and Durability Performance: From Theory to Field Applications
R.J. Torrent, R.D. Neves and K. Imamoto

For more information about this series, please visit: https://www.routledge.com/series-title/book-series/MCT

Concrete Permeability and Durability Performance

From Theory to Field Applications

Roberto J. Torrent
Rui D. Neves
Kei-ichi Imamoto

CRC Press
Taylor & Francis Group
Boca Raton London New York

CRC Press is an imprint of the
Taylor & Francis Group, an **informa** business

First edition published 2022
by CRC Press
6000 Broken Sound Parkway NW, Suite 300, Boca Raton, FL 33487-2742

and by CRC Press
2 Park Square, Milton Park, Abingdon, Oxon, OX14 4RN

© 2022 Taylor & Francis Group, LLC

CRC Press is an imprint of Taylor & Francis Group, LLC

Reasonable efforts have been made to publish reliable data and information, but the authors and publisher cannot assume responsibility for the validity of all materials or the consequences of their use. The authors and publishers have attempted to trace the copyright holders of all material reproduced in this publication and apologize to copyright holders if permission to publish in this form has not been obtained. If any copyright material has not been acknowledged please write and let us know so we may rectify in any future reprint.

Except as permitted under U.S. Copyright Law, no part of this book may be reprinted, reproduced, transmitted, or utilized in any form by any electronic, mechanical, or other means, now known or hereafter invented, including photocopying, microfilming, and recording, or in any information storage or retrieval system, without written permission from the publishers.

For permission to photocopy or use material electronically from this work, access www.copyright.com or contact the Copyright Clearance Center, Inc. (CCC), 222 Rosewood Drive, Danvers, MA 01923, 978-750-8400. For works that are not available on CCC please contact mpkbookspermissions@tandf.co.uk

Trademark notice: Product or corporate names may be trademarks or registered trademarks and are used only for identification and explanation without intent to infringe.

Library of Congress Cataloging-in-Publication Data
Names: Torrent, R., author. | Neves, Rui D., author. | Imamoto, Keiichi, 1966- author.
Title: Concrete permeability and durability performance : from theory to field applications / Roberto J. Torrent, Rui D. Neves, Kei-ichi Imamoto.
Description: First edition. | Boca Raton, FL : CRC Press, 2022. | Series: Modern concrete technology, 1746-2959 ; 21 | Includes bibliographical references and index.
Identifiers: LCCN 2021029833 (print) | LCCN 2021029834 (ebook) |
ISBN 9781138584884 (hbk) | ISBN 9781032039701 (pbk) | ISBN 9780429505652 (ebk)
Subjects: LCSH: Concrete—Testing. | Concrete—Permeability. | Concrete—Deterioration. | Concrete—Service life.
Classification: LCC TA440 .T57 2022 (print) | LCC TA440 (ebook) | DDC 620.1/360287—dc23
LC record available at https://lccn.loc.gov/2021029833
LC ebook record available at https://lccn.loc.gov/2021029834

ISBN: 978-1-138-58488-4 (hbk)
ISBN: 978-1-032-03970-1 (pbk)
ISBN: 978-0-429-50565-2 (ebk)

DOI: 10.1201/9780429505652

Typeset in Sabon
by codeMantra

Contents

Foreword

Concrete as such is a very durable material. There are magnificent examples of concrete structures which have survived 2,000 years without substantial repair measures, and they will survive hundreds of years to come. The Pantheon in Rome and numerous bridges built during the Roman Empire in Italy and Spain that served up to 2,000 years are well-known examples.

Since the large-scale application of reinforced concrete, the construction industry has experienced enormous challenges with respect to achieving the designed service life of concrete structures. According to most standards, reinforced concrete structures are expected to have a service life of at least 100 years. In reality, however, expensive repair and renovation are frequently necessary after not more than 30 years. In recent years, a number of bridges collapsed after less than 50 years. It is estimated that repair of a damaged bridge costs approximately six to eight times more than that of the construction of a new bridge. During repair operation or reconstruction process, the necessary deviation of traffic alone causes additional financial and environmental burdens. Therefore, one major subject in concrete technology research has been to increase repair-free service life of reinforced concrete structures.

Concrete is a porous material with a wide-range distribution of the size of pores, running from a few millimetres down to nanometres. The surface of concrete structures is usually in contact with changing climatic conditions. During a wet period, rain water will be absorbed by capillary action and in humid environment by capillary condensation. The micropores remain water-filled even during dry periods. The humidity in the pore system will initiate corrosion of the steel reinforcement as soon as the carbonation depth exceeds the cover thickness. Another disadvantage of reinforced concrete elements exposed to natural environment is the crack formation due to bending or temperature and humidity gradients. These cracks are preferential pathways for locally deep carbonation and hence early beginning of corrosion of reinforcement.

Service life of reinforced concrete structures depends essentially on the cover thickness and on the permeability of the concrete cover. It is comparatively easy to determine the thickness of the concrete cover. Permeability of the cover, however, is a more complex property. Based on the research findings and experience from practice, the present volume presents various topics related to permeability of concrete. A method to determine permeability of concrete is described in detail, and many possible applications in practice are discussed here. It can be expected that this volume will contribute to our knowledge on how to increase service life of reinforced concrete structures, and the discussions on durability and service life will certainly bring broader awareness of the implications of permeability of concrete.

Durability and service life of reinforced concrete structures, however, do not depend on one dominating parameter. This was shown in a convincing way in a recent publication of RILEM Technical Committee (RILEM TC 246-TDC). Results of this Technical Committee have clearly demonstrated that durability depends on the combination of environmental actions and mechanical load. This volume is an excellent basis for a better understanding of dominating processes which may substantially reduce service life of concrete structures and of steel reinforced concrete structures.

Prof. Dr. Dr. h.c. Folker H. Wittmann

Preface

The genesis of this book originated on August 29, 2016, with a proposal of Prof. Neves to Dr. Torrent on the possibility of writing jointly a book about the permeability of concrete. After some consideration, the proposal was accepted, ending in a first draft of its possible content. Then, finding a suitable interested publisher was required. Believe it or not, on June 1, 2017, an invitation by Tony Moore (Senior Editor of CRC) arrived, asking Dr. Torrent about his willingness to write a book on permeability testing, on advice of Profs. S. Mindess and A. Bentur. This was a fortunate coincidence or superb intelligence services of CRC Press in act... The offer was accepted and, immediately, Prof. Imamoto was invited to join the authors' team, invitation he accepted on July 29, 2017, during an unforgettable exquisite dinner in a small, special sushi restaurant near Tokyo's Narita Airport, agreement possibly helped by a considerable dose of excellent cold sake...

Regarding the subject of this book, it is good to recall that until the early 1980s, the main research efforts on hardened concrete properties were predominantly focused on its mechanical and viscoelastic properties, required for the structural design of reinforced concrete constructions. Since then, a considerable interest arose on durability issues, both in understanding the deterioration mechanisms and in developing suitable test methods and, more recently, in modelling the durability performance of concrete structures.

A quantum leap was made by the work of RILEM TC 116-PCD "Permeability of concrete as a criterion for its durability", chaired by Profs. H.K. Hilsdorf and J. Kropp, that stressed the importance of transport mechanisms, chiefly permeation, on the durability of concrete structures. The results of this work were condensed in a State-of-the-Art Report (RILEM Report 12), published in 1999.

During the ensuing 20 years, several test methods for measuring the permeability of concrete to gases and liquids have been developed and a formidable amount of information has been produced through their application in the laboratory and on site. Part of it was included in RILEM Report 40 (2007) and RILEM State-of-the-Art Report v18 (2016), condensing the work of RILEM TCs 189-NEC and 230-PSC, respectively.

It is the purpose of this book to present the existing knowledge on the permeability of concrete in a consolidated form, describing the available test methods and the effect key technological parameters of concrete have on the measured permeability. It presents a large amount of experimental data from investigations performed on laboratory specimens and full-scale elements and also from real cases of site permeability testing, conducted to solve complex and challenging concrete construction issues (durability, water-tightness, defects, spalling under fire, condition assessment, etc.).

The three authors combine a formidable experience, covering over 30 years of research and testing the permeability of concrete in the lab and on site (they have conducted, with their own hands, permeability tests applying 13 different methods). Thanks to their geographical diversity, they have been active in relevant technical activities in Europe, the Americas, Africa and Asia, thus gaining a good insight into the global situation regarding permeability and durability testing and service life assessment of concrete structures.

This book places a special emphasis on one test method (called *kT*), developed by Dr. Torrent around 1990, that was included in the Swiss Standards in 2003 under the title 'Air-Permeability on Site', with successive updates in 2013 and, recently, in 2019. The credit for this inclusion lies mainly on the initiatives and research work of Prof. E. Brühwiler and Dr. E. Denarié (EPFL, Lausanne), of Dr. F. Jacobs (TFB, Wildegg) and Dr. T. Teruzzi (SUPSI, Lugano). Over 430 documents on the *kT* test method have been recorded to date, out of which some 90 were authored by at least one of this book's authors.

Following Chapter 1, summarizing the fundamentals of durability, the relevance of permeability as a key performance property of concrete is discussed in Chapter 2, already opening the field to the possible applications of its measurement. An understanding of concrete microstructure and of the laws that govern the flow of matter through concrete is considered as essential, aspects that are dealt with in detail in Chapter 3. This is followed by Chapter 4 in which 25 test methods to measure concrete permeability (including capillary suction) are described. Chapter 5 describes in detail the *kT* test method, its fundamentals and the effect of external influences on its results.

Chapter 6 is concerned with the effect of key technological factors on the permeability of concrete to gases and water, tested by various methods.

Chapter 7 reflects the strong conviction of the authors on the relevance of site permeability testing of the end-product to get a realistic assessment of the concrete quality, in particular of its surface layers (the *Covercrete*), of vital importance for the durability of reinforced concrete structures. Having been designed to that end, i.e. to measure the permeability of the *Covercrete* on site, Chapter 8 provides evidence on the suitability of the *kT* test as Durability Indicator, relating its results with other relevant transport properties (sorptivity, diffusion, migration) and with simulation tests (carbonation, freezing/thawing). The same as for any other test, especially when applied on site, the application of *kT* test has not been up to the expectations in a few cases, which are also presented in Chapter 8.

Today, test results are often not enough for designers and owners, who want an assessment of the potential service life of new and existing structures. Chapter 9 presents different service life prediction models with the site permeability of the *Covercrete* as input, often accompanied by a non-destructive evaluation of its thickness.

Chapter 10 presents the relatively new field linking the gas-permeability of concrete to the explosive spalling of the concrete cover during fires. Here, contrary to durability, a not too low permeability is desirable.

Chapter 11 presents a comprehensive series of investigations conducted on site, on full-scale elements and real structures, new and old. Some applications not related to concrete structures are also included.

At the end, in Chapter 12, we draw some conclusions on the present and future of permeability testing of concrete structures, needed developments and unexplored research fields. The book is complemented with Annexes that describe transport tests other than permeation, and a Model Standard on how to conduct *kT* tests in the field and in the laboratory.

The reader is invited to accompany us along this fascinating voyage from the theory of mass transport in concrete to field applications of permeability testing…, fasten your seat belts!!

The Authors

Acknowledgements

R. Torrent: to my wife for her unconditional support, to my children (3) and grandchildren (5) for their love and to my masters in Argentina, late G. Burgoa and R. Kuguel, who arose in me the interest in concrete science and technology and guided me in my first professional steps on these fascinating disciplines. To late Prof. A.M. Neville and to Prof. F. Wittmann for giving me fundamental orientation at turning points of my career.

R. Neves: to my family for enduring my absences. To Senior Researcher A. Gonçalves for changing my mindset from structural design to durability design. To Dr. J. Vinagre and Polytechnic Institute of Setúbal for considering investing in permeability testing equipment and providing me excellent conditions to carry out my research. To Prof. J. de Brito for his friendship, valuable teachings and endless support.

K. Imamoto: to my laboratory students for performing a huge number of permeability tests. To Dr. Hiroshi Tamura, former head of Material Dept., General Building Research Corporation of Japan, for giving me a chance to start research on air-permeability of concrete cover.

All: To so many researchers worldwide that contributed their experiences to the body of knowledge compiled in this book. A thorough literature research has been conducted which, by no means can be considered complete; we apologize to those researchers, the valuable work of whom might have been overlooked when writing this book.

Authors

Dr. Roberto J. Torrent is a researcher, consultant and partner of Materials Advanced Services Ltd. He held positions at the National Institute of Industrial Technology and Portland Cement Institute (Argentina), as well as at Holcim Technology Ltd. (Switzerland). For 30 years, he has been directly involved in durability testing of a large variety of concretes, both in the lab and on site. In the 1990s, he invented the Torrent NDT Method for measuring air-permeability. He is a RILEM Honorary Member.

Dr. Rui Neves was formerly a researcher at the National Laboratory for Civil Engineering (LNEC-Portugal). Currently he is Professor in the Structures and Geotechnics Division at Barreiro School of Technology, Polytechnic Institute of Setúbal, Portugal. His research efforts are mainly devoted to service life of reinforced concrete structures, with special emphasis on investigating and testing the permeability of concrete and rocks. He has carried out relevant consulting activity within the frame of concrete quality control, as well as inspection and appraisal of reinforced concrete structures.

Dr. Kei-ichi Imamoto is a graduate of Tokyo University of Science, Japan. He performed research at Tokyu Construction Co. Ltd. for 9 years and is now Professor at Tokyo University of Science. He received the Young Researcher's award from AIJ (Architectural Institute of Japan) in 2008, and prizes from Japan Society for Finishing Technology, Japan Concrete Institute and Suga Weathering Foundation. He is very active in durability testing and service life assessment of concrete structures.

Chapter 1

Durability performance of concrete structures

1.1 WHAT IS DURABILITY?

Since the title of the book intimately associates permeability with concrete durability, it is worth discussing the latter in this initial chapter.

A good definition of durability has been coined in Section 3.1 of Neville (2003), to which some addenda have been made, resulting in the following tentative definition:

> Durability of a given concrete structure, in its specific exposure environment, is its ability to perform its intended functions, i.e. to maintain its required strength and serviceability, during the specified or traditionally expected service life, without unplanned, extraordinary maintenance or repair efforts.

1.2 DETERIORATION MECHANISMS OF CONCRETE STRUCTURES

Discussing in detail the deterioration mechanisms of concrete structures is beyond the scope of this book; yet, the main ones can be briefly enumerated: steel corrosion induced by carbonation or chlorides, chemical attack (typically by sulphates in the soil and ground water and by acids in sewage systems), Alkali-Silica Reaction (ASR) and frost in cold climates.

All these deterioration mechanisms have two aspects in common:

a. they involve the transfer of mass into or within the concrete member
b. they require the presence of water to take place

The transfer of mass takes place by three physical actions: permeation, diffusion and, to a lesser extent, also by migration (all three thoroughly discussed in Chapter 3) and happens through the interconnected network of pores within the microstructure of concrete (also discussed in Chapter 3).

DOI: 10.1201/9780429505652-1

A succinct analysis will be made in the following sections. For a deeper insight into the problem, the reader can refer to Mehta et al. (1992), Richardson (2002), Dyer (2014), Li (2016), Alexander et al. (2017) and, more specifically for the case of steel corrosion in concrete, to Bertolini et al. (2004), Böhni (2005), Gjørv (2014) and Alexander (2016).

1.2.1 Carbonation-Induced Steel Corrosion

This case of deterioration is due to the penetration (by gas diffusion) of CO_2 from the environment which, in the presence of moisture, reacts preferentially with the reaction product of cement hydration $Ca(OH)_2$ to form $CaCO_3$. From the durability point of view, the main consequence of this reaction is a sharp drop of the pH of the pore solution, displacing the thermodynamic equilibrium of the steel bar, from "passive" to "corrosion". The subsequent corrosion rate is highly dependent on the moisture conditions (as is also the carbonation rate).

According to Mehta et al. (1992), "only porous and permeable concrete products, made with low cement contents, high water/cement ratio (*w/c*), and inadequately moist-cured tend to suffer from serious carbonation".

A tight pore system and a sufficiently thick cover are the main defense strategies against this mechanism, although the cement type (especially the amount of carbonatable material) also plays a role (see Section 9.2.1).

1.2.2 Chloride-Induced Steel Corrosion

This case of deterioration is due to the penetration (by mix modes) of chloride ions from salty solutions in permanent or sporadic contact with the structure. The situation is much more complex than carbonation, due to the overlapping of several physical phenomena taking place, as illustrated in Figure 1.1 for a concrete element in a marine environment (adapted from Hunkeler (2000)).

Chlorides may penetrate by permeation, carried by the saline water solution either under a pressure head for deep parts of the structure or/and due to capillary suction in the critical areas subjected to wetting-drying cycles and, alternatively or complementary, by ion diffusion. Rain washout and evaporation, affecting predominantly the surface layers, add complication to the phenomenon.

When the penetration of the front of a certain elusive critical Cl^- concentration reaches the position of the steel, this is depassivated and metal corrosion may start. The same structure, placed by the sea, will deteriorate much earlier than if exposed to carbonation.

According to Mehta et al. (1992), "The ingress of chlorides into hardened concrete is decisively dependent on and influenced by water transport mechanisms. Substantially greater amounts of chlorides may ingress into the hardened concrete via water transport mechanisms than via pure chloride ion diffusion".

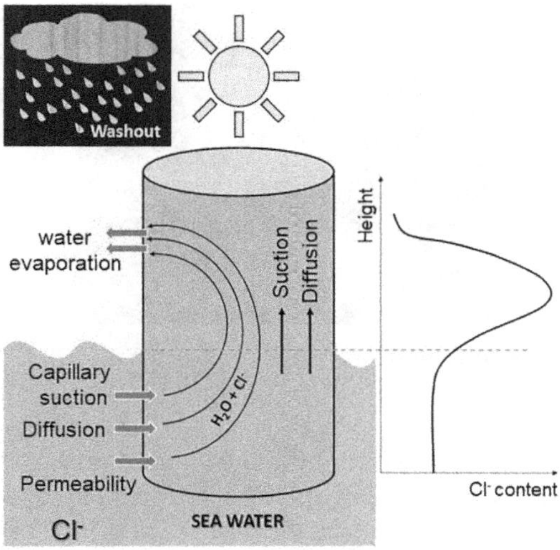

Figure 1.1 Complex combination of physical phenomena in the movement of Cl^- in marine concrete, based on Hunkeler (2000).

1.2.3 External Sulphate Attack

This case of deterioration takes place when sulphate ions from the environment, typically from the soil or ground water, penetrate into concrete, developing deleterious physical–chemical interactions with some minerals in the hydrated cement paste. The main penetration mechanism is permeation of the SO_4^{2-}-rich solution in the form of capillary suction, accompanied by internal redistribution by diffusion.

Salt crystallization, combined with expansive reaction products (e.g. ettringite, gypsum and thaumasite), leads to cracking, loss of mass and/or disintegration of the concrete. There are cements that, due to their composition or performance, are considered as "Sulphate Resistant Cement", although it would be more appropriate to talk of "Sulphate Resistant Concrete", since not just the use of such cements is sufficient to guarantee the immunity of the concrete against sulphate attack.

According to Mehta et al. (1992), "...it can be concluded that, for improved resistance to sulfate attack, a reduction in the porosity and consequently the coefficient of permeability, is more important than modifications in the chemistry of Portland cements".

1.2.4 Alkali-Silica Reaction

This case of deterioration takes place when aggregates containing certain reactive minerals (typically some forms of SiO_2) in sufficient or *pessimum* quantities react, in the presence of moisture, with the alkali (Na^+, K^+) ions in the pore solution, developing expansive reactions. The reaction products,

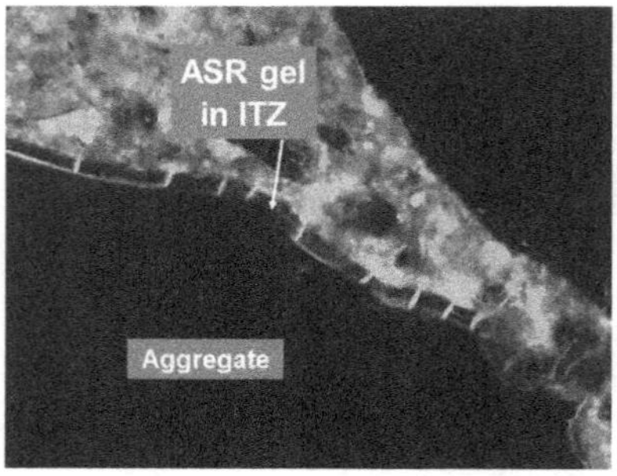

Figure 1.2 ASR gel accommodated along ITZ.

again in the presence of moisture, take the form of an expansive gel which, depending on the circumstances may be innocuous or create enormous deformations of the structure (e.g. dams), cracks, loss of mass and even total disintegration of the concrete.

The expansive gel is sometimes accommodated in microcracks, air voids or along the more porous Interfacial Transition Zone (ITZ), see Section 3.2.3, as shown by the UV light observation of a thin section in Figure 1.2 (Fernández Luco & Torrent, 2003).

The main mechanism of transport of the gel within concrete is permeation, due to expansive pressure, across the system of existing pores and of cracks generated by the expansive action.

The water required to feed the expansive reaction and to swell the ASR gel penetrates the concrete predominantly by permeation (capillary suction) and moves internally by diffusion.

The main defense line against ASR is to avoid the usage of reactive aggregates but, when this is unavoidable, to keep the quantity of alkalis in the concrete sufficiently low (e.g. by using low-alkali cements) or by using adequate types and contents of pozzolanic materials (that compete with advantage in neutralizing the alkalis).

In the case of ASR, the pore structure and permeability of the concrete play a secondary role.

1.2.5 Freezing and Thawing

This case of deterioration takes place when concrete, with a high degree of water-saturation, is exposed to sub-zero temperatures; the saturating water penetrates typically by permeation (capillary suction). The water in the pores freezes, augmenting its volume by 8%, pushing the still unfrozen water along the capillaries, creating damaging pressure on their walls. Successive freeze–thaw cycles continue to accumulate this type of damage, causing scaling and spalling of the surface layers due to internal cracks, typically oriented parallel to the element's surface. The water in the larger pores freezes at higher below-zero temperatures than that in smaller pores. The problem is aggravated if the liquid in the pores contains salts, as typically happens with de-icing compounds, sprayed in winter on roads.

The best-known prevention measure to avoid the freeze–thaw damage is to entrain air bubbles, in quantities and sizes sufficient to relieve the expansion pressures. As discussed in Section 8.3.8, this should be accompanied by a sufficiently tight pore structure (hence the maximum *w/c* ratio typically specified for this case). There is some debate on whether high-strength concrete (HSC), with its reduced porosity and permeability, is resistant to freeze–thaw damage without air-entrainment.

1.3 DETERIORATION PROCESS OF CONCRETE STRUCTURES

When the designer, with the help of codes and standards, defines/specifies the architectural details, the shape and dimensions of the structural elements, the amount, quality and position of the steel bars (including cover thickness) and the quality of the concrete (typically strength and resistance to certain aggressive media), he/she is defining an initial design quality (IDQ), see Figure 1.3 adapted from Beushausen (2014). It is being assumed that, starting with this IDQ, the inevitable degradation process the structure will undergo through the years will follow a certain expected performance such that, when the "traditionally expected" or Design Service Life (DSL) is reached, the structure will still perform at a level above a not very well defined Unacceptable Level of Deterioration (ULD). This is indicated by the full line in Figure 1.3.

Regrettably, in too many cases the True Initial Quality (TIQ) achieved during construction is below that assumed during the design, due to lack of care and/or application of inadequate concrete practices by the Contractor, or to concrete mixes of insufficient quality, or due to lack of zeal of the Inspection or, usually, a combination thereof. Hence, the true decay process (dotted curved line) is faster and the True Service Life (TSL) is reached much earlier than specified or expected (DSL). This requires some Interventions (I1, I2) to restore the condition of the structure to an acceptable level, so as to finally reach the DSL.

This is an expensive solution not only due to the usually high cost of the interventions themselves but also for the lost revenue if the operation of the facility has to be partially or totally interrupted (roads, bridges, tunnels, power stations, cement plants, etc.), not to mention the serious consequences for human lives caused by the deterioration itself (e.g. debris falling from a tall building) or by increased accidents rates caused by traffic restrictions.

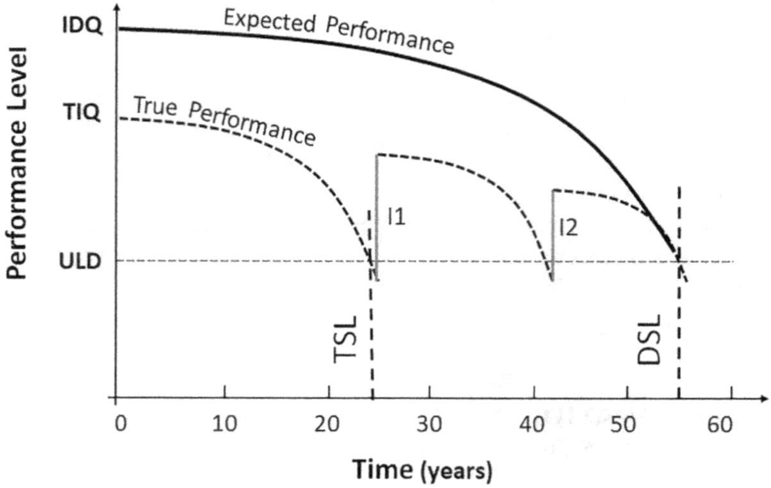

Figure 1.3 Expected and true durability performance of concrete structures (Beushausen, 2014).

Most deterioration processes (steel corrosion, sulphate or chemical attack, frost damage, ASR, etc.) follow a similar pattern, illustrated in Figure 1.4, based on the model proposed by Tuutti (1982), later extended by Nilsson (2012) for steel corrosion.

Initially, there is a period in which no visual damage of the structure is observable. Yet, in this period, some phenomena are taking place internally, such as penetration of the carbonation front or accumulation of enough chloride at the surface of the steel bars (or generation of enough ASR expansive gel), so as to initiate the visible deterioration process. This period

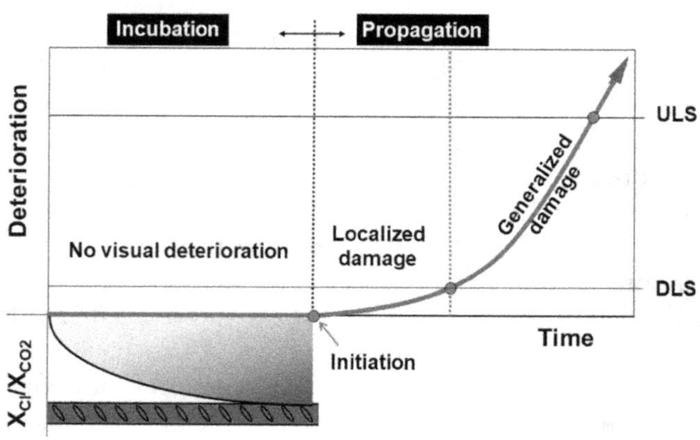

Figure 1.4 Tuutti's model for steel corrosion (Tuutti, 1982), extended by Nilsson (2012).

is defined here as "Incubation" period and the time at which the true damage process starts is called "Initiation" time.

At a certain "Initiation" time, the carbonation front (X_{CO2}) or the penetration of the critical chloride content front (X_{Cl}) has reached the surface of the rebars (see bottom left corner of Figure 1.4), depassivating the steel which under unfavourable conditions will start to corrode. The expansive nature of the corrosion products will produce isolated rust stains and microcracks (which can be considered as localized damage in Figure 1.4), to be followed by spalling of the concrete cover and reduction of the cross section of the steel (generalized damage, see Figure 1.4). This process, if not checked, will lead to a loss of bearing capacity of the element that eventually will reach its Ultimate Limit State (ULS), requiring major retrofitting or simply demolition. Although it is difficult to imagine that a structure would be left deteriorating to such extent, one of the authors was involved in a case in which an important industrial asset had to be stopped and evacuated, due to the risk of collapse caused by extensive steel corrosion damage.

1.4 THE COSTS OF LACK OF DURABILITY

Figure 1.5 shows the deterioration process, after Tuutti's model, as a dotted line referred to the right-hand-side vertical axis. The full line (referred to the left axis) shows the incremental costs of remedial interventions along the service life of the structure.

The full line represents what is called the "Law of Fives" (de Sitter, 1984), by which the cost of intervention grows with time by a factor of 5, law that was confirmed in practice (Wolfseher, 1998); below some further explanations.

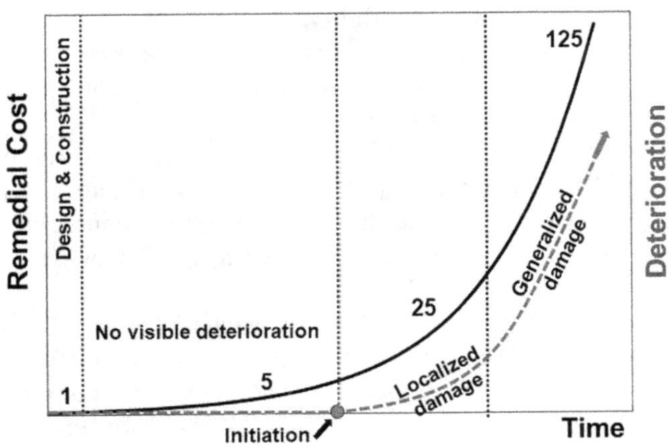

Figure 1.5 Increasing costs of remedial interventions with time, or "Law of Fives" (de Sitter, 1984).

Design and Construction Phase: Here the germs of an unsatisfactory performance are seeded, as a result of a poor design and materials specification or of bad execution. Relative Corrective Cost=1.

Incubation Phase: There is no visible damage yet. If the problem is detected at this stage (NDTs, covermeters, carbonation, chloride profiles, etc.) it is still possible to act preventively, for example, by applying appropriate surface treatments. Relative Corrective Cost=5.

Localized Damage Phase: Deterioration has started in some areas, as revealed by stains, cracks and/or localized spalling. Repair and maintenance work is required. Relative Corrective Cost=25.

Generalized Damage Phase: If repair and maintenance work has not been carried out, the structure will reach a stage in which delicate and complex repair and retrofitting work is required or even the complete replacement of the elements. Relative Corrective Cost=125.

The importance of having things done correctly from the very beginning can be realized from Figure 1.5; it is in the interest of the owners that a good design and construction is achieved. In general, but especially regarding public works, it is in the interest of the whole society and taxpayers that the constructions are durable.

1.5 ECONOMICAL, ECOLOGICAL AND SOCIAL IMPACTS OF DURABILITY

Today it is clear that civil engineers, builders and architects have succeeded in establishing and applying sound criteria to ensure the stability and strength of concrete structures. Fortunately, cases of partial or total collapse of such structures are extremely rare or due to exceptional events.

On the contrary, regarding durability, the situation is not so satisfactory. Indeed, all over the world huge amounts of money are spent in the repair or restoration of concrete structures affected by one or a combination of different degradation mechanisms.

R. Torrent, M. Alexander and J. Kropp have addressed the problem in Chapter 1 of RILEM Report 40 (2007), citing several papers that provide quantitative evidence of the onerous macroeconomic consequences of the problem (Peacock, 1985; Browne, 1989; Mehta, 1997; Neville, 1997; Hoff, 1999; Vanier, 1999; Coppola, 2000).

As the amount of money available for construction in any society is limited, this means that a steady shift of activity from new constructions onto repair and maintenance is taking place. For emerging countries, with a pressing need to improve their infrastructure, this constitutes a serious barrier against development. This has also strong repercussions for the concrete construction industry and all its players (owners, contractors, materials suppliers, engineers, specialized workers, insurance companies, etc.). An example is the recent partial collapse, at an age of 51 years, of an important concrete bridge in Genoa, Italy (Seitz, 2019), in which apparently durability weaknesses might have played a role (Virlogeux, 2019). It is interesting to remark that these weaknesses had already been revealed when the bridge was just 15 years old (Collepardi et al., 2018). Being an essential element in the Italian highways network, it was rebuilt very fast, but as a hybrid steel-concrete structure. Indeed, the deck is a 5 m deep, 30 m wide, hollow hybrid steel concrete structure, with a steel shell and a reinforced concrete slab forming the road surface, supported by 18 reinforced concrete elliptical-shaped piers. The steel shell has been divided into sections which have been prefabricated off-site (Horgan, 2020).

Moreover, ongoing research activities go on to develop solutions aimed at replacing concrete bridge decks by reinforced polymer solutions (Scott, 2010; Rodriguez-Vera et al., 2011; Mara et al., 2014). Construction companies can adapt to building with other materials, but the negative impact on the cement and concrete industry is direct and can be considerable. Repair work involves high costs in diagnosis and design (consulting companies) and uses relatively low volumes of special (usually high-cost) materials that contain little amount of cement/concrete. Hence, the more the activities are shifted from new construction towards repair activities, the higher the negative impact on the cement and concrete industries.

Furthermore, durability and ecology or sustainability go hand by hand. As illustrated in Figure 1.3, a non-durable structure will require one or more interventions during its service life (unnecessary if it had been properly designed, constructed and used). These interventions require partial demolitions and replacement with new materials, with the energy and emissions involved in their production and processing and, in the case of road transport facilities, the extra emissions due to traffic jams caused by the repair work. Quoting de Schutter (2014), it can be stated that "No concrete construction can be sustainable without being durable".

1.6 DURABILITY DESIGN: THE CLASSICAL PRESCRIPTIVE APPROACH

This approach is also known as "deemed-to-satisfy" approach.

The three pillars supposedly supporting the achievement of durable concrete, according to the classical approach adopted by most codes and standards for structural concrete worldwide, are depicted in Figure 1.6 (de Schutter, 2009). The approach is based on specifying maximum limits for the *w/c* or *w/b* (water/binder) ratio and minimum limits for the compressive strength and the cement or binder content (the latter is not always included in codes). The title of the chart is "Parameters for durable concrete?" (de

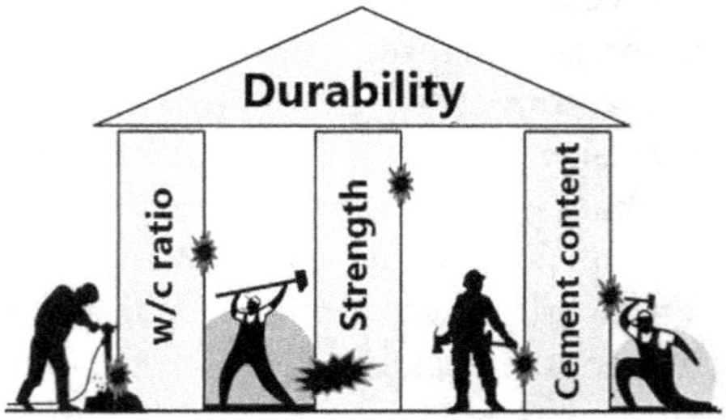

Figure 1.6 Parameters for durable concrete? After EN 206 (de Schutter, 2009).

Table 1.1 Durability requirements for reinforced concrete in Eurocode 2 and EN 206

		Corrosion induced by:									
		Carbonation				*Deicing chlorides*			*Marine chlorides*		
	Exposure class	*XC1*	*XC2*	*XC3*	*XC4*	*XD1*	*XD2*	*XD3*	*XS1*	*XS2*	*XS3*
Row	**Durability indicator**	**Requirements established in Eurocode 2**									
1	$f'c_{min}$ (MPa)[a]	25	30	37	37	37	37	45	37	45	45
2	$c_{min,dur}$ 50 year[b]	15	25	25	30	35	40	45	35	40	45
	Durability indicator	**Requirements established in EN 206**									
3	$f'c_{min}$ (MPa)[a,c]	25	30	37	37	37	37	45	37	45	45
4	w/c_{max}	0.65	0.60	0.55	0.50	0.55	0.55	0.45	0.50	0.45	0.45
5	$Cement_{min}$ (kg/m^3)	260	280	280	300	300	300	320	300	320	340

[a] Compressive strength measured at 28 days on moist-cured concrete cubes.
[b] Minimum cover thickness (mm) for service life of 50 years, structural classes S4.
[c] Optional requirement.

Schutter, 2009), intended to show the weaknesses of each of the three pillars. Regarding steel corrosion, a fourth pillar exists, representing the thickness of the concrete cover.

Table 1.1 shows the three pillars in practice, for the case of steel corrosion induced by carbonation and chlorides, according to Eurocode 2 (EN 1992-1-1, 2004; EN 206, 2013).

In what follows, a brief consideration on the suitability of the four durability indicators in Table 1.1 is provided; for a more detailed discussion of the subject, the reader can refer to Torrent (2018).

1.6.1 Compressive Strength as Durability Indicator

Regarding the suitability of compressive strength as durability indicator, the following comment in CEB-FIP Model Code 1990 (CEB/FIP, 1991), Section d.5.3 "Classification by Durability", is very relevant:

> "Though concrete of a high strength class is in most instances more durable than concrete of a lower strength class, compressive strength per se is not a complete measure of concrete durability, because durability primarily depends on the properties of the surface layers of a concrete member which have only a limited effect on concrete compressive strength."

Similar considerations can be found in p. 156 of Model Code 2010 (*fib*, 2010).

The example in Figure 1.7 illustrates the lack of direct association between strength and durability. Represented in the chart are 18 concretes made with widely different cement types and *w/c* ratios of 0.40 and 0.65 (see Table 5.4); more details on the characteristics of the mixes in Moro and Torrent (2016). In ordinates, the Coefficient of Chloride Migration M_{Cl} (Tang-Nilsson method, described in Section A.2.1.2); in abscissae the compressive strength measured on 150 mm cubes. The samples were moist cured for 28 days, age at which the tests were initiated. Two extreme sets of data are explicitly shown in the chart: those of two different CPCs (triangles) and those of a cement containing 68% of GGBFS (squares). On the right-hand edge of the chart, a classification of resistance to chloride ingress based on 28-day chloride migration results, proposed by Nilsson et al. (1998), is shown with abbreviations: L=Low; M=Moderate; H=High; VH=Very High and EH=Extremely High. A general trend of increasing resistance to chloride ingress with compressive strength can be observed in Figure 1.7. However, for the same strength, the OPC concretes show higher migration coefficients than the rest and, even for strengths above

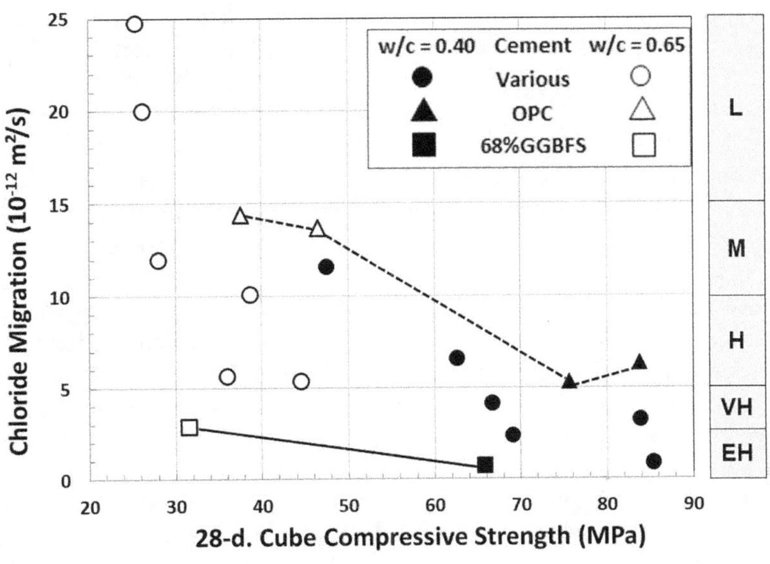

Figure 1.7 Relationship between chloride migration and cube strength at 28 days.

75 MPa, cannot reach a level of VH resistance to chlorides ingress. On the other extreme, the concretes made with a cement containing 68% GGBFS (incidentally made with the same clinker as one of the OPCs) present lower migration coefficients than the rest, for the same compressive strength.

Establishing minimum strength classes as durability requirement is a way to keep *w/c* ratio at low levels, due to the impossibility of measuring the latter. Section R4.1.1 of ACI 318 (2011) openly confesses:

> "Because it is difficult to accurately determine the w/cm of concrete, the f′c specified should be reasonably consistent with the w/cm required for durability. Selection of an f′c that is consistent with the maximum permitted w/cm for durability will help ensure that the maximum w/cm is not exceeded in the field."

1.6.2 Water/Cement Ratio as Durability Indicator

Concrete durability depends, to a large extent, on the resistance of the material to the penetration of aggressive species by a combination of different mechanisms (chiefly permeability and diffusion). This resistance is governed, primarily, by the pore structure of the concrete system, especially that of the cement paste and of the interfacial transition zone around the aggregates (see Chapter 3).

Establishing limits to the composition of the concrete (especially to its *w/c* ratio) constitutes an attempt to regulate the pore structure of the concrete system. However, it implies assuming that all materials (especially cements) perform identically; that is, all concretes of the same *w/c* ratio will perform identically, irrespective of the characteristics of the cement (and other constituents) involved. For the same constituents, it is true that a higher *w/c* ratio means higher "penetrability" of the concrete (Section 3.2.2). However, for the same *w/c* ratio, the "penetrability" of a concrete varies significantly with the type and characteristics of the cement used. Figure 1.8 shows the large range of values of the Coefficient of Chloride Migration M_{Cl} (Jacobs & Leemann, 2007), Tang-Nilsson method (Section A.2.1.2), that can be found for a given *w/c*, when concretes are made with different cements. A similar pattern is shown in Figure 6.5 regarding air-permeability.

These examples show that, technically speaking, *w/c* is not a good durability indicator. This fact is greatly aggravated by the difficulties for the user of the concrete to check compliance with the maximum limits specified.

Two options are offered in EN 206 (2013) for checking compliance with the prescriptive limits for *w/c* ratio:

a. In order to compute the *w/c* ratio of each concrete batch, the contents of cement and added water shall be taken as stated on the print-out of the batch record.

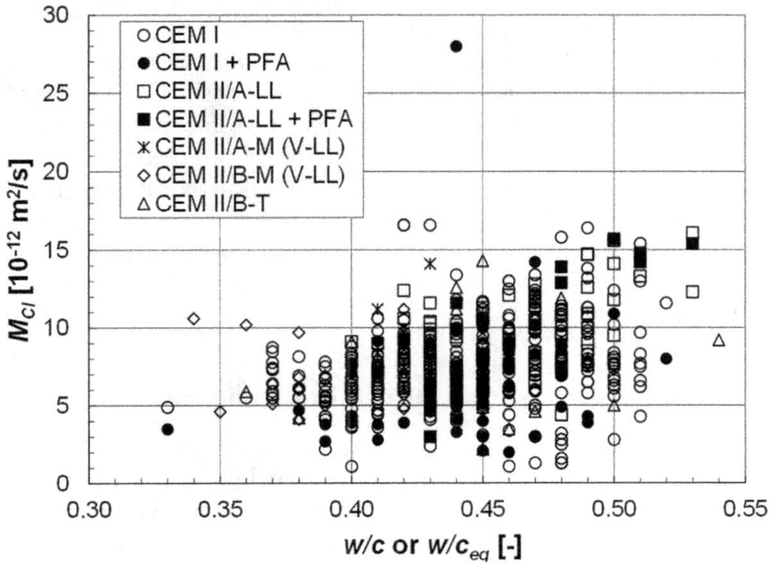

Figure 1.8 Relation of chloride migration with *w/c* ratio.

b. Where the *w/c* ratio of concrete is to be determined experimentally, it shall be calculated on the basis of the determined cement content and the effective water content. The test method and tolerances shall be agreed between the specifier and the producer.

Option (a) is almost exclusively used, despite its grave deficiencies. Indeed, an accidental or deliberate error in the stated sand moisture will end up in a wrong *w/c* ratio reported in the batching print-out. An example is presented in Torrent (2018), where an error in the sand moisture used for the calculation brought the declared *w/c*=0.44 for reported moisture of 0.4% to a true *w/c*=0.59 for the true measured 6.0% of sand moisture.

To this we should add that, sometimes, washing water is left inside the drum when the ready-mixed concrete (r-mc) truck is loaded with a new batch. In addition, "slumping" is a very common practice in the r-mc industry whereas the driver, while washing the loaded truck still in the plant, watches the consistency of the concrete and, if judged too stiff, adds uncontrolled amounts of water into the drum. On arrival to the jobsite, water is sometimes added to *retemper* the mix (in a Western European country, the addition of 30 L/m³ into a truck, was witnessed by one of the authors). All the extra water, discussed in this paragraph, which may be added to the truck, is usually not recorded in the batching protocol which, therefore, underestimates the *w/c* ratio of the concrete delivered, with negative effects on the resulting durability.

Regarding option (b), despite several attempts, no standardized or widely accepted test method to experimentally measure the *w/c* ratio of the freshly delivered concrete has been developed. An overview of such attempts can be found in CR 13902 (2000) that states: "It follows [...] that the problem of measuring water/cement ratio on a sample of fresh concrete about which nothing is known is very difficult and probably impossible".

The fact remains that one of the critical weaknesses of the use of *w/c* ratio as durability indicator is the impossibility of checking compliance by the user. Specifying a characteristic that cannot be measured is clearly meaningless and opens roads to unfair competition by fraudulent practices undetected by the user.

1.6.3 Cement Content as Durability Indicator

The main argument behind the specification of a minimum cement content in the mix is the chemical binding effect that hydrated cement offers to free chlorides and CO_2 that can penetrate the concrete (Wassermann et al., 2009). Along this line of thinking, a higher cement content would imply a higher reservoir of alkalis that need more CO_2 to be carbonated (same for more Cl^- binding), thus delaying the advance of the critical front. But, for the same *w/c* ratio (which is specified in parallel), a higher cement content means also a proportionally higher volume of porous paste that allows more CO_2 (or Cl^-) to penetrate, with a null net result; this has been proved experimentally (Wassermann et al., 2009). Moreover, more paste and more cement mean more susceptibility of the concrete to shrinkage and thermal cracking.

The use of mineral additions batched separately into the concrete mixer adds further complications to establishing the cement content (and also the *w/c* ratio), due to the application of the controversial "k-value concept" of EN 206 (2013) to assess the "cementitious contribution" of the addition used.

1.6.4 Cover Thickness as Durability Indicator

The thickness of the concrete cover is a very important durability indicator for the deterioration of structures due to steel corrosion. A lack of sufficient cover thickness is a recurrent cause of premature corrosion of reinforcing steel (Wallbank, 1989; Neville, 1998; Torrent, 2018).

In theory, both second Fick's diffusion law (through the argument of the error function solution, Section 3.4.1) and capillary suction theory (see Section 3.7) predict a progress of the penetration front of carbonation, chlorides and water with the square root of time. This means that a 10% reduction in cover thickness implies a reduction in service life of 20%, so great is the importance of observing the specified cover thickness. Yet, despite the progress made on electromagnetic instruments capable of assessing non-destructively the cover thickness quite accurately (Fernández Luco, 2005), now largely enhanced by the development of ground penetrating radar (GPR) instruments, their use is not forcibly specified in the standards.

1.7 DURABILITY DESIGN: THE PERFORMANCE APPROACH

In Sections 1.6.1–1.6.3, the inherent weaknesses of the three pillars of the classical durability approach have been revealed. With more or less degree of boldness, codes and standards have been moving, rather timidly, along the P2P (Prescriptive to Performance) road, as discussed in this section and also in Section 12.1.

1.7.1 The "Durability Test" Question

The P2P transit brings to the forefront the question of suitable durability tests, that was discussed in Torrent (2018) and that, given the scope of this book, deserves a revisit.

The durability of concrete is, almost by definition, hardly measurable by testing, as each structure is performing its own durability test, live under its own specific conditions. The prediction of the evolution of the structure condition is uncertain, particularly when based on testing specimens and not the real structure.

Durability involves deterioration processes lasting several years; therefore, it is clear that tests lasting months or even years, although in some cases possibly closer to reality, are not practical for specification and quality control purposes.

Performance specifications need short-term tests, lasting not more than, say, 1 week, including preconditioning of the specimens; otherwise, the approach would not be practical nor acceptable for conformity control purposes, given the current pace of concrete construction.

The durability of concrete structures against deteriorating actions originated from the surrounding environment is strongly related to the resistance of the concrete cover to the penetration, by different transport mechanisms, of external deleterious substances. As a result of 30–40 years of durability research, several test methods have been developed to measure mass transport properties of concrete (RILEM Report 40, 2007; RILEM STAR 18, 2016). They consist typically of tests that measure the resistance of concrete to the transport of matter (gaseous or liquid) by appropriate driving forces (see Chapters 3 and 4 and Annex A). Some of these tests have been standardized in different European countries and in the USA.

All these durability tests have merits and demerits. If we wait until the perfect "durability" test is developed, we will never leave the unsatisfactory and ineffective current prescriptive approach. Indeed, any reasonable durability test will be better than the *w/c* ratio, used today as the durability "panacea".

Drawing a parallel, concrete structural design relies heavily on the compressive strength (measured with a standard test), adopted as the universal, used-for-all property (in codes almost all properties of concrete are derived from its value). And yet, this test can be questioned from different angles: the true stress field is far from uniaxial compression (hence the 20%–25% difference when testing cylinders and cubes); the size is much smaller than that of the structural elements (size effect); the load is statically applied in less than 5 minutes (in bridges, static loads are applied for decades with ≈ 15% strength reduction effect and even cyclically, bringing in also the deleterious effect of fatigue); the specimens are tested saturated (a condition seldom found in reality, which influences the strength by ≈ 20%), and so on and so forth. Yet, despite all these limitations, the standard compressive strength is accepted by civil engineers as a suitable indicator of the bearing capacity of concrete and is used, without objections, in the structural design of concrete structures.

A similar, pragmatic approach is required for durability, that is, the adoption of well proved standard tests to measure relevant "Durability Performance Indicators", focusing on the merits and positive contribution of the tests and less on their demerits.

1.7.2 Canadian Standards

Canadian Standards specify limiting values of the result (electrical charge passed Q in Coulombs) in a migration test (ASTM C1202, 2019) (see description in Section A.2.1.1).

Canadian Standard (CSA, 2004) specifies a maximum limit of $Q = 1{,}500$ Coulombs for exposure classes C-1 (structurally reinforced concrete exposed to chlorides with or without freezing and thawing conditions) and A-1 (structurally reinforced concrete exposed to severe manure and/or silage gases, with or without freeze-thaw exposure. Concrete exposed to the vapour above municipal sewage or industrial effluent, where hydrogen sulphide gas may be generated). That limit is reduced to $Q = 1{,}000$ Coulombs for exposure class C-XL (structurally reinforced concrete exposed to chlorides or other severe environments with or without freezing and thawing conditions, with higher durability performance expectations). The testing age should not exceed 56 days.

1.7.3 Argentine and Spanish Codes

The Codes for Structural Concrete of Argentina (CIRSOC 201, 2005) and Spain (EHE-08, 2008) rely on a water-permeability test for durability specifications. The test is known as Water Penetration under Pressure (EN 12390-8, 2009), described in Section 4.1.1.2, and the result is the maximum (sometimes also the mean) depth of penetration of water W_p reached under pressure onto the surface of a concrete specimen. Table 1.2 summarizes the requirements of the Argentine and Spanish Codes referred to this test.

The Argentine Code also specifies, for all aggressive environments, a maximum water sorptivity of 4.0 g/m²/s½ (see Section 4.2.1) which looks quite demanding, especially if applied to all exposure classes.

1.7.4 Japanese Architectural Code

The Japanese Architectural Code "Recommendations for Durability Design and Construction Practice of Reinforced Concrete Buildings" (Noguchi et al., 2005) includes, in its Chapter 2, the principles of durability design.

Regarding performance-based design of carbonation-induced steel corrosion, a probabilistic approach is adopted, based on Eq. (1.1) of carbonation progress.

$$C_t = k \cdot \alpha_1 \cdot \alpha_2 \cdot \alpha_3 \cdot \beta_1 \cdot \beta_2 \cdot \beta_3 \cdot \sqrt{t} \tag{1.1}$$

Table 1.2 Performance laboratory tests and limiting values specified in some national standards

Country	*Standard*	*Indicator*	*Test method*	*Exposure*	*Limit*
Canada	CSA A23.1/ A23.2	Electric charge passed (Coulomb)	ASTM C1202	Chlorides, manure gases	≤1 500
				Extended service life	≤1 000
Argentina and Spain	CIRSOC 201, EHE 08	Maximum water penetration under pressure (mm)	EN 12390-8	Moderate	≤50
				Severe	≤30
Switzerland	SIA 262/1	Water sorptivity (g/m²/h)	SIA 262/1: Annex A	Mild chlorides	≤10
		Chloride migration coefficient (10^{-12} m²/s)	SIA 262/1: Annex B	Chlorides	≤10
		Carbonation rate (mm/y½)	SIA 262/1: Annex I	Carbonation	≤5.0
		Frost-thaw-salts, mass loss (g/m²)	SIA 262/1: Annex C	Mild frost	≤200
				Severe frost	≤1 200
		Sulphate resistance expansion (‰)	SIA 262/1: Annex D	Sulphates	≤1.2

where

C_t=carbonation depth at time t
k=coefficient (1.72 after Kishitani or 1.41 after Shirayama)
α_1=coefficient function of concrete and aggregate type
α_2=coefficient function of cement type
α_3=coefficient function of mix proportions (w/c ratio)
β_1=coefficient function of air temperature
β_2=coefficient function of relative humidity of air
β_3=coefficient function of CO_2 concentration

In Chapter 7 of AIJ (2016), "Practice and quality management", the coefficient of air-permeability *kT* (Torrent method, described in Chapter 5) is used to predict concrete carbonation with consideration of moisture effect. Sampling method follows Annex E of Swiss Standard (SIA 262/1, 2019).

The main purpose of this code is not to establish durability specifications but to predict the carbonation progress in concrete structures.

1.7.5 Portuguese Standards

In Portugal, the performance-based durability design is possible through the application of LNEC E 465 (2007). This standard addresses the deterioration by reinforcement corrosion, induced by carbonation and sea chlorides. Its major features are summarized as follows:

- applies a semi-probabilistic approach, where the reliability analysis is carried out in the service life format
- the end of service life is defined as the occurrence of corrosion-induced cracking
- the service life is broken down in two periods (initiation and propagation) following Tuutti's model (Figure 1.4)
- comprises one analytical model for the propagation period, based on Faraday's law and on the empirical expression proposed by Rodriguez et al. (1996)
- comprises three analytical models for the initiation period, one for chloride penetration and two for concrete carbonation
- the analytical model for chloride penetration is based on the model proposed by Mejlbro (1996) and uses chloride migration coefficient from NT Build 492 test (see A.2.1.2), as durability indicator
- one of the analytical models for concrete carbonation is also based on "CEB Task Group V, 1+2 model" (DuraCrete, 1998) and uses concrete resistance to accelerated carbonation (LNEC E 391, 1993) as durability indicator
- the other model for concrete carbonation uses the oxygen-permeability coefficient from CEMBUREAU test (see 4.3.1.2) as durability indicator, adapting Parrott's model (Parrott, 1984), see Section 9.3.1.

Further, the input parameters for the analytical models vary according to the exposure conditions and these are grouped in exposure classes according to EN 206 (2013). Three safety factors for service life are defined, one for each of the reliability classes identified in Eurocode 0 (EN 1990, 2002).

This methodology allows the user to define a combination of nominal cover thickness and performance requirement (chloride migration coefficient. oxygen-permeability coefficient or carbonation resistance), to ensure the intended service life.

1.7.6 South African Standards

For many years, thanks to the continuous and persistent work of several distinguished researchers such as Alexander et al. (1999), Alexander (2004) and Beushausen and Alexander (2009), an original performance concept was introduced and consolidated in South Africa, crowned with its acceptance in South African Standards (CO3-2, 2015; CO3-3, 2015). It consists in measuring "Durability Indices" (oxygen-permeability and chloride conductivity) in the laboratory, on cores drilled from the finished structure. These indices, coupled with the assumed or measured cover thickness allow, via modelling, the assessment of the service life of reinforced concrete structures exposed to carbonation or chlorides. This approach and its test methods are described in more detail in Sections 4.3.1.3, 9.3.2 and A.2.2.3.

1.7.7 Swiss Standards

The Swiss Codes and Standards for Concrete Construction have taken decisively the road to performance specifications, based primarily on three separate standards, namely:

- SIA 262 (2013) based on Eurocode 2 is the Swiss Concrete Construction Code, defining exposure classes and corresponding cover thicknesses
- SIA 262/1 (2019) describes special, non-EN Standard tests and sets performance requirements associated with the exposure classes, to be fulfilled for laboratory and site tests (NDT or drilled cores)
- SN EN 206 (2013), prescriptive, is the Swiss version of EN 206
- Table 1.3 shows the evolution of Swiss Standards' requirements for exposures that promote steel corrosion, moving from purely Prescriptive to Performance-based; more details can be found in Torrent and Jacobs (2014). In Switzerland, the following has been achieved:
- for each exposure class, a suitable "Durability Performance Indicator" test has been adopted, for example:
  - accelerated carbonation for XC3 and XC4
  - water capillary suction for XD1 and XD2a
  - chloride migration for XD2b and XD3

- a standard for conducting each of these tests has been issued (SIA 262/1, 2019)
- limiting values and conformity rules have been established for the test results (average of different samples) in each exposure class (see Table 1.2 and Table 1.3, rows 5–7)

The concrete producer, supplying concrete for structures under a given exposure classes, shall design the mixes complying with the prescriptive requirements of rows 1 and 2 in Table 1.3. In addition, the concrete producer shall cast specimens ("*Labcrete*", see Section 7.1.2) from samples taken during the regular production with a frequency that is function of the volume produced, but at least four times per year. The averages of the test results on these samples must comply with the maximum requirements of Table 1.3, rows 5–7. More important, perhaps, now the user can take samples during delivery and check compliance of the received concrete with the specifications. It is envisaged that, once enough experience has been accumulated with the performance requirements, the prescriptive requirements will be removed from the standard or kept as recommended values.

One of the most innovative aspects of the Swiss Standards is the recognition that tests made on cast samples are not truly representative of that of the cover concrete *Covercrete*, see Section 7.1.4) of the real structure.

SIA 262 Code (SIA 262, 2013) describes the measures to be adopted in order to ensure durability and, acknowledging the importance of the role of the *Covercrete*, specifically states (free translation from German into English):

- "with regard to durability, the quality of the cover concrete is of particular importance", Section 5.2.2.7 of SIA 262 (2013)
- "the tightness of the cover concrete shall be checked, by means of permeability tests (e.g. air-permeability measurements), on the structure or on cores taken from the structure", Section 6.4.2.2 of SIA 262 (2013)

Therefore, since 2013, the air-permeability kT of the *Covercrete* of structural elements exposed to the most severe environments shall be checked on site, with the "Air-Permeability on the Structure" test, according to Annex E of SIA 262/1 (2019).

The requirements for site air-permeability are indicated in Row 8 of Table 1.3, the specified kT_s values being "characteristic" upper limits, having their own conformity rules (SIA 262/1, 2019; Torrent et al., 2012), see Section 8.5.1.

Table 1.3 Evolution of Swiss Standards requirements (for corrosion exposure classes)

Year	Row	Exposure class	*Carbonation-induced corrosion*				*Chloride-induced corrosion*			
			XC1	*XC2*	*XC3*	*XC4*	*XD1*	*XD2a*	*XD2b*	*XD3*
2003		**Durability indicator**	*PRESCRIPTIVE*							
	1	w/c_{max}	0.65	0.65	0.60	0.50	0.50	0.50	0.45	0.45
	2	C_{min} (kg/m³)	280	280	280	300	300	300	320	320
	3	$f'c_{min}$ (MPa)	25	25	30	37	30	30	37	37
	4	d_{nom} (mm)	20	35	35	40	40	40	55	55
2008		**Durability indicator**	*LABCRETE*							
	5	$q_{w\,max}$ (g/m²/h)	-	-	-	-	10	10	-	-
	6	$D_{Cl\,max}$ (10^{-12} m²/s)	-	-	-	-	-	-	10	10
	7	$K_{N\,max}$ (mm/y$^{1/2}$)	-	-	5.0/4.0	5.0/4.5	-	-	-	-
2013		**Durability indicator**	*REALCRETE*							
	8	kT_s (10^{-16} m²)	-	-	-	2.0	2.0	2.0	0.5	0.5

Note: EN 206 Class XD2 was subdivided in 2008 into XD2a and XD2b, for chloride contents of the solution in contact with the concrete of up to or over 0.5 g/L, respectively.

w/c, water/cement ratio by mass; *C*, cement content, including SCM with corresponding factors *k*; *f'c*, strength class (cube); q_w, water conductivity coefficient, Annex A of SIA 262/1 (2019). Rather complex indicator, closely related to water absorbed in 24 hours w_{24} (g/m²): $w_{24}=217+326\times q_w$; D_{Cl}=chloride migration coefficient, measured after Tang-Nilsson method (Section A.2.1.2); K_N=carbonation resistance=0.136 K_S, with K_S measured in an accelerated test after 7, 28 and 63 days exposure to CO_2 concentration of 4%-vol. (Annex I of SIA 262/1 (2019)). The values indicated correspond to expected service lives of 50/100 years; *kT*, coefficient of air-permeability, measured after Torrent method (Annex E of SIA 262/1 (2019)); value not be exceeded by more than 1 test out of 6; d_{nom}, nominal cover depth, values indicated are for reinforced concrete (values for prestressed concrete are 10 mm higher); typical tolerance ± 10 mm.

1.8 CONCRETE PERMEABILITY AS "DURABILITY INDICATOR"

In Section 1.2, we could see that most relevant mechanisms of deterioration of concrete structures have a close relation to the permeability of concrete.

It is not surprising, then, that several performance-based standards and codes (see Sections 1.7.3–1.7.7) select water-permeability (in the form of penetration under pressure or of capillary suction) or gas-permeability as durability indicator. In the particular case of the South African and Swiss

Standards, based on core testing and non-destructive measurements conducted on site, respectively, the end-product is tested, which is more representative than laboratory tests performed on cast specimens, as discussed in Chapter 7.

Being the main topic of this book, the suitability of concrete permeability as a durability indicator is broadly and deeply dealt with.

1.9 BEYOND 50 YEARS: MODELLING

Most requirements described in Sections 1.5 and 1.6 correspond to an expected service life of 50 years. Nowadays, important infrastructure constructions are intended for service lives that largely exceed the 50 years expected by the application of the prescriptive EN standards or the performance Swiss standards. Examples are the Alp Transit Tunnel in Switzerland (100 years) (Alp Transit, 2012), the new Panama Canal (100 years) (Cho, 2012), the Chacao Bridge in Chile (100 years) (Valenzuela & Márquez, 2014), the Hong Kong-Zhuhai-Macao link in China (120 years) (Li et al., 2015), the Port of Miami Tunnel in USA (150 years) (Torrent et al., 2013) and the second Brisbane Gateway Bridge in Australia (300 years) (Gateway, 2009), all of them exposed to very aggressive environments.

Due to the lack of experience with such longevous structures (reinforced concrete is a rather "recent" building system) from which to draw learnings, the solution lies on the judicious use of predictive models.

The most widespread model used today in Europe is Duracrete (DuraCrete, 2000), later partially adopted by *fib* (2006), dealing with steel corrosion induced by carbonation or chlorides, whilst in North America (Life-365, 2012) model (only for chloride-induced corrosion) is the preferred one. These models are based on the assumption that the penetration of chlorides (and carbonation) is a purely diffusive process governed by Fick's second law (see Chapter 3), with the main input durability indicators being the cover thickness and the coefficient of chloride-diffusion (or migration) of the concrete. In Chapter 9, several service life design models, based on the use of concrete permeability as input, are presented.

REFERENCES

ACI 318 (2011). "Building code requirements for structural concrete". ACI.

AIJ (2016). "Recommendations for durability design and construction practice of reinforced concrete buildings". Architectural Institute of Japan (in Japanese).

Alexander, M.G. (2004). "Durability indexes and their use in concrete engineering". International RILEM Symposium on Concrete Science and Engineering: 'A Tribute to Arnon Bentur', 9–22.

Alexander, M.G. (2016). *Marine Concrete Structures: Design, Durability and Performance*. Woodhead Publishing, Duxford, UK, 485 p.

Alexander, M.G., Ballim, Y. and Mackechnie, J.R. (1999). "Concrete durability index testing manual". Research Monograph No.4, Univs. Cape Town & Witwatersrand, South Africa, 33 p.

Alexander, M.G., Bentur, A. and Mindess, S. (2017). *Durability of Concrete: Design and Construction*. CRC Press, Boca Raton, FL, 345 p.

Alp Transit (2012). "Alp Transit Gotthard – New traffic route through the heart of Switzerland". *Brochure*, 48 p.

ASTM C1202 (2019). "Standard test method for electrical indication of concrete's ability to resist chloride ion penetration".

Bertolini, L., Elsener, B., Pedeferri, P. and Polder, R. (2004). *Corrosion of Steel in Concrete*. Wiley-VCH, Weinheim, Germany, 391 p.

Beushausen, H. (2014). "RILEM TC230-PSC: Performance-based specification and control of concrete durability". RILEM Week, São Paulo, Brazil, September, 42 slides.

Beushausen, H. and Alexander, M. (2009). "Application of durability indicators for quality control of concrete members – A practical example". *Concrete in Aggressive Aqueous Environments – Performance, Testing, and Modeling*, Alexander, M.G. and Bertron, A. (Eds.). RILEM Publications, Bagneaux, 548–555.

Böhni, H. (2005). *Corrosion in Reinforced Concrete Structures*. Woodhead Publishing, Cambridge, UK, 241 p.

Browne, R. (1989). "Durability of reinforced concrete structures". New Zealand Concrete Construction, September 2–10 and October 2–11.

Cho, A. (2012). "Dramatic digs mark panama canal expansion progress". *Engineering News-Record*, July 18.

CIRSOC 201 (2005). "Reglamento Argentino de Estructuras de Hormigón". Argentine Concrete Code.

Collepardi, M., Troli, R. and Collepardi, S. (2018). "Ponte di Genova: alcune considerazioni sul calcestruzzo e gli agenti esterni che ne hanno ridotto la durabilità". Ingenio-web.it, August 22, 5 p.

Coppola, L. (2000). "Concrete durability and repair technology". 5th *CANMET/ACI International Conference On 'Durability of Concrete'*, Barcelona, Spain, June 4–9, 1209–1220.

CO3-2 (2015). "Civil engineering test methods. Part CO3-2: Concrete durability index testing – Oxygen permeability test". South African Standard SANS 3001-CO3-2:2015.

CO3-3 (2015). "Civil engineering test methods. Part CO3-3: Concrete durability index testing – Chloride conductivity test". South African Standard SANS 3001-CO3-3:2015.

CR 13902 (2000). "Test methods for determining the water/cement ratio of fresh concrete". CEN Report, European Committee for Standardization, May, 7 p.

CSA (2004). Canadian Standard A23.1/A23.2: "Concrete materials and methods of concrete construction/methods of test and standard practices for concrete".

de Schutter, G. (2009). "How to evaluate equivalent concrete performance following EN 206-1? The Belgian approach". *PRO 66: Concrete Durability and Service Life Planning – ConcreteLife'09*. Kovler, K. (Ed.). RILEM Publications, Haifa, Israel, 1–7. ISBN: 978-2-35158-074-5.

de Schutter, G. (2014). "No concrete is sustainable without being durable!". XIII DBMC, São Paulo, Brazil, 49–56.

de Sitter, W.R. (1984). "Costs for service life optimization: The Law of Fives". Durability of Concrete Structures, Workshop Report, Rostam, S. (Ed.), Copenhagen, Denmark, May 18–20, 131–134.
DuraCrete (1998). "Modelling of degradation". The European Union–Brite EuRam III, BE95–1347/R4–5, CUR, Gouda, The Netherlands.
DuraCrete (2000). "Probabilistic performance based durability design of concrete structures". The European Union–Brite EuRam III, BE95–1347/R17, CUR, Gouda, The Netherlands.
Dyer, T. (2014). *Concrete Durability*. CRC Press, Boca Raton, FL, 402 p.
EHE-08 (2008). "Instrucción de Hormigón Estructural". Spanish Concrete Code.
EN 206 (2013). "Concrete – Specification, performance, production and conformity", December.
EN 1990-1-1 (2002). "Basis of structural design".
EN 1992-1-1 (2004). "Eurocode 2: Design of concrete structures – Part 1-1: General rules and rules for buildings", December.
EN 12390-8 (2009). "Testing hardened concrete – Part 8: Depth of penetration of water under pressure".
Fernández Luco, L. (2005). "RILEM recommendation of TC 189-NEC: Comparative test – Part II – Comparative test of *Covermeters*". *Mater. & Struct.*, v38, 907–911.
Fernández Luco, L. and Torrent, R. (2003). "Diagnosis of a case of harmless alkali-silica reaction in a cracked concrete pavement". 6th *CANMET/ACI International Conference on Durability of Concrete*, Thessaloniki, Greece, June 1–7, 521–536.
fib (2006). "Model code for service life design". *fib Bulletin 34*, February, 112 p.
fib (2010). "Model code 2010". 1st Complete Draft, v1, March.
Gateway (2009). "A second Gateway Bridge". Gateway Upgrade Project, Fact Sheet 4, February.
Gjørv, O.E. (2014). *Durability Design of Concrete Structures in Severe Environments*. 2nd ed., CRC Press, Boca Raton, FL, 254 p.
Hoff, G.C. (1999). "Integrating durability into the design process". *Controlling Concrete Degradation*. Dhir, R.K., Newlands, M.D. (Eds.). Thomas Telford, London, 1–14.
Horgan, R. (2020). "Polcevera viaduct | Fatal collapse replacement bridge to be completed this month". New Civil Eng., April 07.
Hunkeler, F. (2000). "Corrosion in reinforced concrete: Processes and mechanisms". *Corrosion in Reinforced Concrete Structures*. Böhni, H. (Ed.). CRC Press, Cambridge, UK, 1–45.
Jacobs, F. and Leemann, A. (2007). "Betoneigenschaften nach SN EN 206-1". ASTRA Report VSS Nr. 615, Bern.
Li, K. (2016). *Durability Design of Concrete Structures*. Wiley, Singapore, 280 p.
Li, Q., Li, K.F., Zhou, X., Zhang, Q. and Fan, Z. (2015). "Model-based durability design of concrete structures in Hong Kong–Zhuhai–Macau sea link project". *Structural Safety*, v53, March, 1–12.
Life-365 (2012). "Service Life Prediction Model™ and computer program for predicting the service life and life-cycle cost of reinforced concrete exposed to chlorides". *Life-365 Consortium II*, 80 p.
LNEC E 391 (1993). "Concrete. Determination of carbonation resistance", May.

LNEC E 465 (2007). "Concrete: Methodology for estimating the concrete performance properties allowing to comply with the design working life of the reinforced or prestressed concrete structures under the environmental exposures XC and XS", November.

Mara, V., Haghani, R. and Harryson, P. (2014). "Bridge decks of fibre reinforced polymer (FRP): A sustainable solution". *Constr. & Build. Mater.*, v50, January, 190–199.

Mehta, P.K. (1997). "Durability – Critical issues for the future". *Concr. Intern.*, July, 27–33.

Mehta, P.K., Schiessl, P. and Raupach, M. (1992). "Performance and durability of concrete systems". 9th *Congress on the Chemistry of Cement*, New Delhi, India, 571–659.

Mejlbro, L. (1996). "The complete solution of Fick's second law of diffusion with time-dependent diffusion coefficient and surface concentration". Durability of Concrete in Saline Environment, Cementa AB, Danderyd, Sweden, 127–158.

Moro, F. and Torrent, R. (2016). "Testing fib prediction of durability-related properties". *fib* Symposium 2016, Cape Town, South Africa, 21-23 Nov.

Neville, A. (1997). "Maintenance and durability of structures". *Concr. Intern.*, November, 52–56.

Neville, A. (1998). "Concrete cover to reinforcement — or cover up?" Concr. Intern., v20, n11, November, 25–29.

Neville, A. (2003). *Neville on Concrete.* ACI, Farmington Hill, MI.

Nilsson, L.-O. (2012). "Transport processes in the microstructure of concrete and their relevance for durability". Keynote paper, Microdurability, Amsterdam, April 11–13.

Nilsson, L., Ngo, M.H. and Gjørv, O.E. (1998). "High performance repair materials for concrete structures in the port of Gothenburg". *CONSEC* 1998, Tromsø, Norway, June 21–24, v.2, 1193–1198.

Noguchi, T., Kanematsu, M. and Masuda, Y. (2005). "Outline of recommendations for durability design and construction practice of reinforced concrete buildings in Japan". ACI SP 234, 347–372.

Parrott, P.J. (1984). "Design for avoiding damage due to carbonation-induced corrosion". 3rd *International Conference on Durability of Concrete*, Nice, France, May, 283–298.

Peacock, W.J. (1985). "The maintenance of buildings and structures – The problem, some causes and remedies". *Proceedings on Thomas Telford Seminar*, "Improvement in concrete durability". Institution Civil Engs., London, May, 131–161.

Richardson, M.G. (2002). *Fundamentals of Durable Reinforced Concrete.* Spon Press, London, 254 p.

RILEM Report 40 (2007). "Non-destructive evaluation of the penetrability and thickness of the concrete cover". State-of-the-Art Report of RILEM TC 189-NEC, Torrent, R. and Fernandez Luco, L. (Eds.), 223 p.

RILEM STAR 18 (2016). "Performance-based specifications and control of concrete durability". *State-of-the-Art Report* Vol. 18, RILEM TC 230-PSC, Beushausen, H. and Fernandez Luco, L. (Eds.), 373 p.

Rodriguez, J., Ortega, L.M., Casal, J. and Diez, J.M. (1996). "Corrosion of reinforcement and service life of concrete structures". *DBMC 7*, Stockholm, Sweden, May 19–23, v.1, 117–126.

Rodriguez-Vera, R.E., Lombardi, N.J., Machado, M.A., Liu, J. and Sotelino, E.D. (2011). "Fiber reinforced polymer bridge decks". *Publication FHWA/IN/JTRP*-2011/04. Joint Transportation Research Program, Indiana DoT and Purdue Univ., West Lafayette, Indiana, 113 p.

Scott, R. (2010). "FRP bridge decking – 14 years and counting". Reinf. Plastics, January/February.

Seitz, P. (2019). "System Nummer 9 – Einsturz der Morandi-Brücke in Genua". *TEC21*, Zürich, 7–8, 20–25.

SIA 262 (2013). "Betonbau".

SIA 262/1 (2019). "Betonbau – Ergänzende Festlegungen".

SN EN 206 (2013). "Beton – Teil 1: Festlegung, Eigenschaften, Herstellung und Konformität".

Torrent, R.J. (2018). "Bridge durability design after EN standards: Present and future". *Struct. & Infrastruct. Engng.*, DOI: 10.1080/15732479.2017.1414859, 14 p.

Torrent, R., Alexander, M. and Kropp, J. (2007). "Introduction and problem statement". *RILEM Report 40*, May, 1–11.

Torrent, R., Armaghani, J. and Taibi, Y. (2013). "Evaluation of port of Miami tunnel segments: Carbonation and service life assessment made using on-site air permeability tests". *Conc. Intern.*, May, 39–46.

Torrent, R., Denarié, E., Jacobs, F., Leemann, A. and Teruzzi, T. (2012). "Specification and site control of the permeability of the cover concrete: The Swiss approach". *Mater. & Corrosion*, v63, n12, December, 1127–1133.

Torrent, R. and Jacobs, F. (2014). "Swiss standards 2013: World's most advanced durability performance specifications". 3rd *All-Russian Conference on Concrete and Reinforced Concrete*, Moscow, May 12–16.

Tuutti, K. (1982). "Corrosion of steel in concrete". *Research report No.4.82*. Swedish Cement and Concrete Research Institute (CBI), Stockholm.

Valenzuela, M.A. and Márquez, M.A. (2014). "Consideraciones para la Inspección y Mantenimiento del Puente Chacao". CINPAR, Santiago, Chile, June 4–6.

Vanier, D.J. (1999). "Why industry needs asset management tools". NRCC Seminar Series 'Innovations in Urban Infrastructure', 11–25.

Virlogeux, M. (2019). "Damals glaubte man, Beton halte ewig". *TEC21*, Zürich, 7–8, 26–29.

Wallbank, E.J. (1989). "The performance of concrete in bridges". HMSO, London, April 1989.

Wassermann, R., Katz, A. and Bentur, A. (2009). "Minimum cement content requirements: A must or a myth?. *Mater. & Struct.*, v42, 973–982.

Wolfseher, R. (1998). "Economical aspects of repair and maintenance". 5th *International Conference on Materials Property & Design, 'Durable Reinforced Concrete Structure'*, Weimar, Germany, October, 33–48.

Chapter 2

Permeability as key concrete property

2.1 FOUNDATIONS OF PERMEATION LAWS

The foundations of today's knowledge on the permeation of fluids through porous media were laid down by the work of Jean Léonard Marie Poiseuille (1799–1869), a French physician and physiologist, who was interested in the conditions of the flow of liquids through narrow tubes, basically associated with the arterial system of blood circulation. He conducted a series of experiments, from which he established that the flow rate of a fluid through a tube of radius r is proportional to r^4. Independently, the German civil engineer Gotthilf Heindrich Ludwig Hagen (1797–1884) arrived at the same result by conducting experiments in brass tubes of different diameters, concluding in what is now known as Hagen-Poiseuille law (see Section 3.5 1).

More or less simultaneously, the French engineer Henry Darcy (1803–1858) was studying the laminar flow of water through sand beds, finding that the flow rate was proportional to the energy loss (water head loss), inversely proportional to the length of the flow path and proportional to a coefficient K that depended on the type of sand and also on the type of fluid.

The combination of these discoveries led to the general law of permeation of liquids through porous media (viscous laminar flow of Newtonian liquids):

$$Q = K \cdot \frac{A}{\mu} \cdot \frac{\Delta P}{\Delta L} \tag{2.1}$$

where

Q = flow rate (m^3/s)
K = (intrinsic) coefficient of permeability (m^2)
A = cross-sectional area traversed by the fluid (m^2)
μ = viscosity of the fluid (Pa.s)
$\Delta P/\Delta L$ = gradient of pressure across the element (Pa/m)

DOI: 10.1201/9780429505652-2

In the case of an ideal impermeable solid body traversed by parallel capillary tubes of radius r, the coefficient of permeability is (see derivation of formulae in Section 3.5.1):

$$K = \frac{\varepsilon \cdot r^2}{8} \tag{2.2}$$

where ε is the porosity of the body (area of tubes/total cross-sectional area of the body).

2.2 RELATION BETWEEN PERMEABILITY AND PORE STRUCTURE OF CONCRETE

Equation (2.2) indicates that the coefficient of permeability of concrete, recognized as a porous medium, must be closely related to the pore structure of the material.

One of the main investigations on the permeability of cementitious material was due to the researcher who, possibly, did more to establish studies on concrete as a scientific, rather than an empirical discipline: Treval C. Powers (1900–1997).

During his fundamental research on the microstructure of hardened cement paste (h.c.p.), still valid today, and its effect on key properties, he could establish a relationship between the coefficient of water-permeability of h.c.p. and its capillary porosity, quite independent of the cement types investigated at the time (Powers, 1958). He also found that, due to the extremely low size of the gel pores, flow through h.c.p. takes place primarily through its capillary pores.

He also established the approximate hydration time required for the capillary pores of h.c.p. of different *w/c* ratios to become segmented, i.e. connected between them through the gel pores, resulting in very low permeability (Powers et al., 1959). They found that for *w/c* ratios above 0.70 that segmentation is impossible, as shown in Table 1.6 of Neville (1995).

Since that pioneer work, the permeability of concrete received growing attention by researchers worldwide, which resulted in a consolidated knowledge on that property, on how it is influenced by different factors and on how it can be measured, both in the laboratory and on site. These aspects are dealt with in detail in the rest of this book.

2.3 PERMEABILITY AS KEY CONCRETE PROPERTY

Water-permeability of concrete is relevant to structures that contain or transport water (or other liquids), in particular dams, tanks containing water or other liquids, retaining walls, canals, culverts, pipes, etc.

Similarly, gas-permeability of concrete is relevant to structures that contain or transport gases, in particular tanks and pipes, underground gas reservoirs to store/release energy, evacuated tunnels for high speed trains, etc.

Gas-permeability plays an important role in the release of water vapour under fire, thus decreasing the risk of explosive spalling in the event of fire, topic that is discussed in detail in Chapter 10.

In this section, some engineering applications in which concrete permeability plays a key role, not specifically associated with durability, are presented.

2.3.1 Permeability for Liquids' Containment

2.3.1.1 ACI Low Permeability Concrete

Section 4.3 of ACI 318 (2019) includes exposure class P1 "Low Permeability Requirement", assigned on the basis of the need for concrete to have a low permeability to water, when the permeation of water into concrete might reduce durability or affect the intended function of the structural member. An example is an interior water tank. Requirements: $w/b \leq 0.50$ and cylinder compressive strength class ≥ 28 MPa.

2.3.1.2 Dams

Conventional concrete for dams is usually sufficiently water-tight to avoid leakage across the thick body of the dam. For conventional concrete dams, built in lifts, construction joints as well as expansion joints are the weak points regarding water-tightness. An interesting research was reported by Görtz et al. (2021), in which the water-tightness of the joints was measured experimentally and modelled numerically. The coefficient of permeability was measured on Ø64.5 mm cores, drilled from a 90-year-old dam in Germany, so as to obtain specimens without joints and with horizontal and vertical joints. The measured water-permeability of the specimens without joints was in the range $0.5-3.0\times10^{-9}$ m/s, whilst for those containing horizontal and vertical joints it climbed to the ranges $5-100\times10^{-9}$ m/s and $1-30\times10^{-6}$ m/s, respectively (i.e. one and three orders of magnitude higher, respectively). Two numerical models were applied, that successfully fit to the experimental results, especially the 'dual-permeability model'.

In the case of two concrete-face rockfill dams (Barrancosa and Condor Cliff) on the River Santa Cruz, Patagonia, Argentina, a maximum value for the water-permeability of 2×10^{-9} cm/s was specified for the upstream concrete slab (Di Pace, 2021). Due to difficulties in measuring that property, an equivalent value of the coefficient of air-permeability (Torrent method) of 0.2×10^{-16} m² was proposed, applying the relation:

$$K_w = 6.24 \cdot kT^{0.68} \tag{2.1}$$

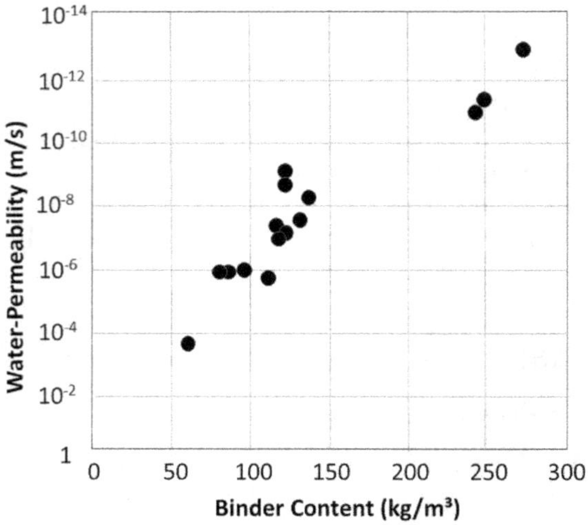

Figure 2.1 Effect of binder content on the water tightness of several RCC dams, data from Dunstan (1988).

where K_w=water-permeability (10^{-9} cm/s) and kT=Torrent air-permeability (10^{-16} m^2).

The relation in Eq. (2.1) was derived from the investigation reported by Sakai et al. (2013), discussed in Section 3.5.5 (Figure 3.12).

In the case of roller-compacted concrete (RCC) dams, the material's characteristics and the construction techniques present some challenges in this respect. The permeability of the RCC mass and that of the horizontal lift surfaces are key elements for the performance of hydraulic RCC structures (USACE, 2000). The permeability of RCC depends on the mix proportioning (especially on the binder content); a survey conducted by Dunstan (1988) showed that the water-tightness of RCC dams (measured *in situ*) increased with the amount of binder in the mix, see Figure 2.1.

The permeability of the dam depends also on the placing techniques and the use of bedding mortar along lift surfaces, sometimes complemented by the use of impermeable membranes in the upstream face, as used for Urugua-í dam in Argentina, the RCC of which contained the record low 60 kg/m^3 of cement (one of the authors was involved in its mix design) (Dam Search, 2021). Yet, the dam was not 100% watertight (Schrader, 2003).

2.3.1.3 Pervious Concrete

In all the cases discussed in this chapter, a performing concrete is one having low permeability to either gases or liquids. In the case of pervious concrete, the opposite is true, as the goal is to achieve a pavement concrete that is sufficiently strong to resist the applied loads, but highly permeable to allow the drainage of water from its surface through the body of the material.

Pervious concrete is a special type of concrete which has an open structure which allows water to freely percolate through the pavement into the ground. The concrete is manufactured with uniform, open-graded coarse aggregate, cement and water and little or no fines. Elimination of fines creates a void structure in the finished concrete which allows fluids to rapidly pass through the pores and into the subgrade. This characteristic is very useful for reducing the rate and quantity of storm water runoff. The void structure, when coupled with an aggregate subbase, will slow down the rate of runoff and store substantial quantities of storm water.

Pervious concrete is aimed for flatwork concrete applications which have a direct impact on storm water management. These include pedestrian walkways, parking areas, residential streets and areas with light traffic. Its open structure makes pervious concrete a good sound insulator and is also used for noise barrier construction.

The void content of pervious concrete can range from 15% to 35%, with typical compressive strengths of 3.0–30 MPa. The drainage rate of pervious concrete pavement will vary with aggregate size and density (and strength) of the mixture, but will generally fall into the range from 0.15 to 1.25 cm/s (ACI 522R, 2010).

Test methods exist to measure the draining rate of a pervious concrete, typically the constant head and the falling head permeameters (Sandoval et al., 2017). One of the latter has been standardized as ASTM C1701 (2009), which can be applied in the field.

2.3.1.4 Liquid Gas Containers

Storage containers for liquified gases, typically liquified natural gas (LNG), are made with prestressed concrete, with or without an alloy steel liner. Avoiding the liner is more cost effective and safer (precluding the risk of rupture of the steel liner), but the challenge is to place a concrete that is sufficiently liquid-tight at cryogenic temperatures (Hanaor, 1985). An important aspect of the mix design is the thermal compatibility of aggregates and cement paste, so as to minimize differential internal thermal stresses and microcracks formation. In ACI 376 (2011), a maximum intrinsic concrete permeability to cryogenic fluids (typically liquid N_2) of 0.01×10^{-16} m^2 is specified, remarking that the cryogenic permeability of even partially-dried concrete is approximately half that obtained at ambient temperature. This reduction in permeability is associated with the formation of ice in the concrete pores, blocking them and also improving the mechanical performance (higher strength, E-modulus, tensile strain capacity, etc.) (Kogbara et al., 2013).

2.3.2 Permeability for Gas Containment

2.3.2.1 Evacuated Tunnels for High-Speed Trains

The "Swissmetro" project was launched in Switzerland in the early 1990s, aimed at establishing a high-speed Maglev (magnetic levitation) rail system. Trains, under magnetic sustentation, would run at speeds of up to 500 km/h, exclusively through small Ø5 m tunnels. To achieve such speeds economically, it was planned to operate the trains along partly evacuated tunnels. Details on the project can be found in Cassat et al. (2003) and Cassat and Espanet (2004); the project was abandoned, reportedly due to lack of political and financial support (RTS, 2010).

However, the concept was revived around 2012 in the USA, with the name "Hyperloop", with the more ambitious goal of trains running at twice the speed of planes, even at hypersonic speeds. To operate efficiently, the system requires magnetically levitated trains to run on tunnels evacuated to air pressures of the order of 1 mbar (100 Pa).

In the case of the Swissmetro, the trains would operate along concrete-lined tunnels, whilst Hyperloop considers steel tubes. It is clear that, for the former, a low air-permeability concrete is required to achieve and maintain the required low air-pressure inside the concrete tunnels efficiently. For that purpose, air-permeability tests (Cembureau method, described in Section 4.3.1.2) on cores drilled from panels simulating the Swissmetro tunnels walls were performed at TFB laboratory in Switzerland (Badawy & Honegger, 2000), as well as using the Torrent method (see Chapter 5) (Badoux, 2002). The role of cracks on the air-permeability was also investigated (Badoux, 2002), a topic dealt with in Section 6.11.

Probabilistic numerical modelling of the air-tightness of concrete-lined tunnels has been performed in the context of the Korean Super-Speed Tube Transport (SSTT), designed for trains running at 700 km/h. The modelling involved parameters such as concrete permeability and wall thickness, as well as the diameter of the tube structure, complemented by experimental investigation on the effect of joints and connections of the concrete tube (Park et al., 2015). The modelling was extended to include the effect of cracks on the performance of the evacuated tunnels (Devkota & Park, 2019).

2.3.2.2 Underground Gas "Batteries"

It is of economic interest to store the energy surplus, generated in periods of low demand, in such a way that it can be recovered in periods of high demand. This has been traditionally done, in mountainous geographical regions, by pumping up water to hydropower reservoirs, to be used later as hydroelectric energy.

Compressed air energy storage (CAES) is a concept in which the surplus energy generated by wind turbines and solar cells is used to compress air, stored in underground caverns in solid bedrock (SINTEF, 2017). At the appropriate time, the air pressure is released through a gas turbine that generates electricity. The cavern wall structure is capable, in interaction with the surrounding rock, to resist pressures of over 20 MPa and high-frequency operations. Although the structural function of the reservoir is fulfilled by concrete elements, some such systems rely on steel lining inside the gas reservoir for gas tightness (Tengborg et al., 2014).

In other cases, concrete linings of low permeability that are appropriately reinforced against potential tensile fracturing due to high air storage pressure, allow the more flexible realization of underground CAES and result in significant construction cost reduction. The system relies on the permeability of the concrete lining and of the surrounding rock for the design of long-term air-tightness performance. Numerical modelling indicated that a concrete lining with air-permeability less than 0.01×10^{-16} m^2 would result in an acceptable air leakage rate of less than 1%, with the operation pressure ranging between 5 and 8 MPa at a depth of 100 m (Kim et al., 2012). In order to get experimental data to feed the model, large-scale mock-up permeability tests were conducted, comprising not just the concrete itself, but also the joints and their treatment, the latter becoming critical for the air-tightness of the system (Kim et al., 2014).

2.3.3 Permeability for Radiation Containment

2.3.3.1 Radon Gas

Radon (Rn) is a colourless, odourless and tasteless radioactive noble gas which occurs naturally in soils in amounts depending on the local geology (CIP 18, 2000). Some concerns on the presence of this gas exist due to its association with the development of lung cancer. This happens when it accumulates, due to its high density, in low areas like basements or crawl spaces. Radon decays to other radioactive elements in the uranium series, called "radon progeny" that exists as solid particles that can become attached to dust particles in the air. If inhaled, they can lodge in the lung, where they can cause cancerous tissue damage due to energy emitted during radioactive decay. Concrete constitutes an effective barrier to radon penetration if cracks and openings are sealed. The World Health Organization (WHO) regards Rn as a health hazard, constituting the second cause of lung cancer, after tobacco.

To quantify the problem, in Canton Ticino (Switzerland), a huge campaign of site measurements was run, collecting data from ca. 50,000 dwellings by 2010 (LC, 2010). Out of them, 91% had Rn concentration values below 400 Bq/m^3, 7% within 400–1,000 Bq/m^3 and 2% above 1,000 Bq/m^3. The Swiss Federal Bureau of Public Health establishes a limit of 1,000 Bq/m^3 for dwellings. For new constructions or refurbishing, the applicable limit is 400 Bq/m^3, with a recommendation of not exceeding 300 Bq/m^3, on the basis of WHO guidelines (LC, 2010).

The main transport mechanism of penetration of Rn from the soil into the dwellings is gas-permeability, through the so-called "chimney-effect", by which the warm air moving upwards generates a small depression in the ground floor and basement, that aspirates gas from the surrounding ground. This is aggravated in winter time by heating.

Calculations of the indoor radon entry rate by pressure-driven flow rely on the permeability coefficient of the concrete. The coefficient of air-permeability is the key transport parameter; in typical radon problems, the pressure difference across the foundation walls may range between 1 and 20 Pa (Abu-Irshaid & Renken, 2002). A test method to measure the coefficient of air-permeability, especially for that purpose, has been developed (Ferguson et al., 2001).

The relation between Rn concentration and the coefficients of gas diffusion and air-permeability is discussed in Rogers et al. (1995), based on experimental measurements performed on 25 samples of new residential concretes in Florida (USA). It is claimed that the low permeability of concrete, even microcracked, compared with that of soils, makes it a very suitable material to protect from Rn emissions (Piedecausa García & Chinchón-Payá, 2011). An attempt to estimate the coefficient of radon-diffusion in concrete from pore structure measurements was presented in Linares (2015).

A special laboratory test method to measure the permeability to Rn of cementitious composites has been developed at SUPSI (Teruzzi & Antonietti, 2019). The same authors are investigating the relation between Rn permeability and air-permeability using the Torrent method (Chapter 5) (Teruzzi, 2020).

2.3.3.2 Nuclear Waste Disposal Containers

The adoption of nuclear power energy for the production of electricity includes, among the formidable challenges it poses, that of the disposal and storage of radioactive waste. The increasing amount of low and medium radioactive waste needs a serious concept of a long-term policy in radioactive waste management. Periods in the range of 300–1,000 years are considered in which the storing facilities have to guarantee the safety of human population and environment against radiation.

The design and construction of many of the facilities and structures for long-term storage of radioactive waste materials employ reinforced concrete for support, containment, and environmental protection functions. During the desired very long service life of these structures, the reinforced concrete is to provide both physical and chemical barriers to isolate the waste from the environment.

Citing Naus (2003):

> "Although a number of deteriorating influences can affect these structures, corrosion of embedded steel, leaching, elevated temperature, and irradiation (depending of the application) probably represent the

greatest initial threats. Over the long term, leaching and cracking have increased importance as water will provide the transport medium for radionuclides should the other engineered barriers fail. In nearly all chemical and physical processes influencing the durability of concrete structures, dominant factors involved include transport mechanisms within the pores and cracks, and the presence of a fluid. Concrete permeability therefore is of significant importance relative to the long-term durability of radioactive waste facilities. Concrete permeability will vary according to such things as the proportions of constituents, degree of cement hydration, cement fineness, aggregate gradation, and moisture content. One of the most important factors affecting ionic transport through concrete is the presence of cracks. Cracking not only controls the quantity of ions transported, but can also control whether there will be any convective transport. Factors that contribute to increases in concrete permeability or cracking therefore are of importance to the durability of the radioactive waste management facilities."

To achieve the required performance, the use of high-performance steel fibre reinforced concrete was advocated in Slovakia, as reported by Hudoba (2007), who presents detailed drawings and pictures of the underground chambers with containers position and the act of placing a container in the storage chamber.

Within this context, Kubissa et al. (2018) investigated the Autoclam air-permeability index (API), see Section 4.3.2.7, and moisture distribution on cores drilled from heavy-weight concretes (density around 3,250 kg/m^3) of large-scale mock elements.

In Argentina, O_2-Permeability tests (Cembureau method, see Section 4.3.1.2) and site air-permeability tests (Torrent method, see Chapter 5) were performed on a concrete container to assess its long-term durability (Ramallo de Goldschmidt, 2004). Similar investigations were developed on cement-based materials for radioactive waste repositories in Japan (Kurashige et al., 2009 and Niwase et al., 2012).

A report of the International Atomic Energy Agency (IAEA, 2013) emphasizes the relevance of concrete permeability for the durability of nuclear waste storage facilities, stating that porosity and permeability of Cement Waste Products, function of their composition, are directly related to diffusion and leaching characteristics of incorporated radionuclides. In Table 13 of IAEA (2013), three test methods applicable to measure the permeability of cementitious materials are listed, namely: water-permeability (ASTM D5084, 2010), gas-permeability (ASTM C577, 2019) and air-permeability (Torrent method, described in detail in Chapter 5). In Table 17 of IAEA (2006), three NDT methods for measuring air-permeability are listed: Surface airflow method (Whiting & Cady, 1992) and, again, the Torrent method. In addition, test methods to measure water sorptivity are also included (all described in Chapter 4): ISAT, Figg and Autoclam (the latter, also for air-permeability).

Unless sensors are left inside the containers, concrete permeability can be measured only before they are put into service. There are indications that the gas and water-permeability of concrete exposed to radiation may increase with age, based on tests performed on cores drilled from the recovery of a nuclear reactor after 25 years of service (Mills, 1990).

Monitoring of air-permeability (Torrent method), together with carbonation, electrical resistivity and electrochemical tests, was performed on concrete of the El Cabril disposal containers (Andrade et al., 2003).

With regard to nuclear waste disposal, not just the permeability of the container matters, but also that of the surrounding rock, see Section 11.5.2.2.

2.4 PERMEABILITY AND DURABILITY

As discussed in Sections 1.2 and 1.9, most deterioration actions affecting the durability of concrete structures (carbonation, chlorides and sulphates ingress, chemical attack and even frost) are related to the penetration of aggressive agents from the environment into the concrete. The mechanisms by which this penetration takes place are basically two: permeation (that includes capillary suction) and diffusion, which are described in detail in Chapter 3. Whatever the mechanism, a concrete that has an open pore structure (i.e. more and larger pores) or presents micro-cracks will deteriorate at a higher rate than a concrete with a tighter pore structure. The former will show a higher permeability than the latter, see Eq. (2.2), suggesting that the coefficient of permeability (to gases or liquids) is a sensitive indicator of the "penetrability" of a given concrete; this is discussed in detail in Section 3.8. Hence, a direct relation exists between the permeability of concrete and its durability performance (as this book is titled).

In this respect, it is worth quoting Prof. K. Mehta (Mehta, 1991) who, in his review of the durability situation and research progress made at that time, expressed the following:

> "It seems that, in spite of some important discoveries valuable from the standpoint of durability enhancement, today more concrete structures seem to suffer from lack of durability than was the case 50 years ago. In order of decreasing importance, the major causes of concrete deterioration today are as follows: corrosion of reinforcing steel, frost action in cold climates, and physico-chemical effects in aggressive environments. There is a general agreement that the permeability of concrete, rather than normal variations in the composition of Portland cement, is the key to all durability problems."

This statement has full validity even today.

The rest of this book is dedicated to studying the permeability of concrete and the relation of this important property with the durability of concrete structures.

REFERENCES

Abu-Irshaid, E. and Renken, K.J. (2002). "Relationship between the permeability coefficient of concrete and low-pressure differentials". *2002 International Radon Symposium Proceedings*, 13 p.

ACI 318 (2019). "Building code requirements for structural concrete".

ACI 376 (2011). "Code requirements for design and construction of concrete structures for the containment of refrigerated liquefied gases". American Concrete Institute, 170 p.

ACI 522R (2010). "Report on pervious concrete", 43 p.

Andrade, C., Sagrera, J.L., Martínez, I., García, M. and Zuloaga, P. (2003). "Monitoring of concrete permeability, carbonation and corrosion rates in the concrete of the containers of El Cabril (Spain) disposal". *Trans. SMiRT 17*, Prague, Czech Rep., August 17–22, Paper #O03-2, 8 p.

ASTM C577 (2019). "Standard test method for permeability of refractories", 5 p.

ASTM C1701 (2009). "Standard test method for infiltration rate of in place pervious concrete", 3 p.

ASTM D5084 (2010). "Standard test methods for measurement of hydraulic conductivity of saturated porous materials using a flexible wall permeameter", 23 p.

Badawy, M. and Honegger, E. (2000). "Swissmetro – Tests on air permeability of concrete". *16th IABSE Congress*, Lucerne, Switzerland, 8 p.

Badoux, M. (2002). "Experimental research for the liners of the Swissmetro tunnels". *MAGLEV 2002*, Lausanne, Switzerland, September, Paper PP05107, 8 p.

Cassat, A., Bourquin, V., Mossi, M., Badoux M., Vernez, D., Jufer, M., Macabrey, N. and Rossel, P. (2003). "SWISSMETRO – Project development status". *STECH '03*, Tokyo, Japan, 8 p.

Cassat, A. and Espanet, C. (2004). "SWISSMETRO: Combined propulsion with levitation and guidance". *MAGLEV 2004*, Shanghai, China, October 26–28, 10 p.

CIP 18 (2000). "Radon resistance buildings". Concrete in Practice, NRMCA, 2 p.

Dam Search (2021). "Database of 906 RCC Dams", http://www.rccdams.co.uk/dam-search/.

Devkota, P. and Park, J. (2019). "Analytical model for air flow into cracked concrete structures for super-speed tube transport systems". *Infrastructures*, v4, n76, 13 p.

Dunstan, M.R.H. (1988). "Wither roller compacted concrete for dam construction?". *Roller Compacted Concrete II*. Hansen, K.D. (Ed.). A.S.C.E., New York, pp. 294–308.

Di Pace, G. (2021). Private communication, June.

Ferguson, L.J., Daoud, W.Z. and Renken, K.J. (2001). "Further measurements on the permeability coefficient of a concrete sample under low pressure differences". *The* 2001 *International Radon Symposium*, Pre-Prints, 78–86.

Görtz, J., Wieprecht, S. and Terheiden, K. (2021). "Dual-permeability modelling of concrete joints". Constr. & Buildg. Mater., v302, October, 124090.

Hanaor, A. (1985). "Microcracking and permeability of concrete to liquid nitrogen". *ACI J.*, March–April, 147–153.

Hudoba, I. (2007). "Utilization of concrete as a construction material in the concept of Radioactive Waste Storage in Slovak Republic". *Acta Montanistica Slovaca*, v12, n1, 157–161.

IAEA (2006). *Ageing Management of Concrete Structures in Nuclear Power Plants*. Series No. NP-T-3.5. International Atomic Energy Agency, Vienna, 372 p.

IAEA (2013). *The Behaviours of Cementitious Materials in Long Term Storage and Disposal of Radioactive Waste*. Report IAEA-TECDOC-1701. International Atomic Energy Agency, Vienna, 75 p.

Kim, H.-M., Lettry, Y., Ryu, D.-W., Synn, J.-H. and Song, W.-K. (2014). "Mock-up experiments on permeability measurement of concrete and construction joints for air tightness assessment". *Mater. & Struct.*, v47, n1, 127–140.

Kim, H., Rutqvist, J., Ryu, D., Choi, B., Sunwoo, C. and Song, W. (2012). "Exploring the concept of compressed air energy storage (CAES) in lined rock caverns at shallow depth: A modeling study of air tightness and energy balance". *Appl. Energy*, v92, April, 653–667.

Kogbara, R.B., Iyengar, S.R., Grasley, Z.C., Masad, E.A. and Zollinger, D.G. (2013). "A review of concrete properties at cryogenic temperatures: Towards direct LNG containment". *Constr. & Bldg. Mater.*, v47, October, 760–770.

Kubissa, W., Glinicki, M.A. and Dąbrowski, M. (2018). "Permeability testing of radiation shielding concrete manufactured at industrial scale". *Mater. & Struct.*, v51, Article 83, 15 p.

Kurashige, I., Hironaga, M., Yoshinori, M. and Kishi, T. (2009). "Quality inspection system of cement-based materials supporting performance confirmation for radioactive waste repository. Part 1. Non destructive evaluation of rebound number and air permeability of surface concrete of in-situ structures and laboratory specimens". *Manag. Radioact. Wastes Non-Radioact. Wastes Nucl. Facilit.* (S12), v41, n15, August, 1–15 (in Japanese).

LC (2010). "Radon nelle abitazioni ticinesi: resoconto campagna 2009–2010 (Lugano campagna) e conclusione del programma di misurazione a tappeto 2005–2010". Labor. Cantonale, Bellinzona, January 5, 17 p.

Linares, P. (2015). "Caracterización del hormigón en relación a la difusión de gases y su correlación con el Radón". PhD Thesis, Univ. Polit. Madrid, Spain, 220 p.

Mehta, P.K. (1991). "Durability of concrete – Fifty years of progress?" ACI SP-126, 1–32.

Mills, R.H. (1990). "Gas and water permeability tests on 25 years old concrete from the NPD nuclear generation station". Report Info-0356, Atomic Energy Control Board, Project No. 2.147.1, May, 56 p.

Naus, D.J. (2003). "Use of concrete in radioactive waste disposal facilities". *RILEM TC 202-RWD*, General Information, 5 p.

Neville, A.M. (1995). *Properties of Concrete*. 4th ed., Longman Group Ltd., Harlow, 844 p.

Niwase, K., Sugihashi, N., Edamatsu, Y. and Sakai, E. (2012). "Study on the influence of fly-ash quality on the properties of cementitious materials and the applicability of non destructive test for quality control at radioactive waste disposal facility in Japan". Concr. Res. and Technol., v23, n1, 13–24 (in Japanese).

Park, C.-H., Synn, J.-H. and Park, J. (2015). "Probabilistic performance assessment of airtightness in concrete tube structures". *KSCE J. Civil Eng.*, v20, 1443–1451.

Piedecausa García, B. and Chinchón-Payá, S. (2011). "Radiactividad natural de los materiales de construcción. Radiación interna: el gas radón". *Cemento-Hormigón*, n946, September–October, 34–50.

Powers, T.C. (1958). "Structure and physical properties of hardened portland cement paste". J. Amer. Ceramic Soc., v41, n1, January 1, 6 p.

Powers, T.C., Copeland, L.E. and Mann, H.M. (1959). "Capillary continuity or discontinuity in cement pastes". *J. Portl. Cem. Assoc. Res. and Development Labs.*, v1, n2, 38–48.

Ramallo de Goldschmidt, T.R. (2004). "Low and intermediate level waste package assessment under interim storage and final disposal conditions". Report IAEA-TECDOC-1397, International Atomic Energy Agency, June, 33–50.

Rogers, V.C., Nielson, K.K. and Holt, R.B. (1995). "Radon generation and transport in aged concrete". EPA/600/SR-95/032 Report, Project Summary, March, 2 p.

RTS (2010). "Le projet Swissmetro abandonné faute d'argent". RTS Info, June 28.

Sakai, Y., Nakamura, C. and Kishi, T. (2013). "Correlation between Permeability of Concrete and Threshold Pore Size obtained with Epoxy-Coated Sample". J. Adv. Concr. Technol., v11, August, 189–195.

Sandoval, G.F.B., Galobardes, I., Teixeira, R.S. and Toralles, B.M. (2017). "Comparison between the falling head and the constant head permeability tests to assess the permeability coefficient of sustainable Pervious Concretes". *Case Studies Constr. Mater.*, v7, 317–328.

Schrader, E.K. (2003). "Performance of roller compacted concrete (RCC) dams – An honest assessment". *Roller Compacted Concrete Dams*, Berga, L., Buil, J.M., Jofré, C. and Chonggang, S. (Ed.). A.A. Balkema Publishers, Lisse, NL, 91–102.

SINTEF (2017). "Air could be the world's next battery". PhysOrg, March 28, 5 p.

Tengborg, P., Johansson, J. and Durup, J.G. (2014). "Storage of highly compressed gases in underground Lined Rock Caverns – More than 10 years of experience". *Proceedings on World Tunnel Congress 2014* – Tunnels for a better Life. Foz do Iguaçu, Brazil, 8 p.

Teruzzi, T. (2020). Personal Communication, June 2.

Teruzzi, T. and Antonietti, S. (2019). "PERMEA – Permeabilità al gas radon dei conglomerati cementizi". Clickin, SUPSI, Lugano, Switzerland, Giugno, 2 p.

USACE (2000). "Roller-compacted concrete". Engng. Manual, U.S. Army Corps of Engineers, EM1110-2-2006, January 15, 77 p.

Whiting, D. and Cady, P.D. (1992). "Condition evaluation of concrete bridges relative to reinforcement corrosion. Volume 7: Method for field measurement of concrete permeability". SHRP-S/FR-92–109 Report, National Res. Council, Washington DC, September, 93 p.

Chapter 3

Theory: concrete microstructure and transport of matter

3.1 CEMENT HYDRATION

3.1.1 Main Hydration Reactions and Resulting Changes

The main component and “heart” of Portland cements is the Portland cement clinker, the product obtained as the main output of the chemical process developed in the cement kiln, at temperatures reaching ≈ 1,450°C. Clinker is later ground into a very fine powder, together with a suitable amount of a set regulator (typically some form of calcium sulphate), to obtain the so-called ordinary Portland cement (OPC). Often, Portland cement incorporates other components of various degrees of cementitious contribution, such as granulated blast-furnace slag (GBFS), pulverized fly ash (PFA), natural or artificial pozzolans (POZ), limestone filler (LF), and microsilica such as silica fume (SF). Depending on the nature of the added component(s), the incorporation may happen in the cement plant (interground in the cement mill or blended in powder form) or directly at the concrete plant as a separate powder. In the first case, it is usual to designate the added component(s) as mineral components (MIC) whilst, in the second case, as supplementary cementitious materials (SCM).

Clinker is an assemblage of different minerals, the main ones being[1]:

- alite (main component), based on C_3S
- belite, based on C_2S
- aluminate and ferrite, based on C_3A and C_4AF, respectively

The “pure” compounds C_3S, C_2S, C_3A, thus, do not occur in this simple form, but as solid solutions containing a whole series of oxides such as MgO, Na_2O, K_2O, Fe_2O_3, and Al_2O_3 (Holcim, 2011).

[1] In cement chemistry notation: $C = CaO$; $S = SiO_2$; $A = Al_2O_3$; $F = Fe_2O_3$; $H = H_2O$.

DOI: 10.1201/9780429505652-3

In cement chemistry, the term "hydration" denotes the totality of the changes that occur when an anhydrous cement, or one of its constituent phases, is mixed with water (Taylor, 1997).

With respect to the formation of the microstructure of the hydrated Portland cement paste and concrete, the hydration of the two calcium silicate components – alite and, to a lesser extent, belite – plays a key role. The hydration of both minerals leads to the formation of calcium silicate hydrates (C-S-H) and calcium hydroxide (CH), as summarized in Eq. (3.1).

$$\begin{matrix} C_3S \\ C_2S \end{matrix} + \text{Water} \rightarrow C-S-H+CH \tag{3.1}$$

In the case of cements containing MIC, other reactions take place, such as the hydration of ground GBFS (GGBFS) in the alkaline environment created by clinker hydration or the pozzolanic reaction of the reactive silica in low-calcium PFA, SF or natural or artificial POZ with the calcium hydroxide created during clinker hydration. Both reactions follow that of clinker hydration, ending up in further formation of C-S-H as reaction product. Both secondary reactions have the beneficial effect of increasing the amount of C-S-H but, on the negative side, are delayed and slower than clinker hydration reaction.

The cement hydration reaction has the following consequences:

- the volume occupied by the hydration products becomes approximately twice that originally occupied by the anhydrous cement
- the specific surface of the hydration products is about 1,000 times that of the anhydrous cement
- the reactions are exothermic, with each cement gram generating about 380 J of heat
- radical changes in the microstructure, leading to a strong and stable material

3.1.2 Hydrothermal Conditions for Hydration (Curing)

The chemical reactions involved in cement hydration (Eq. 3.1) can only take place under certain favourable conditions of temperature and moisture.

The hydration reactions accelerate at higher temperatures, a fact made practical use of in the precast industry or whenever a high early strength is required (although often the "quality" of the microstructure at later ages, say 28 days, may suffer). At the other extreme, hydration slows down at low temperatures, a fact to be taken into account in what is called "cold weather concreting". It is generally accepted that hydration virtually stops at temperatures below –10°C.

By definition, the hydration reactions need the presence of water. Water added to concrete during mixing becomes chemically bound and physically bound (adsorbed), both bonding processes being relevant for the progress of hydration (ACI 308R, 2016). About 0.25 g of water is required to completely hydrate 1 g of cement, water that is removed from the pores in a process called "self-desiccation" (of relevance for concretes of low *w/c* ratios). According to Powers et al. (1954), for cement hydration to proceed, the relative humidity in the pores should be at least 80%.

This relative humidity can be achieved, either by providing extra water externally (ACI 308R, 2016) or internally (Zhutowsky & Kovler, 2012; Sensale & Gonçalves, 2014; Weiss & Montanari, 2017) or, alternatively, by preventing the non-reacted water to evaporate by protecting the concrete surface using plastic sheets, wet burlaps or curing compounds. These processes are restrictedly known as "curing".

According to Xue et al. (2015), once water is lost, the hydration of the cement will cease, reason why it is important to keep sufficient moisture in concrete while the cement is actively hydrating, especially at early ages (Ye et al., 2005).

For mixes with low *w/c* ratio, preventing the evaporation of water may not be sufficient and supplying water, at least in an initial stage, will improve concrete performance (Maslehuddin et al., 2013). According to Al-Gahtani (2010), water curing is the most effective method of curing.

When using SCM, water is required for the pozzolanic reaction between such materials and calcium hydroxide to take place in the later stages of hydration of cement (Bentur & Goldman, 1989; Sensale & Gonçalves, 2014; Maslehuddin et al., 2013). Actually, the pozzolanic reaction requires higher relative humidity than that required for the cement hydration (Ye et al., 2005). Therefore, cements containing MIC or concretes containing SCM are generally more sensitive to the lack of adequate curing (evaporation, low temperatures), due to the delayed occurrence of the secondary reactions involved in the hydration of the mineral additions.

3.2 MICROSTRUCTURE OF HARDENED CONCRETE

3.2.1 Overview

To study the transfer of matter through concrete, it is important to have a clear understanding of the microstructure of the material. Here, an overview of this rather complex topic is provided, that is more extensively treated in several excellent classical books on concrete, e.g. in Chapter 1 of Neville (1995), in Chapter 4 of Mindess et al. (2003) and in Chapter 2 of Mehta and Monteiro (2006).

First of all, hardened concrete is a heterogeneous material composed of two main macroscopic phases, well differentiated morphologically and behaviourally: the aggregates and the hardened cement paste (h.c.p.) matrix.

As shown in Figure 3.1, the aggregates are embedded in the h.c.p. matrix, such that – for conventional concrete – there is no physical contact between the aggregate particles, as they are all coated by a thin layer of cement paste, which also fills the voids between the individual particles.

The h.c.p. matrix constitutes a continuous phase, while the aggregate particles are a discrete phase. Hence, it would be possible for a "microorganism" to traverse the entire concrete sample travelling exclusively along the h.c.p. This indicates that the h.c.p. has a key role in the transfer of matter through concrete, with isolated aggregate particles playing just a secondary role. It is worth mentioning that the permeability of aggregates (rocks) covers the same range as that of concrete, as discussed in Section 11.5.2.

A synonym for "aggregate" in Italian, Portuguese and Spanish languages is "inert", indicating a totally passive role of the particles. Alkali-aggregate reaction is just one example that aggregates are not inert but that they can react chemically with the h.c.p. Moreover, since long it has been recognized that the contact zone between aggregates and h.c.p. is influenced by both phases, in what nowadays is called the "Interfacial Transition Zone" (ITZ). The ITZ consists of a 10–50 µm thick "rim" around the aggregates and can be considered as a third distinctive phase of concrete microstructure (together with h.c.p. and the aggregates) (Mehta & Monteiro, 2006).

The ITZ is the result of many factors, such as the accumulation of bleeding water under the coarse particles, of unwashed aggregates coated with "dirt" or, more generally, of selective deposition of $Ca(OH)_2$ crystals closer to silicate aggregates, to chemical reactions between paste and aggregate, etc.

The fact is that, typically, the ITZ is more porous than the bulk h.c.p., as can be seen in the example presented in Figure 3.2, which shows a thin section of concrete (20–25 µm thick) taken from a sample that was previously

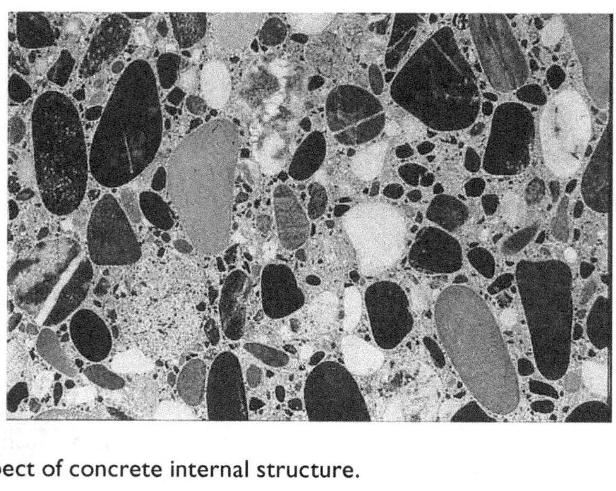

Figure 3.1 Aspect of concrete internal structure.

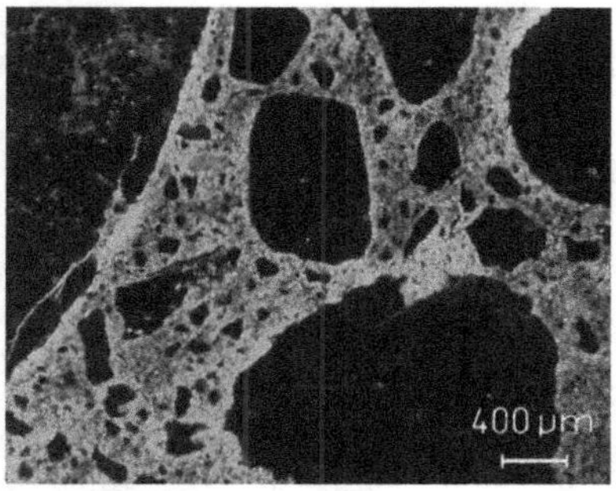

Figure 3.2 More porous rim around aggregates.

vacuum-impregnated with a low-viscosity resin containing a fluorescent dye. The microscopy observation of the thin section under transmitted UV light reveals very clearly the microstructure of the concrete, in dark what is solid (typically aggregate particles) and in light colour what is porous, which is brighter the more porous the phase. It can be seen quite clearly that a rim of more porous h.c.p. contours the aggregates, not just the coarse, but also the fine particles. The characteristics and role of the ITZ is further discussed in Section 3.2.3; those more interested in the topic may consult (Alexander et al., 1999).

3.2.2 Microstructure of Hardened Cement Paste

Let us try to understand the microstructure of h.c.p. During hydration of cement (Eq. 3.1), the predominant calcium silicate crystals in clinker (C_3S and C_2S) react with water to form calcium silicate hydrates (C-S-H) and $Ca(OH)_2$. C-S-H makes up to 50%–60% of the volume of h.c.p., in the form of tiny fibrous crystals (<0.2 µm thick and 0.5–2.0 µm long (Neville, 1995)) which, due to their fineness and extremely high specific surface (≈1,000 times that of the original cement grains), develop strong inter-particle attraction forces between them and with the aggregate particles, making concrete a strong monolithic material. Often, this fine crystals' system of colloidal size is known as "cement gel".

The hydration products occupy a bulk volume that is about twice the volume of the original unhydrated cement crystals. This larger volume is only ≈72% solid, as the resulting C-S-H crystals are arranged in a disordered pattern that leaves a characteristic porosity of ≈28%. These extremely tiny pores (about 3 nm in diameter) are called "gel pores" and are an inherent feature of the microstructure of cement hydration products that cannot be altered.

Figure 3.3 shows a sketch of the evolution of hydration of cement particles in permanent contact with water, from time 0 (unhydrated grains) to several weeks, say 28 days, when some 70% of the hydration is achieved.

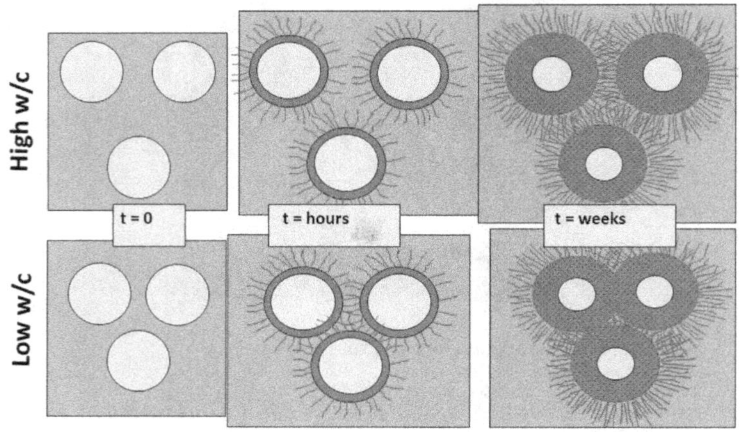

Figure 3.3 Evolution of cement hydration.

The case of two pastes shown in Figure 3.3, one of high and the other of low *w/c* ratio, illustrates that the *w/c* ratio actually measures the original degree of dispersion of cement particles in water. Thus, the distance between grains of the same cement will be greater the higher the *w/c* ratio of the mix. After a few hours, the hydration products of low *w/c* ratio paste (or concrete) will start making contact and bonding with those of neighbouring particles, leading to stiffening and eventually to setting, whilst the system with high *w/c* ratio will still remain soft. After several weeks of hydration and hardening, the hydrated particles would have grown to a larger volume, filling the space originally occupied by water. The resulting voids (remaining space occupied by water) are called "capillary pores" and play a significant role in the ability of concrete to transport matter. Figure 3.3 also shows that the paste or concrete with high *w/c* ratio will contain larger capillary pores than that of low *w/c* ratio, the latter being less permeable and, at the same time, stronger. Therefore, Figure 3.3 illustrates well the beneficial effect of lowering the *w/c* ratio and maintaining the conditions for an extended development of hydration (curing) on reducing the permeability of concrete. It has to be mentioned that the resulting pore structure will depend also on the physical-chemical characteristic of the cement, affected by the raw materials used for production, the raw meal grinding and clinker burning processes, the form and dosage of $CaSO_4$ set regulator used and the type and intensity of the cement grinding process.

Figure 3.4 is based on one of the best-known models describing the microstructure of h.c.p. (Feldman & Sereda, 1968). It represents a h.c.p. after partial hydration, with the still unhydrated cement particles in grey shade. The thick lines represent the original boundaries of the unhydrated

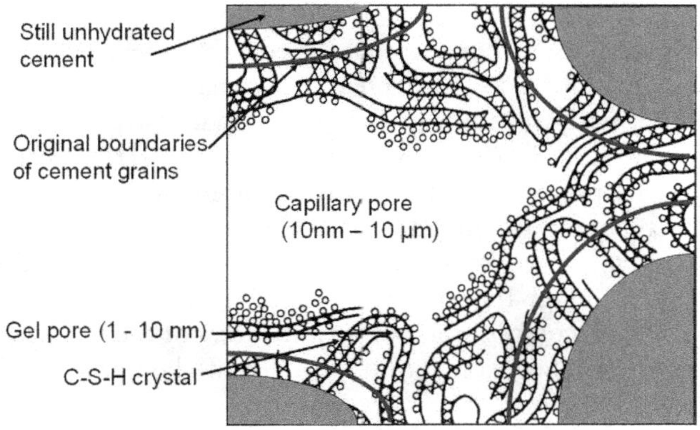

Figure 3.4 Model of h.c.p. microstructure, adapted from Feldman and Sereda (1968).

cement particles. The complex system of curved lines represents the irregular array of C-S-H crystals, leaving between them very tiny pores (the "gel pores"). Finally, the space between the original particles' boundaries, not yet filled with hydration products represents a capillary pore, which is considerably larger than the gel pores.

All what has been described above refers to the microstructure of CPC, i.e. to a cement obtained by inter-grinding Portland cement clinker with some form of $CaSO_4$ (set regulator), with only minor additional components. This type of cement is becoming rarer nowadays, due to the pressure to reduce CO_2 emissions (largely associated with clinker production). As a result, the worldwide trend is towards the use of composite cements, in which increasing quantities of clinker are substituted by mineral additions that provide widely different hydraulic contributions to the cement. The most used mineral additions are, in increasing order of hydraulic contribution:

- LF
- pozzolans: natural POZ, low-calcium PFA, thermally activated clays (CC), SF
- high-calcium fly ash and GGBFS

The presence of these – now usual – cement components formidably complicates the hydration process with interactions and secondary reactions, making the modelling of the resulting microstructure a scientific discipline in itself. Suffice here to say that, in terms of mass transfer through h.c.p., the presence of active mineral additions has, in general, a beneficial effect, due to the achievement of denser matrices, refinement of pore structure and mitigation of the porosity of the ITZ.

3.2.3 Interfacial Transition Zone

Since long, it has been recognized that the contact zone between aggregates and h.c.p. is influenced by both phases, in what nowadays is called the "ITZ". Indeed, early research in the URSS (Lyubimova & Pinus, 1962), cited in Alexander et al. (1965), Swamy (1971), showed that a mutual interaction aggregate-paste exists, leading to a softening of the first 30 μm on both sides of the interface boundary (see Figure 3.5).

According to many works in this field, the ITZ thickness ranges between 10 and 40 μm, although the structure and the thickness are affected by several factors such as composition of the binder and of the inclusions and age (Uchikawa, 1988). There seems to be a preferential deposition of $Ca(OH)_2$ crystals in the ITZ.

The change of ITZ thickness of mortar and concrete (*w*/*c*=0.50, kept at 20°C) with age is illustrated in Figure 3.6, based on the results from Uchikawa (1988). For concrete, the ITZ thickness reached 12 μm at 3 days and exhibited the maximum value of 29 μm at 7 days. Thereafter, the value decreased gradually to 22 μm at 91 days as a result of the densification of the paste due to cement hydration.

Figure 3.7 (l.) shows the amount of $Ca(OH)_2$ formed by diffusion of ions into the ITZ, calculated on the basis of measurement of the distribution of $Ca(OH)_2$ in the ITZ (Scrivener, 1999). This is accompanied by higher

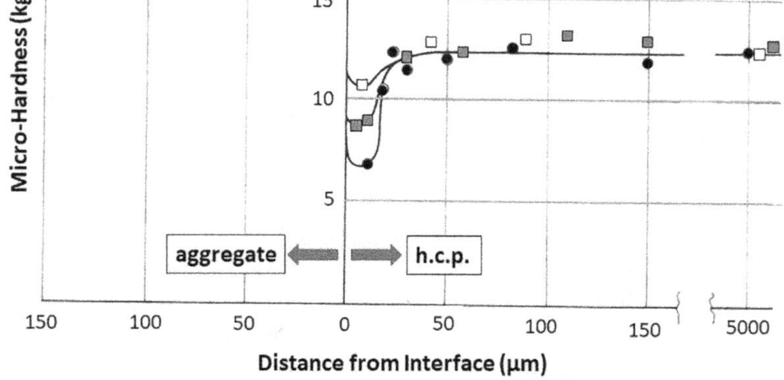

Figure 3.5 Interaction paste-aggregate leading to softening of the ITZ, adapted from Alexander et al. (1965).

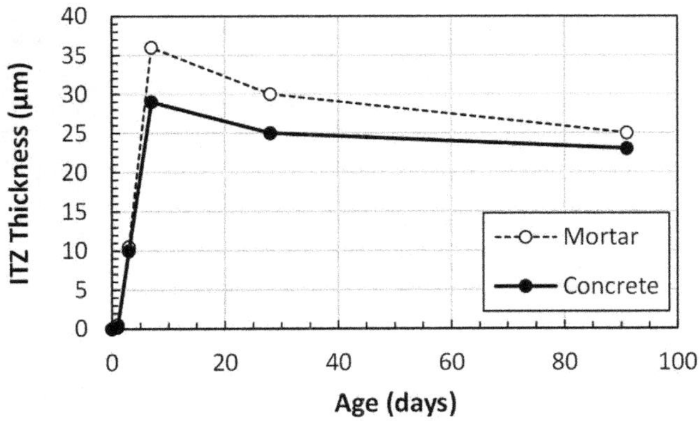

Figure 3.6 Age-dependence of ITZ thickness of hardened mortar and concrete, data from Uchikawa (1988).

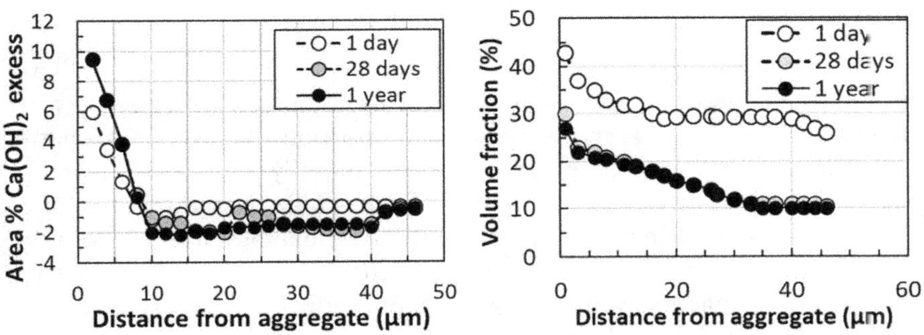

Figure 3.7 $Ca(OH)_2$ and porosity profiles within ITZ, data from Scrivener (1999).

h.c.p. porosity in the vicinity of the aggregate, as shown in Figure 3.7 (r.) (Scrivener, 1999).

The presence of the more porous ITZ can contribute significantly to the transfer of matter, since it is estimated that ITZ can account for about 20%–50% of the total h.c.p. volume in concrete (Garboczi, 1995). The influence of ITZ on concrete permeability is discussed in Section 6.4.2. It can be concluded, then, that extrapolation of transport test results, obtained on cement paste and mortars, to concrete has to be taken with caution.

3.2.4 Pore Structure of Hardened Concrete

Figure 3.8 presents the different types of pores that can be found in concrete. Capillary pores and ITZ pores, just described, constitute a continuous network which accounts for a very large proportion of the flow of

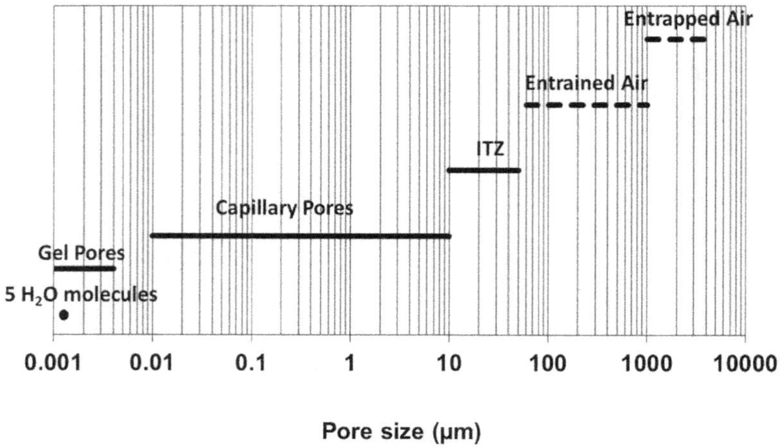

Figure 3.8 Pores and voids found in concrete and their size ranges.

matter through concrete. Gel pores, despite being connected to the capillary pores, are of such tiny size (of the same order of magnitude of 5–10 water molecules) that they carry a very small proportion of the flow of matter through concrete. The same applies to intentionally entrained air and/or naturally entrapped air (dotted lines in Figure 3.8), consisting of isolated air bubbles. Matter, in order to traverse these air bubbles, has first to travel through the continuous network of pores that predominantly dictate the flow rate. Compaction voids (not shown in Figure 3.8) are an intolerable defect to be avoided by all means.

The formation factor, which is the ratio between the electrical conductivity of the pore solution alone and that of the bulk solid, is a measure of the intricacy of the pore structure of the solid, with some application to concrete (Snyder, 2001; Spragg et al., 2013).

A final comment regarding the heterogeneous microstructure within a concrete specimen, not to mention within real structural elements (to be discussed in Chapter 7). There is a microstructural gradient near a cast surface (Bentz et al., 1999) and also near the top, finished surface, leading to the concept of "skin of concrete" coined by Kreijger (1984). Due to the "wall effect", the cast surfaces are richer in cement paste and leaner in aggregates content within a distance of about 5 mm. The higher paste content of the skin means that, in general, the penetration resistance to gases, liquids and ions will be lower than in the bulk concrete (Buenfeld & Okundi, 1988). Bleeding and finishing operations of the top surface also lead to a more porous layer, of higher "penetrability". As discussed in Chapter 5, the existence of a more permeable skin is a challenge for surface permeability test methods.

3.2.5 Binding

A final, important element to discuss here is the chemical interactions that take place between relevant substances that penetrate concrete from the outside (e.g. CO_2 from the air and/or Cl^- from salty solutions in contact with the structures) and the h.c.p.

The reaction of CO_2 from the air with the hydrated cement paste, in the presence of moisture, is known as "carbonation". CO_2 reacts primarily with $Ca(OH)_2$ but also with the C-S-H. As a result, a certain densification of the surface layers occurs, combined with a decrease of the alkalinity of the pore solution, the latter a phenomenon that may lead to carbonation-induced corrosion of the embedded steel. It is said that cement "binds" the CO_2, with some cements having more "binding" capacity than others. Indeed, OPCs have a higher CO_2 binding capacity and usually resist better carbonation than cements containing additions, because of the fixation and consumption of $Ca(OH)_2$ in the secondary reactions of the latter. For deeper insight into the topic, the reader is advised to consult (Li, 2016).

Similarly, Cl^- ions also react chemically with the hydrated cement paste, in particular with hydrated C_3A and C_4AF phases to form what is known as "Friedel's salt". Chloride binding is important, because it subtracts free Cl^- ions from the pore solution, which are responsible for chloride-induced steel corrosion. Some mineral additions, typically GBFS, increase the chloride binding capacity and, contrary to carbonation-induced corrosion, show a better performance regarding chloride-induced corrosion. However, if concrete carbonates, this binding effect is reversed. For a deeper insight into the topic, the reader is advised to consult (Justnes, 1997; Nilsson, 2002; Yoon, 2010).

Carbonation- and chloride-induced steel corrosion are discussed in more detail in Chapter 9.

3.3 WATER IN THE PORES OF HARDENED CONCRETE

Concrete exposed to natural drying contains some water in its pore system. This has two main consequences for durability:

- the transport mechanisms dictating the penetration of aggressive agents are strongly affected by the degree of saturation of the concrete pores
- most durability damaging actions involve some kind of chemical (carbonation, sulphate attack, ASR, etc.) or electrochemical (steel corrosion) reactions, in which the presence of water in the pores plays a significant role. Frost damage, being predominantly a physical phenomenon, requires a near-saturation condition of concrete to take place

After mixing, concrete is a concentrated suspension of solids in water, which means that – with the exception of entrained and entrapped air voids – the system is fully saturated with "water". The inverted commas warn that, since the very beginning and along the whole service life of the structure, the liquid phase is not just water, but a complex solution containing a variety of ions, called the "pore solution". The ionic composition and pH of this pore solution play an important role in durability.

As discussed in 3.1.2, curing is a vital process intended to delay as much as possible the loss of moisture from the concrete pores to the environment, thus favouring hydration and reducing the risk of drying shrinkage cracking.

Except for permanently submerged structures, sooner or later concrete will be exposed to an environment causing evaporation of water. Figure 3.4 shows, besides solids and voids, the presence of water, in particular that which is adsorbed onto the C-S-H crystals surfaces (little circles) or inside the gel pores, called the "interlayer water" (x symbols). These water molecules are strongly bound to the solid surfaces and will hardly be removed from a concrete exposed to a natural environment. On the other hand, the water filling the larger capillary pores is beyond the interaction range with the C-S-H crystals and can be considered as "free water" that evaporates rather easily into a drier atmospheric environment.

Water in the pores has a strong effect on all transport mechanisms, since full saturation is needed for water permeation and migration, whilst low degree of saturation is required for capillary suction and gas-permeability.

3.4 MECHANISMS OF TRANSPORT OF MATTER THROUGH CONCRETE

The majority of concrete deterioration processes involves the penetration of gases, water or solutions containing aggressive agents from the environment. In this context, concrete transport mechanisms are of high relevance. These mechanisms and the properties that characterize them are discussed in this section. Often, more than one mechanism is involved in the penetration and movement of deleterious substances in concrete. In what follows, each mechanism is described in some detail, with special emphasis on permeability.

3.4.1 Diffusion: Fick's Laws

Diffusion is a transport mechanism taking place whenever a gradient of concentration of a given substance exists across a porous body.

The movement of atoms and molecules, reflected as Brownian movement, generates collisions with neighbouring particles. These collisions are more frequent in regions of higher concentration than in those of lower concentration, which results in a net flux of mass from the high concentration to the low concentration regions. This type of transport is known as diffusion.

Adolf Fick (1829–1901) was a German scientist who developed the theory of diffusion, applying it to physics and medicine. His work is condensed into two laws, known as the first and second Fick's laws of diffusion.

Fick's first law applies to the case of steady-state diffusion regime, i.e. when the concentrations of the transported substance are constant, and reads:

$$-J = D \cdot \frac{\partial C}{\partial x} \tag{3.2}$$

where J = flux of ions or molecules (g/m²/s). The minus indicates that the flux is opposite to the gradient

D = coefficient of diffusion of the material, expressed in [L^2/T], typically (m^2/s)

$\partial C/\partial x$ = gradient of concentration across the element (g/m^4)

Equation (3.2) is used to measure gas-diffusion but rarely for chloride-diffusion, because it requires relatively thin specimens and long times to achieve the steady state. A description of the diffusion cell test used to measure the coefficient of diffusion to chlorides, under steady-state conditions, can be found in Section 3.2.1 of Tang et al. (2012). The measured value is designated as D_s (for steady).

In concrete having its surface(s) exposed to the environment, the penetration of substances by diffusion (CO_2, Cl^-, SO_4^{2-}, etc.) happens in a non-steady-state regime. In this case, Fick's second law applies; Collepardi et al. (1970) seem to have been the first to apply it to the diffusion of chlorides in concrete.

Fick's second law, for unidirectional flow and assuming no chemical reactions taking place, reads:

$$\frac{\partial C}{\partial t} = \frac{\partial}{\partial x}\left(D \frac{\partial C}{\partial x}\right) \tag{3.3}$$

which relates the concentration C of a substance at time t and depth x with the gradient of concentration and the material property coefficient of diffusion of concrete D, expressed typically in m^2/s or mm^2/y.

If D does not vary with C or x, Eq. (3.3) becomes

$$\frac{\partial C}{\partial t} = D \frac{\partial^2 C}{\partial x^2} \tag{3.4}$$

If we assume now that D (m^2/s) and the concentration of the substance at the concrete surface are constant, Eq. (3.4) has an explicit solution:

$$C(x,t) = C_0 + (C_s - C_0) \cdot \left[1 - erf\left(\frac{x}{\sqrt{4 \cdot D \cdot t}}\right)\right] \quad (3.5)$$

where

$C\ (x,\ t)$ = concentration (kg/m^3) at distance x (m) from the surface and time t (s)

C_s = concentration (kg/m^3) at the surface (x=0)

C_0 = initial concentration (kg/m^3) in the concrete, before being exposed to the diffusing substance

erf = error function

Equation (3.5) is often used to calculate parameters C_s, D and C_0 by curve fitting to chloride profiles obtained after immersion or ponding tests in which one face of a concrete specimen has been in contact for several months with a solution with a defined Cl^- concentration. Some diffusion tests are discussed in Section A.1.

Equation (3.5) is also widely used for service life prediction, as discussed in Chapter 9.

3.4.2 Migration: Nernst-Planck Equation

Nernst-Planck equation is an extension of Fick's first law of diffusion of ions in a fluid, see Eq. (3.3), to the case where an electric field is superimposed to the system and reads, assuming the fluid at rest – only the ions move, static electromagnetic conditions and uniaxial flow:

$$-J = D\left(\frac{\partial C}{\partial x} + C\ \frac{z.F}{R.T}\ \frac{\partial E}{\partial x}\right) \quad (3.6)$$

where

z = electric charge of the ion (–)

C = ion concentration in the fluid (g/m^3)

F = Faraday constant = 96,485 J/V/mol

R = universal gas constant = 8.3144598 J/mol/K

T = absolute temperature (K)

$\partial E/\partial x$ = gradient of electric potential across the element (V/m)

The first term within the parenthesis in Eq. (3.6) corresponds to the component of the flow of ions driven by the concentration gradient (diffusion) and the second term by the electric potential gradient. This second component is known as migration. Notice that, according to Eq. (3.6), both diffusion and migration are in theory governed by the same property of concrete D. Some migration tests are discussed in Section A.2; in some of them, it is assumed that the current measured in a migration test is carried exclusively by Cl^-.

It is, therefore, important to mention that the solution filling the concrete pores constitutes a complex ionic system (Andrade, 1993). When an electric potential field is applied, the anions (e.g. OH^-, Cl^-, SO_4^{2-}) move towards the anode, whilst the cations (e.g. Ca^{2+}, Na^+, K^+) move in the opposite direction (see Figure 3.9).

The ions present in the aqueous pore solution have different electrical charges and mobilities (mobility is the velocity reached by the ion under an electric field of 1 V/m) and, therefore, contribute differently to the electric current generated by the electric field. Table 3.1 lists the electrical mobility of the more abundant ions in the pore solution.

The ionic composition of the pore solution in concrete varies with the degree of hydration (time) and is strongly dependent on the type of binder used. In concrete made with OPC, the more abundant ions are OH^-, K^+ and Na^+, in that order (Page & Vennesland, 1983; Lothenbach & Winnefeld, 2006; Clariá et al., 2012). The inclusion of 10% SF reduces the presence of cations and anions alike from ≈ 900 mg equivalent/L to just ≈ 320 at 28 days (Page & Vennesland, 1983). Then, it is to be expected that the electrical conductivity of the pore solution and, thus, of concrete will be reduced when clinker is substituted by active mineral additions, a fact observed in practice and of relevance in the interpretation of migration test results described in Section A.2.

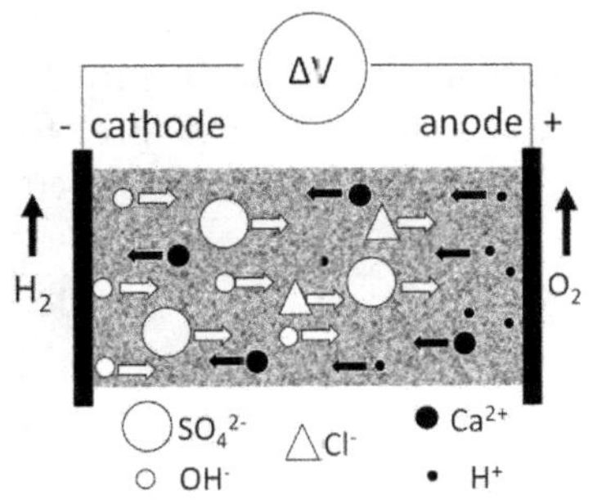

Figure 3.9 Sketch of ion migration under electric field, adapted from Andrade (1993).

Table 3.1 Ionic mobilities in water at 25°C, taken from Atkins and de Paula (2013)

Ion	*OH^-*	*SO_4^{2-}*	*Cl^{-a}*	*K^+*	*Ca^{2+}*	*Na^+*
Electrical mobility (10^{-8} m²/s V)	20.64	8.29	7.92	7.62	6.17	5.19

[a] Not abundant in pore solution unless naturally or deliberately introduced from external source.

3.5 PERMEABILITY

3.5.1 Laminar Flow of Newtonian Fluids. Hagen-Poiseuille Law

When a liquid is placed on an inclined slope, as illustrated in Figure 3.10 (l.), the liquid will flow due to the shear stress τ resulting from gravity. In the case of laminar flow, the liquid moves in laminae, with external layers moving faster than the internal ones, with the fluid at rest at the contact with the solid surface below. The velocity of the laminar flow is measured by the change of angle γ with time or $\dot{\gamma}$.

In the case of a Newtonian fluid, there is a linear relation between τ and $\dot{\gamma}$ expressed by Eq. (3.7).

$$\tau = \mu \cdot \frac{dV}{dy} = \mu \cdot \dot{\gamma} \tag{3.7}$$

where τ is the shear stress and μ is the dynamic coefficient of viscosity of the fluid.

Water and honey are Newtonian liquids, with honey having a viscosity about 2,000–10,000 times that of water. The viscosity of water drops significantly with temperature, as shown in Table 3.2. The viscosity of gases is about 2 orders of magnitude smaller than that of water; moreover, the viscosity of both air and CO_2 does not change significantly with temperature.

Now, let us study the steady laminar flow of a Newtonian liquid through a cylindrical tube of radius r (a simplified version of a capillary pore) subjected to a gradient of pressure dP/dL, see Figure 3.10 (r.).

For a steady flow (from left to right), the three forces acting on the plug of liquid (of radius y and length ΔL) must be in equilibrium, as described by Eqs. (3.8) and (3.9).

$$F_1 - F_2 - R_v = 0 \tag{3.8}$$

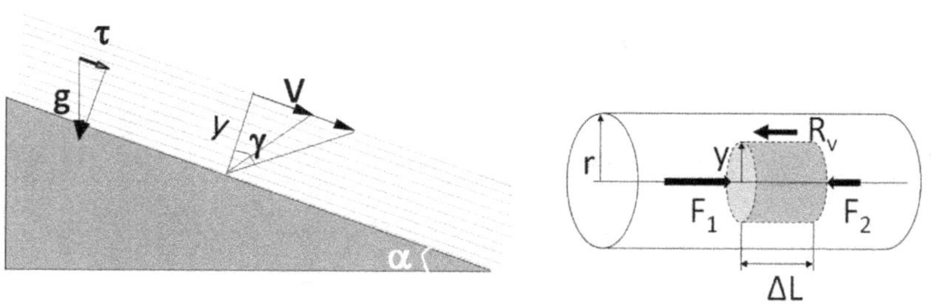

Figure 3.10 Sketch of liquid flow on an inclined slope (l.) and plug of fluid under a pressure gradient (r.)

Table 3.2 Coefficient of viscosity of various fluids and its temperature dependence (Weast, 1975)

Liquid	**T** *(°C)*	*μ (N.s/m²)*	*Gas*	**T** *(°C)*	*μ (N.s/m²)*
Water	0	1.787×10^{-3}	Air	0	1.708×10^{-5}
	10	1.307×10^{-3}		18	1.827×10^{-5}
	20	1.002×10^{-3}		40	1.904×10^{-5}
	30	0.798×10^{-3}		54	1.958×10^{-5}
	40	0.653×10^{-3}	CO_2	0	1.390×10^{-5}
	50	0.547×10^{-3}		19	1.499×10^{-5}
	60	0.467×10^{-3}		30	1.530×10^{-5}
Honey		2–10		40	1.570×10^{-5}

or

$$P\pi y^2 - \left(P - \frac{dP}{dL}\,\Delta L\right)\pi y^2 - 2\pi y\;\Delta L\,\tau = 0 \tag{3.9}$$

where the first two terms are the forces due to the pressure at the left- and right-hand surfaces of the plug, respectively, and the third term is the viscous resistance to flow offered by the surrounding liquid. Operating on Eq. (3.9), one gets:

$$\tau = \frac{dP}{dL}\,\frac{y}{2} \tag{3.10}$$

and, considering Eq. (3.7):

$$\mu\,\frac{dV}{dy} = \frac{dP}{dL}\,\frac{y}{2} \tag{3.11}$$

since dP/dL is not function of y, we can integrate Eq. (3.11):

$$V = \frac{1}{4\,\mu}\,\frac{dP}{dL}\,y^2 + \text{Constant} \tag{3.12}$$

Assuming a "no-slip" condition, i.e. $V=0$ for $y=r$, Eq. (3.13) describes the well-known parabolic profile of velocities of a liquid through a tube of radius r:

$$V(y) = -\,\frac{r^2 - y^2}{4\mu}\,\frac{dP}{dL} \tag{3.13}$$

To calculate the flow Q we have to integrate the dQ that flows through a "ring" of radius y and thickness dy with velocity $V(y)$:

$$Q = \int_0^r -\frac{r^2 - y^2}{4\mu} \frac{dP}{dL} 2\pi y dy \tag{3.14}$$

$$Q = -\frac{\pi r^4}{8\mu} \frac{dP}{dL} \tag{3.15}$$

Equation (3.15) is the Hagen-Poiseuille equation indicating that the flow of a liquid through a tube, under pressure, is proportional (but in opposite sense) to the pressure gradient and to the fourth power of the tube radius and inversely proportional to the viscosity of the liquid.

Now, let us consider the flow of a liquid through an impermeable body of cross section area A, traversed by n capillary tubes of radius r, as illustrated in Figure 3.11.

The flow rate will be the contribution of that carried by each individual tube, or from Eq. (3.15):

$$Q = -\frac{n\pi r^4}{8\mu} \frac{dP}{dL} \tag{3.16}$$

introducing the porosity of the body

$$\varepsilon = \frac{n\pi r^2}{A} \tag{3.17}$$

we have

$$Q = -\frac{\varepsilon r^2}{8} \frac{A}{\mu} \frac{dP}{dL} \tag{3.18}$$

Now, the general equation of the permeability of liquids through a porous body is Eq. (2.8) of Kropp et al. (1995):

$$Q = -K_l \frac{A}{\mu} \frac{dP}{dL} = -K_l \frac{A}{\mu} \frac{P_2 - P_1}{L} \tag{3.19}$$

where K_l is the coefficient of permeability of the material to liquids, expressed in $[L^2]$, typically in m^2. In petroleum industry, the Darcy unit is often used to quantify permeability (1 Darcy$=9.869\times10^{-13}$ m^2). P_2 and P_1 are the inlet and outlet pressures across the fluid traversing a material of length L.

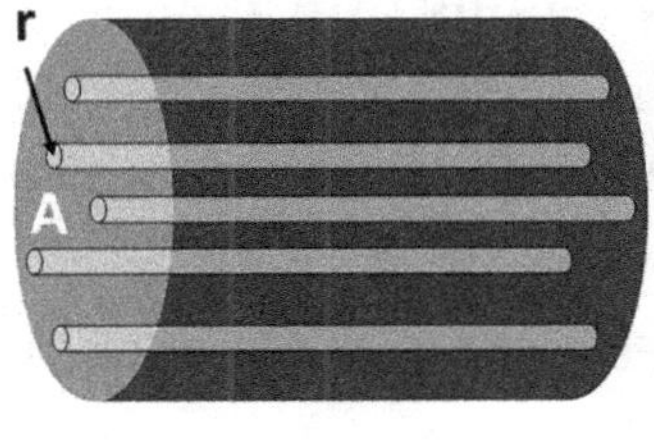

Figure 3.11 Sketch of an idealized porous body.

By comparing Eqs. (3.18) and (3.19) we can see that the coefficient of permeability of the body sketched in Figure 3.11 is

$$K_I = \frac{\varepsilon r^2}{8} \tag{3.20}$$

The permeability, so expressed, is often known as the "intrinsic coefficient of permeability", as it depends exclusively on the pore structure and should be the same irrespective of the fluid, the influence of which is considered explicitly through its viscosity μ.

3.5.2 Water-Permeability: Darcy's Law

A law was formulated in 1895 by the French engineer Henry Darcy on the basis of experiments on the flow of water through sand beds that reads:

$$Q = -\ K_D A\ \frac{\Delta h}{\Delta L} \tag{3.21}$$

where $\Delta h/\Delta L$ is the gradient of pressure expressed in water head per linear distance [L/L] and K_D is Darcy's coefficient of permeability of the material to water, expressed in $[L/T]$, typically in m/s.

For most engineers in touch with hydraulics, the familiar units of the coefficient of permeability of a material are m/s, helped by the fact that they are also "velocity" units. However, it shall be stressed that these units correspond, exclusively, to the flow of water and cannot be used for other fluids as they are derived from Eq. (3.21) where non-SI units have been applied to express the water pressure and in which the viscosity of the liquid is hidden.

Dividing Eqs. (3.21) and (3.19), one has the following conversion between K and K_D, using the density and viscosity of water δ_w and μ_w, respectively,

$$K_D = \frac{K}{\mu}\frac{\Delta P}{\Delta h} = K\ \frac{g\ \delta_w}{\mu_w} \tag{3.22}$$

where g is the acceleration of gravity; at 20°C is K_D (m/s) = 9.77×10^6 K (m²).

From now on, unless otherwise explicitly stated, when a coefficient of permeability of a material is referred to, it will be K expressed in m².

3.5.3 Permeation of Liquids through Cracks

If the flow across a prismatic tube of cross section $a \times w$ is analyzed instead of the cylindrical one shown in Figure 3.11, the flow rate through a smooth crack can be calculated. The length of the slot or "crack" would be a and its opening would be w (with $w \ll a$).

It can be shown that Eq. (3.10) still holds for this case, with Eq. (3.13) becoming:

$$V(y) = -\frac{(w/2)^2 - y^2}{2\mu} \frac{dP}{dL} \tag{3.23}$$

where y=distance from the centre of the slot. Similar as for the circular tube, there is a parabolic distribution of velocities, with its maximum at the centre of the slot and V=0 on its walls.

Equation (3.14) becomes:

$$Q = -\int_{-w/2}^{+w/2} \frac{(w/2)^2 - y^2}{2\mu} \frac{dP}{dL} a \cdot dy \tag{3.24}$$

which, integrating, leads to

$$Q = \frac{1}{12} \cdot \frac{a \cdot w^3}{\mu} \cdot \frac{\Delta P}{L} \tag{3.25}$$

Equation (3.25) is used in Section 6.11 to discuss the permeability of cracked concrete.

3.5.4 Hagen-Poiseuille-Darcy Law for Gases

The main difference between the flow of gases and liquids through a porous body is that liquids are incompressible (the density virtually does not change with pressure), whilst gases are compressible (their density increases with pressure).

In the case of gas flow, a correction for compressibility has to be included, multiplying the second member of Eq. (3.19) by the mean pressure and dividing it by the outlet pressure, or:

$$Q = -K_g \frac{A}{\mu} \frac{P_2 - P_1}{L} \frac{P_2 + P_1}{2\,P_1} = -K_g \frac{A}{\mu L} \frac{P_2^2 - P_1^2}{2\,P_1} \tag{3.26}$$

Equation (3.26) is the Hagen-Poiseuille-Darcy equation for gases, where K_g is the coefficient of permeability of the material to gases, expressed in m^2. Equation (3.26) is used to measure K_g experimentally, case in which P_1 is the pressure at which the gas flow Q is measured.

3.5.5 Relation between Permeability to Gases and Liquids

In theory, the intrinsic coefficient of permeability of a material, expressed in m², should be the same if measured with liquids, applying Eq. (3.19), or gases, applying Eq. (3.26), as it would depend exclusively on the pore structure of the material, Eq. (3.20).

It has to be borne in mind that, to compare gas vs. water-permeability, the latter should be measured on a material with its pores saturated with water (100% saturation), whereas the former should be measured on a material which is completely dry (0% saturation), which complicates the experimental conditions of the comparison. Normally, gas-permeability tests are seldom conducted at fully dried conditions, but on sufficiently dry specimens (see Chapters 4 and 5).

Figure 3.12 presents data of air-permeability *kT* (Torrent method) and of water-permeability K_w, reported by Sakai et al. (2013). Two series of laboratory data are plotted in Fig. 3.12 (site tests also reported in the original paper are not included). One series ('Lab Japan') was obtained on 15 concrete specimens made with OPC (in some cases with the addition of GGBFS or PFA) with *w/b* ratios between 0.40 and 0.70, subjected to different curing conditions up to 28 days, later stored in a room until the testing age of 2.75 years. The air-permeability *kT* tests were performed on Ø150×300 mm cylinders. The water-permeability K_w tests were performed on 38 mm thick slices saw-cut from the centre of Ø100×200 mm cylinders, which were vacuum-saturated for 24 h prior to the test. The applied water pressure was 2.5 MPa; the original K_w results were reported in cm/s, which were converted into m², applying the conversion factor in Equation (3.22).

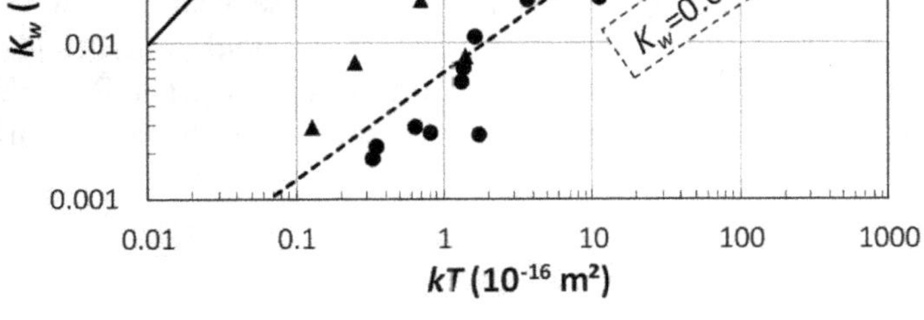

Figure 3.12 Relation between the coefficients of permeability to water (K_w) and to air (*kT*) (Sakai et al. 2013).

The other series ('Lab Korea') corresponds to tests performed with Korean materials, involving six concretes with *w/b* ratios of 0.45 and 0.53, with and without PFA, subjected to different curings (only four results are plotted in Figure 3.12 because in two cases no water permeation was reached).

Two aspects are worth mentioning from the results presented in Figure 3.12. The first one is that the coefficient of water-permeability K_w is two to three orders of magnitude smaller than the coefficient of air-permeability *kT*. This cannot be attributed to the applied test methods, since Jung [1969] obtained similar differences between water- and air-permeability applying different test methods, see Figure 3.12a of Nilsson and Tang, (1995). The second relevant aspect is the good correlation ($R=0.90$) existing between K_w and *kT*, with the regression equation shown in Figure 3.12 (broken line). The regression shown in Figure 3.12 has a practical value, as it allows to estimate the coefficient of water-permeability of concrete K_w for concretes of which the coefficient of air-permeability *kT* is known (see Eq. (2.1)).

The reasons for the huge difference between water- and air-permeability are quite intriguing, since both molecules (water and main gases in air) are of about the same size, around 0.3–0.4 nm. This is sufficiently small for both molecules to be capable of permeating through the h.c.p. pores system, but the concentration of molecules is much larger in a liquid which increases the resistance to flow (Nilsson & Tang, 1995). Perhaps the high adhesion properties of water, due to its polar nature, increases the resistance to flow at or near the pore wall surfaces. According to Zhou et al. (2017), the anomalous low water-permeability, compared to gas-permeability, is physically ascribed to the swelling of C-S-H gel specific to water, which makes the pore structure finer at water-saturated state.

Section 3.6 provides another explanation for such a difference, in line with what was proposed by Bamforth (1987).

3.6 KNUDSEN AND MOLECULAR GAS FLOW: KLINKENBERG EFFECT

The flow of gas under pressure may take different forms, see Figure 3.13.

The "viscous flow" of gases under pressure happens by collision between the gas molecules with a preferential (exclusive in the case of laminar flow) component along the direction of flow; depending on Reynold's number R_e, viscous flow can be turbulent or laminar. Laminar flow happens when Reynold's number (Eq. 3.27) $R_e<2{,}200$ (Britannica, 1980); otherwise, the flow is turbulent.

$$R_e = \frac{d \cdot V \cdot \delta}{\mu} \tag{3.27}$$

where d=tube diameter (m), V=fluid velocity (m/s), δ=density of the fluid (g/m^3) and μ=viscosity of the fluid (g/m/s).

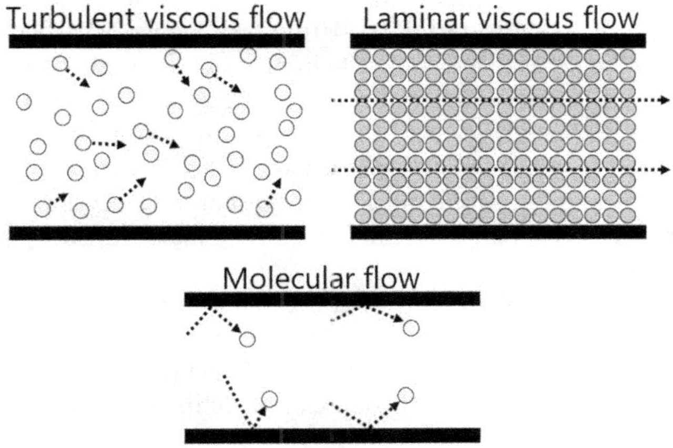

Figure 3.13 Types of gas flow under pressure.

Viscous flow takes place when the free path of the molecules is small with respect to the size of the pores. Under a vacuum, the distance that a molecule has to travel to collide with another one increases and, in small pores, the gas molecules collide more frequently with the pore walls than with a companion, resulting in the so-called "molecular flow".

The mean free path of air molecules λ (mean distance travelled before colliding with a companion) can be calculated as a function of the applied pressure p (Umrath, 1998):

$$\lambda(\mu\text{m}) = 66.7/p(\text{mbar}) \tag{3.28}$$

When λ is much larger than the size of a pore, "molecular flow" takes place whereby the molecules can move freely without mutual interference. The ratio between the mean free path λ and the diameter of the tube (pore) d, known as Knudsen number K_n, defines the predominant type of flow:

$$K_n = \frac{\lambda}{d} \tag{3.29}$$

Although not unanimously, viscous flow is considered to take place when $K_n < 0.01$; molecular flow takes place when $K_n > 0.5$ and transitional or Knudsen flow takes place for intermediate values of K_n (Umrath, 1998; Livesey, 1998).

By combining Eqs. (3.28) and (3.29) it is possible to relate the pressure of the air flow with the tube diameter, for the boundary conditions of viscous and molecular flow. This has been done in Figure 3.14 indicating the regions where the three flow regimes prevail, as function of p and d.

The shaded box represents the area where the diameter of the capillary pores of h.c.p. is likely to fall (0.01–10 μm, see Figure 3.8). It can be seen that viscous flow takes place, and only in pores larger than 10 μm, for pressures above 1,000 mbar (≈atmospheric pressure at sea level). For flow under vacuum ($p<1,000$ mbar) most of the flow is non-viscous. This means that the assumption of viscous flow for calculating the coefficient of permeability of concrete to gases, applying Hagen-Poiseuille-Darcy Eq. (3.26), may not represent the actual physical phenomenon of mass transport for systems under vacuum.

Therefore, the flow of gas can be conceived as a combination of viscous and molecular flow, suggesting that the no-slip assumption of the derivation of Hagen-Poiseuille law (Sections 3.5.1 and 3.5.4) is not really valid.

The flow under molecular diffusion can be defined as follows:

$$-N_M = D_K \frac{\Delta C}{\Delta L} \tag{3.30}$$

where

N_M = molar flux due to molecular diffusion (mol/m²/s)

D_K = coefficient of Knudsen diffusion (m²/s)

$\Delta C/\Delta L$ = gradient of concentration of gas moles (mol/m⁴)

or, expressing the gas concentration in terms of the applied pressure difference ΔP:

$$-N_M = D_K \frac{1}{R \cdot T} \frac{\Delta P}{\Delta L} \tag{3.31}$$

For the case of viscous flow of a gas, from Eq. (3.26) we can write:

$$-J_v = \frac{K_g}{\mu} \frac{\overline{P}}{P} \frac{\Delta P}{\Delta L} \tag{3.32}$$

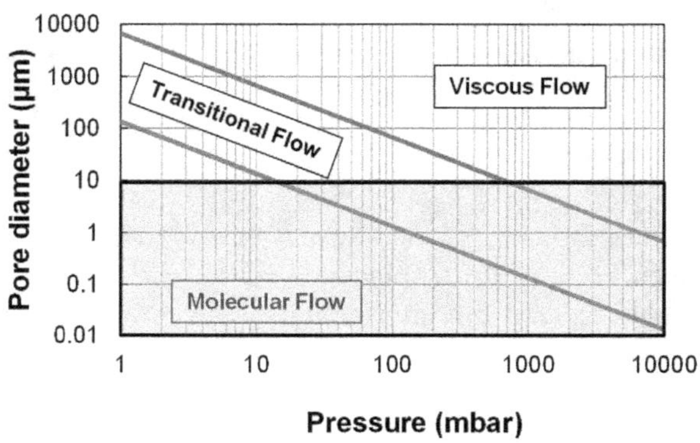

Figure 3.14 Regions of different gas flow regimes.

where $\overline{P}$ is the mean pressure acting on the gas and P (N/m^2) the pressure at which the volumetric flux J_v is measured.

The volumetric and molar fluxes are related by the fundamental equation of ideal gases:

$$P \cdot J_v = N_v \cdot R \cdot T \tag{3.33}$$

Combining Eqs. (3.32) and (3.33), we get

$$-N_v = \frac{K_g}{\mu} \frac{\overline{P}}{R \cdot T} \frac{\Delta P}{\Delta L} \tag{3.34}$$

where N_v is the molar flux due to viscous flow.

The total molar flux N will be the contribution of both the molecular and viscous flows; hence, adding Eqs. (3.31) and (3.34):

$$-N = \left(\frac{K_g}{\mu} + \frac{D_K}{\overline{P}}\right) \cdot \frac{\overline{P}}{R \cdot T} \frac{\Delta P}{\Delta L} \tag{3.35}$$

or

$$-N = \frac{K_g}{\mu}\left(1 + \frac{D_K \cdot \mu}{K_g \cdot \overline{P}}\right) \cdot \frac{\overline{P}}{R \cdot T} \frac{\Delta P}{\Delta L} \tag{3.36}$$

where an "apparent" coefficient of gas-permeability K_g^{app} can be defined as follows:

$$K_g^{app} = K_g\left(1 + \frac{D_K \cdot \mu}{K_g \cdot \overline{P}}\right) = K_g\left(1 + \frac{b}{\overline{P}}\right) \tag{3.37}$$

Equation (3.37) expresses the known fact that the apparent coefficient of gas-permeability K_g^{app}, measured in a test, decreases as the mean applied pressure increases. The third member of Eq. (3.37) has the form of the empirical correction proposed by Klinkenberg (1941), where parameter b can be experimentally obtained by plotting the measured K_g^{app} against the reciprocal of the mean pressures $\overline{P}$ applied. The slope b and the intersect K_g allow also the calculation of the Knudsen coefficient of diffusion D_K, applying the second member of Eq. (3.37).

Figure 3.15, built with data from Neves (2012), shows results of the apparent coefficient of oxygen-permeability K_g^{app}, measured on five different concrete mixes at different pressures, plotted against the reciprocal of the mean test pressure. The data in Figure 3.15 confirm the validity of Eq. (3.37).

An analysis of concrete mix D is shown in Figure 3.15, describing how parameters K_g and b are obtained from the regression line fitted to the experimental results. Table 3.3 presents the results of K_g, b and the Knudsen coefficient of diffusion D_K, Eq. (3.37), for the five concrete mixes shown in Figure 3.15.

Sakai and Kishi (2016) modelled the response of Torrent air-permeability test method, assuming molecular flow instead of viscous flow. Sakai (2020) developed a theoretical relation between the coefficient of gas-permeability k and the coefficient of molecular diffusion of gases D, which was adjusted to fit experimental results. This relation enables to convert the coefficient of gas-permeability obtained experimentally into an equivalent coefficient of gas diffusion or vice versa.

An interesting theoretical and experimental study on the Klinkenberg effect on gas-permeability, measured under pressure (Cembureau test) and under vacuum (special test), can be found in Sogbossi et al. (2019). A more

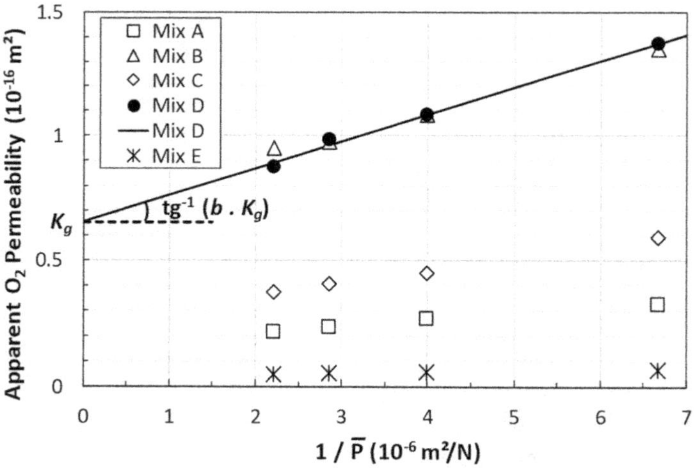

Figure 3.15 Evidence of Klinkenberg effect and calculation of parameters K_g and b (Neves, 2012).

Table 3.3 Parameters K_g and b of Klinkenberg effect and Knudsen diffusion D_K of data in Figure 3.15

Mix	*A*	*B*	*C*	*D*	*E*
K_g (10^{-16} m^2)	0.162	0.722	0.262	0.652	0.039
b (N/m^2)	151,906	128,943	187,200	165,510	115,102
D_K (10^{-8} m^2/s)	12.3	46.5	24.5	54.0	2.2

detailed analysis of the gas flow theory, particularly in relation to permeability tests, including Cembureau and Torrent methods, can be found in Sogbossi (2017).

3.7 CAPILLARY SUCTION AND WATER VAPOUR DIFFUSION

As discussed in Section 3.3, knowledge of the movement of water through concrete is of importance. However, the phenomenon of transport of water through the concrete pores is quite complex, involving several mechanisms such as capillary suction, surface diffusion, effusion and diffusion (Alvaredo, 1994). An explanation of these mechanisms, which depend on the moisture content of the concrete and the boundary conditions, can be found in Roelfstra (2001).

In what follows, two important mechanisms of water ingress (capillary suction), movement and evaporation (water vapour diffusion) will be discussed in some detail.

3.7.1 Capillary Suction: A Special Case of Water-Permeability

If the impermeable body of cross section area A, traversed by n capillary tubes of radius r, illustrated in Figure 3.11 is placed vertically in contact with water at its base (typical water sorptivity test), water will rise along the tubes due to the capillary pressure P_{cap} (N/m²) established at the interface between air and water:

$$P_{cap} = \frac{2 \cdot \sigma \cdot \cos\theta}{r} \tag{3.38}$$

where

σ = surface tension/energy at the interface, ≈0.073 N/m for concrete
θ = contact angle water/concrete, ≈0

when the water has risen a height z from the bottom, the gradient of pressure along the tube is

$$\frac{\Delta P}{z} = \frac{2 \cdot \sigma \cdot \cos\theta}{r \cdot z} \tag{3.39}$$

The gravity effect of the column of water is negligible compared to the capillary force in a tube of capillary pore size and has been neglected in Eq. (3.39).

Now, if we consider Eq. (3.15) relating the flow rate in a tube of radius r (given by the area of the tube multiplied by the rate of water rise) under a gradient of pressure given by Eq. (3.39) for the case of capillary suction, we have

$$\pi \cdot r^2 \frac{dz}{dt} = \frac{\pi \cdot r^4}{8 \cdot \mu} \frac{2 \cdot \sigma \cdot \cos\theta}{r \cdot z} \tag{3.40}$$

rearranging

$$z \cdot dz = \frac{\sigma \cdot \cos\theta}{4 \cdot \mu} \cdot r \cdot dt \tag{3.41}$$

and integrating

$$z = \sqrt{\frac{\sigma \cdot \cos\theta}{2 \cdot \mu}} \cdot \sqrt{r} \cdot \sqrt{t} = S \cdot \sqrt{t} \tag{3.42}$$

If we want to express the suction in terms of the increase in mass Δm of the body, we have to multiply z by the area of the tube ($\pi\, r^2$), by the number of tubes n and by the density of water δ_w, or

$$\Delta m = \pi \cdot r^2 \cdot n \cdot \delta_w \sqrt{\frac{\sigma \cdot \cos\theta}{2 \cdot \mu}} \cdot \sqrt{r} \cdot \sqrt{t} \tag{3.43}$$

or, expressing the mass increase per unit area of the body:

$$\frac{\Delta m}{A} = \frac{\pi \cdot r^2 \cdot n \cdot \delta_w}{A} \sqrt{\frac{\sigma \cdot \cos\theta}{2 \cdot \mu}} \cdot \sqrt{r} \cdot \sqrt{t} \tag{3.44}$$

and, entering the porosity of the body, Eq. (3.17):

$$\frac{\Delta m}{A} = \delta_w \cdot \varepsilon \sqrt{\frac{\sigma \cdot \cos\theta}{2 \cdot \mu}} \cdot \sqrt{r} \cdot \sqrt{t} = a \cdot \sqrt{t} \tag{3.45}$$

S ([L]/[T]^{½}, typically mm/s^{½}) is often known as "coefficient of water capillary rise", whilst a ([M]/[L]^2/[T]^{½}, typically g/m^2/s^{½}) as "coefficient of water suction". Both are common parameters used to express the rate of water suction of a porous body, which can be indistinctly designated as "Sorptivity".

Equations (3.42) and (3.45) are telling us that the water capillary rise and the corresponding increase in mass of the idealized body follow a "square root of time" law, something acknowledged in the units of S and a. In general, experimental evidence indicates that the exponent of t for concrete tends to be lower than ½, but it should not be forgotten that concrete is composed of an intricate network of pores of quite different sizes, not just of parallel tubes of the same radius. The "square root of time" law reflects the decreasing driving force (pressure gradient) as the water level in the tube rises. Some researchers (Villagrán Zaccardi et al., 2017) propose that a better fit to the capillary suction test results is achieved relating the water absorption with $t^{0.25}$ instead of $t^{0.5}$.

Equations (3.42) and (3.45) are also telling us that the sorptivity grows only with the square root of the radius of the tube, reflecting the combined effects of higher permeability and lower capillary pressure for larger r, that tend to partially compensate each other.

3.7.2 Water Vapour Diffusion

Diffusion through concrete is such a slow process that various phases of water may coexist in thermodynamic equilibrium in the pores (vapour, capillary water and adsorbed water) (Bažant & Najjar, 1972).

If the gradient of (evaporable) water content per unit volume of concrete w is considered to be the driving force of the diffusion of water through concrete, the first Fick's law of diffusion is applicable, Eq. (3.1), in the form:

$$-J = D_0 \cdot \frac{\partial w}{\partial x} \tag{3.46}$$

The problem is that the coefficient of diffusion D_0 is not constant for concrete. As concrete drying proceeds, the water remaining in the pores is lost with increasing difficulty and drying becomes much slower than predicted by the linear diffusion theory; so, the problem becomes non-linear (Bažant & Najjar, 1972).

The moisture flux J can also be expressed in terms of H, the pore relative humidity, that is the ratio of the partial pressure of water vapour to the equilibrium vapour pressure of water at a given temperature. A gradient of H is, therefore, a gradient of pressure and, therefore, the phenomenon can be seen as a permeation case.

$$-J = k \cdot \frac{\partial H}{\partial x} \tag{3.47}$$

where k is the permeability.

Now, the relation between w and H at constant temperature (and fixed degree of hydration) is given by the desorption or adsorption isotherms that can be expressed as:

$$dH = \varphi \cdot dw \tag{3.48}$$

where φ is the cotangent of the slope of the isotherm $w = w(H)$.

Equations (3.49) and (3.50) were adopted by Roelfstra (2001) and Saetta et al. (1993) to describe adsorption and desorption isotherms, respectively, for modelling purposes.

$$\frac{w}{w_{sat}} = H\left(1.16H^3 - 1.05H^2 - 0.11H + 1\right) \quad \text{Adsorption} \tag{3.49}$$

$$\frac{w}{w_{sat}} = H \quad \text{Desorption} \tag{3.50}$$

where w_{sat} is the content of water (kg/m³) of the saturated concrete.

The rate of change of mass of water per unit volume of porous material w is determined by the flux field as follows:

$$\frac{\partial w}{\partial t} = -\frac{\partial J}{\partial x} \tag{3.51}$$

Combining Eqs. (3.47), (3.48) and (3.51) we get, assuming a constant φ (the desorption isotherm is quasi-linear in the interval $0.15 < H < 0.95$) (Alvaredo, 1994),

$$\frac{\partial H}{\partial t} = \frac{\partial}{\partial x}\left(k \cdot \varphi \cdot \frac{\partial H}{\partial x}\right) = \frac{\partial}{\partial x}\left[D(H) \cdot \frac{\partial H}{\partial x}\right] \tag{3.52}$$

Equation (3.52) is shown as Eq. (5.1– 130) in *fib* Model Code 2010 (*fib*, 2012) and has been used in the "TransChlor" service life model for chloride-induced corrosion developed at EPFL (École Polytechnique Fédérale de Lausanne, Switzerland) (Conciatori, 2005), described in Section 9.4.1.

3.8 TRANSPORT PARAMETERS AND PORE STRUCTURE

3.8.1 Relationship between Transport Parameters and Pore Structure

The study of the relation between transport parameters and pore structure of h.c.p. probably started in 1958 (Powers, 1958), who established a good correlation between water-permeability and capillary porosity, stating that the flow through the extremely tiny gel pores is negligible.

Many studies in the same direction followed (Garboczi, 1990; Reinhardt & Gaber, 1990; Olivier & Massat, 1992; Tang & Nilsson, 1992; Holly et al., 1993; Roi et al., 1993; El-Dieb & Hooton, 1994; Christensen et al., 1996; Breysse & Gérard, 1997), particularly after the advent of mercury intrusion porosimetry (MIP), to study the pore structure of cementitious materials. Due to advantages in terms of homogeneity and size of samples, most studies focused on pastes and mortars, with few data produced for concretes. These studies established relations between transport properties, predominantly permeability, and pore structure of the h.c.p.; however, most of these studies ignored the important contribution of ITZ around coarse aggregates to the permeability of the system.

In what follows, an analysis of the main transport mechanisms will be made, based on the simplest possible model, i.e. that of an impervious body of cross section *A*, traversed by a single rectilinear tube of radius *r*.

From Eqs. (3.15) and (3.19), the permeability *K* of such body is:

$$K = \frac{\pi \cdot r^4}{8 \cdot A} \tag{3.53}$$

The diffusion coefficient *D* of such material can be described as (Nilsson & Tang, 1995):

$$D = D_0 \frac{\pi \cdot r^2}{A} \tag{3.54}$$

where D_0 is the bulk diffusion coefficient of the specie in the fluid

We can summarize by stating that, theoretically, the coefficient of permeability *K*, of diffusion *D* and the water sorptivity *S* respond to the following functions of the tube pore radius *r*:

$$K = C_K \cdot r^4 \tag{3.55}$$

$$D = C_D \cdot r^2 \tag{3.56}$$

From Eq. (3.42) and from Eq. (3.44) with $n=1$:

$$S = C_S \cdot r^{0.5} \tag{3.57}$$

$$a = C_a \cdot r^{2.5} \tag{3.58}$$

where C_k, C_D, C_S and C_a are constants.

Figure 3.16 shows data of air-permeability coefficient kT (SIA 262/1, Annex E), chloride-diffusion D_{Cl} (AASHTO T259–80) and coefficient of water suction at 24 hours a_{24} (SIA 262/1, Annex A) of 12 concrete mixes, of which the pore size distribution was measured by MIP. The data come from Tables 3.2-IV, a–c of Torrent and Ebensperger (1993); the kT_3 data reported were increased by a factor 1.846 to get kT, as discussed in Section 5.3.4. The above-mentioned properties are plotted as a function of the reported mean pore radii, fitting power trend lines; two D_{Cl} outliers, shown as empty triangles in Figure 3.16, were not included in the regression.

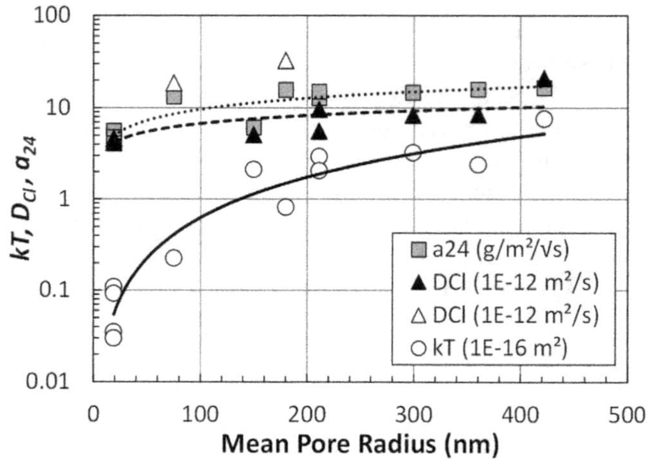

Figure 3.16 Variation of air-permeability *kT*, Cl^- diffusion D_{Cl} and water sorptivity a_{24} with pore radius.

Table 3.4 Min., max. and its ratio for r_m, *kT*, D_{Cl} and a_{24}, data from Torrent and Ebensperger (1993)

Property	r_m *(nm)*	**kT** *(10^{-16} m²)*	D_{Cl} *(10^{-12} m²/s)*	a_{24} *(g/m²/s½)*
Minimum	19	0.03	4.2	4.1
Maximum	422	7.61	32.6	16.5
Max/min ratio	22	250	8	4

The results confirm the much higher sensitivity of kT to the pore size, compared with both D_{Cl} and a_{24}, as predicted theoretically by Eqs. (3.55), (3.56) and (3.58). Table 3.4 presents the minimum and maximum values and their ratios of the following variables: mean pore radius r_m, kT, D_{Cl} and a_{24}.

The decreasing sensitivity of the transport properties $kT \rightarrow D_{Cl} \rightarrow a_{24}$, with the pore radius corresponds, at least qualitatively, to the theoretical predictions.

3.8.2 Permeability Predictions: Theory vs Experiments

3.8.2.1 Gas- and Water-Permeability vs Pore Structure

A method for impregnating concrete samples with epoxy before applying MIP test was proposed by Sakai et al. (2013, 2014a); it is claimed that, in this manner, a more realistic threshold pore radius r_t can be obtained. This r_t is the peak value of the derivative of the intruded volume with respect to the pore radius. In Sakai et al. (2013) parallel measurements of the coefficient of air-permeability kT (Torrent method) and the water-permeability K_w were reported. MIP, kT and K_w were measured on 15 concretes of w/c ratios within the range 0.40–0.70, made with different binders and containing entrained air. The permeability tests were conducted as described in connection with Figure 3.12. Figure 3.17 presents the results of kT and K_w as function of the theoretical permeability $\varepsilon\, r_t^2/8$, Eq. (3.20). The threshold pore radius r_t is taken as characteristic of the pore system and ε is the MIP-measured porosity.

Figure 3.17 displays also the equivalence line, showing that the agreement between theoretical prediction and experimental measurement of air-permeability kT is very good. The K_w values, although showing a good correlation with $\varepsilon\, r_t^2/8$, are some two orders of magnitude lower than those predicted by Eq. (3.20). The same data have been plotted in Figure 3.12, showing the same effect.

3.8.2.2 *Water Sorptivity vs Pore Structure*

Water sorptivity, assumed as a special case of permeability, is governed by Eq. (3.42). Operating on it, it is possible to write:

$$\frac{t}{z^2/r} = \frac{2\cdot\mu}{\sigma\cdot\cos\theta} = \frac{2\cdot 1\,10^{-3}}{0.073\cdot 1} = 0.027\,\mathrm{s/m} \qquad (3.59)$$

Therefore, if we could plot the time t required for water to penetrate a distance z by capillary suction (knowing the pore radius r), as function of z^2/r,

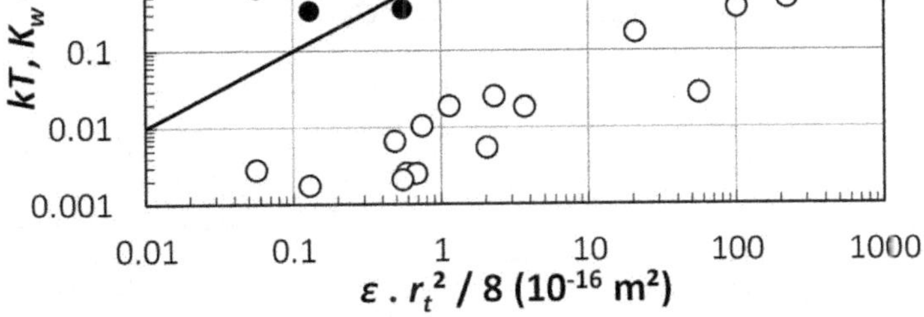

Figure 3.17 Measured air- and water-permeability (*kT* and *Kw*) vs. theoretical value based on MIP. Data from Sakai et al. (2013).

we should get a straight line passing through the origin of coordinates with a slope 0.027 s/m.

This plot was made, based on an experimental research by Sakai et al. (2014b), involving six concrete mixes prepared with different binders and *w/c* ratios between 0.40 and 0.70 (which were part of the 15 mixes used to build Figure 3.17). They measured the threshold pore radius r_t of each concrete with the same technique used to build Figure 3.17. In parallel, they cut slices of 2.75 years old Ø100×200 mm cylinders, slices that were dried for 3 months in a chamber at 40°C/20% RH. A water-sensitive paper was attached to the bottom surface of the slices, which were "loaded" with 30 mm of water on their top surface, allowing to measure the time t taken for the water to traverse the whole slice thickness z (which varied between 5 and 40 mm). Figure 3.18 (l.) shows the plot t vs. z^2/r for the 20 slices investigated, in a log-log scale, showing a good correlation between both variables of the form:

$$t = 1.65 \cdot \left(\frac{z^2}{r_t} \right)^{0.97} \tag{3.60}$$

The full line in Figure 3.18 (l.) corresponds to the theoretical relation expressed by Eq. (3.59). We can see that the experimental results follow the same trend as the theoretical line, but displaced one and a half orders of magnitude upwards. That means that the experimental "sorptivity" is significantly lower than the one predicted by Eq. (3.42).

Another evaluation of the validity, in this case, of Eq. (3.45), can be done analysing the data presented in Torrent and Frenzer (1995), where the coefficient of water absorption at 24 hours (a_{24}) was measured on eight concrete cubes of four different *w/c* ratios and two curing conditions, as described in Section 5.6.2. The results are displayed in the bottom line of Table 5.3.

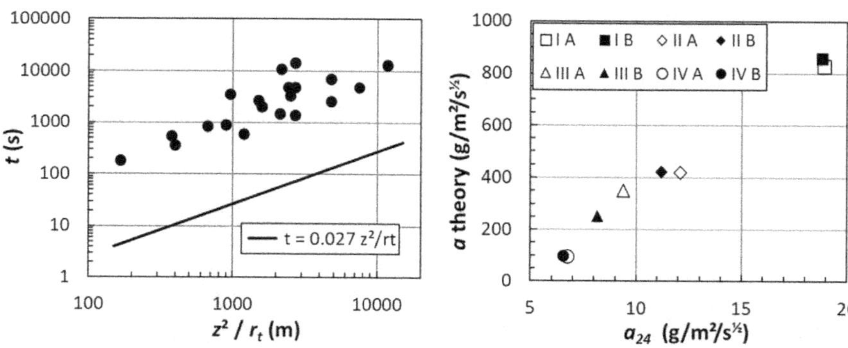

Figure 3.18 (l.) Permeation time t vs z^2/r_t: experimental results (●) and theory (line); (r.) theoretical a vs. experimental a_{24} coefficients of water absorption.

In parallel, two measurements of pore structure by MIP were conducted on samples extracted from each cube, as described in Section 11.4.2.1. A theoretical value of the coefficient of water absorption can be calculated from Eq. (3.45) as follows:

$$a = \delta_w \cdot \varepsilon \sqrt{\frac{\sigma \cdot \cos\theta}{2 \cdot \mu}} \cdot \sqrt{r} \tag{3.61}$$

where ε and r are the MIP total porosity and median pore radius, respectively.

Figure 3.18 (r.) presents in abscissae the experimental a_{24} values and in ordinates the theoretical value of a after Eq. (3.61). We can see a very good qualitative relation between both variables, but with Eq. (3.61) predicting values of the coefficient of water absorption that are between 15 and 45 times higher than the experimental ones, the ratio growing with the value of a_{24}.

The data presented in this Section 3.8 illustrate clearly the close relation existing between permeability and pore structure of concrete. In the case of gas-permeability, there is a quantitative agreement between theory and practice, whilst for water-permeability and sorptivity, the agreement is just qualitative, with the theoretical predictions overestimating the water-permeability/sorptivity by over one order of magnitude.

3.9 THEORETICAL RELATIONSHIP BETWEEN TRANSPORT PARAMETERS

It can be concluded that a material designed with a tight pore system will be resistant to the penetration of aggressive species via permeability (including capillary suction) or diffusion mechanisms, thus resulting in more durable structures. The relations between transport parameters are of practical interest, because short-term transport tests (e.g. permeability or migration) can be used as proxies to predict transport properties that would require long-term tests to be measured, typically chloride-diffusion and, to a lesser extent, gas diffusion.

Operating with Eqs. (3.55)–(3.58), we can derive the following theoretical relations between transport parameters D, S and a, with respect to K:

$$D = C_{D,K} \cdot K^{0.50} \tag{3.62}$$

$$S = C_{S,K} \cdot K^{0.125} \tag{3.63}$$

$$a = C_{a,K} \cdot K^{0.625} \tag{3.64}$$

where $C_{D,K}$, $C_{S,K}$ and $C_{a,K}$ are constants.

We will see in Section 8.3 that good correlations exist between different transport parameters and, also, that the experimental relations are not far from the theoretical predictions.

REFERENCES

ACI 308R (2016). "Guide to external curing of concrete".

Alexander, M.G., Arliguie, G., Ballivy, G., Bentur, A. and Marchand, J. (1999). "Engineering and transport properties of the interfacial transition zone in cementitious composites – State-of-the-art report of RILEM TC 159-ETC and 163-TPZ". *RILEM Report 20*, 404 p.

Alexander, K.M., Wardlaw, J. and Gilbert, D.J. (1965). "Aggregate-cement bond, cement paste strength and the strength of concrete". *The Structure of Concrete and Its Behaviour under Load, Proceedings on International Conference*. Brooks, A.E. and Newman, K. (Eds.), Cem. & Concrete Assoc., London, September, 59–81.

Al-Gahtani, A.S. (2010). "Effect of curing methods on the properties of plain and blended cement concretes". *Constr. Build. Mater.*, v24, n3, 308–314.

Alvaredo, A.M. (1994). "Drying shrinkage and crack formation". PhD Thesis, ETHZ, Switzerland, 102 p.

Andrade, C. (1993). "Calculation of chloride diffusion coefficients in concrete from ionic migration measurements". *Cem. & Concr. Res.*, v23, 724–742.

Atkins, P. and de Paula, J. (2013). *Elements of Physical Chemistry*. 6th Ed., Oxford University Press, Oxford, 624 p.

Bamforth, P.B. (1987). "The relation between permeability coefficients for concrete obtained using liquid and gas". *Mag. Concr. Res.*, v39, n138, March, 3–11.

Bažant, Z.P. and Najjar, L.J. (1972). "Nonlinear water diffusion in nonsaturated concrete". *Mater. & Struct.*, v5, n25, 3–20.

Bentur, A. and Goldman, A. (1989). "Curing effects, strength and physical properties of high strength silica fume concretes". *J. Mater. Civ. Eng.*, v1, n1, 46–58.

Bentz, D.P., Clifton, J.R., Ferraris, C.F. and Garboczi, E.J. (1999). "Transport properties and durability of concrete: Literature review and research plan". NISTIR 6395, Nat. Inst. of Stand. and Technol., Gaithersburg (MD), September, 49 p.

Breysse, D. and Gérard, B. (1997). "Modelling of permeability in cement-based materials: Part 1- uncracked medium". *Cem. & Concr. Res.*, v27, n5, 761–775.

Britannica (1980). *Encyclopedia Britannica*, 15th Ed., v11, *Macropedia*, Chicago, IL. 788.

Buenfeld, N.R. and Okundi, E. (1998). "Effect of cement content on transport in concrete". *Mag. Concr. Res.*, v50, n4, 339–351.

Christensen, B.J., Nason, T.O. and Jennings, H.M. (1996). "Comparison of measured and calculated permeabilities for hardened cement pastes". *Cem. & Concr. Res.*, v26, n9, 1325–1334.

Clariá, M.A., Irassar, E.F., López, R. and Bonaveti, V. (2012). "Cementos". *Ese material llamado hormigón*, Asoc. Arg. Tecnol. Hormigón (AATH), Maldonado, N.G. and Carrasco, M.F. (Eds.), Buenos Aires, Argentina, 366 p.

Collepardi, M., Marcialis, A. and Turriziani, R. (1970). "La cinetica di penetrazione degli ioni cloruro nel calcestruzzo". *Il cemento*, v67, 157–164.
Conciatori, D. (2005). "Effet du microclimat sur l'initiation de la corrosion des aciers d'armature dans les ouvrages en béton armé". PhD Thesis, EPFL, Switzerland, 264 p.
El-Dieb, A.S. and Hooton, R.D. (1994). "Evaluation of the Katz-Thompson model for estimating the water permeability of cement-based materials from mercury intrusion porosimetry data". *Cem. & Concr. Res.*, v24, n3, 443–455.
Feldman, R.F. and Sereda, P.J. (1968). "A model for hydrated Portland cement paste as deduced from sorption-length change and mechanical properties". *Matériaux et Constructions*, v1, 509–520. https://doi.org/10.1007/BF02473639.
fib (2012). *Model code 2010*, Final Draft, v1, fib Bulletin 65, March, 357 p.
Garboczi, E.J. (1990). "Permeability, diffusivity, and microstructural parameters: A critical review". *Cem. & Concr. Res.*, v20, 591–601.
Garboczi, E.J. (1995). "Microstructure and transport properties of concrete". RILEM report 12, *Performance Criteria for Concrete Durability*. Kropp, J. and Hilsdorf, H.K. Eds., E&FN Spon, London, 198–212.
Holcim (2011). *Holcim Cement Course – V2: Materials Technology II*. Holderbank, Switzerland.
Holly, J., Hampton, D. and Thomas, M.D.A. (1993). "Modelling relationships between permeability and cement paste pore microstructures". *Cem. & Concr. Res.*, v23, 1317–1330.
Jung, M. (1969). "Beiträge zur Gütewebertung korrosions- und wasserdichter Betone". Dissertation Weimar.
Justnes, H. (1997). "A review of chloride binding in cementitious systems". November, 17 p. https://www.danskbetonforening.dk/media/ncr/publication-no-21-4.pdf.
Klinkenberg, L.J. (1941). "The permeability of porous media to liquids and gases". *API Drilling and Production Practice*, 200–213.
Kreijger, P.C. (1984). "The skin of concrete. Composition and properties". *Mater. & Struct.*, v17, n100, 275–283.
Kropp, J., Hilsdorf, H.K., Grube, H., Andrade, C. and Nilsson, L.-O. (1995). "Transport mechanisms and definitions". RILEM report 12, *Performance Criteria for Concrete Durability*. Kropp, J. and Hilsdorf, H.K. (Eds.), E&FN Spon, London, 4–14.
Li, K. (2016). *Durability Design of Concrete Structures*. John Wiley & Sons, Singapore, 280 p.
Livesey, R.G. (1998). "Flow of gases through tubes and orifices". *Foundation of Vacuum Science and Technology*. Lafferty, J.M. (Ed.), John Wiley & Sons, New York, Chichester, Weinheim, Brisbane, Singapore, Toronto, 760 p.
Lothenbach, B. and Winnefeld, F. (2006). "Thermodynamic modelling of the hydration of Portland cement". *Cem. & Concr. Res.*, v36, n2, 209–226.
Lyubimova, T.Y. and Pinus, E.R. (1962). "Crystallization structure in the contact zone between aggregate and cement in concrete". *Kolloidnyi Zhurnal*, v24, n5, September–October, 578–587 (in Russian).
Maslehuddin, M., Ibrahim, M., Shameem, M., Ali, M.R. and Al-Mehthel, M.H. (2013). "Effect of curing methods on shrinkage and corrosion resistance of concrete". *Constr. & Build. Mater.*, v41, 634–641.

Mehta, P.K. and Monteiro, P.J.M. (2006). *Concrete. Microstructure, Properties and Materials.* 3rd Ed., McGraw-Hill, New York, Chicago, San Francisco, Athens, London, Madrid, Mexico City, Milan, New Delhi, Singapore, Sydney, Toronto, 660 p.

Mindess, S., Young, J.F. and Darwin, D. (2003). *Concrete.* 2nd Ed., Pearson Education Inc., Upper Saddle River, NJ, 644 p.

Neves, R.D. (2012). "A Permeabilidade ao Ar e a Carbonatação do Betão nas Estruturas". PhD Thesis, Universidade Técnica de Lisboa, Instituto Superior Técnico, Portugal, 502 p.

Neville, A.M. (1995). *Properties of Concrete.* 4th Ed., Longman Group Ltd., Harlow, Essex, 844 p.

Nilsson, L.-O. (2002). "Concepts in chloride ingress modelling". *RILEM PRO 38*, 29–48.

Nilsson, L.-O. and Tang, L. (1995). "Relation between different transport parameters". RILEM report 12, *Performance Criteria for Concrete Durability.* Kropp, J. and Hilsdorf, H.K. (Eds.), E&FN Spon, London, 15–32.

Olivier, J.P. and Massat, M. (1992). "Permeability and microstructure of concrete: A review of modelling". *Cem. & Concr. Res.*, v22, 503–514.

Page, C.L. and Vennesland, Ø. (1983). "Pore solution composition and chloride binding capacity of silica-fume cement pastes". *Mater. & Struct.*, v16, n1, 19–25.

Powers, T.C., Copeland, L.E., Hayes, J.C., and Mann, H.M. (1954). "Permeability of Portland Cement Paste". *Proc.* ACI, v51: 285, Portland Cement Assn., Research Dept., Bull. 53.

Powers, T.C. (1958). "Structure and physical properties of hardened portland cement paste". *J. Amer. Ceramic Soc.*, v41, January, 1–6.

Reinhardt, H.W. and Gaber, K. (1990). "From pore size distribution to an equivalent pore size of cement mortar". *Mater. & Struct.*, v23, 3–15.

Roelfstra, G. (2001). "Modèle d'évolution de l'état des ponts-routes en béton". PhD Thesis, EPFL, Switzerland, 175 p.

Roi, D.M., Brown, P.W., Shi, D., Scheetz, B.E. and May, W. (1993). "Concrete microstructure, porosity and permeability". SHRP C-628 Report, Washington, 90 p.

Saetta, A.V., Scotta, R.V. and Vitalini, R.V. (1993). "Analysis of chloride diffusion into partially saturated concrete". *ACI Mater. J.*, v9, n5, 441–451.

Sakai, Y. (2020). "Relationship between air diffusivity and permeability coefficients of cementitious materials". *RILEM Techn. Lett.*, v5, 26–32.

Sakai, Y. and Kishi, T. (2016). "Numerical simulation of air permeability in covercrete assuming molecular flow in circular tubes". *Constr. & Build. Mater.*, v125, 784–789.

Sakai, Y., Nakamura, C. and Kishi, T. (2013). "Correlation between permeability of concrete and threshold pore size obtained with epoxy-coated sample". *J. Adv. Concr. Technol.*, v11, August, 189–195.

Sakai, Y., Nakamura, C. and Kishi, T. (2014a). "Threshold pore radius of concrete obtained with two novel methods". *International Workshop on Performance-based Specifications and Control of Concrete Durability*, Zagreb, Croatia, June 11–13, 109–116.

Sakai, Y., Nakamura, C. and Kishi, T. (2014b). "Evaluation of mass transfer resistance of concrete based on representative pore size of permeation resistance". *Constr. & Build. Mater.*, v51, 40–46.

Scrivener, K. (1999). "Characterization of the ITZ and its quantification by test methods". Engineering and Transport Properties of the Interfacial Transition Zone in Cementitious Composites – State-of-the-Art Report of RILEM TC 159-ETC and 163-TPZ". RILEM Report 20.

Sensale, G.R. de and Gonçalves, A.F. (2014). "Effects of fine LWA and SAP as internal water curing agents". *Int. J. Concr. Struct. Mater.*, v8, n3, 229–238.

Snyder, K.A. (2001). "The relationship between the formation factor and the diffusion coefficient of porous materials saturated with concentrated electrolytes: theoretical and experimental considerations". *Concr. Sci. & Eng.*, v3, December, 216–224.

Sogbossi, H. (2017). "Etude de l'évolution de la perméabilité du béton en fonction de son endommagement: transposition des résultats de laboratoire à la prédiction des débits de fuite sur site". *Matériaux composites et construction*. Univ. Paul Sabatier, Toulouse, France.

Sogbossi, H., Verdier, J. and Multon, S. (2019). "New approach for the measurement of gas permeability and porosity accessible to gas in vacuum and under pressure". *Cem. & Concr. Composites*, v103, 59–70.

Spragg, R., Bu, Y., Snyder, K., Bentz, D. and Weiss, J. (2013). "Electrical Testing of Cement Based Materials: Role of Testing Techniques, Sample Conditioning, and Accelerated Curing". Publication FHWA/IN/JTRP-2013/28. Joint Transportation Res. Program, Indiana Dept. of Transportation and Purdue Univ., West Lafayette, Indiana, USA, 2013, 20 p.

Swamy, N. (1971). "Aggregate-matrix interaction in concrete systems". *Structure, Solid Mechanics & Engineering Design*, Part 1. Te'eni, M. (Ed.), John Wiley & Sons (Interscience), Bristol, 301–315.

Tang, L. and Nilsson, L.-O. (1992). "A study of the quantitative relationship between permeability and pore size distribution of hardened cement paste". *Cem. & Concr. Res.*, v22, 541–550.

Tang, L., Nilsson, L.-O. and Basheer, M. (2012). *Resistance of Concrete to Chloride Ingress. Testing and Modelling.* Spon Press, Abingdon and New York, 246 p.

Taylor, H.F.W. (1997). *Cement Chemistry.* 2nd Ed., Thomas Telford, London, 459 p.

Torrent, R. and Ebensperger, L. (1993). "Studie über Methoden zur Messung und Beurteilung der Kennwerte des Überdeckungsbetons auf der Baustelle". *Office Fédéral des Routes*, Rapport No. 506, Bern, Suisse, Januar, 119 p.

Torrent, R. and Frenzer, G. (1995). "Methoden zur Messung und Beurteilung der Kennwerte des Ueberdeckungsbetons auf der Baustelle -Teil II". Office Fédéral des Routes, Rapport No. 516, Bern, Suisse, October, 106 p.

Uchikawa, H. (1988). "Similarities and discrepancies of hardened cement paste, mortar and concrete from standpoints of composition and structure". Engng. Foundation Confer. *Advances in the Production and Utilization of Cement Based Materials*, Trout Lodge, Potosi, Missouri, v1, 271–310.

Umrath, W. (1998). "Fundamentals of vacuum technology". *Leybold Vacuum Products and Reference Book*, Cologne, Germany, 188 p.

Villagrán Zaccardi, Y.A., Alderete, N.M. and De Belie, N. (2017). "Improved model for capillary absorption in cementitious materials: Progress over the fourth root of time". *Cem. & Concr. Res.*, v100, October, 153–165.

Weast, R.C. (1975). *Handbook of Chemistry and Physics*. 56th Ed., CRC Press, Cleveland, OH.

Weiss, W.J. and Montanari, L. (2017). "Guide specification for internally curing concrete". Part of InTrans Project 13–482, Iowa State University, 40 p.

Xue, B., Pei, J., Sheng, Y. and Li, R. (2015). "Effect of curing compounds on the properties and microstructure of cement concretes". *Constr. & Build. Mater.*, v101, 410–416.

Ye, D., Zollinger, D., Choi, S. and Won, M. (2005). "Literature review of curing in portland cement concrete pavement". Report FHWA/TX06/0–5106-1, The Univ. of Texas, Austin, 80 p.

Yoon, I-S. (2010). "Reaction Experimental Study on chloride binding behavior in cement composition". *2nd International Symposium Service Life Design for Infrastructure*, Delft, The Netherlands, October 4–6, 631–635.

Zhou, C., Ren, F., Wang, Z., Chen, W. and Wang, W. (2017). "Why permeability to water is anomalously lower than that to many other fluids for cement-based material?" *Cem. & Concr. Res.*, v100, 373–384.

Zhutowsky, S. and Kovler, K. (2012). "Effect of internal curing on durability-related properties of high-performance concrete". *Cem. & Concr. Res.*, v42, 20–26.

Chapter 4

Test methods to measure permeability of concrete

In Chapter 3, the main mechanisms and laws governing mass transport through concrete were discussed, including their links to the pore structure of the material. Since the resistance of the material to the ingress and movement of aggressive substances is crucial to ensure durability, developing test methods capable of measuring that resistance is of key importance, associated with what was discussed in Section 1.7.1. The durability-related properties measured by those test methods are often designated as Durability Indicators (Alexander, 2004; Baroghel-Bouny, 2006).

Permeability of concrete is the main subject of this book; hence, this chapter presents in some detail the main test methods available to measure it, both in the laboratory and on site. In particular, methods that are standardized and/or commercially available are treated more extensively. Some test methods, referred to in various chapters of this book, involving other mechanisms of mass transport such as diffusion and migration, are described in Annex A.

For more comprehensive information on mass transport tests in concrete, the following reports of RILEM TCs (in which all three authors were actively involved) can be consulted (RILEM TC 116-PCD, 1995; RILEM TC 189-NEC, 2007; RILEM TC 230-PSC, 2016).

This chapter is structured presenting first the tests to measure water-permeability and water suction, in the laboratory and on site, followed by tests to measure gas-permeability, also in the laboratory and on site. Since the transport of liquids and gases under pressure is strongly affected by the degree of saturation of concrete's pore system, some pre-treatment of the specimens prior to laboratory testing is required, with the intention of testing the concrete samples under adequate and similar conditions.

4.1 WATER-PERMEABILITY

Many aggressive agents are dissolved in water when they ingress into concrete, hence the interest in measuring water-permeability. The water-permeability of concrete may be assessed by providing a constant flow rate

DOI: 10.1201/9780429505652-4

through concrete, under an inlet pressure. Achieving a constant flow rate, i.e. steady state, when feasible, takes a long time in practice (weeks/months), which makes the attractiveness of water-permeability tests rather limited. Moreover, the contact of concrete with water for such long periods alters the imperviousness of the sample during the test due to further hydration of the binder. However, it is a relevant technological property concerning hydraulic structures, such as dams, liquid-containers, pipes, canals, etc., and, therefore, some test methods have been proposed to characterize the water-permeability of concrete.

4.1.1 Laboratory Water-Permeability Tests

4.1.1.1 Steady-State Flow Test

First of all, it is required that the specimens are saturated before testing, see pp. 135–179 of RILEM TC 116-PCD (1995).

Sealing all specimens' surfaces but two identical areas destined for inlet and outlet flow is also necessary. The applied pressure varies from one test to another, according to the expected water-permeability of the tested concrete, in order to achieve steady-state flow within reasonable time (Coutinho & Gonçalves, 1994). The range of common test durations lies typically between days and weeks. Avoidance of thick specimens is advised, so as to keep the test duration within reasonable limits and to minimize deviations from Darcy's law. Sometimes it is also advisable to apply a confining pressure to the specimen. Figure 4.1 depicts an example of a steady-state water-permeability test set-up existing in Holcim laboratory in Toluca, México.

The outlet flow rate is monitored from the collected water and, when constant, Darcy's water-permeability coefficient can be determined applying Eq. (3.21) in the form:

$$K_D = \frac{Q \cdot L}{A \cdot H} \tag{4.1}$$

where

K_D=Darcy's water-permeability coefficient (m/s)
Q=water flow (m^3/s)
L=thickness of the specimen (m)
A=area of inlet/outlet concrete surface (m^2)
H=applied pressure, converted into height of water column (m)

For average quality concrete, Darcy's water-permeability coefficients in the order of 10^{-12} m/s shall be expected. Eq. (5.1–122) of *fib* (2012) allows a rough estimation of this coefficient on the basis of the concrete compressive strength. Darcy's water-permeability coefficient can be converted into an intrinsic permeability coefficient applying Eq. (3.22).

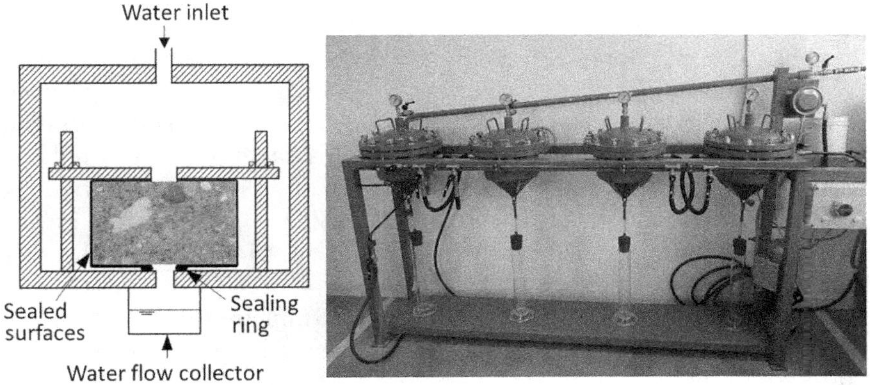

Figure 4.1 Experimental set-up for steady-state water-permeability testing.

4.1.1.2 Non Steady-State Flow Test: Water-Penetration under Pressure

To overcome the relatively long duration of steady-state testing of water-permeability, the non steady state condition has been adopted. If a through-flow is not established, then the measured variable is the amount of water that penetrates the sample, which can be evaluated in terms of penetration depth or volume. There are several methods to measure water-penetration in concrete through permeation. In their essence they are like the steady-state water permeation test, but in this case the test pressure and duration are predefined and, instead of measuring a flow, which is not supposed to occur within test duration, the volume of penetrating water is monitored, or the specimen is broken when the test ends and the water penetration depth is measured. There is a European Standard to test the water penetration depth by permeation (EN 12390-8, 2009), derived from German Standard (DIN 1048, 1978), the layout of which is shown in Figure 4.2.

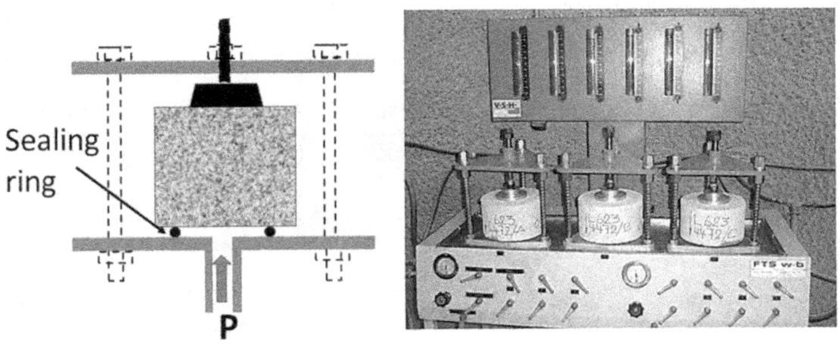

Figure 4.2 Set-up for the water penetration under pressure test (EN 12390-8).

A previously saturated cube or cylinder (150 mm size) is placed in a frame (see Figure 4.2), such that the bottom flat surface is in contact with water at 0.5 MPa (≈5 atm) of pressure, which is kept constant for 72 hours (variable between 0.1 and 0.7 MPa for DIN 1048 (1978)). At this time, the specimen is split, and the contour of the visible water penetration depth is drawn, recording the maximum value W_p in mm. This value is the reported test result (in some cases the mean value of the penetration is also reported). A classification of concrete quality in terms of this test was reported in The Netherlands (Van Eijk, 2009), included in Table 4.1 together with other water-permeability tests.

Typically, maximum W_p values of 50 and 30 mm are specified for environments of moderate and severe aggressiveness, respectively (EHE-08, 2008), see Table 1.2.

A similar test, but applied on a truncated cone-shaped specimen (Ø175/185 mm, height 150 mm), exists as Chinese Standard (GBJ 82-85, 1986) that is regularly used, especially for concrete dams.

Although the result of the water penetration under pressure test is expressed in terms of the maximum penetration W_p, a Darcy's water-permeability coefficient can be computed from test data using a relation originally developed by Valenta (1970), see also Claisse et al. (2009), requiring knowledge of the available porosity ε of the concrete:

$$K_D = \varepsilon \frac{W_p^2}{2 \cdot H \cdot t} \tag{4.2}$$

where

K_D=Darcy's water-permeability coefficient (m/s)
ε=available porosity (-)
W_p=water penetration depth (m)
H=applied pressure, converted to height of water column (m)
t=test duration (s)

Once the water penetration W_p has been measured by applying a pressure head H during a period t, Eq. (4.2) can be used to calculate the time t'

Table 4.1 Concrete quality rating based on results of various water-permeability tests

Quality class	*Max. W_p (mm)*	*KFPT (10^{-13} m/s)*	*ISA @10 minutes (mL/m²/s)*	*Figg's WAR (10^3 s/mL)*
Very high	-	<2.5	-	>100
High	<30	2.5–15	<0.25	30–100
Moderate	30–50	15–75	0.25–0.50	10–30
Low	50–80	75–150	>0.50	3–10
Very low	>80	>150	-	<3

required for the water front, under a pressure head H', to reach a certain depth x:

$$t' = \left(\frac{x}{W_p}\right)^2 \frac{H}{H'} \cdot t \tag{4.3}$$

4.1.2 Site Water-Permeability Tests

There are some test methods which were developed to assess the water-permeability of concrete on site; in particular, three methods that ended up in commercial instruments are described in this Section. The first two test methods are based on the same principle: the apparatus have a chamber in contact with the tested concrete surface, that is filled with water, kept under constant pressure while monitoring the volume of water penetrating the concrete.

4.1.2.1 Germann Test

In GWT (Germann Water permeation Test) method, a pressure chamber containing a watertight gasket is secured tightly onto the surface by two anchored clamping pliers or by means of a suction plate (GWT, 2014). The gasket may optionally be glued to the surface.

The chamber is filled with water and the valves closed. The top lid of the chamber is turned until a desired water pressure is achieved. The pressure selected is maintained by means of a micrometer gauge pressing a piston into the chamber, substituting the water penetrating into the material. The travel of the piston over time measures the amount of water penetrated and is used for characterizing the permeation of the surface tested.

A commercial instrument is produced by Germann Instruments.

4.1.2.2 Autoclam System

The Autoclam permeability system allows the measurement of water-permeability and sorptivity, combined with air-permeability. The major difference between the water-permeability and sorptivity test is the level of applied relative pressure, 0.5 or 0.02 bar for permeability and sorptivity test, respectively. As complete saturation of concrete cannot be assured, sorptivity almost always contributes to concrete water intake; it is claimed that testing pressures $\geq$0.5 bar are sufficient to make permeation dominant in relation to sorptivity (Basheer, 1993).

The fundamentals of the method were developed in the 1980s (Montgomery & Adams, 1985; Montgomery & Basheer, 1989), ending up in a commercial instrument branded "Autoclam" (Amphora, n.d.).

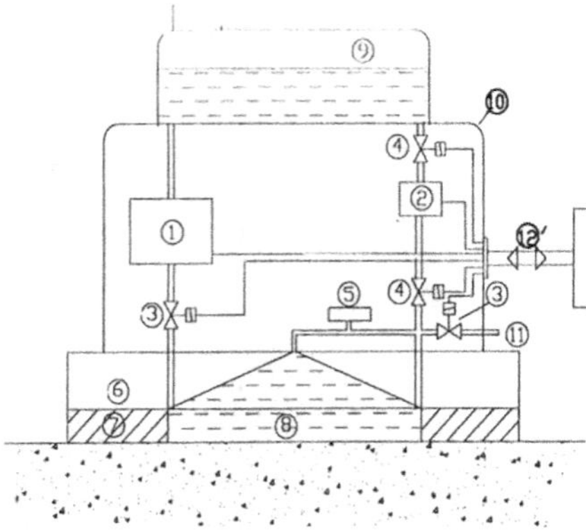

Figure 4.3 "Autoclam" system: sketch (RILEM TC 189-NEC, 2007) and view of two units at work.

The layout of the test is shown in Figure 4.3. Water from the reservoir (9) is admitted into the test area (8) through a priming pump (1) with air escaping through the bleed pipe (11). A pressure transducer (5) measures the water pressure.

When the test chamber is completely filled with water the priming pump automatically switches off and a micro pump pressurizes the test area to 0.5 bar (0.02 bar for water sorptivity) above atmospheric, at which stage the test starts. As water permeates (is also absorbed by capillary action), the pressure inside tends to decrease, but it is maintained constant by the micro pump and the control system. The volume of water delivered is measured and recorded every minute for a test duration of 15 minutes, so that the total quantity of water permeated/absorbed during the test is accurately known. Testing has shown that a plot of the quantity of water permeated/absorbed with the square root of time elapsed is linear. The slope of this graph is reported as the water-permeability/sorptivity index with units of $m^3/min^{1/2}$.

4.1.2.3 Field Water-Permeability

The method described here is the Field Permeability Test (FPT), developed by Meletiou (1991) and reported by Meletiou et al. (1992). The basic concept of the FPT apparatus is to drill a hole (23 mm in diameter and 152 mm deep) in the concrete. Then, a cylindrical probe is introduced into the hole; tightening the top nut causes the two neoprene packers to expand and seal-off a central chamber. First, a full vacuum is applied to the chamber for 5–10 minutes.

The probe is then connected to the instrumentation unit. Water, pressurized by a nitrogen bottle, is introduced into the chamber through the hollow probe, at a pressure within the range from 10 to 35 bar (normally 17 bar), which flows radially into the surrounding concrete. It is claimed that a steady-state flow is achieved (at about 30 minutes), at which time the flow rate is recorded from 5 to 15 minutes intervals for about 2 hours by means of a capillary flow-meter. The pressure and the flow rate allow the calculation of the apparent coefficient of permeability by Field Permeability Test or KFPT (cm/s) according to Darcy's law (Meletiou et al., 1992). The total duration of a single determination is 2 hours, which is the duration of the test itself, to which the time to drill the hole, insert the probe and preconditioning the zone should be added, making a total of about 3 hours.

An excellent correlation was reported between KFPT and the coefficient of water-permeability determined by a laboratory method KLAB (Soongswang et al., 1988; Meletiou et al., 1992), both applied on concrete blocks made of the same batches of concrete.

Table 4.1 presents a proposed classification of water-permeability of concrete, based on FPT results. Some results of the application of FPT in the laboratory, on 1 m cubes, can be found in Table 2.2.3.3 of Torrent and Frenzer (1995), who reported KFPT values in the range of $8–29 \times 10^{-13}$ m/s for a high-strength concrete (80 MPa 28-day cube strength) and in the range of $16–55 \times 10^{-13}$ m/s for a 51 MPa concrete.

4.2 SORPTIVITY: SPECIAL CASE OF WATER-PERMEABILITY

As discussed in Section 3.7.1, an important mechanism of water (and the aggressive ions it may carry in dissolution, e.g. Cl^-, SO_4^{2-}, Na^+, K^+) ingress in concrete is capillary suction, that can be considered a special case of water-permeability, in which the driving force is capillary pore pressure. This type of permeation occurs whenever concrete, that is not fully saturated, is put in contact with small heads of water. Typical examples are concrete surfaces exposed to cycles of dry and rainy weather, concrete in marine spray, splash and tidal zones or exposed to cyclic spray and splash of de-icing salts solution. Capillary suction is a much more powerful driver than diffusivity.

Capillary suction tests or, in short, sorptivity tests, consist in simulating in the laboratory or on site, this major transport mechanism. The test consists in exposing a sufficiently dry concrete surface to contact with water and in monitoring the rate of penetration of water into the material.

Under laboratory conditions, where a proper pre-conditioning of the sample is feasible (see Section 4.3.1.1, almost identically applicable to sorptivity tests), it is a simple-to-assess property and the results have a relatively low scatter in a single laboratory. Thus, it may constitute an interesting durability indicator.

It shall be mentioned that the sorptivity test has a double dependence on the moisture content of the concrete tested, as not only the permeability (blockage of pores) is affected by it, but also the force (capillary pressure) driving the water into the pores. Hence, differences in moisture conditions of the specimens (that depend not only on the drying temperature and duration, but also on the relative humidity, ventilation, packing of the oven, etc.) have a strong effect on the test result. Therefore, although the repeatability of the test in a single laboratory may be very good, the reproducibility of the test among different labs can be very poor, mainly due to difficulties in achieving the same saturation degree of the samples at the initiation of the test.

4.2.1 Laboratory Sorptivity Tests

Concerning laboratory testing, there are several methods to assess concrete sorptivity. Among them, the most popular are RILEM TC 116-PCD (1999d), ASTM C 1585 (2013) and the South African water sorptivity test (Alexander et al., 1999). A sorptivity test has existed in the Swiss Standards since 1989 (SIA 162/1, 1989), nowadays under Annex A of SIA 262/1 (2019), which is used for specification and control of the concrete supplied by producers for certain exposure classes. All these methods share the same basic principle, introduced by Fagerlund (1977): immersion of the bottom end of the specimen into a shallow head of water (typically 3–5 mm) and monitor the water absorption through periodical weighing of the specimen.

Following these methods, the lateral faces of the specimens are sealed (not in the Swiss Standards), as the objective is to promote a unidirectional water flow into concrete, driven by capillary forces. The specimens shall lay upon a pile of filter papers or a pair of rollers or pins, so the water absorption into the tested surface is not disturbed. All of them also recommend care in removing the free water from the concrete surface prior to weighing, with a dampened paper towel or cloth. Swiss Standard (SIA 262/1, 2019) requires cores to be drilled from cast specimens or from site concrete to conduct the test, to avoid the "wall effect" ("skin of concrete"). The curved surfaces are not sealed and the test box is covered by a sliding transparent lid.

There are slight differences in test procedures, as that of the top end of the specimen being covered to prevent evaporation of water from the specimen (RILEM TC 116-PCD, 1999d; ASTM C 1585, 2013) or not (Alexander et al., 1999; SIA 262/1, 2019), or the fluid being calcium hydroxide solution (Alexander et al., 1999) or tap water (RILEM TC 116-PCD, 1999d; ASTM C 1585, 2013; SIA 262/1, 2019). However, the major differences concern the test output. The RILEM recommendation mentions that the validity of Washburn (1921) equation is not expected, i.e. there is no linear relationship between mass gain and the square root of time, as theoretically

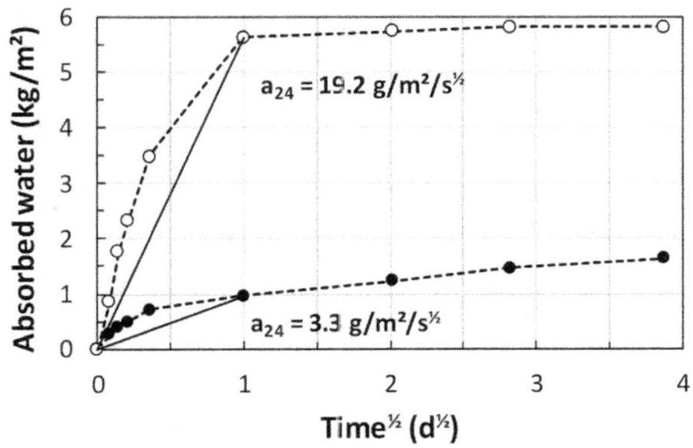

Figure 4.4 Mass gain with the square root of time for two concretes of different sorptivities, measured after SIA 162/1(1989).

predicted, see Eqs. (3.42) and (3.45). This is clearly shown in Figure 4.4, presenting the water absorption with the square root of time t for two concretes of different sorptivities; Villagrán Zaccardi et al. (2017) suggest linearity with $t^{0.25}$ instead of with $t^{0.5}$.

Therefore, the test result often consists in the amount of water absorbed per surface area in contact with water at the end of defined periods. Usually, it is calculated by the following equation:

$$A_t = \frac{m_t - m_0}{A} \tag{4.4}$$

where

A_t=absorption of water at time t (g/m²)
m_t=mass of the specimen at time t (g)
m_0=initial mass of the specimen (g)
A=area of the specimen surface in contact with water (m²)

Sometimes, the coefficient of water absorption a_t is reported, which is A_t divided by the square root of time; typical units of a_t are g/m²/s½ (see Figure 4.4). Since the mass gain is not strictly proportional to the square root of time, the time at which A_t and a_t are determined shall be reported, 24 hours being a common test duration to discriminate concretes of low, medium or high sorptivity.

ASTM C 1585 (2013) method yields both a sorptivity coefficient and water absorbed at a given time. The water absorbed after a given time is expressed in length units, converting the mass increase into volume of absorbed water and assuming that the specimen has a prismatic shape (constant cross-section):

$$I_t = \frac{\left(\dfrac{m_t - m_0}{\rho_w}\right)}{A} \tag{4.5}$$

where

I_t=water absorption at time t (mm)
m_t=mass of the specimen at time t (g)
m_0=initial mass of the specimen (g)
ρ_w=density of water (1×10^{-3} g/mm^3)
A=area of the specimen surface in contact with water (mm^2)

Parameter I_t may give the wrong impression that it measures the height reached by the water in the sample at time t, which is not the case. In fact, I_t measures the height of water placed on the sample that is absorbed in time t. To convert mass gain to capillary rise in the sample, the porosity of the concrete has to be known, as described below (Eq. 4.7).

The sorptivity coefficient is defined as an initial rate of water absorption S (mm/s$^{1/2}$), set by the slope of the straight line that fits best the I_t versus the square root of time plot.

$$S = \left(n \times \sum_{i=1}^{n} \sqrt{t_i} \times I_i - \sum_{i=1}^{n} \sqrt{t_i} \times \sum_{i=1}^{n} I_i \right) \Big/ \left(n \times \sum_{i=1}^{n} t_i - \sum_{i=1}^{n} \sqrt{t_i} \times \sum_{i=1}^{n} \sqrt{t_i} \right) \tag{4.6}$$

where

S=initial rate of water absorption (mm/s$^{1/2}$)
n=number of time intervals
t_i=time at a specific weighing (s)
I_i=water absorption at time t_i, *viz.* Eq. (4.5) (mm)

The validity of this rate requires achieving a minimum correlation coefficient of 0.98 between the analysed variables, at least, in the first 6 hours of testing. In concrete, values of S in the order of magnitude 10^{-3} to 10^{-4} mm/s$^{1/2}$ are expected.

The test result according to the South African method (Alexander et al., 1999) is a sorptivity coefficient, calculated as per ASTM C1585, although the allowed testing time to achieve a minimum correlation coefficient of 0.98 can be as low as 25 minutes and the coefficient computation is affected by concrete porosity and expressed in mm/h$^{1/2}$:

$$H = \frac{S}{\varepsilon} \cdot 60 \tag{4.7}$$

where

H=sorptivity coefficient (mm/h$^{1/2}$)

S=initial rate of water absorption (mm/s$^{1/2}$), from Eq. (4.6)
ε=open porosity of concrete (-)

The determination of H, besides the capillary absorption test, requires also the vacuum water absorption test. Interestingly, this test output theoretically corresponds to a rate of an average height of capillary rise, as the average height of absorbed water in the concrete capillaries, neglecting tortuosity, can be estimated as follows:

$$h_t = \frac{V_{w,t}}{A_c} = \frac{\left(\dfrac{m_t - m_0}{\rho_w}\right)}{\varepsilon \times A} = \frac{I_t}{\varepsilon} \tag{4.8}$$

where
h_t=height of capillary rise at time t (mm)
$V_{w,t}$=volume of absorbed water at time t (mm^3)
A_c=cross-section of capillaries in contact with water (mm^2)
m_t=mass of the specimen at time t (g)
m_0=initial mass of the specimen (g)
ρ_w=density of water (1×10^{-3} g/mm^3)
ε=open porosity of concrete (-)
A=area of the specimen surface in contact with water (mm^2)
I_t=water absorption at time t *viz.* Eq. (4.5) (mm)

4.2.2 Site Sorptivity Tests

There are several test methods to assess concrete sorptivity on site, described in the following sections. All of them consist in attaching a chamber on or drilling a hole in the concrete surface to be tested, where water (at low pressure) is put into contact with concrete. The rate at which water is absorbed is measured, which is – per se – an indication of the sorptivity of the concrete surface. As discussed in Section 4.2, the initial moisture condition of the concrete has a strong effect on the results of this test.

One shortcoming of many site sorptivity tests is that they cannot be applied on the underside of structural elements (bottom side of slabs and beams).

4.2.2.1 ISAT

The Initial Surface Absorption Test (ISAT) was originally developed by Glanville (1931) and later modified by Levitt (1969a, 1969b, 1971) in the early 1970s. The ISAT has the merit of being probably the first site "penetrability" test ever covered by a standard, BS 1881-5 (1970), now superseded by BS 1881-208 (1996).

The method consists in placing a cap, preferably transparent, onto the concrete surface and making it water-tight by clamping it (the cap carries a rubber O-ring for this purpose). The concrete surface under the cap is subject to a slight water head pressure of 200 mm (approx. 0.02 bar) by allowing the water to flow from the reservoir into the cap, opening the inlet valve (Figure 4.5). Once the cap is filled, water continues flowing and fills also the horizontal calibrated capillary tube; then, the water inlet valve is closed. The rate of water absorption by the concrete under the cap is measured by the movement of the water meniscus along the capillary tube.

After 10, 30 and 60 minutes of opening the valve to allow the initial contact between water and concrete, absorption readings are taken. When a measurement is finished, the inlet valve is reopened. Based on the measurements, an index of initial surface absorption can be calculated as follows:

$$ISA = \frac{\pi \cdot \phi^2 \cdot L}{4 \cdot A \cdot t} \tag{4.9}$$

where

ISA = initial surface absorption (mL/m²/s)
ϕ = diameter of the capillary tube (mm)
L = length travelled by the water meniscus in the capillary tube (mm)
A = area of concrete in contact with water (m²)
t = time elapsed in L measurement (s)

Table 4.1 shows a rating of concrete cover quality, based on *ISA* values, proposed by Concrete Society (1987). ISA values at 30 minutes of a range of concrete qualities, reported in Table 3.1-IIIj of Torrent and Ebensperger (1993), ranged between 0.1 and 0.8 mL/m²/s.

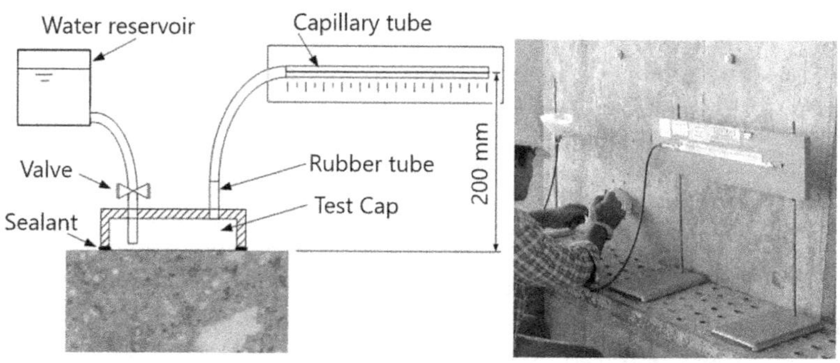

Figure 4.5 Sketch and view of site application of ISAT.

4.2.2.2 Karsten Tube

The Karsten tube method, developed in the early 1960s by Rudolf Karsten (Hendrickx, 2013), is one of the simplest (and cheapest) methods to measure water sorptivity on site. It consists of a glass cap, fixed onto the concrete surface with putty (mastic), to which a calibrated tube is attached (see Figure 4.6 with different configurations for vertical or horizontal surfaces). The tube is filled with water and the rate at which the water is absorbed is measured by the drop of the meniscus along the tube.

The test has found more acceptance for testing façades of stone and brick or renderings than for concrete (Pereira Apps, 2011; Duarte et al., 2011). Indeed, RILEM Recommendations (RILEM TC 25-PEM, 1980) were issued to evaluate on site the quality and degradation of stones as well as the effectiveness of protective treatments. Measurements are recorded after 5, 10 and 15 minutes, extending the test until 1 hour, depending on the permeability of the rock.

An application to concrete, with and without surface sealant treatments, can be found in Nwaubani (2018). In Table 3.2-IV of Torrent and Ebensperger (1993) Karsten results ranging, for a variety of concrete qualities, between 3 and 20 g/m^2/s$^{1/2}$ are reported.

4.2.2.3 Figg

This method was originally proposed by Figg (1973) and later modified by Cather et al. (1984), whose recommendations are considered in the latest commercial version of the test equipment (Porosiscope, formerly Poroscope Plus). Each individual measurement requires the drilling of a Ø10 and

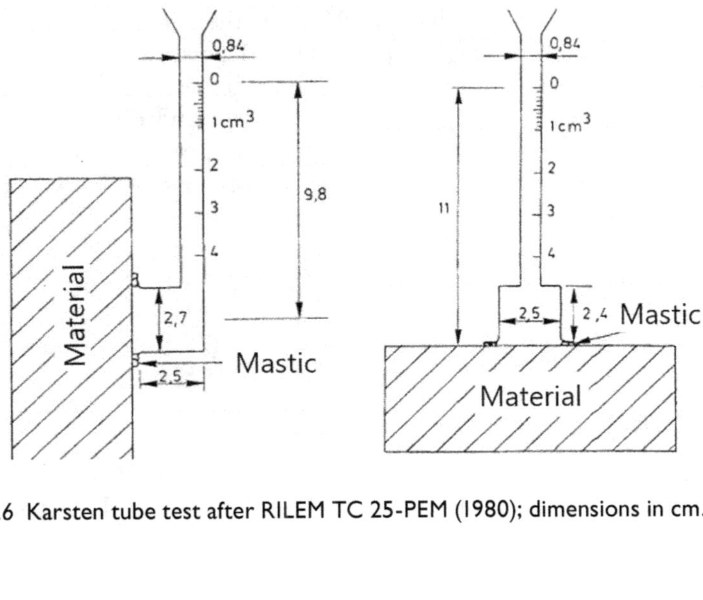

Figure 4.6 Karsten tube test after RILEM TC 25-PEM (1980); dimensions in cm.

40 mm deep hole. The hole is then partially sealed by means of a rubber plug, resulting in a cylindrical test chamber of Ø10 and 20 mm height at the bottom of the hole. Then a needle, connected to two concentric tubes, perforates the seal plug. The outer tube ends at the connection and the inner tube is longer and is inserted into the hole through the needle. The inner tube, in the opposite end, is connected to a syringe filled with water. Pressing the plunger of the syringe will drive water into the hole. When the hole is filled, the water starts to return through the remaining space in the needle and will proceed into the outer tube, which ends in a capillary tube. When the water that is coming out of the capillary tube is free of air bubbles, the connection between the syringe and the inner tube is interrupted by closing a stopcock. As the concrete of the hole walls absorbs water, the meniscus inside the capillary tube starts to move. The time required for the water meniscus to travel a length that corresponds to a volume of absorbed water of 0.01 mL is measured. The Porosiscope apparatus is equipped with two sensors that detect the water meniscus movement and a stopwatch, which is activated when the meniscus crosses the first sensor and stopped when it crosses the second sensor.

The test result is denominated Water Absorption Rate, which is given by:

$$WAR = \frac{t}{0.01} \tag{4.10}$$

where

WAR = water absorption rate (s/mL)

t = measured time (s)

Actually, the test result is the reciprocal of a water absorption rate, as higher values of WAR correspond to low rates of water absorption by concrete. In regular quality concretes, values of WAR in the range of $3–100 \times 10^3$ s/mL shall be expected (James, 2007). Table 4.1 presents a classification of concrete qualities based on the WAR values, proposed for the Porosiscope.

WAR data reported in Table 3.2-IV of Torrent and Ebensperger (1993) obtained on a wide variety of concrete qualities, range between 1×10^3 and over 500×10^3 s/mL, which are consistent with the ranges indicated by the manufacturers.

4.2.2.4 *Autoclam System*

See Section 4.1.2.2.

4.2.2.5 *SWAT*

The Surface Water Absorption Test (SWAT) was developed at the University of Yokohama, Japan (Akmal et al., 2011; Hayashi & Hosoda, 2013).

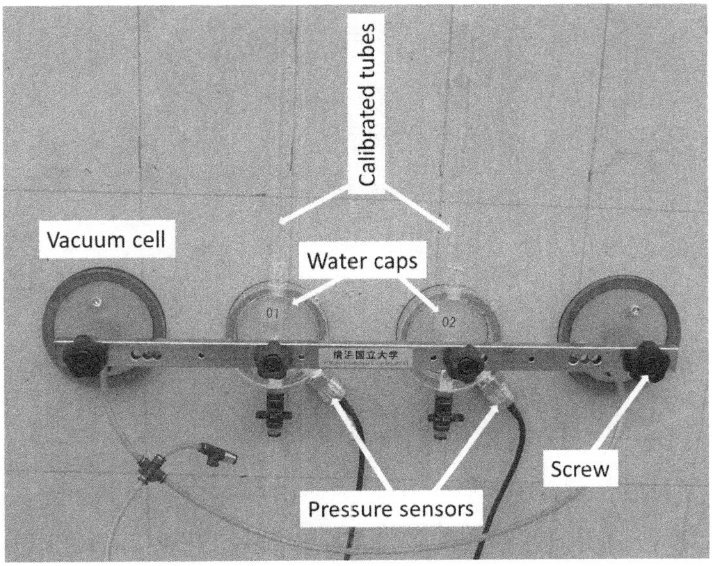

Figure 4.7 SWAT method set-up on vertical wall.

It consists of a steel frame, attached onto the concrete surface by means of a small vacuum pump operating on two vacuum cells located at its ends (Figure 4.7). The frame holds two or more water caps with rubber gaskets that are fixed individually onto the concrete surface by means of screws.

Each water cap is filled with water through a tube until reaching a small head of water (300 mm). The absorbed water reduces the water head, reduction that is monitored automatically by a pressure sensor and data recording system.

A SWAT can be conducted within around 5 seconds after starting filling water; thus, initial absorbing behaviour can be measured. Through readings of SWAT after 10 minutes, the water absorption capability of the concrete can be measured.

4.2.2.6 WIST

The Water Intentionally Spraying Test (WIST) is possibly the simplest indirect method of measuring the water absorption capability of a vertical concrete surface (it cannot be applied on horizontal surfaces). It was developed by S. Nishio of Japan Railway Technical Research Institute (Nishio, 2017; Nguyen et al., 2019).

It consists of spraying a controlled small amount of water, through a mask, so as to create a "smile"-shaped wet pattern on the concrete surface (Figure 4.8).

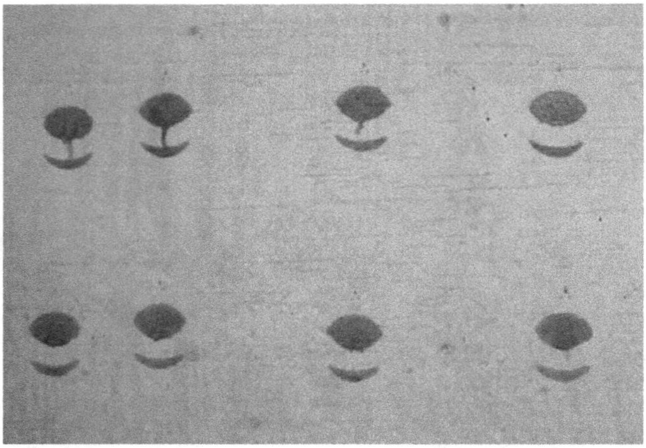

Figure 4.8 "Smile"-shaped wet patterns on the concrete surface for WIST test.

Up to 4 or 8 shapes can be produced at about the same time; the sprays are repeated at 60 seconds intervals at the same place, until water drips down from the oval wet area reaching the crescent-shaped wet area (two top-left shapes in Figure 4.8). The test result is the number of sprays required for this to happen, which typically ranges between 3 and 15.

Although extremely ingenious and simple, the test result is dependent on a number of factors that influence the rate of evaporation (temperature of concrete in relation to ambient, relative humidity and wind velocity, solar radiation) and, possibly, on the wall roughness as well. Nevertheless, it may constitute a fast, semi-quantitative way of identifying areas of low and high permeability, where other, more accurate tests can be applied later.

4.3 GAS-PERMEABILITY

As discussed in Section 3.5, the intrinsic permeability of concrete to gases is several orders of magnitude higher than to water. This easier gas flow through the intricate pore structure of concrete means that a steady-state flow is more rapidly reached and the duration of a gas-permeability test can be much shorter (minutes or hours) compared with a water-permeability test (days or weeks).

Regarding the significance of gas-permeability of concrete, it is worth quoting H. Hilsdorf and J. Kropp, in p. 287 of RILEM TC 116-PCD Report (1995):

> "The present knowledge on suitable test methods and correlations between the individual transport parameters justifies the extensive use of gas permeability measurements to characterize concrete transport

properties. Although not directly relevant to degradation, this parameter offers close correlations with the diffusion coefficient for gases, the diffusion of aggressive ions in the liquid phase, the rate of water absorption as well as with the permeability to water or diluted solutions. Therefore, this single parameter may characterize the penetrability of concrete in a variety of different cases, thereby covering various corrosion mechanisms".

There are several test methods to assess concrete gas-permeability, both for laboratory and site applications, as described in the following sections.

4.3.1 Laboratory Gas-Permeability Test Methods

4.3.1.1 Influence of Moisture and the Need for Pre-Conditioning

Contrary to water-permeability tests that require complete pore saturation, gas-permeability tests require that at least the larger pores (remember that permeability is proportional to the fourth power of the pore size, Eq. (3.15)) are sufficiently free of water so as to allow the passage of the gas molecules through them. Same kind of treatment is required by water suction (sorptivity) tests, see Section 4.2, because an interface air-water is required in the pores of concrete to develop capillary forces capable of driving the water into the pore system.

A great advantage of laboratory over site testing (of gas-permeability and water sorptivity) is that the specimens can be preconditioned in a standard manner prior to testing, to minimize the highly significant effect of moisture on the results of such tests (see Section 6.9). In the case of laboratory tests, this requires a pre-conditioning of the specimens consisting of a drying process, be it in a controlled drying room (takes longer) or in an oven (shorter).

One approach, possibly the most scientifically correct, consists in oven-drying the specimen to a level that corresponds to its equilibrium saturation degree for a given temperature and relative humidity. The problem of this approach is that it requires measuring the desorption isotherm of the material, which is a lengthy and delicate process.

Along that line, Recommendation RILEM TC 116-PCD (1999b) describes a process by which a saturated 5 mm-thick slice of the concrete is kept at 20°C/75% RH (in a CO_2-free environment to avoid carbonation) until equilibrium is reached, from which the corresponding equilibrium moisture content is obtained. The specimen to be tested for permeability (typically Ø150×50 mm disc) is dried in a ventilated oven at 50°C until the same degree of saturation as that established in the desorption experiment is reached. Then, a moisture homogenization phase follows, by which the now sealed disc is kept in the oven at 50°C for at least a further 14 days. This procedure is recommended for both gas-permeability and water sorptivity tests. Spanish Standard UNE 83966 (2008) follows the same procedure as the RILEM Recommendation, considered as 'reference' conditioning.

The above described pre-conditioning process is clearly impractical for quality control purposes, due to its complexity and long duration. Hence, faster and simpler pre-conditioning processes are prescribed, as illustrated by the following examples:

- the same Spanish Standard UNE 83966 (2008) proposes an "accelerated" pre-conditioning, by which the Ø150×50 mm disc is dried in a ventilated oven at 50°C during 4 days, followed by a homogenization stage (specimen sealed) of 3 extra days at 50°C, to be completed by 21 days in a dry room at 20°C/65% RH
- a large number of very consistent data of Cembureau gas-permeability tests, on a variety of concrete mixes, was produced in Holcim laboratories in Switzerland, whereby the Ø150×50 mm discs were dried in an oven at 50°C for 6 days, followed by 24 hours cooling in a desiccator (Torrent & Ebensperger, 1993; Torrent & Frenzer, 1995)
- similarly, the South African gas-permeability test, designated as Oxygen-Permeability Index (OPI), described in 4.3.1.3 requires the Ø70×30 mm cores to be dried in an oven at 50°C for 7 days, followed by a cooling stage of 2–4 hours in a desiccator
- Annex A of Swiss Standard SIA 262/1 (2019) dealing with a water sorptivity test prescribes a pre-conditioning consisting of drying Ø50×50 mm cores for 2 days in a ventilated oven at 50°C, followed by 24 hours cooling at 20°C
- a more extreme preconditioning is applied in France for the Cembureau gas-permeability test, where drying at 105°C until constant mass is applied (Baroghel-Bouny, 2006) citing AFPC-AFREM (1997). Such severe drying is likely to induce cracking in the concrete and is usually not recommended (see Section 5.7.2.3)
- recommendations for measuring air-permeability (Torrent method) in the lab, call for 3–6 days oven drying at 50°C of specimens (typically 150 mm cubes), until the surface moisture, measured by an electrical impedance-based device, shows an indication between 4.0% and 5.5% (see Annex B)
- in Portugal (LNEC), Ø150×50 mm discs are dried in a ventilated oven at 50°C during 3 days, followed by a homogenization stage (specimen sealed), also at 50°C, until 24 hours prior to the test, when the specimens are taken from the oven and left (still sealed) at 20°C/ 65% RH

As seen from the above examples, there is a general consensus that drying at a temperature around 50°C is a good compromise between a sufficiently fast drying rate and avoidance of damage of the concrete due to too high temperature drying.

4.3.1.2 Cembureau Gas-Permeability Test

There is a general consensus on the validity of the fundamentals of the Cembureau method, and on the usefulness of the results provided by it to evaluate gas-permeability. It is a laboratory test method originally proposed by Grube and Lawrence (1984), later improved (Kollek, 1989), based on the laminar flow of a compressible gas in steady-state regime. Due to its good acceptance among researchers and practitioners, the method is standardized in some countries, e.g. Portugal (LNEC E 392, 1993), Spain (UNE 83981, 2008) and experimentally in France (XP P 18–463, 2011), and is object of a RILEM technical recommendation (RILEM TC 116-PCD, 1999c). It has also been appointed as reference test method for benchmarking purposes in the Comparative Test of RILEM TC 189-NEC (2007) and the Application Test of RILEM TC 230-PSC (2016).

The test consists in measuring the volume of gas that passes through a concrete sample (typically Ø150×50 mm discs) in a unit time (flow rate), under the application of an inlet gas pressure. To provide a unidirectional flow and to ensure the accuracy of its measurements, a tight cell that only allows gas passing through the testing surfaces is required. The tightness is achieved by an inflatable rubber tube that is pressurized against the curved wall of the disc. Devices to regulate the inlet pressure P and to measure the outgoing gas flow rate Q are provided (see Figure 4.9). As the gas to be used in this test shall be inert to the concrete components, the most commonly used gases are oxygen and nitrogen.

The gas flow meter may be of the "soap bubble" type, wherein the gas volume flow rate is calculated from the gas-driven advance of a soap bubble in graduated tubes with time. The flow rate is computed by dividing the outlet volume (read on the graduated tube) by the time taken by the bubble to travel the distance between measuring marks. Knowing the flow rate Q,

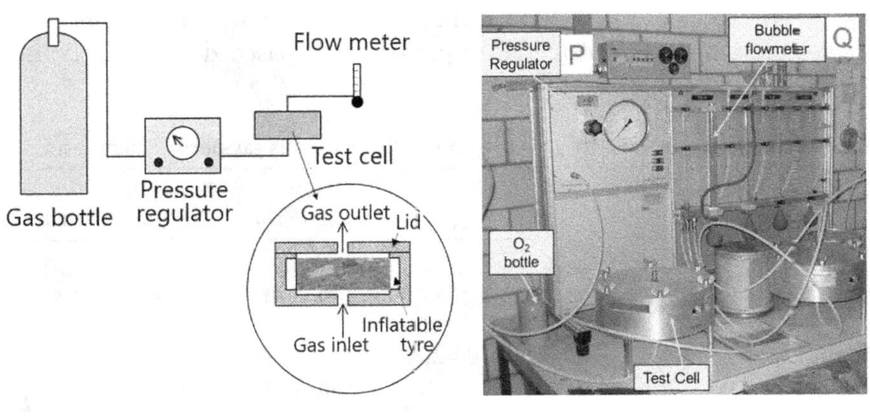

Figure 4.9 Set-up and equipment for Cembureau gas-permeability test.

the applied pressure P, the specimen cross-section A and thickness L and the viscosity μ of the gas used, it is possible to calculate the permeability coefficient kO through Eq. (3.26).

The usual range of permeability coefficients, which obviously depends on the concrete quality, lies between 0.01×10^{-16} and $1.0\times10^{-16}m^2$. A high scatter has been reported, pages 135–179 of RILEM TC 116-PCD Report (1995) and in RILEM TC 116-PCD (1999a), but it seems attributable to the variability of the measured property rather than to the test method.

As the flow rate is measured when a steady-state condition has been reached, it is necessary to monitor the flow rate variation with time, until no variation is detected. It is considered that the steady state has been reached when two successive flow rate measurements, within 5 minutes, do not differ by more than 3%. The time needed to reach the steady-state condition can vary from 15 minutes to a few hours, depending on the concrete permeability and the applied pressure. Considering that it is usual to apply three to four pressure levels (two may be enough), the Cembureau method may be quite time-consuming for high-quality concrete.

A classification of concrete quality according to kO is shown in Table 4.2, together with other test methods.

Recently, a study was made (Sogbossi et al., 2019) to adapt the Cembureau test method (eventually adding a double-cell) to measure the porosity and permeability of concrete under vacuum or under pressure.

4.3.1.3 South African Oxygen-Permeability Index Test

Another laboratory method to measure concrete gas-permeability in the laboratory, developed in South Africa (Alexander et al., 1999), has been gaining acceptance due to the research efforts placed on its improvement and application (Hooton et al., 2001; Alexander, 2004; Stanish et al., 2006; Alexander et al., 2008; Nganga et al., 2013; Kessy et al., 2015; Muigai et al., 2012; Starck et al., 2017). In fact, the result of this test, called by its acronym OPI, is already considered in performance-based design concerning

Table 4.2 Concrete quality rating based on results of various gas-permeability tests

Concrete quality	*Cembureau (10^{-16} m^2)*	*OPI*	*Figg t (s)*	*Autoclam (ln(bar)/min)*	*Torrent kT (10^{-16} m^2)*	*Paulini n*
Very good	-	>10	>1,000	<0.10	<0.01	>2.0
Good	<0.01	9.5–10	300–1,000	0.10–0.50	0.01–0.1	1.5–2.0
Normal	0.01–1.0	-	100–300	-	0.1–1.0	-
Poor	>1.0	9.0–9.5	30–100	0.50–0.90	1.0–10	1.0–1.5
Very poor	-	<9.0	<30	>0.90	>10	<1.0

concrete durability (Salvoldi et al., 2015; Beushausen & Alexander, 2009), see Section 9.3.2, and the test is standardized in South Africa (SABS, 2015).

The test specimens are Ø70×30mm discs drilled and saw-cut from the structure (or from larger cast specimens). Hence, it is a laboratory test meant for testing the quality of the as-built structure.

The disc is placed in between an air-tight vessel pressurized with O_2 (100kPa) and a perforated lid that acts as gas outlet. The test consists in monitoring the pressure decay in the vessel due to the gas passing through the concrete sample (Figure 4.10).

A quasi-linear relation exists between the natural logarithm of the measured pressure and time (see example in Figure 4.11), the slope of which, z, allows to calculate the coefficient of permeability k_{OPI} through:

$$k_{OPI} = \frac{w \cdot V \cdot L \cdot z}{R \cdot T \cdot A} \tag{4.11}$$

where

k_{OPI}=Darcy's coefficient of oxygen-permeability (m/s)
w=molecular weight of oxygen gas (w=0.313 N/mol)
V=volume of oxygen vessel (m^3)
L=thickness of the specimen (m)
z=relative pressure decay rate *viz.* Figure 4.11 (s^{-1})
R=gas constant (R = 8.314 J/K/mol)
T=mean temperature (K)
A=area of specimen cross-section (m^2)

Values of k_{OPI} in the range of 10^{-12} to 10^{-9}m/s shall be expected for regular concrete mixes. For practical purposes, the output of the test is an index,

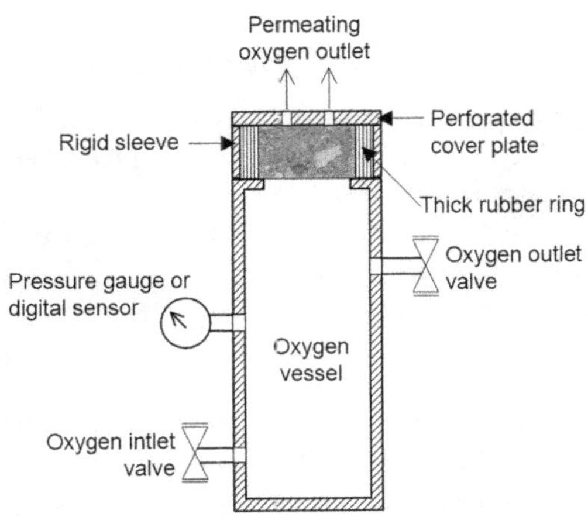

Figure 4.10 Set-up and equipment for OPI test.

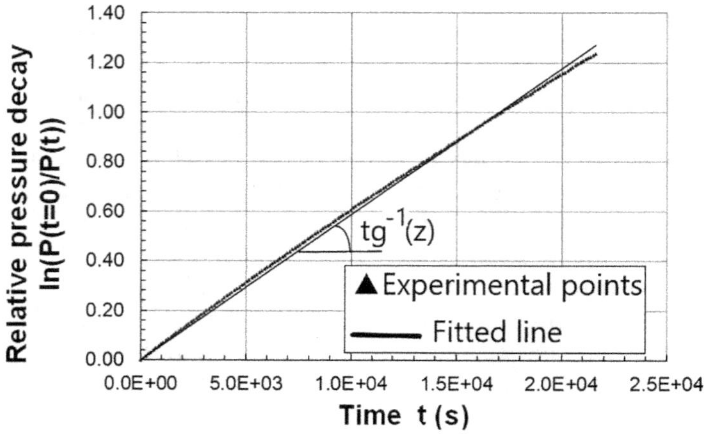

Figure 4.11 Example of data from OPI test (OPI=9.45).

instead of a permeability coefficient. In fact, to avoid scientific notation, the permeability coefficient is converted into an OPI through:

$$\mathrm{OPI} = -\log_{10} k_{\mathrm{OPI}}\,(\mathrm{m/s}) \qquad (4.12)$$

Accordingly, OPI values between 9 and 12 are foreseen for regular concrete mixtures. In relation to the permeability coefficient, the OPI should be in fact an oxygen "impermeability" index because higher values of OPI correspond to concretes of lower permeability.

A classification of concrete quality according to k_{OPI} and OPI is shown in Table 4.2 (Starck et al., 2017).

4.3.2 Site Gas-Permeability Test Methods

Gas-permeability has provided a fertile ground for the development of site testing methods, which can be grouped according to different criteria, e.g.:

- non-destructive, semi-destructive (hole-drilling) or destructive (core extraction for laboratory testing)
- non-intrusive (surface methods) or intrusive (applied on drilled holes)
- positive or negative (vacuum) working pressure

The evolution of some of these methods is presented in Figure 4.12, showing the growing interest in them within the decade of 1984–1994, accompanied by several RILEM TCs dealing with the topic, some involving round-robin experiments, discussed in Section 4.4.

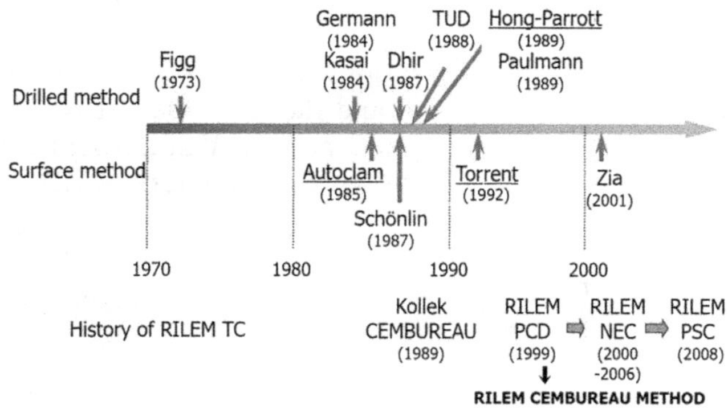

Figure 4.12 Evolution of site test methods to measure gas-permeability of concrete.

4.3.2.1 Figg

The Figg intrusive test method evaluates concrete air-permeability around a drilled hole, exactly as described in Section 4.2.2.3. A vacuum is created in the chamber through a hypodermic needle that perforates the rubber plug, which is connected to a manual vacuum pump (Figure 4.13). When an absolute pressure below 0.45 bar is achieved, a valve is closed, insulating the pneumatic system. The pressure inside the hole rises due to air penetrating the test chamber through the concrete porous structure surrounding it. The commercial instrument Porosiscope, formerly Poroscope (Figure 4.13), is equipped with a pressure cell and an embedded stopwatch, that is activated when the pressure inside the hole reaches 0.45 bar and stopped when that pressure rises to 0.50 bar. The time *t* recorded by the stopwatch is used as an air-permeability index.

Table 4.2 presents a classification of concrete qualities, based on Figg (Porosiscope) time *t*.

Data reported in Tables 3.1-III and 3.2-IV of Torrent and Ebensperger (1993), obtained with the Poroscope on a wide variety of concrete qualities,

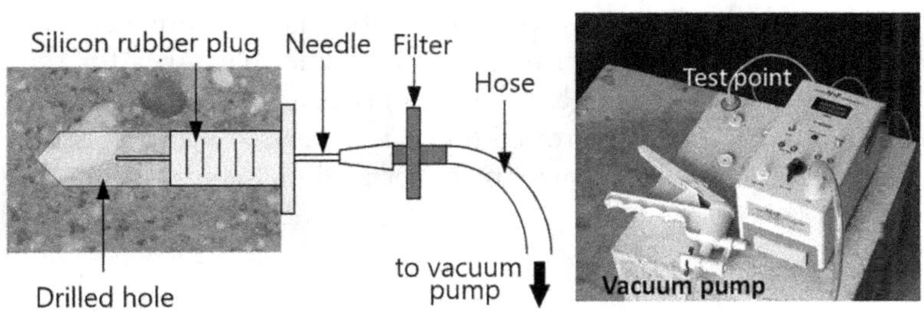

Figure 4.13 Sketch of the Figg method and commercial equipment (Porosiscope).

ranged between 19 and >18,000 seconds, which fit well to the range in Table 4.2.

Clearly, the hole and tube dimensions (volume of depressed air V) influence the rate of pressure variation and therefore the test result. Hence, a durability indicator which is independent from V and from the initial and final pressure may be computed. The air exclusion rate (AER) is calculated through:

$$AER = \frac{t}{\frac{p_i \cdot V}{p_f} \cdot \frac{(p_i + p_f)}{200}} \tag{4.13}$$

where

AER = air exclusion rate (s/mL)
V = volume of depressed air (mL)
p_i = initial pressure (kPa)
p_f = final pressure (kPa)
t = time for pressure variation between p_i and p_f (s)

The common range of AER in concrete is 25–75 s/mL (Concrete Society, 1987).

The Figg test enjoys some popularity and some modifications concerning the hole dimensions were proposed by Kasai et al. (1984) and Neves and Gonçalves (2006).

4.3.2.2 Hong-Parrott

This intrusive method developed by Hong and Parrott (1989) and Parrott and Hong (1991) assesses air-permeability by means of a Ø20 and 35 mm deep hole, partially sealed with an expansive plug, leaving a test chamber at the bottom of the hole. The hole may be cast, drilled or cored. The test consists in pressurizing the test chamber by pumping air until reaching a relative pressure around 0.75 bar. The air inlet valve is closed, isolating the test chamber, the pressure of which will decrease with time due to the air flowing through the concrete around the hole. The pressure inside the hole is monitored and a stopwatch measures the time that takes for the pressure to decay from 0.6 to 0.4 bar. For less permeable concrete, the pressure decay in 5 minutes, starting from 0.6 bar, is measured instead.

Based on these measurements, an air-permeability coefficient can be calculated as follows:

$$K = \frac{c}{t} \cdot \frac{(p_i - p_f)}{(p_i + p_f)} \tag{4.14}$$

where

K=air-permeability coefficient (m^2)
c=factor that depends on the volume of pressurized air
p_i=initial pressure (bar)
p_f=final pressure (bar)
t=time for pressure variation between p_i and p_f (s)

This permeability coefficient usually varies between 0.01×10^{-16} and 1.0×10^{-16} m^2. Besides the measurements being carried out under a non-steady-state flow of air, the concrete surface crossed by the air flow (estimated by applying a soap solution around the cell) is not constant during the test.

A model to predict carbonation-induced corrosion risk has been developed, based on the result of this test (Parrott, 1994); it is described in Section 9.3.1.

4.3.2.3 Paulmann

The intrusive Paulmann method is named after its main author (Paulmann & Rostasy, 1989). A Ø10×40–45 mm hole is drilled in the concrete surface whereas the deeper 25 mm of the hole is subjected to pressurized (2 bar) N_2 by means of an injection packer that seals the remaining part of the hole. Then the pressurized gas starts to permeate through concrete and part of the flow is collected and measured on a predefined area by means of a guard ring located around the hole.

As the length travelled by the flow can be estimated, a permeability coefficient can be calculated from the flow rate measurement in the guard ring and the pressure applied in the central hole. The usual range of values for the permeability coefficient according to the Paulmann method is reportedly like that of the Cembureau method.

4.3.2.4 TUD

The TUD (Technical Univ. Delft) intrusive test method was developed by Reinhardt and Mijnsbergen (1989) and later modified by Dinku (1996). Like the Hong-Parrott method, it assesses concrete permeability in a drilled hole with pressurized gas. However, in this case, the hole is Ø14 and 45 mm long, the gas is nitrogen and much higher pressures are applied. A cylindrical hollow probe with a rubber ring near its tip is introduced in the hole. Tightening a screw nut at the upper end of the probe expands the ring against the hole wall to seal the test chamber cavity. In this way, the disposable rubber plugs, required in the Figg method, are avoided.

Then the cavity is filled with nitrogen, through the probe hole, at approximately 12 bar. After waiting a few seconds for the pressure to stabilize, a valve in the pressure line is closed and the pressure starts to decrease. The time interval between the pressure levels of 11.0 and 10.5 bar is recorded and used as an indicator of concrete permeability.

Values of TUD time ranging from 5 to 200 seconds are reported by Reinhardt and Mijnsbergen (1989), concerning non-cured and well-cured concrete mixes with *w/c* ratios of 0.6 and 0.4.

Reported data in Tables 3.1-III and 3.2-IV of Torrent and Ebensperger (1993), obtained on a wide variety of concrete qualities, yielded TUD values in the range between 6 and 3,250 seconds.

A simple equation, based on the Hagen-Poiseuille law, that is valid for a total volume of 94 cm³ of pressurized gas, was suggested by Dinku and Reinhardt (1997) to calculate a gas-permeability coefficient:

$$K = \frac{105.09 \cdot 10^{-17}}{t} \tag{4.15}$$

where K is the gas-permeability coefficient in m^2 and t is the time elapsed between pressure levels of 11 and 10.5 bar, in seconds. According to the same research, K values from 7×10^{-18} to $2\times10^{-17}\,m^2$ correspond to average permeability concrete.

4.3.2.5 GGT

The intrusive Germann Gas Test (GGT) is based on the work by Hansen et al. (1984). It requires drilling a Ø18 mm hole at a small angle to the concrete surface, so that the end of the hole is at an approximate depth of 25 mm from the concrete surface; a pressure sensor is located at the end of the hole that is sealed. Exactly above the sensor, a test rig is clamped with a sealing gasket. Compressed (1–4 bar) carbon dioxide is applied to the concrete surface inside the rig. Then, part of the pressurized CO_2 flows into the drilled hole increasing the pressure inside it. The pressure rise inside the hole is monitored by the pressure sensor.

Following a theoretical study carried out by Hansen et al. (1984), it is possible to calculate a coefficient of gas-permeability, from the test data and concrete porosity. The values of the permeability coefficient obtained by the GGT method are reportedly comparable with those obtained by the Cembureau method.

A commercial GGT was once produced by Germann Instruments, but apparently not anymore.

4.3.2.6 Paulini

The intrusive Paulini method was developed by Paulini and Nasution (2007), initially for laboratory testing, later modified to allow for site testing (Paulini, 2010). The site test requires drilling a Ø30 mm hole in

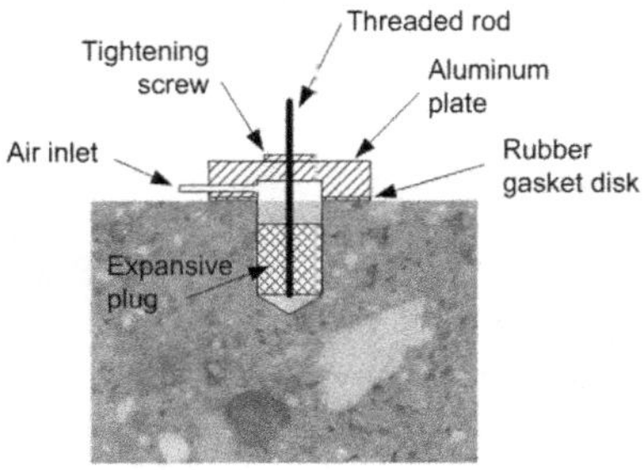

Figure 4.14 Sketch of the Paulini method.

the concrete surface, where a packer will be lodged (the bore dust is collected for humidity measurement and eventual chemical analysis), see Figure 4.14. The packer contains a plug that is expanded by means of a threaded rod, thus sealing the lowest part of the hole and anchors the reaction to the force that will compress an aluminium plate towards a rubber gasket disk that seals the concrete surface around the hole. The first 10 mm of the drilled hole are left unsealed to allow air permeation through surrounding concrete.

The air inlet is connected to a bottle of compressed air, equipped with a control valve to allow pressure regulation. The air pressure inside the hole is monitored, and the pressure drop is used to calculate the flow rate Q. Based on test data, an air-permeability coefficient may be computed through the following equation, developed by Paulini (2010).

$$K = \ln\left(\frac{\phi_d}{\phi_h}\right)\frac{Q \cdot \mu \cdot p_a}{2 \cdot \pi \cdot h \cdot \left(p_i^2 - p_a^2\right)} \tag{4.16}$$

where

K = permeability coefficient (m^2)
ϕ_d = outside diameter of the rubber gasket disk (m)
ϕ_h = diameter of the hole (m)
h = unsealed depth of the hole (m)
Q = computed air flow (m^3/s)
μ = viscosity of air (Pa.s)
p_a = atmospheric pressure (Pa)
p_i = applied gas pressure (Pa)

The test is run by applying a minimum of three pressure levels, controlled by means of the bottle's valve, starting with the higher one (6 bar). A time derivative of air pressure inside the hole is computed to check if constant flow has been reached. Data for Q computation are only considered after steady state being achieved at each applied pressure level.

Furthermore, Paulini and Nasution (2007) suggested a power law for gas permeation in concrete:

$$v = v_{\text{ref}} \left(\frac{p}{L} \right)^n \tag{4.17}$$

where

v=air flow velocity (m/s)
v_{ref}=reference air flow velocity (m/s)
p=applied absolute gas pressure (MPa)
L=length travelled by the air flow inside concrete (m)
n=permeability exponent (-)

If flow velocities are plotted against pressure gradients (p/L) in a log-log scale a nearly linear relationship shall be expected. The slope of that line is the permeability exponent, whereas the intersection with the vertical axis is the reference air flow velocity. Higher permeability exponents and lower reference velocities correspond to denser pore structures of concrete. Therefore, both parameters can be used as permeability indexes, particularly the permeability exponent, as it has a positive association with concrete quality, and therefore stands for the "Permeability Exponent" method designation. When assessing concrete permeability through the Paulini method, values of air-permeability coefficient in the order of magnitude of 10^{-16}m^2 shall be expected, while n values between 1 and 2 and v_{ref} values between 10^{-4} and 10^{-6}m/s correspond to regular/low concrete permeability (see Table 4.2).

4.3.2.7 Autoclam System

The same Autoclam system device, already described in detail for water-permeability and sorptivity measurement (Sections 4.1.2.2 and 4.2.2.4), serves also to measure the air-permeability of concrete. It is a surface method but requires drilling three small holes per testing spot, to attach the measuring cell onto the tested concrete, ensuring a tight connection. After attaching the cell, the air inside its chamber is pressurized and the pressure decay, due to the air from the chamber flowing through the concrete, is monitored. When plotting the natural logarithm of pressure vs. time, a quasi-linear relation is expected, the slope of which is taken as the test result, named air-permeability index (API). The API is usually expressed in ln(bar)/minute and, for concrete, values in the order of magnitude of 10^{-1} ln(bar)/minute shall be expected.

Table 4.2 presents a classification of concrete qualities, based on Autoclam API.

4.3.2.8 Single-Chamber Vacuum Cell

The surface single-chamber method was developed, almost simultaneously, by Bérissi et al. (1986) and Schönlin and Hilsdorf (1987) and was further developed by Imamoto et al. (2006). The test principle is to create a vacuum on the concrete surface by means of a single chamber cell. Then, the pressure gradient between the pores of the surrounding concrete (at atmospheric pressure) and the evacuated chamber drives air through concrete into the chamber. The vacuum inside the chamber also holds it "stuck" onto the concrete surface by means of a rubber ring (Figure 4.15). A tight connection between the chamber and the concrete surface must be ensured, to avoid air flowing through the connection. Therefore, a ring made of convenient rubber and with appropriate thickness is necessary between the chamber borders and concrete surface. The basic setup is shown in Figure 4.15.

The depression is created by means of a vacuum pump, or a syringe, connected to the chamber, where a valve is closed when the intended vacuum (e.g. 20 mbar, absolute pressure) is reached. Afterwards, the pressure rise inside the chamber is monitored and used to assess concrete permeability. Schönlin and Hilsdorf (1987) proposed an equation to calculate a permeability index, later amended in its units in Chapter 3 of RILEM TC 189-NEC (2007):

$$M = \frac{2(p_f - p_i)V_c}{(t_f - t_i)p_a(2p_a - p_f - p_i)} \quad (4.18)$$

where

M = permeability index (m³/s/mbar)
p_f = absolute chamber pressure at time t_f (mbar)
p_i = absolute chamber pressure at time t_i (mbar)

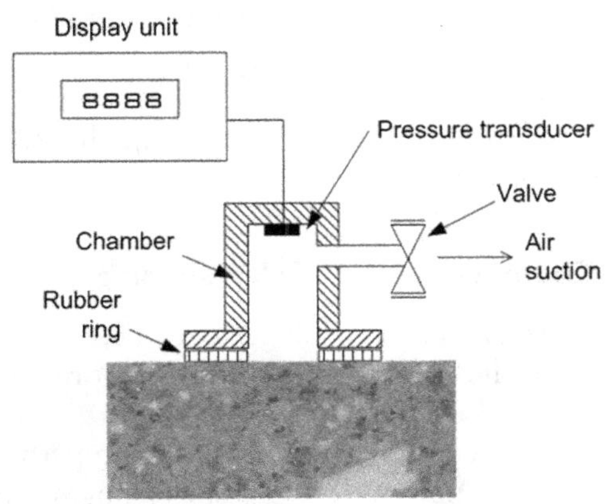

Figure 4.15 Sketch of single-chamber vacuum cell test method.

p_a=atmospheric pressure (mbar)
V_c=inner volume of chamber and accessories (m^3)
t_f=time at measurement of p_f (s)
t_i=time at measurement of p_i (s)

For regular concretes, values of the permeability index within the range from 10^{-9} to $10^{-4} m^3/s/mbar$ are expected. When testing trowelled surfaces, overestimation of concrete permeability was reported and attributed to a possible preferential flow path along the thin top superficial layer (Torrent, 1992). To overcome this potential shortcoming, Gabrijel et al. (2008) suggested spraying the near 100 mm of concrete surface around the chamber with a transparent car coating.

4.3.2.9 Double-Chamber Vacuum Cell (Torrent)

The surface, double-chamber cell method, developed by Torrent (1992) can be considered an evolution of the single chamber method, as it is also non-intrusive and operates under vacuum. Indeed, this method comprises two concentric chambers that are permanently kept at the same pressure. The external chamber, besides preventing possible air flow through the concrete skin near the inner (measurement) chamber, ensures that the air flow into the latter is basically unidirectional. This last feature allows the derivation of a formula to calculate a coefficient of air-permeability of the material. This was the first test method for site measurement of air-permeability to be standardized (2003), now updated in Annex E of Swiss Standard (SIA 262/1, 2019). The above-mentioned standard not only prescribes how the test must be performed (age, sampling, temperature and moisture limits, calibration and testing procedure) but also specifies maximum limits to the values obtained on site, depending on the environmental conditions to which the element is exposed and defines conformity criteria for acceptance of the end-product. This test method is thoroughly addressed in Chapter 5.

Table 4.2 presents a classification of concrete qualities, based on Torrent air-permeability coefficient kT.

4.3.2.10 Triple-Chamber Vacuum Cell (Kurashige)

This test method, developed by Kurashige (2015), consists in adding a third concentric chamber to the Torrent method's cell, but removing the pressure regulator.

After 60 seconds of evacuation by the vacuum pump, the three concentric cells are isolated from the pump, their pressure rising naturally as function of the permeability characteristics of the underlying concrete. A numerical model of the gas flow, complemented by assumptions on the relation between permeability and porosity and between permeability and depth from the surface, allows the determination of the parameters that provide a best fit to the experimentally obtained pressure-time curves for the three chambers.

A Japanese patent has been filed covering this test method.

4.3.2.11 Zia-Guth

The Zia-Guth method, named after its authors (Guth & Zia, 2001), operates like the Torrent method, as it is a surface test and the probe has two concentric chambers that are evacuated by means of a vacuum pump. Then, the external chamber valve is opened, allowing air entrance at atmospheric pressure that flows into the inner chamber through a near surface path. The pressure increase in the inner chamber is monitored. The pressure vs. time data are compared with predefined curves, corresponding to specific permeability coefficients, established by numerical modelling. The 'best fit' to the theoretical predictions provides a coefficient of air-permeability expressed in m² which, for normal quality concrete, ranges between 3×10^{-16} and 6×10^{-16} m² (Guth & Zia, 2000).

4.3.2.12 "Seal" Method

The proposed surface test method called the "Seal" method (Ujike et al., 2009), aimed at measuring the air-permeability coefficient of the cover concrete on site, is sketched in Figure 4.16 (l.).

An easily removable rubber latex resin is applied on the concrete surface forming a circle of radius r_2 leaving an uncovered inner circle of radius r_1 (Figure 4.16).

The air-permeability of the concrete is measured by suctioning air out of the inner circle using a vacuum pump, with the area sealed by the airtight resin preventing sucking air from the surrounding concrete (Figure 4.16). A surface moisture meter is used to check the surface moisture of the concrete and the seal can be easily removed when the test has ended.

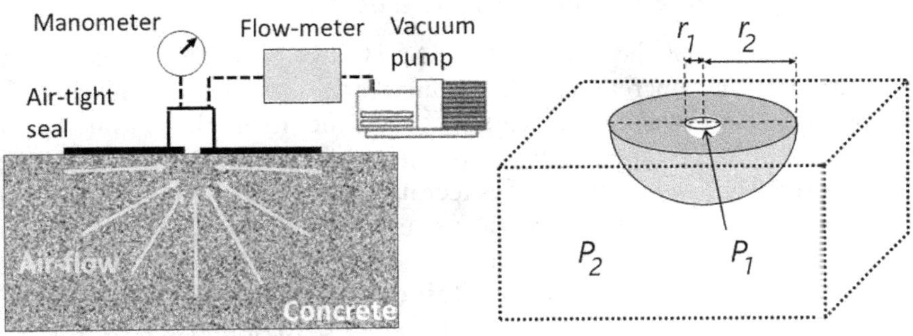

Figure 4.16 Air-permeability "Seal" test sketch (l.) and theoretical air hemisphere (r.).

A model is applied, based on the sketch shown in Figure 4.16 (r.), assuming that the air-permeability coefficient is constant at any depth from the concrete surface and that the air-permeability zone forms a hemisphere, whose centre is the suction port of radius r_1, immediately below the vacuum chamber.

The air-permeability coefficient is computed using the following equation (Kawaai & Ujike, 2016):

$$k = \frac{Q_1 \cdot P_1 \cdot \mu}{\pi \cdot \left(P_2^2 - P_1^2\right)} \left(\frac{\sqrt{2}}{r_1} - \frac{1}{r_2}\right) \tag{4.19}$$

where

k=coefficient of air-permeability (m^2)
Q_1=air flow rate (m^3/s), measured by the flow meter
P_1=pressure in vacuum chamber (N/m^2)
P_2=atmospheric pressure (N/m^2)
μ=viscosity of air (N.s/m^2)
r_1 and r_2=inner and outer radii of the seal (m)

4.3.3 Assessment of Concrete Quality by Gas-Permeability Test Methods

Table 4.2, an extension of the one included in RILEM TC 230-PSC (2016), presents a comparison on how several gas-permeability test methods rate concrete quality, based on their test results. It is interesting to note that four to five quality categories can be established to classify concrete quality by means of gas-permeability test methods.

4.4 COMPARATIVE TEST RILEM TC 189-NEC

4.4.1 Objective and Experiment Design

Within the frame of the work of RILEM TC 189-NEC "Non-destructive evaluation of the concrete cover", a Comparative Test was organized with the declared objective of establishing whether the site methods available at the time (2003), designed to measure the "penetrability" of the concrete cover in the field, were capable of detecting differences in the *w/c* ratio and curing conditions of different concretes. In addition, their correlation with so-called "Reference" laboratory methods was evaluated.

A full report on this Comparative Test can be found in RILEM TC 189-NEC (2007) and a condensed report in Romer (2005), updated by Torrent (2008).

Table 4.3 Test conditions investigated in the comparative test

	Test condition #									
Variable	*1*	*2*	*3*	*4*	*5*	*6*	*7*	*8*	*9*	*10*
w/c ratio	0.40	0.55	0.60	0.40	0.55	0.55	0.40	0.55	0.40	0.55
Cement type	OPC			BFSC[a]		OPC	OPC		OPC	
Moist curing (days)	7	7	7	7	7	1	7	7	7	7
Temperature (°C)	20	20	20	20	20	20	20	20	10	10
Storage condition	Normal						Moist		Cold	

[a] Blast-Furnace Slag cement, containing nominally 63% of slag.

The Comparative Test was a typical "blind" test in which a series of concrete panels (0.3×0.9×0.12 m), depicted in right-hand picture in Figure 4.3, were cast with different concrete mixes by an independent laboratory (EMPA, Switzerland, in this case), moist cured and stored under prescribed conditions, leading to ten test conditions, summarized in Table 4.3.

The tests were performed in the rooms where the panels were stored, with the following ambient conditions: Normal (20°C/70% RH); Moist (20°C/90% RH) and Cold (10°C). The age of the slabs at the initiation of the tests, which lasted 5 days, ranged between 54 and 69 days.

Later, four cores were drilled from each of the ten slabs, cut to size, dried at 50°C, weighed and shipped to LNEC (Portugal) for testing, applying standardized or RILEM-recommended tests, under controlled laboratory conditions. These tests are referred to as "Reference Tests". Another set of cores was drilled and shipped to the University of Cape Town for the determination of the South African Durability Indices. Both during the application of the site tests directly on the slabs as well as of the "Reference Tests" on the cores, the identity of the samples was coded. So, the tester did not know to which test condition the slabs or specimens belonged.

Table 4.4 summarizes the main site tests applied directly on the panels and the laboratory "Reference Tests" applied on the cores drilled from the panels.

4.4.2 Evaluation of Test Results

4.4.2.1 Significance of Test Method

This evaluation was aimed at establishing whether the test methods were capable of differentiating the "penetrability" of concretes of different *w/c* ratios (sets 1–2–3 for OPC and 4–5 for BFSC), of the same *w/c* ratio but different curing (sets 2–6) and of different *w/c* ratios for measurements conducted in moist room (sets 7–8) and cold room (sets 9–10). The capability of the methods was tested applying a Student's t-statistical test of the

Table 4.4 Comparative Test: main "penetrability" tests applied on the panels

Property measured	*Test method*	*Described in*	*Laboratory*	*Measurements per test condition*
Site tests				
Gas-permeability	Autoclam	4.3.2.7	QUB (UK)	3
	Torrent	4.3.2.9 + Chapter 5	TFB (CH)[a]	6
			IETcc (E)[a]	8
	Hong-Parrott	4.3.2.2	LNEC (P)	4
Water sorptivity	Autoclam	4.1.2.2/4.2.2.4	QUB (UK)	3
Electrical resistivity	Wenner	A 2.2.2	TNO (NL)	20
Laboratory Reference tests				
O_2 – permeability	Cembureau	4.3.1.2	LNEC (P)	4
Water sorptivity	RILEM 116-PCD	4.2.1		
Chloride migration	ASTM C1202	A 2.1.1		
	Tang-Nilsson	A 2.1.2		
Electrical resistivity	Wenner	A 2.2.2		

QUB, Queen's Univ. Belfast; IETcc, Instituto Eduardo Torroja.[a]The results obtained by both participants were very similar; only those of TFB are discussed here.

difference between the means of the pairs of sets of results under comparison, as shown in Table 4.5.

The null hypothesis H0 was that both sets of results come from populations having the same mean "penetrability". The alternative hypothesis H1 was that one set has a mean "penetrability" higher than the other as indicated in the second row of Table 4.5, for which a one-tailed test is applicable.

The outcome of the statistical test was evaluated as follows:

- if the result of the statistical test allows to reject the null hypothesis H0 at a level of significance < 1%, the differentiation capability of the test, for the particular sets compared, is "highly significant" (++)
- if the result of the statistical test allows to reject the null hypothesis H0 at a level of significance between 1% and 5%, the differentiation capability of the test, for the particular sets compared, is "significant" (+)
- if the result of the statistical test does not allow to reject the null hypothesis H0 at a level of significance of 5%, the differentiation

Table 4.5 Variables tested, differentiation capabilities of "Reference" and Site tests and correlations between them

Compared sets	*1–2*	*2–3*	*1–3*	*4–5*	*2–6*	*7–8*	*9–10*	
Variable tested	*w/c* OPC			*w/c* BFSC	Curing	*w/c* moist	*w/c* cold	
Expected "penetrability"	2>1	3> 2	3>1	5>4	6>2	8>7	10>9	
Reference test	**Differentiation capability**							
O_2 permeability	++	++	++	++	++	++	++	
Water sorptivity	++	++	++	++	++	++	++	
Cl^- migration ASTM C1202	++	++	++	++	++	++	++	
Cl^- migration Tang-Nilsson	++	++	++	+	++	++	++	
Electrical resistivity	++	++	++	++	--	++	++	
Site test	**Differentiation capability**							*R*
Autoclam air[a]	--	++	++	o	++	++	++	0.67 / 0.90[b]
Torrent air-permeability	++	++	++	o	++	++	++	0.97
Hong-Parrott	o	++	++	++	+	++	++	0.92
Autoclam water sorptivity[a]	++	o	++	+	++	+	++	0.47
Wenner resistivity	++	--	++	++	--	++	++	0.83

[a] The internal RH reached by all panels exceeded the maximum 80% required by the test (Torrent, 2008).

[b] After removal of outliers (Torrent, 2008).

capability of the test, for the particular sets compared, is "not significant" (**o**)

- if the results are in reverse order than expected, the response of the test is "wrong" (--)

Table 4.5 shows that all the selected "Reference Tests" were capable of correctly differentiating the "penetrability" of all sets at a highly significant (++) or significant (+) level, with the exception of the Wenner electrical resistivity test, that assessed wrongly the effect of curing on the "penetrability" (Sets 2–6).

With regard to the Site Tests the picture is not as good but yet quite positive. All test methods were capable of differentiating, at highly significant (++) level, the "penetrability" of the OPC concretes with *w/c* ratios 0.40 and 0.60 (Sets 1–3). Yet, the only test method capable of differentiating the "penetrability" of OPC concretes with *w/c* ratios between 0.40 and 0.55

(Sets 1–2) and between 0.55 and 0.60 (Sets 2–3) was the Torrent method and at a highly significant level (++). Autoclam Air and Torrent methods failed to differentiate the "penetrability" of the BFSC concretes of *w/c* ratios 0.40 and 0.55 (Sets 4–5), where the other methods succeeded.

All test methods but Wenner electrical resistivity assessed correctly the positive effect of extended moisture curing on the "penetrability" (Sets 2–6). This is most likely due to the strong influence of the moisture on the resistivity; the effect of the higher moisture of the well cured specimens probably prevailed upon the beneficial effect of moist curing in reducing the "penetrability". Finally, all site test methods performed satisfactorily under the Moist (Sets 7–8) and Cold (Sets 9–10) testing conditions.

All Site Test methods, but Torrent and Wenner resistivity left some traces on the surface (holes of different sizes), as reported in RILEM TC 189-NEC (2007) and Romer (2005).

4.4.2.2 Correlation between Site and "Reference" Tests

The degree of correlation between the results obtained on each set by each site test and its corresponding "Reference" test was investigated. The results of the three gas-permeability site tests (Autoclam, Torrent and Hong-Parrott) were correlated with the Cembureau O_2-Permeability "Reference Test"; the Autoclam Water Sorptivity test with the RILEM water sorptivity test and the Wenner electrical resistivity measured on the panels with those measured on the drilled cores. The correlation coefficients R obtained for each test are shown in the last column of Table 4.5.

It can be seen that in the case of Torrent and Hong-Parrott site tests the correlation with the "Reference" test is excellent, it is acceptable for Autoclam Air and Wenner Resistivity and poor for Autoclam Water Sorptivity (see the footnote of Table 4.5).

Correlations between the South African Indexes and other "Reference" and Site tests can be found in Beushausen and Alexander (2008).

4.4.2.3 Conclusions of the Comparative Test

Transcribing from Romer (2005):

> "It can be concluded that the Comparative Test at EMPA was well designed, planned and executed to provide meaningful and objective results. The fact that the testers involved, both on site and at the laboratories, did not know the identity of the slabs or cores they were testing, guarantees the objectivity of the results obtained.
>
> Although to a varying degree, the Comparative Test proved that there are methods capable of evaluating the "penetrability" of the concrete cover on site, in a reliable and statistically significant manner. In five or six out of seven cases, the test methods were capable of detecting

correctly the expected differences in "penetrability" at a significant or highly significant level. Moreover, some of the site methods showed very good correlations with corresponding relevant Reference Test methods.

This opens good perspectives for the application of such methods in practice, for the specification and *in situ* compliance control of the "penetrability" of the vital concrete cover, aiming at performance-oriented criteria regarding the durability of concrete structures."

ACKNOWLEDGEMENTS

Some sections of this chapter took Chapters 3 and 4 from RILEM TC 189-NEC (2007) as reference. The contribution of P.A.M. Basheer and A.F. Gonçalves to those chapters is duly acknowledged.

REFERENCES

AFPC-AFREM (1997). "Durabilité des Bétons, Méthodes recommandées pour la mesure des grandeurs associées à la durabilité". Compte-rendu des journées techniques de l'AFPC-AFREM, 11 et 12 décembre, Toulouse, France, 284 p.

Akmal, U., Hosoda, A., Hayashi, K. and Fujiwara, M. (2011). "Analysis of quality of covercrete subjected to different curing conditions using new surface water absorption test". *Proceedings of the 13th International Summer Symposium*, JSCE, 287–291.

Alexander, M.G. (2004). "Durability indexes and their use in concrete engineering". *International RILEM Symposium on Concrete Science and Engineering:* A Tribute to Arnon Bentur, 9–22.

Alexander, M.G., Ballim, Y. and Mackechnie, J.R. (1999). "Concrete durability index testing manual". Research Monograph No.4, Univs. Cape Town & Witwatersrand, South Africa, 33 p.

Alexander, M.G., Ballim, Y. and Stanish, K. (2008). "A framework for use of durability indexes in performance-based design and specifications for reinforced concrete structures". *Mater. & Struct.*, v41, n5, June, 921–936.

Amphora (not dated). "Autoclam ()". Brochure, Amphora Technologies Ltd., Belfast, 4 p.

ASTM C 1585 (2013). "Standard test method for measurement of rate of absorption of water by hydraulic – cement concretes".

Baroghel-Bouny, V. (2006). "Durability indicators: relevant tools for performance-based evaluation and multi-level prediction of RC durability". RILEM PRO *047*, 3–30.

Basheer, P.A.M. (1993). "Technical Note. A brief review of methods for measuring the permeation properties of concrete in situ". *Proc. Inst. Civ. Eng. Struct. Build.*, v99, n1, February, 74–83.

Bérissi, R., Bonnet, G. and Grimaldi, G. (1986). "Mesure de la porosité ouverte des bétons hydrauliques". Bull. Liaison des Lab. des Ponts et Chaussée, n142, 59–67.

Beushausen, H. and Alexander, M. (2008). "The South African durability index tests in an international comparison". *J. South African Institution Civil Eng.*, v50, n1, 25–31.
Beushausen, H. and Alexander, M. (2009). "Application of durability indicators for quality control of concrete members – A practical example". Conc*rete* in *Aggressive Aqueous Environments – Performance, Testing, and* Modeling. Alexander, M.G. and Bertron, A. (Eds.). RILEM Publications, Bagneaux, 548–555.
BS 1881-5 (1970). "Testing concrete. Methods of testing hardened concrete for other than strength". British Standards Institution (Withdrawn).
BS 1881-208 (1996). "Methods of testing hardened concrete for other than strength". BS 1881, Part 208, British Standards Institution.
Cather, R., Figg, J.W., Marsden, A.F. and O'Brien, T.P. (1984). "Improvements to the Figg method for determining the air-permeability of concrete". *Mag. Concr. Res.*, v36, n129, December, 241–245.
Claisse, P.A., Elsayad, H.I. and Ganjian, E. (2009). "Water vapour and liquid permeability measurements in cementitious samples". *Adv. Cem. Res.*, v21, n2, April, 83–89.
Concrete Society (1987). "Permeability testing of site concrete – A review of methods and experience". Technical Report No. 31, London, 95 p.
Coutinho, A. de S. and Gonçalves, A. (1994). *Fabrico e Propriedades do Betão. Volume III.* 2nd ed., LNEC, Lisboa, 368 p.
DIN 1048 (1978). "Prüfverfahren für Beton – Bestimmung der Wassereindringtiefe".
Dinku, A. (1996). "Gas-permeability as a means to assess the performance properties of concrete". IWB, Stuttgart, 236 p.
Dinku, A. and Reinhardt, H.W. (1997). "Gas-permeability coefficient of cover concrete as a performance control". *Mater. & Struct.*, v30, n7, 387–393.
Duarte, R., Flores-Colen, I. and Brito, J. (2011). "In situ testing techniques for evaluation of water penetration in rendered facades – The portable moisture meter and Karsten tube". XII DBMC, Porto, Portugal, April 12–15, 8 p.
EHE-08 (2008). "Instrucción de Hormigón Estructural". Spanish Concrete Code.
EN 12390-8 (2009). "Testing hardened concrete – Part 8: Depth of penetration of water under pressure".
Fagerlund, G. (1977). "The critical degree of saturation method of assessing the freeze/thaw resistance of concrete". *Matériaux & Constr.*, v10, n4, 217–229.
fib (2012). Model Code 2010, Final Draft, v1, fib Bulletin 65, March, 357 p.
Figg, J.W. (1973). "Methods of measuring the air and water-permeability of concrete". *Mag. Concr. Res.*, v25, n85, December, 213–219.
Gabrijel, I., Mikulic, D., Bjegovic, D. and Stipanovic-Oslakovic, I. (2008). "In-situ testing of the permeability of concrete". SACoMaTiS, 337–346.
GBJ 82-85 (1986). "Standard for test methods of long-term performance and durability of ordinary concrete". Chinese Standard.
Glanville, W.H. (1931). "The permeability of portland cement concrete". Building Research Establishment, Technical paper, No.3, 62 p.
Grube, H. and Lawrence, C.D. (1984). "Permeability of concrete to oxygen". *Proceedings on RILEM Seminar* Durability Conc*rete* Struct*ures* under Normal Out-door Exposure, Univ. Hanover, March, 68–79.

Guth, D.L. and Zia, P. (2000). "Correlation of air-permeability with rapid chloride permeability and ponding tests". *PCI/FHWA/FIB International Symposium on High Performance Concrete*, Orlando, September 25–27, 304–315.

Guth, D.L. and Zia, P. (2001). "Evaluation of new air-permeability test device for concrete". *ACI Mater. J.*, v98, n1, 44–51.

GWT (2014). "NDT systems. Bridging theory and practice". Germann Instruments.

Hansen, A.J., Ottosen, N.S. and Petersen, C.G. (1984). "Gas-permeability of concrete in situ: Theory and practice". ACI SP 82, 543–556.

Hayashi, K. and Hosoda, A. (2013). "Fundamental study on evaluation method of covercrete quality of concrete structures by surface water absorption test". *J. J.S.C.E.*, Ser E2 (Materials and Concrete Structures), v69, n1, 82–97. (in Japanese).

Hendrickx, R. (2013). "Using the Karsten tube to estimate water transport parameters of porous building materials". *Mater. & Struct.*, v46, 1309–1320.

Hong, C.Z. and Parrott, L.J. (1989). "Air-permeability of cover concrete and the effect of curing". British Cement Assoc. Report C/5, October, 25 p.

Hooton, R., Griesel, E. and Alexander, M. (2001). "Effect of controlled environmental conditions on durability index parameters of portland cement concretes". *Cem. Concr. & Aggr.*, v23, n1, 44–49.

Imamoto, K., Shimozawa, K., Nagayama, M., Yamasaki, J. and Nimura, S. (2006). "Evaluation of air-permeability of cover concrete by single chamber method". *31st Conference on* 'Our World in Concrete & Structures' 2006, 16–17.

James (2007). James Instruments Inc., "P-6050 & P-6000 porosiscope plus operating instructions".

Kasai, Y., Nagano, M. and Matsui, L. (1984). "On site rapid air-permeability test for concrete". ACI SP 082, 525–542.

Kawaai, K. and Ujike, I. (2016). "Influence of bleeding on durability of horizontal steel bars in RC column specimen". *Proceedings of IALCCE2016*, Delft, The Netherlands, October 16–19, 839–846.

Kessy, J.G., Alexander, M.G. and Beushausen, H. (2015). "Concrete durability standards: International trends and the South African context". *J South African Inst. Civ. Eng.*, v57, n1, 47–58.

Kollek, J.J. (1989). "The determination of the permeability of concrete to oxygen by the Cembureau method – A recommendation". *Mater. & Struct.*, v22, n3, May, 225–230.

Kurashige, I. (2015). "Novel non-destructive test method to evaluate air-permeability distribution in depth direction in concrete – Development of triple-cell air-permeability tester (TCAPT)". In*ternational* Symp*osium on* Non-Destructive Testing in Civil Eng*ineering*. (NDT-CE), Berlin, Germany, September 15–17.

Levitt, M. (1969a). "Non-destructive testing of concrete by the initial surface absorption method". *Proceedings on Symposium* Non-destructive *Testing* of *Concrete* and *Timber*, Inst. of Civil Engs., London, June 11–12, 23–28.

Levitt, M. (1969b). "An assessment of the durability of concrete by ISAT". *Proceedings on RILEM Symposium* Durability of Concrete, Prague.

Levitt, M. (1971). "The ISAT: A non-destructive test for the durability of concrete". *Br. J. NDT*, July, 106–112.

LNEC E 392 (1993). "Betões. Determinação da permeabilidade ao oxigénio". Laboratório Nacional de Engenharia Civil, May, 4 p.

Meletiou, C.A. (1991). "Development of a field permeability test for assessing the durability of concrete in marine structures". PhD Thesis, Univ. of Florida, 184 p. https://ufdc.ufl.edu/AA00037920/00001/4.

Meletiou, C.A., Tia, M. and Blooquist, D. (1992). "Development of a field permeability test apparatus and method for concrete". *ACI Mater. J.*, v89, n1, January–February, 83–89.

Montgomery, F.R. and Adams, A. (1985). "Early experience with a new concrete permeability apparatus". *Proceedings on 2nd International Conference on Structural Faults and Repair*, Forde and Topping (Eds.), 359–363.

Montgomery, F.R. and Basheer, M. (1989). "Durability assessment of concrete bridges by in-situ testing, early results". The Life of Structures. Physical Testing. Armer, G.S.T., Clarke, J.L., Garas, F.K. (Eds.). Butterworths, London, 352–359.

Muigai, R., Moyo, P. and Alexander, M. (2012). "Durability design of reinforced concrete structures: A comparison of the use of durability indexes in the deemed-to-satisfy approach and the full-probabilistic approach". *Mater. & Struct.*, v45, n8, August, 1233–1244.

Neves, R. and Gonçalves, A. (2006). "Concrete durability evaluation based on modified Figg test". RILEM PRO 47, 249–256.

Nganga, G., Alexander, M. and Beushausen, H. (2013). "Practical implementation of the durability index performance-based design approach". *Constr. & Build. Mater.*, v45, August, 251–261.

Nguyen, N.H., Nakarai, K., Kuboria, Y. and Nishio, S. (2019). "Validation of simple non destructive method for evaluation of cover concrete quality". *Constr. & Build. Mater.*, v201, March, 430–438.

Nishio, S. (2017). "Simple evaluation of water-permeability in cover concrete by water spray method". *Q. Rep. RTRI*, v58, n1, 36–42.

Nwaubani, S.O. (2018). "Non-destructive testing of concrete treated with penetrating surface sealant using a Karsten-tube". ICCRRR, Cape Town, South Africa.

Parrott, L. (1994). "Design for avoiding damage due to carbonation-induced corrosion". ACI SP-145, 283–298.

Parrott, L. and Hong, C.Z. (1991). "Some factors influencing air permeation measurements in cover concrete". *Mater. & Struct.*, v24, 403–408.

Paulini, P. (2010). "A laboratory and on-site test method for air-permeability of concrete". *2nd International Symposium on Service Life Design for Infrastructure*, Delft, October 4–6, 995–1002.

Paulini, P. and Nasution, F. (2007). "Air-permeability of near-surface concrete". *CONSEC'07*, Tours, France, 8 p.

Paulmann, K. and Rostasy F.S. (1989). "Praxisnahes Verfahren zur Beurteilung der Dichtigkeit oberflächennäher Betonschichten im Hinblick auf die Dauerhaftigkeit". Institut für Baustoffe, Massivbau und Brandschutz, Techn. Univ. Braunschweig.

Pereira Apps, C.A.C. (2011). "Evaluation of the variability of the Karsten tube in-situ test technique on measuring liquid water-permeability of renders and ceramic tile coatings". IST, Univ. Técnica Lisboa, October.

Reinhardt, H.W. and Mijnsbergen, J.P.G. (1989). "In-situ measurement of permeability of concrete cover by overpressure". The Life of Structures, Physical Testing. Armer, Clarke and Garas (Eds.), Butterworth-Heinemann, 243–254.

RILEM TC 25-PEM (1980). "Recommended tests to measure the deterioration of stone and to assess the effectiveness of treatment methods". *Matériaux et Constructions*, v13, 175–253. https://doi.org/10.1007/BF02473564.

RILEM TC 116-PCD (1995). "Performance criteria for concrete durability". Kropp, J. and Hilsdorf, H.K. (Eds.), *RILEM Report 12*, E&FN Spon, 316 p.

RILEM TC 116-PCD (1999a). "Concrete durability – An approach towards performance testing". Final Report, *Mater. & Struct.*, v32, April, 163–173.

RILEM TC 116-PCD (1999b). "Preconditioning of concrete test specimens for the measurement of gas-permeability and capillary absorption of water". *Mater. & Struct.*, v32, April, 174–176.

RILEM TC 116-PCD (1999c). "Measurement of the gas-permeability of concrete by the RILEM-Cembureau method". *Mater. & Struct.*, v32, April, 176–178.

RILEM TC 116-PCD (1999d). "Determination of the capillary absorption of water of hardened concrete". *Mater. & Struct.*, v32, April, 178–179.

RILEM TC 189-NEC (2007). "Non-destructive evaluation of the penetrability and thickness of the concrete cover". Torrent and Fernández Luco (Eds.), *RILEM Report 40*, May, 223 p.

RILEM TC 230-PSC (2016). "Performance-based specifications and control of concrete durability". Beushausen and Fernández Luco (Eds.), *RILEM Report 18*, 373 p.

Romer, M. (2005). "Comparative test – Part I – Comparative test of penetrability methods". *Mater. & Struct.*, v38, December, 895–906.

SABS (2015). "Civil engineering test methods. Part CO3-2: Concrete durability index testing – Oxygen permeability test". SABS/TC 081/SC 01, SANS-3001-CO3-2.

Salvoldi, B.G., Beushausen, H. and Alexander, M.G. (2015). "Oxygen permeability of concrete and its relation to carbonation". *Constr. & Build. Mater.*, v85, 30–37.

Schönlin, K.S. and Hilsdorf, H.K. (1987). "Evaluation of the effectiveness of curing of concrete structures". ACI SP 100, 207–226.

SIA 162/1 (1989). "Test No. 5- Water conductivity". Swiss Society of Engineers and Architects.

SIA 262/1 (2019). "Concrete construction – Complementary specifications". Swiss Society of Engineers and Architects.

Sogbossi, H., Verdier, J. and Multon, S. (2019). "New approach for the measurement of gas permeability and porosity accessible to gas in vacuum and under pressure". *Cem. & Concr. Composites*, v103, 59–70.

Soongswang, P., Tia, M., Blooquist, D., Meletiou, C.A. and Sessions, L. (1988). "Efficient test set-up for determining the water-permeability of concrete". Transp. Res. Rec., n1204, 77–82.

Stanish, K., Alexander, M.G. and Ballim, Y. (2006). "Assessing the repeatability and reproducibility values of South African durability index tests". *J. South African Inst. Civ. Eng.*, v48, n2, 10–17.

Starck, S., Beushausen, H., Alexander, M. and Torrent, R. (2017). "Complementarity of in situ and laboratory-based concrete permeability measurements". *Mater. & Struct.*, v50, n3, June, 177–191.

Torrent, R.J. (1992). "A two-chamber vacuum cell for measuring the coefficient of permeability to air of the concrete cover on site". *Mater. & Struct.*, v25, n6, July, 358–365.

Torrent, R. (2008). "Update of article RILEM TC 189-NEC *Comparative test – Part I – Comparative test of penetrability methods, Mater. & Struct.*, v38, Dec 2005, pp. 895–906". *Mater. & Struct.*, v41, April, 443–447.

Torrent, R. and Ebensperger, L. (1993). "Methoden zur Messung und Beurteilung der Kennwerte des Überdeckungsbetons auf der Baustelle". Office Fédéral des Routes, Rapport No. 506, Bern, Switzerland, January, 119 p.

Torrent, R. and Frenzer, G. (1995). "Methoden zur Messung und Beurteilung der Kennwerte des Ueberdeckungsbetons auf der Baustelle -Teil II". Office Fédéral des Routes, Rapport No. 516, Bern, Switzerland, October, 106 p.

Ujike, I., Okazaki, S. and Nakamura, T. (2009). "A study on improvement of in-situ air-permeability test for concrete structures". *Cem. Sci. & Concr. Technol.*, v63, 189–195.

UNE 83966 (2008). "Acondicionamiento de probetas de hormigón para los ensayos de permeabilidad a gases y capilaridad". Norma Española, 6 p.

UNE 83981 (2008). "Determinación de la permeabilidad al oxígeno del hormigón endurecido". Norma Española, 10 p.

Valenta, O. (1970). "The permeability and the durability of concrete in aggressive conditions". Comm. Intern. des Grandes Barrages (ICOLD), Montreal, Q. 39, R.6, 103–117.

Van Eijk, R.J. (2009). "Evaluation of concrete quality with Permea-TORR, Wenner Probe and Water Penetration Test". KEMA Report, Arnhem, July 8, 2009 (in Dutch), 48 p.

Villagrán Zaccardi, Y.A., Alderete, N.M. and De Belie, N. (2017). "Improved model for capillary absorption in cementitious materials: Progress over the fourth root of time". *Cem. & Concr. Res.*, v100, October, 153–165.

Washburn, E.W. (1921). "The dynamics of capillary flow". *Phys. Rev.*, v17, n3, March, 273–283.

XP P 18–463 (2011). "Essai de perméabilité aux gaz sur béton durci". Experimental French Standard, 15 p.

Chapter 5

Torrent NDT method for coefficient of air-permeability

5.1 INTRODUCTION: WHY A SEPARATE CHAPTER?

The reader may ask why is this method treated differently from the other methods described in Chapter 4?; the answer is:

- the authors are among the top experts in the application of this test method, created by one of them
- the authors share the conviction on the importance of measuring the air-permeability, not just in the laboratory like other methods do, but – more essentially – on site, for which this test method is especially suitable
- is a test method that can be equally applied on cast specimens and *in situ*, allowing a direct comparison between those results and, thus, discerning responsibilities along the concrete construction chain (Neves et al., 2015)
- the large and growing number of users of the test method worldwide, for whom this detailed chapter (and the whole book) should be extremely helpful

5.2 THE ORIGIN

The creator of the test method, subject of this chapter, explains the origin and development of the idea as follows:

> In 1989, R. Torrent (RT) who in 1987 had joined the Materials Dept. of "Holderbank" Management & Consulting Ltd. (HMC, later Holcim Technology Ltd.) was asked to prepare an R&D project on the general topic of "Durability". RT always acknowledged a relatively weak background in Chemistry but a strong one in Physics. During an intensive literature investigation, in order to elaborate the R&D proposal requested, RT came across with the concept of *Covercrete*, discussed in Section 7.1.4, particularly through Kreijger (1984), Newman (1987), Mayer

DOI: 10.1201/9780429505652-5

(1987). As a result, RT proposed a R&D project named *Covercrete*, aimed at studying the influence of the cement type, mix composition and curing on the quality ("penetrability") of the *Covercrete*, measured through predominantly physical tests. The proposal was endorsed by HMC's Research Council, resulting in funds and time available for RT to implement the test methods in HMC laboratory and to carry out the project (Torrent & Jornet, 1990). Within this context, RT joined RILEM TC 116-PCD (Permeability of Concrete as a Criterion of its Durability), attending a meeting in Göteborg, Sweden, that happened to be inspirational. In this meeting, different test methods to measure the "penetrability" of the *Covercrete* were presented and discussed, among them two on-site gas-permeability tests (described in Sections 4.3.2.3 and 4.3.2.8). At some point along the Bruneggerstrasse (Canton Aargau, Switzerland), while driving towards his office, RT got the idea of somehow combining the approaches of both test methods. On arrival, he immediately started to develop the concept which, after getting the needed components, assembling a prototype and a few trials, evolved into the test method called double-chamber vacuum cell, or Torrent test method to measure, non-destructively, the coefficient of air-permeability of concrete, described in this chapter.

Some brief acknowledgements by R.Torrent:

It is worth, at this point, to thank my Holderbank bosses at the time: T. Dratva (who asked me to submit the R&D project) and Drs. J. Gebauer and H. Braun (for useful discussions and for giving me the chance to freely develop my ideas). Also, to Drs. A. Jornet, L. Ebensperger and to G. Frenzer, former Holderbank colleagues for their still lasting friendship and for accompanying me during my first steps in the field of permeability testing. To ASTRA (Swiss Federal Highways Administration) for financial support and strict control of the R&D projects (Torrent & Ebensperger, 1993; Torrent & Frenzer, 1995) by a top team of advisors: M. Donzel, P. Wüst, F. Wittmann and late C. Menn. To the Swiss researchers that, through their competent and dedicated work, paved the road to the standardization of the method in Swiss Standard SIA 262/1: E. Brühwiler, E. Denarié, F. Jacobs, F. Hunkeler, A. Leemann, M. Romer and T. Teruzzi and, finally, to those who helped converting a crude prototype into a user-friendly, practical instruments: M. Fischli and K. Baumann (Proceq S.A.) and J. Szychowski, G. Zino and V. Bueno (Materials Advanced Services Ltd.).

5.3 FUNDAMENTALS OF THE TEST METHOD

5.3.1 Principles of the Test Method

The so-called Torrent test method for measuring the coefficient of air-permeability of the *Covercrete* (see Chapter 7) is an improvement on the single-chamber vacuum cell test method already described in Section 4.3.2.8.

One of the limitations of the single-chamber vacuum cell method is that the geometry of the air flow into the cell is undefined and cannot be controlled, see Figure 5.1. This prevents the calculation of a proper coefficient of air-permeability, which has to be replaced by some kind of air-permeability index.

A more serious limitation is the fact that the air that flows into the vacuum chamber, raising its pressure, can take a preferential path along the usually more permeable concrete "skin", artificially increasing the value of permeability measured; this was experimentally proved by Torrent (1991, 1992).

To overcome this problem, the idea of creating a guard-ring around the measurement vacuum chamber, keeping the pressure of both chambers always balanced, was developed (Torrent, 1991, 1992). Hence the designation became double-chamber vacuum cell test method, or shorter Torrent method.

The scheme of the air-flow in this test method is shown in Figure 5.2. Since the pressure P_e in the external guard-ring is kept permanently balanced with that in the inner test chamber (P_i), a unidirectional flow of a cylinder of air into the latter can be assumed. Now, under this controlled flow of air into the measurement chamber, the coefficient of air-permeability kT can be calculated, as derived below. So, the test method has two distinctive features: two circular rubber rings that seal the two chambers on the concrete surface, dividing the flow of air into the two concentric chambers and a pressure regulator that keeps the air pressure in both chambers permanently balanced ($P_e=P_i$). Figure 5.3 shows the aspect of the concentric rings and the book's cover picture the vacuum cell stuck onto the concrete surface, pressed by the external atmospheric pressure.

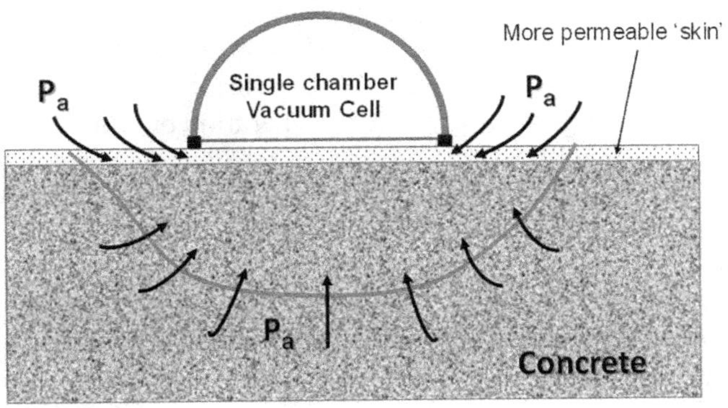

Figure 5.1 Undefined air-flow geometry and preferential surface path in the single chamber vacuum cell test method.

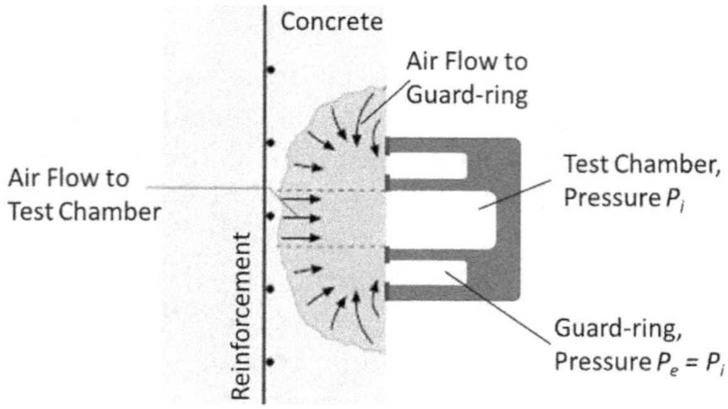

Figure 5.2 Scheme of air-flow in the Torrent method, with the assumed cylindrical flow of air into the central, test chamber.

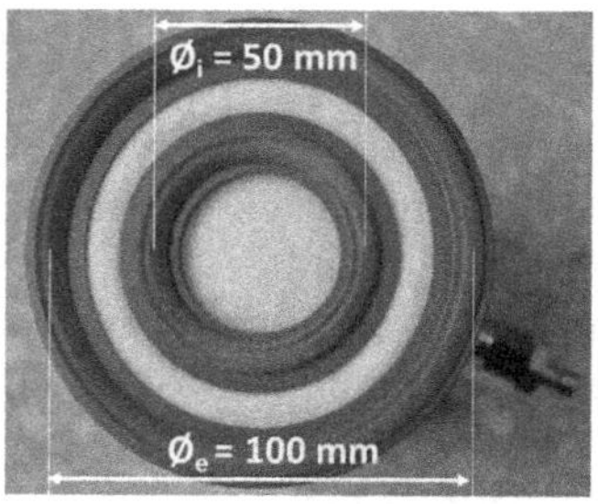

Figure 5.3 Aspect and dimensions of the vacuum cell's concentric rings.

5.3.2 Historical Evolution

Five relevant steps can be identified in the development and evolution of the instrument (Torrent & Szychowski, 2016, 2017), as summarized in Table 5.1.

Only the second- and fifth-generation instruments are currently commercially available. The lay-out and components of these commercial instruments are shown in Figures 5.4 and 5.5. Figure 5.4 corresponds to the second-generation instrument branded *Torrent Permeability Tester* (*TPT*) (Proceq, 2019) and Figure 5.5 to the fifth-generation branded *PermeaTORR AC+* (Active Cell) (M-A-S, 2019). Instruments up to the fourth generation relied on a mechanical regulator (membrane type) to control the evacuation of the external chamber by a constant regime vacuum pump. The fifth-generation instrument relies on an electronic pressure regulator that controls the speed of a variable regime embedded vacuum pump. More details on the evolution and improvements of the instrument from the first- to the fourth-generation instruments can be found in Torrent and Szychowski (2016, 2017).

Table 5.1 Evolution of five generations of the Torrent method to measure the air-permeability coefficient

Generation	*Description/brand*	*Developer*	*Features*
First (1990)	Research Prototype	R. Torrent (Switzerland)	Manual operation + calculations
Second (1995)	*TPT*	Proceq (Switzerland)	Manual operation + automatic calculations
Third (2008)	*PermeaTORR*	Materials Advanced Services Ltd. (Argentina)	Automatic operation + calculations + graphic plot
Fourth (2016)	*PermeaTORR AC(Active Cell)*		Ibid + Active Cell + embedded pump
Fifth (2021)	*PermeaTORR AC+ (Active Cell)*		Ibid + Electronic pressure regulation + remote control

Generations 1 (prototype), 2 (*TPT*) and 3 (*PermeaTORR*) work on the principle of a "passive", hollow vacuum cell, the role of which was merely to divide the flow of air into that sucked by the inner chamber from that sucked by the outer chamber. The pressure measurement and regulation are conducted remotely at a control unit, connected to the cell chambers by means of a pair of rubber hoses (see Figure 5.4) that are part of the pneumatic system.

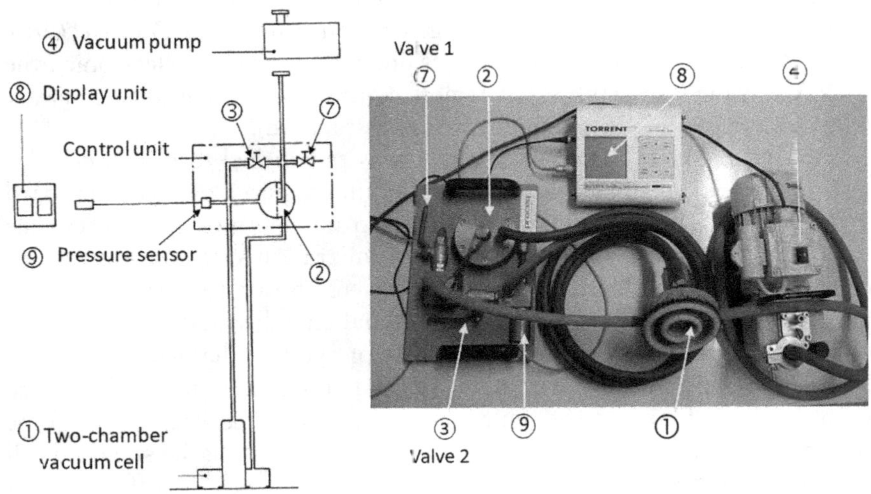

Figure 5.4 Sketch (Proceq, n.d.) and components of the second-generation instrument *TPT*.

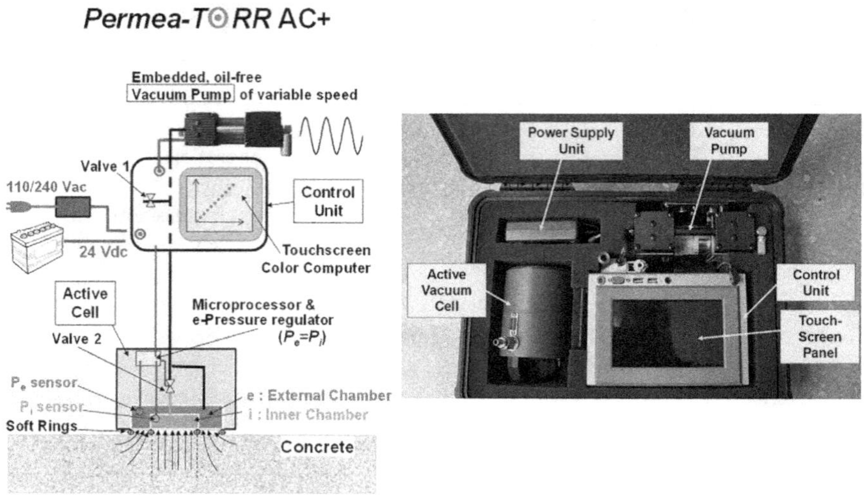

Figure 5.5 Sketch and components of the automatic fifth-generation instrument *PermeaTORR AC+*.

The fourth- and fifth-generation instrument *PermeaTORR AC* "Active Cell" (Figure 5.5) presents as a novelty the fact that the vacuum cell now houses several "active" components inside it: two pressure sensors (one for each chamber), the pressure regulator, valve 2 and a microprocessor that reads the pressure signals, operates valve 2 and communicates with the control unit. It also incorporates a small oil-free vacuum pump. The advantages of the new "Active Cell" design are described in Torrent and Szychowski (2016, 2017). The fifth-generation instrument *PermeaTORR AC+* "Active Cell" presents a different pressure regulation (electronic), the option of using a smartphone as remote control and an upgraded software (allowing voice and visual inputs into the test record).

The dimension of the inner chamber was established to have a circular testing area of Ø50 mm, which is the minimum required by Swiss Standard (SIA 262/1, 2019) for testing cores, drilled from laboratory cast specimens or on site, for capillary suction and chloride migration tests. The dimensions of the guard-ring (external Ø=100 mm) were chosen to ensure a higher natural increase in pressure than the internal chamber and to fit the most common standard specimens (cube length or cylinder diameter=150 mm). There was some concern that aggregate particles of 32 mm could interfere with the flow of air into the central chamber (Romer, 2005c). To elucidate this matter, some comparative tests were performed using the standard cell and one of bigger dimensions (not disclosed) on concretes of different compositions (maximum size of the aggregate=32 mm). Both cells produced very similar *kT* results (Romer, 2005c), confirming the suitability of the cell with internal and external diameters of 50 and 100 mm, respectively.

5.3.3 Operation of the Instrument

The principles of the instruments' operation are basically the same, irrespective of the different models:

a. the two chambers are initially evacuated by the vacuum pump, which presses the cell onto the concrete surface and seals the two chambers by means of the soft concentric rubber rings (Figure 5.3)
b. after 60 seconds, the central, test chamber (then at a pressure typically of 0–50 mbar) is isolated from the pump, by closing the connecting valve (valve 2 in Figures 5.4 and 5.5), moment at which its pressure starts to rise due to the air (at atmospheric pressure ≈ 1,000 mbar) in the concrete pores flowing into the chamber through the concrete (Figure 5.2)
c. the pump continues to run, operating exclusively on the external chamber, extracting just the necessary amount of air for its pressure P_e to balance the pressure of the central chamber P_i (this is achieved by a high-precision pressure regulator); so that, at any time, $P_e = P_i$
d. the rate of increase in pressure in the central chamber during the test, that is higher the more permeable the concrete, is recorded, allowing the calculation of the coefficient of air-permeability *kT*, applying the formula (Eq. 5.13) derived below
e. after storing the results in the instrument's memory, a venting valve (valve 1 in Figures 5.4 and 5.5) is opened to restore the whole system to atmospheric pressure, moment in which the cell can be detached from the concrete surface, ready for a new test

5.3.4 Model for the Calculation of the Coefficient of Air-Permeability *kT*

The following assumptions are made in deriving the equation for the calculation of *kT*:

- initially, all the pores in the concrete are at atmospheric pressure (hence, a repetition of a test at or near the place where one was previously made, can only be performed after ≈30 minutes waiting period)
- the air flows into the inner chamber by viscous laminar flow (see Section 3.6) as a unidirectional cylindrical plug (Figure 5.2); this hypothesis is possibly valid only for penetrations of the vacuum front L not beyond the inner chamber diameter, i.e. ≈50 mm; to be verified by numerical modelling
- the characteristics of the concrete (permeability and porosity) and its temperature and moisture condition are constant within the volume of area A of the inner chamber and penetration L affected by the test

- although the conditions are non steady, the distribution of pressure is regarded as linear between the atmospheric pressure front and the surface of the concrete beneath the test chamber
- the air pressure in the test chamber (normally 10–50 mbar) is much lower than the atmospheric pressure P_a (≈1,000 mbar)
- the cell is placed on a semi-infinite body; even when not, the penetration of the test does not exceed the thickness of the element/specimen

Figure 5.6 sketches the conditions in the concrete pores before and during the application of the Torrent test method. At the left, there is a sketch indicating the inner test chamber (i) and outer guard-ring (e); the inner test chamber has a cross sectional area *A* and a total volume V_c.

At the right-hand side of Figure 5.6, the distribution of pressure of the air in the concrete pores is shown, at different instants of the test, from (a) to (d).

Initially, at time $t=0$, all the pores in the concrete contain air at atmospheric pressure P_a (≈1,000 mbar), situation (a) in Figure 5.6. As soon as vacuum is being made in the inner chamber, its pressure drops drastically, reaching at $t_0=60$ seconds a value $P_0 \approx 0$–50 mbar, situation (b) of Figure 5.6; at this moment, valve 2 in Figures 5.4 and 5.5 is closed and the inner test chamber is isolated from the pump. The pressure in the inner test chamber *P* starts to grow steadily, due to the air flowing from the concrete

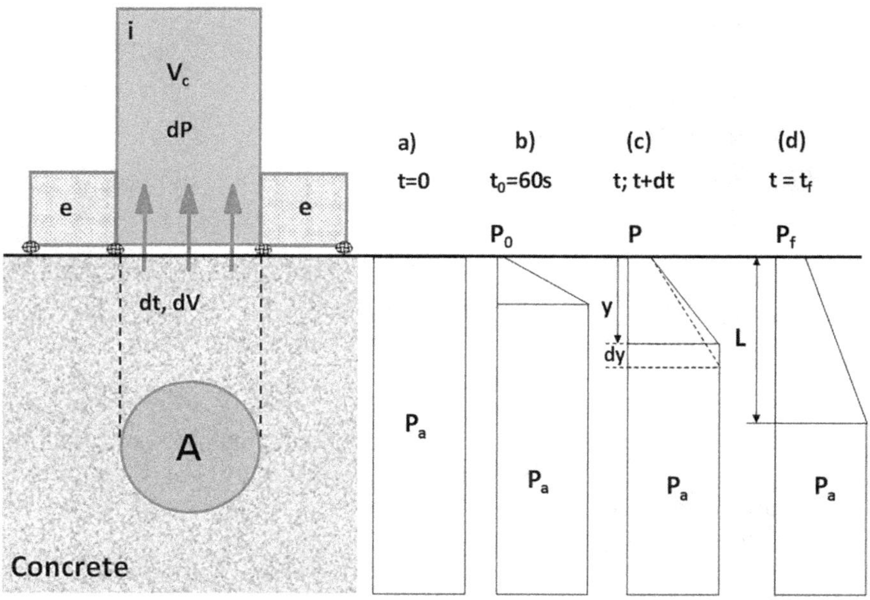

Figure 5.6 Evolution of pressure profiles with depth during the test.

pores into the inner chamber, driven by the gradient of pressure that has been established, situation (c) in Figure 5.6.

Applying the Hagen-Poiseuille-Darcy law for gases, Eq. (3.26), to the situation (c) of Figure 5.6, we can calculate the volume of air dV (m^3) that flows into the inner chamber in an interval dt (s) as follows:

$$dV = \frac{kT \cdot A}{2 \cdot \mu \cdot y} \cdot \frac{P_a^2 - P^2}{P} \cdot dt \tag{5.1}$$

where

kT=coefficient of air-permeability, measured by the Torrent method (m^2)
A=cross-sectional area of the inner test chamber (m^2)
μ=dynamic viscosity of air ($N.s/m^2$)
y=penetration depth of atmospheric pressure front (m)
P=pressure in the inner chamber at time t (N/m^2)
P_a=atmospheric pressure (N/m^2)

Applying the general gas equation, we can calculate the number of air moles corresponding to a volume dV at a pressure P and absolute temperature T as follows:

$$dn = \frac{P}{R \cdot T} \cdot dV \tag{5.2}$$

where R is the universal gas constant.

Substituting dV in Eq. (5.2) by its value in Eq. (5.1) we get

$$dn = \frac{kT \cdot A \cdot \left(P_a^2 - P^2\right)}{2 \cdot \mu \cdot y \cdot R \cdot T} \cdot dt \tag{5.3}$$

By mass conservation, the number of moles of air dn entering the vacuum chamber must correspond to the same number affected by a differential deepening dy of the atmospheric pressure front y, dotted lines in situation (c) in Figure 5.6. The affected volume of air is

$$dV = \varepsilon \cdot A \cdot dy \tag{5.4}$$

where

ε=open porosity of the concrete (-)

This volume can be assumed to be at the mean pressure $(P_a+P)/2$, which allows us to compute, from Eq. (5.2), the number of moles affected as follows:

$$dn = \frac{(P_a + P)}{2 \cdot R \cdot T} \cdot dV = \frac{(P_a + P)}{2 \cdot R \cdot T} \cdot \varepsilon \cdot A \cdot dy \tag{5.5}$$

Equating dn in Eqs. (5.5) and (5.3), we have:

$$\frac{(P_a + P)}{2 \cdot R \cdot T} \cdot \varepsilon \cdot A \cdot dy = \frac{kT \cdot A \cdot \left(P_a^2 - P^2\right)}{2 \cdot \mu \cdot y \cdot R \cdot T} \cdot dt \tag{5.6}$$

$$y \cdot dy = \frac{kT}{\mu \cdot \varepsilon} \cdot P_a \cdot \left(1 - \frac{P}{P_a}\right) \cdot dt \tag{5.7}$$

integrating and neglecting the term P/P_a, we can calculate the position y of the atmospheric front at time t

$$y = \sqrt{\frac{2 \cdot kT \cdot P_a \cdot t}{\varepsilon \cdot \mu}} \tag{5.8}$$

now, looking at the situation in the cell, we know that the volume of air dV, entering the inner test chamber (of volume V_c) at pressure P, will produce a differential increase in its pressure dP given by

$$dP = \frac{P \cdot dV}{V_c} \tag{5.9}$$

now, introducing Eqs. (5.1) and (5.8) into (5.9) and rearranging, we get

$$\frac{1}{P_a^2 - P^2} dP = \frac{A}{2 \cdot V_c} \cdot \sqrt{\frac{kT \cdot \varepsilon}{2 \cdot \mu \cdot P_a}} \cdot \frac{1}{\sqrt{t}} dt \tag{5.10}$$

integrating both members between conditions (b) and (d) in Figure 5.6:

$$\int_{P_0}^{P_f} \frac{1}{P_a^2 - P^2} dP = \frac{A}{2 \cdot V_c} \cdot \sqrt{\frac{kT \cdot \varepsilon}{2 \cdot \mu \cdot P_a}} \cdot \int_{t_0}^{t_f} \frac{1}{\sqrt{t}} dt \tag{5.11}$$

leads to

$$\frac{1}{2 \cdot P_a} \cdot \ln\left[\frac{(P_a + P_f) \cdot (P_a - P_0)}{(P_a - P_f) \cdot (P_a + P_0)}\right] = \frac{A}{V_c} \cdot \sqrt{\frac{kT \cdot \varepsilon}{2 \cdot \mu \cdot P_a}} \cdot \left(\sqrt{t_f} - \sqrt{t_0}\right) \tag{5.12}$$

from where the value of the coefficient of air-permeability can be obtained as

$$kT = \left(\frac{V_c}{A}\right)^2 \cdot \frac{\mu}{2 \cdot \varepsilon \cdot P_a} \cdot \left[\frac{\ln\left(\frac{P_a + \Delta P}{P_a - \Delta P}\right)}{\sqrt{t_f} - \sqrt{t_0}}\right]^2 \tag{5.13}$$

Equation (5.13) is the one used nowadays to compute the coefficient of air-permeability kT (m²) by the Torrent method, where

V_c=volume of the inner test chamber pneumatic system (m³)

A=area of the inner test chamber (m²)

μ=dynamic viscosity of air (N s/m²)

ε=open porosity of the concrete (-) which, by default is taken as 0.15

P_a=atmospheric pressure (N/m²)

ΔP=increase of effective pressure in the inner chamber ($\Delta P = P_f - P_0$) between time t_0 and t_f (N/m²)

t_0=once valve 2 was closed, time from which the increase in pressure is measured (60 seconds)

t_f = time at which the test is finished (s)

The test ends when the increase of effective pressure ΔP_{eff}[1] exceeds 20 mbar or, for concretes of low permeability, when t_f=720 seconds. However, in the automatic instruments, the test may also be optionally stopped at t_f=360 seconds provided an approximately linear relationship between ΔP_{eff} and $\left(\sqrt{t} - \sqrt{t_0}\right)$ is observed during the test (see justification in Section 5.3.5).

It is worth mentioning that the equation to calculate kT in the original publications of the test method (Torrent, 1991, 1992; Torrent & Ebensperger, 1993) was just an approximation; the correct Eq. (5.13) was developed later (Torrent & Frenzer, 1995). Therefore, the values of kT published before 1995 need to be converted by multiplying them by a factor 1.846, as shown in p. 60 of Torrent and Frenzer (1995).

The final penetration of the atmospheric pressure front L (m), i.e. the depth of concrete affected by the vacuum, can be calculated from Eq. (5.8) as

$$L = \sqrt{\frac{2 \cdot kT \cdot P_a \cdot t_f}{\varepsilon \cdot \mu}} \tag{5.14}$$

Notice that Eq. (5.14) is identical to Eq. (4.2), if the terms are rearranged. If the penetration of the test L exceeds the thickness of the element e, a correction is required, see Section 5.3.6.2.

5.3.5 Relation between ΔP and $\sqrt{t}$

5.3.5.1 Theoretical Linear Response

Let us justify the stopping of the test when t_f = 360 seconds if the plot ΔP_{eff} vs. $\left(\sqrt{t} - \sqrt{t_0}\right)$ is approximately linear. Equation (5.13) can be rewritten as

[1] The effective pressure rise is the measured pressure rise during the test minus the pressure rise observed when the instrument is applied on an impermeable plate (calibration plate).

$$\ln\left(1+\Delta P/P_a\right)-\ln\left(1-\Delta P/P_a\right)=C\ .\left(\sqrt{t_f}-\sqrt{t_0}\right) \quad (5.15)$$

with

$$C=\left(\frac{A}{V_c}\right)\sqrt{\frac{2\ .\ kT\ .\ \varepsilon\ .\ P_a}{\mu}} \quad (5.16)$$

developing $\ln(1+x)$ for $|x|<1$ as a Taylor series:

$$\ln\left(1+x\right)=\sum_{n=1}^{\infty}\frac{\left(-1\right)^{n+1}}{n}\ x^n \quad (5.17)$$

Considering just the first four terms of the series:

$$\ln\left(1+x\right)=x-\frac{x^2}{2}+\frac{x^3}{3}-\frac{x^4}{4} \quad (5.18)$$

$$\ln\left(1-x\right)=-x-\frac{x^2}{2}-\frac{x^3}{3}-\frac{x^4}{4} \quad (5.19)$$

$$\ln\left(1+x\right)-\ln\left(1-x\right)=2x-0+\frac{2\ .\ x^3}{3}-0 \quad (5.20)$$

Making $x=\Delta P/P_a$ and disregarding the third term in Eq. (5.20) ($\Delta P << P_a$):

$$\ln\left(1+\frac{\Delta P}{P_a}\right)-\ln\left(1-\frac{\Delta P}{P_a}\right)=2\ \frac{\Delta P}{P_a} \quad (5.21)$$

Considering Eqs. (5.21), (5.15) and (5.16):

$$\Delta P=\left[\left(\frac{A}{V_c}\right)\sqrt{\frac{kT\ .\ \varepsilon\ .\ P_a^3}{2\ .\ \mu}}\right]\left(\sqrt{t_f}-\sqrt{t_0}\right) \quad (5.22)$$

So, if the model described above is correct and kT and ε are constant in the cylinder of area A and length L affected by the test, the plot ΔP_{eff} vs. $\left(\sqrt{t}-\sqrt{t_0}\right)$ should be linear (which is the usual outcome of most tests), with a slope proportional to $\sqrt{kT}$. In that case, no matter when you measure kT, you will get the same result. Instruments of third and later generations display the relation ΔP_{eff} vs. $\left(\sqrt{t}-\sqrt{t_0}\right)$ graphically, allowing the user to stop the test after 6 minutes (360 seconds) if the plot is quasi-linear, thus shortening the test duration. The above-mentioned generations of instruments display the actual plot on a chart containing reference lines, built from Eq. (5.22), which allows the user to anticipate at a very early stage of the test the approximate kT value under measurement.

5.3.5.2 Lack of Linear Response: Possible Causes

In the majority of cases, the response of the instrument indicates a linear relation ΔP_{eff} vs. $\left(\sqrt{t}-\sqrt{t_0}\right)$. The departure from this linearity can be due to a gradient of permeability with depth ($\partial K/\partial y$). Indeed, the curvature of the ΔP_{eff} vs. $\left(\sqrt{t}-\sqrt{t_0}\right)$ plot is related to that gradient such that, if $(\partial K/\partial y)<0$, the curvature is also negative whilst, if $(\partial K/\partial y)>0$ the curvature is positive and, if $(\partial K/\partial y)=0$ the plot is linear (Dobel et al., 2010; Dobel & Fernández Luco, 2012). The more usual departure from linearity happens when a gradient of moisture with depth exists in the concrete (e.g. after a rapid severe drying) leading to a negative curvature. A rare example of a positive curvature is presented in Torrent (2012) when testing a surface treated with a permeable formwork liner (see Section 7.2.3). The action of the liner is to reduce the *w/c* ratio of the near-surface layers leading to a positive permeability gradient (higher permeability at increasing depths).

Another reason for the lack of linearity of the ΔP_{eff} vs. $\left(\sqrt{t}-\sqrt{t_0}\right)$ relation is caused by the evaporation of water from the concrete under test into the vacuum chamber, that produces an initial "jump" of ΔP_{eff} which increases artificially the computed value of kT, especially for concretes of low permeability (Romer, 2005a). To avoid this effect, automatic instruments work always at pressures above 30 mbar (water vapour pressure at about 25°C). To this effect the different results obtained by the *TPT* instrument and the *PermeaTORR* instruments' family are attributed (Torrent, 2012), as discussed in Section 5.6.4.2 and shown in Figure 5.13.

5.3.6 Relation between *L* and *kT*. Thickness Correction

5.3.6.1 Relation between Test Penetration *L* and *kT*

There seems to be a direct relation between L and kT in Eq. (5.14); however, the relation is not unique, as t_f depends on the mode in which the test is stopped. Mode 1 happens for moderate to high-permeability materials, for which t_f corresponds to an increase in (effective) pressure in the inner chamber of $\Delta P_f = 20$ mbar. For concretes of lower permeability, Mode 2, t_f happens when it reaches 720 seconds (optionally 360 seconds for the third- and fourth-generation instruments). Using Eq. (5.22), Szychowski (2010) found that the function $L(kT)$ is composed of two stages, depending on the mode of ending the test:

For Mode 2 and $t_f = 720$ or 360 seconds, Eq. (5.14) holds.

For Mode 1, Eq. (5.23) is applicable, with $\Delta P_f = 20$ mbar; so, in this case, L is instrument-dependent as it is function of the ratio between the volume and cross-section area of the inner chamber.

$$L = \sqrt{\frac{2 \cdot kT \cdot P_a \cdot 60}{\varepsilon \cdot \mu} + \frac{V_c}{A \cdot \varepsilon} \cdot \frac{2 \cdot \Delta P_f}{P_a}} \tag{5.23}$$

In the case of the *PermeaTORR* family, $V_c = 16 \times 10^{-5}$ m³ and $A = 196 \times 10^{-5}$ m²; for the *TPT* is $V_c = 22 \times 10^{-5}$ m³. Figure 5.7 shows the relation between L and kT for both instruments, assuming $P_a = 1{,}000$ mbar (100 kPa), for the different stopping modes. The turning points are shown in the graph. It can be seen that the computed L ranges from a few mm for kT below 0.01×10^{-16} m² to over 100 mm for kT above $\approx 10 \times 10^{-16}$ m.

The validity of Eq. (5.14) was experimentally investigated in the laboratory by Kato (2013), using six concrete mixes of *w/c* in the range 0.30–0.70, resulting in compressive strengths between 23 and 90 MPa. He conducted kT measurements on intact 150 mm cubes, as well as in others in which he drilled Ø6 mm holes of variable lengths between 100 and 130 mm, i.e. arriving to depths between 20 and 50 mm below the surface where kT was to be measured. By analysing the changes between the kT measured on the intact cubes and on those perforated, he could guess when the vacuum front reached the holes (resulting in significantly higher kT). The experimental research showed that the measurable L is 15 mm for $kT < 0.05 \times 10^{-16}$ m²; 30 mm when 0.05×10^{-16} m² $< kT < 2.0 \times 10^{-16}$ m² and 40 mm when $kT > 2.0 \times 10^{-16}$ m², in very good agreement with the calculation through Eq. (5.14), displayed in Figure 5.7.

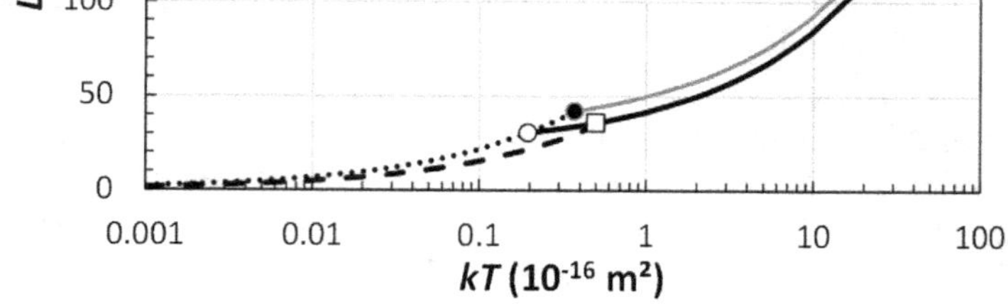

Figure 5.7 Relation between L and kT for different stoppage modes and instruments (*TPT* and *PermeaTORR*), assuming $P_a = 100$ kPa.

5.3.6.2 Correction of kT for Thickness

The derivation of Eq. (5.13) assumes that the test is performed on a semi-infinite body. Invariably, the test is performed in practice on specimens or elements that have finite dimensions. As Figure 5.7 shows, the penetration of the test can be considerable; for instance, for kT values above 40×10^{-16} m^2, the penetration of the test given by Eq. (5.23) exceeds 150 mm. In case that the thickness of the specimen e is lower than L, Eq. (5.13) is no longer valid and a correction is required, as described below.

In the case that $L>e$, two successive time intervals have to be considered, separated by the time t_e required for the penetration front y to reach the thickness of the element e:

t_0–t_e, where the flow takes place across a growing depth y, which corresponds to the general case
t_e–t_f, where the flow takes place across a constant depth $y=e$

Therefore, Eq. (5.10) has to be modified to take into consideration the gas flow happening in both intervals, becoming, where kT is the true coefficient of permeability of the material:

$$\frac{1}{P_a^2 - P^2}\,dP = \frac{A}{2\cdot V_c}\cdot\sqrt{\frac{kT\cdot\varepsilon}{2\cdot\mu\cdot P_a}}\cdot\frac{1}{\sqrt{t}}\,dt + \frac{kT\cdot A}{2\cdot V_c\cdot\mu\cdot e}\,dt \tag{5.24}$$

Now, if we integrate the first member of Eq. (5.24) between t_0–t_f, we will get:

$$\int_{t0}^{tf}\frac{1}{P_a^2 - P^2}\,dP = \frac{1}{2\cdot P_a}\cdot\ln\left[\frac{(P_a+P_f)\cdot(P_a-P_0)}{(P_a-P_f)\cdot(P_a+P_0)}\right] = \frac{A}{V_c}\cdot\sqrt{\frac{kT_i\cdot\varepsilon}{2\cdot\mu\cdot P_a}}\cdot\left(\sqrt{t_f}-\sqrt{t_0}\right) \tag{5.25}$$

where kT_i is the value indicated by the instrument, assuming $L=\infty$, Eq. (5.12).

The first term of the second member of Eq. (5.24) must be integrated for the time interval t_0–t_e, whilst the second term for the interval t_e–t_f which, considering the first member (Eq. 5.25) results in

$$\frac{A}{V_c}\cdot\sqrt{\frac{kT_i\cdot\varepsilon}{2\cdot\mu\cdot P_a}}\cdot\left(\sqrt{t_f}-\sqrt{t_0}\right) = \frac{A}{V_c}\cdot\sqrt{\frac{kT\cdot\varepsilon}{2\cdot\mu\cdot P_a}}\cdot\left(\sqrt{t_e}-\sqrt{t_0}\right) + \frac{kT\cdot A}{2\cdot V_c\cdot\mu\cdot e}\cdot\left(t_f - t_e\right) \tag{5.26}$$

Now, applying Eq. (5.14) to compute the time t_e necessary for the vacuum front to reach the thickness e, we have

$$t_e = \frac{e^2 \cdot \varepsilon \cdot \mu}{2 \cdot kT \cdot P_a} \tag{5.27}$$

Substituting t_e in Eq. (5.26) and reorganizing, we get

$$a \cdot kT + b \cdot \sqrt{kT} + c = 0 \tag{5.28}$$

where

$$a = \frac{P_a \cdot t_f}{\mu \cdot e}; \quad b = -\sqrt{\frac{2 \cdot P_a \cdot \varepsilon \cdot t_0}{\mu}};$$

$$c = \frac{e \cdot \varepsilon}{2} - \left[\sqrt{\frac{2 \cdot P_a \cdot \varepsilon \cdot kT_i}{\mu}} \cdot \left(\sqrt{t_f} - \sqrt{t_0}\right)\right] \tag{5.29}$$

With all variables in Eq. (5.25) expressed in the SI system of units [m, kg, s] and solving the second degree Eq. (5.28)

$$kT = \left(\frac{-b + \sqrt{b^2 - 4 \cdot a \cdot c}}{2 \cdot a}\right)^2 \quad \text{valid for } L \geq e \tag{5.30}$$

The corrected value kT is always smaller than the value indicated by the instrument kT_i, because in the latter a smaller gradient of pressure is assumed ($y>e$) than the real one ($y=e$), for the same gas flow. Instruments of third and later generations allow an automatic thickness correction after entering the value of e.

5.4 RELEVANT FEATURES OF THE TEST METHOD

The most relevant features of the Torrent test method are

- it is entirely non-destructive (no traces left on the concrete surface after application) and suitable for both laboratory and site applications
- it is fast, taking from 1.5 to 6 or 12 minutes for high- and low-permeability concrete, respectively
- the known and controlled gas flow into the inner chamber allows the calculation, through Eq. (5.13), of an important physical property as the coefficient of permeability of concrete to air kT is, in SI units (m^2) and not just a technological index, as for the single chamber test (Section 4.3.2.8)

- the penetration of the vacuum front L can be calculated from Eq. (5.14), which allows knowing the concrete depth affected by the test and applying a correction in case it exceeds the thickness of the element
- it is not affected by the "skin" or "wall" effects since, thanks to the outer guard ring, the spurious air entering the vacuum cell through the permeable "skin" (Figure 5.1) is now sucked by the outer chamber, not affecting the unidirectional flow into the inner, test chamber. This was demonstrated in Torrent (1991, 1992)

5.5 INTERPRETATION OF TEST RESULTS

5.5.1 Permeability Classes

It has been shown that, in theory, the coefficient of permeability is proportional to the porosity and to the pore size squared, Eq. (3.20). As the interconnected pores (capillary pores and ITZ) cover a size range of four orders of magnitude (Figure 3.8), it is expected that the air-permeability would cover around twice that range.

Table 5.2 (first three columns) presents a classification of concrete, based on the air-permeability kT, that confirms that expectation.

A similar classification, based on kT, was already proposed in p. 31 of Torrent and Frenzer (1995), extended now to the one presented in Table 5.2. The lower limit of measurable permeability is $kT = 0.001 \times 10^{-16}$ m²; below that value the flow of air is so small that the induced pressure rise cannot be accurately measured by the pressure sensors.

As shown in Table 5.2, the classification follows a logarithmic scale of kT, a fact to be taken into consideration when interpreting the test results. As discussed in more detail in Chapter 8, the durability performance of a concrete is related to the logarithm of kT rather than to kT directly.

Table 5.2 Estimated porosity ε and characteristic radius r of the concrete pore system as function of kT

			Expected porosity ε *(%)*	*Expected pore radius* r *(nm)*	
kT $(10^{-16}$ m²$)$	*Permeability Class*		*Eq. (5.31)*	*Eq. (5.32)*	*Eq. (5.33)*
<0.001	PK0	Negligible	<6	<5	<2
0.001 – 0.01	PK1	Very low	6 – 9	5 – 10	2 – 5
0.01 – 0.1	PK2	Low	9 – 12	10 – 25	5 – 15
0.1 – 1.0	PK3	Moderate	12 – 16	25 – 70	15 – 45
1.0 – 10	PK4	High	16 – 19	70 – 200	45 – 150
10 – 100	PK5	Very high	19 – 23	200 – 600	150 – 450
>100	PK6	Ultra high	>23	> 600	>450

5.5.2 Microstructural Interpretation

The results of *kT* can be interpreted also from a microstructural perspective.

Several investigations produced parallel results of mercury intrusion porosimetry or MIP (porosity ε_{MIP} and characteristic pore radius r) on samples taken from specimens where *kT* had previously been measured (Torrent & Ebensperger, 1993; Torrent & Frenzer, 1995; Sakai et al., 2013). In the last two investigations, data from site concrete are included.

In Torrent and Ebensperger (1993), the relation between ε_{MIP} and the capillary ε_c and total porosity ε_t was investigated for a wide range of concretes. Porosities ε_c and ε_t were measured after the old Swiss Standard (SIA 162/1,1989), very similar to current Swiss Standard (SIA 262/1-K, 2019). The volume of pores was obtained gravimetrically by measuring the mass after saturation under water (ε_c) and after water saturation under vacuum (ε_t), referred to the mass after drying at 105°C. The total porosity is meant to include not just the capillary pores but also the air voids. The results of Torrent and Ebensperger (1993) indicate that ε_{MIP} measures a porosity somewhere in between the capillary and the total porosity, as determined gravimetrically by water saturation referred to a sample dried at 105°C.

Figure 5.8 (l.) shows the relation obtained between the measured air-permeability *kT* and the porosity measured by MIP ε_{MIP} in the three investigations.

The regression of Eq. (5.31) is fitted to all 62 test results (black line in Figure 5.8, left), with $R = 0.80$.

$$\varepsilon_{MIP} = 0.16 + 0.015 \ln(kT) \tag{5.31}$$

According to the correlation coefficient test (Urdan, 2011), these results present a highly significant correlation between ε_{MIP} and *kT*, as the test returned a near zero p-value.

Figure 5.8 (r.), in turn, shows the relation between the measured air-permeability *kT* and K_{MIP}, the latter computed from Eq. (3.20) as follows:

$$K_{MIP} = \frac{\varepsilon_{MIP} \cdot r^2}{8} \tag{5.32}$$

where the radius r is equal to the mean pore radius for the data from Torrent and Ebensperger (1993) and Torrent and Frenzer (1995) and to the threshold value for the data from Sakai et al. (2013). It is worth mentioning the different degrees of saturation of the samples in Figure 5.8, as *kT* was measured on specimens kept 21–28 days in a dry room at 20°C/50% RH for the first two investigations, compared to 6 months at 20°C and ≈60% RH for Sakai et al. (2013). The MIP analysis was performed on thoroughly dried small samples.

The black line in Figure 5.8 (r.) is the equality line. It can be seen that, despite the large scatter in results, especially from site tests, the calculated

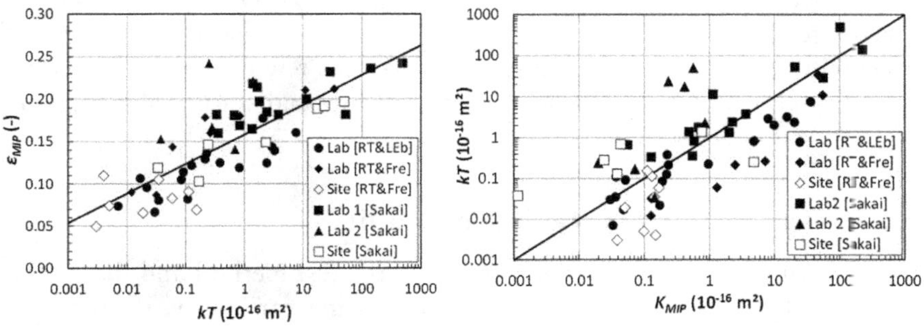

Figure 5.8 Relation between ε_{MIP} and *kT* (l.) *and* between *kT* and K_{MIP} (r.).

K_{MIP} matches reasonably well the experimental *kT*. Given the different preconditioning of the *kT* tests and the differences in MIP techniques and instruments used to establish the values of ε_{MIP} and *r*, the agreement in Figure 5.8, (l. and r.), is quite remarkable.

From Eqs. (5.31) and (5.32), it is possible to obtain an estimate of the porosity ε and characteristic pore radius *r* of the concrete pore system from the measured *kT*, see Table 5.2.

An estimate of the so called "representative pore size of permeation resistance" (RPSPR), as function of *kT*, was proposed by Sakai et al. (2014) as

$$RPSPR = 46 \cdot \sqrt{kT} \tag{5.33}$$

with *RPSPR* in (nm) and *kT* in (10^{-16} m²). The calculated values of *RPSPR* are included in the last column of Table 5.2, resulting slightly lower than those obtained from Eqs. (5.31) and (5.32).

Table 5.2, although providing just gross estimates of ε and *r*, adds a physical meaning to the measured *kT* values in terms of the pore structure of the concrete tested.

For a more detailed analysis, from a different and more general perspective, the reader can refer to Sakai (2019).

5.6 REPEATABILITY AND REPRODUCIBILITY

When the repeatability and reproducibility of the Torrent test method are assessed, the fact that the *kT* values range over six orders of magnitude (Table 5.2) has to be borne in mind. Indeed, *kT* values within a ratio of 10:1 actually belong to the same Permeability Class. This means that *kT* is a variable very sensitive to the microstructure (especially to the pore structure) of the material under test. For instance, in the case of concrete, if a large aggregate particle (say, 38 mm size) happens to lay below the concrete surface, just where the central chamber (Ø 50 mm) is placed, a different *kT* value will be obtained than in a place displaced a few mm from that point. Therefore, the scatter of *kT* values requires a special statistical treatment, as discussed in Section 5.8.

5.6.1 Testing Variability: Repeatability

This section deals with the repeatability of the test method and the instruments used to measure *kT*. It refers to the testing variability, i.e. the variability of test results obtained with the same instrument applied repeatedly on the same sample (if possible, exactly on the same spot).

There are few data on the repeatability of *kT* measurements. Some data were produced by González Gasca, at the Instituto Eduardo Torroja, Madrid, Spain with a *TPT* (Torrent, 1997). She repeated five measurements at approximately the same place, on the same face of a concrete disc (Ø 150×50 mm), with a waiting time of 30 minutes between successive measurements. She reported a mean value of 0.057×10^{-16} m^2, with a coefficient of variation CoV of 19%.

Ten successive *kT* test results were obtained by R. Torrent, with a *PermeaTORR AC* instrument, on the same location of a reference sample used to calibrate the instruments. A waiting time of 15 minutes was observed between the end of one measurement and the initiation of the next one. The mean value of the ten measurements was 0.106×10^{-16} m^2, with a standard deviation of 0.004×10^{-16} m^2, yielding a CoV of 3.8%.

The repeatability and stability of readings can be judged by repeating tests at intervals in the same spots on specimens that are stable in terms of hydration and of moisture and temperature. One investigation of this kind was reported by Adey et al. (1998), who repeated air-permeability tests with a *TPT* instrument (Proceq S.A.) on seven well-defined spots (#1–#7) of slabs removed from a bridge and stored in EPFL Lausanne laboratory for 6 months before the measurements. Two of these spots (#4 and #7) were located in coincidence with steel bars with 20 mm cover thickness (assessed by a covermeter), whilst the rest were located away of reinforcing bars. Measurements of *kT* were repeated five times over a period of 5 weeks. Except for the second reading on spot #3, the stability and repeatability of the measurements are very good. It is worth mentioning that most of the readings fall within the "High" Permeability Class (Table 5.2), including those obtained in coincidence with the steel bars, with spots #5 and #2 yielding values corresponding to the "Very High" and "Moderate" Permeability Classes.

The testing variability of the Torrent test method is recognized as very low; hence, the recommendation of performing just one reading per measuring spot; it is better to spend time measuring several points within the elements or specimen (i.e. different lateral faces of a cube) rather than repeating readings on the same spot.

5.6.2 Within-Sample Variability

Here, the variability of *kT* within, nominally, the same sample is dealt with. This variability reflects the effect the heterogeneity of concrete has on the measured coefficient of air-permeability. For this, we will refer to an experiment conducted at the Swiss Federal Polytechnic University, Zürich (ETHZ), within the frame of a project financed by the Swiss Federal Highway Administration (Torrent & Frenzer, 1995). Concrete cubes (0.5 m) were cast at ETHZ with four different concrete mixes made with the same constituents, but with a wide range of characteristics (w/c=0.3–0.75; OPC=200–450 kg/m³; $f'c$=14–66 MPa). Two cubes were cast with each mix, one of which was moist cured during 7 days (B), whilst the other was totally deprived of moist curing (A). At the age of 28 days, the cubes were tested by now Holcim Technology personnel, without knowing the preparation conditions of any of the eight cubes (blind test). The tests were conducted, using a *TPT*, on two opposite faces of each cube, five tests on each face following a pattern similar to number five of a dice. The results obtained are presented in Table 5.3 (the results in the two bottom rows are not relevant here, but will be discussed in Section 6.2.1.2).

Table 5.3 Within sample variability of *kT* measured on 0.5 m cubes of different concretes

Cube	*IA*	*IB*	*IIA*	*IIB*	*IIIA*	*IIIB*	*IVA*	*IVB*
Curing (d)	0	7	0	7	0	7	0	7
w/c	0.75		0.50		0.40		0.30	
OPC (kg/m³)	200		300		335		450	
Air (%)	8.0		7.5		7.5		2.3	
28-day $f'c_{cyl}$ (MPa)	13.8		25.4		38.3		65.7	
kT (10^{-16} m²) Face 1	37.2	15.3	0.771	0.992	0.208	0.093	0.028	0.018
	19.2	12.8	5.05	0.443	0.215	0.091	0.041	0.008
	18.7	6.12	1.08	0.121	0.150	0.036	0.043	0.018
	33.3	15.3	0.310	0.103	0.159	0.077	0.032	0.007
	20.1	10.4	0.812	0.116	0.066	0.06	0.017	0.009
kT (10^{-16} m²) Face 2	35.7	10.3	1.09	0.765	0.094	0.062	0.075	0.014
	37.1	10.6	1.19	0.570	0.242	0.073	0.059	0.027
	43.6	8.49	1.05	0.347	0.603	0.071	0.023	0.018
	71.3	20.5	0.483	0.146	1.186	0.024	0.023	0.007
	60.6	7.36	0.261	0.109	0.187	0.053	0.020	0.010
Mean (10^{-16} m²)	37.7	11.7	1.21	0.371	0.311	0.064	0.036	0.014
Std. dev. (10^{-16} m²)	18.5	4.3	1.39	0.317	0.341	0.022	0.019	0.007
CoV (%)	48.9	37.1	114.9	85.4	109.7	34.5	51.9	48.4
kT_{gm} (10^{-16} m²)	34.0	11.0	0.84	0.262	0.215	0.060	0.032	0.012
s_{LOG}	0.20	0.16	0.36	0.39	0.36	0.18	0.21	0.21
kO (10^{-16} m²)	22.6	16.5	2.00	1.74	0.463	0.346	0.173	0.194
a_{24} (g/m²/s^{½})	18.9	18.8	12.1	11.2	9.39	8.15	6.73	6.53

The cubes were cast by the same personnel, using same forms, tools and techniques, and the tests were performed by the same person with the same instrument, following the same routine. Hence, in principle, one would expect the within sample variability to be very similar for the different specimens.

Looking at the standard deviation of the results (sixth row from bottom of Table 5.3), it is clear that it changes enormously with the permeability level of the sample. The CoV (fifth row from the bottom) shows more stability. The fourth and third rows (from the bottom) show the parameters obtained assuming that the results of *kT*, especially on site, follow reasonably well a log-normal distribution (see Section 5.8). The kT_{gm} value is the geometric mean of the test results which, as expected, is somewhat smaller than the arithmetic mean (seventh row from the bottom). Remember that the geometric mean corresponds to the antilogarithm of the average of the logarithms of the test results. The s_{LOG} value reported corresponds to the standard deviation of the $\log_{10}$ of the *kT* results, which is quite stable for the different cubes, demonstrating to be a fair indicator of the variability of the results. In the future, the results of *kT* measurements will be described in terms of their kT_{gm} and s_{LOG} statistical parameters, as recommended by Jacobs et al. (2009).

Now, if we compare the CoV reported in Table 5.3 with the 3.8% or even 19% (Section 5.6.1) obtained when applied on the same specimen, we see the much larger Within-Sample Variability compared with the Test Variability. This reflects the heterogeneity of concrete within the sample, even when the specimen (0.5 m cubes in this case) has been prepared in a laboratory under strict casting, compaction and curing procedures.

5.6.3 Global Variability

The Global Variability encompasses the Testing and Within-Sample Variability, discussed in Sections 5.6.1 and 5.6.2, plus the Between-Samples Variability. The latter is the result of variations in quality of the *Covercrete* resulting from batch-to-batch concrete quality variations, changes in the conditions of placement, compaction, curing, etc., differences in the skills of the personnel involved in the construction process, variable weather conditions during construction and testing, different exposures of the surfaces to sun, wind, rain/snow, etc.

Data on the Global Variability, expressed as s_{LOG}, were reported by Jacobs et al. (2009), based on site investigations of 52 concrete structures in Switzerland (Jacobs, 2006), reproduced in Figure 5.9.

This gives an indication of the large differences in s_{LOG} that can be encountered when testing *kT* on site concrete, reflecting the large number

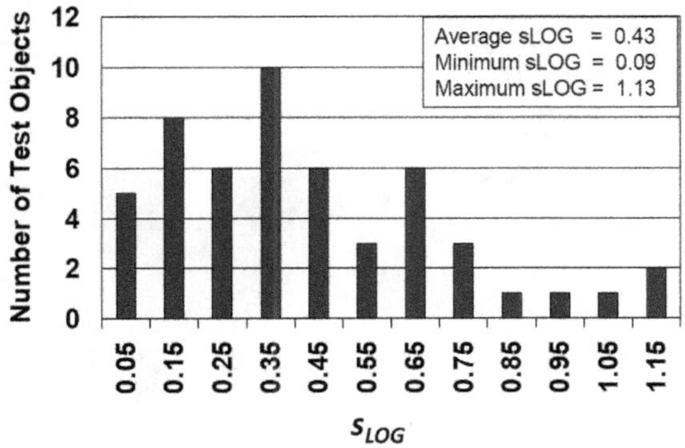

Figure 5.9 Distribution of *in situ* global variability s_{LOG} obtained on 52 structures in Switzerland.

of factors involved in the Global Variability (compared them with those reported in Table 5.3 for the laboratory cubes).

5.6.4 Reproducibility

On several occasions, the same samples were tested, in the lab and on site, using different instruments, of the same and/or different brands.

5.6.4.1 Reproducibility for Same Brand

Within the frame of a project, sponsored by the Swiss Federal Highways Administration (ASTRA) (Jacobs et al., 2009), several Swiss institutions participated in an interlaboratory test, conducted in two different construction sites, a bridge and a tunnel.

Two segments of the bridge were tested (D-E and XI), with the total test surfaces divided into test boxes, each box being assigned to each one of the five Swiss laboratories taking part in the experiment (see Figure 5.10): EMPA, TFB (organizer of the inter-laboratory test), EPFL, SUPSI and Holcim.

The test results are summarized in Figure 5.11, where the kT_{gm} and s_{LOG} of the 15 values obtained by each laboratory on each of the two bridge segments are plotted; the central dot represents the kT_{gm} value and the length of the segment represents $\pm s_{LOG}$ (see Section 5.8.3). It has to be borne in mind that each laboratory tested different samples (test areas) of a presumably same population (the bridge segment), so a certain difference between the results is to be expected. Figure 5.11 shows quite clearly that all laboratories rated the permeability of the bridge segments as "Moderate", with a lower variability for Segment XI than for Segment DE.

Figure 5.10 Dr. F. Jacobs and five instruments at comparative field test.

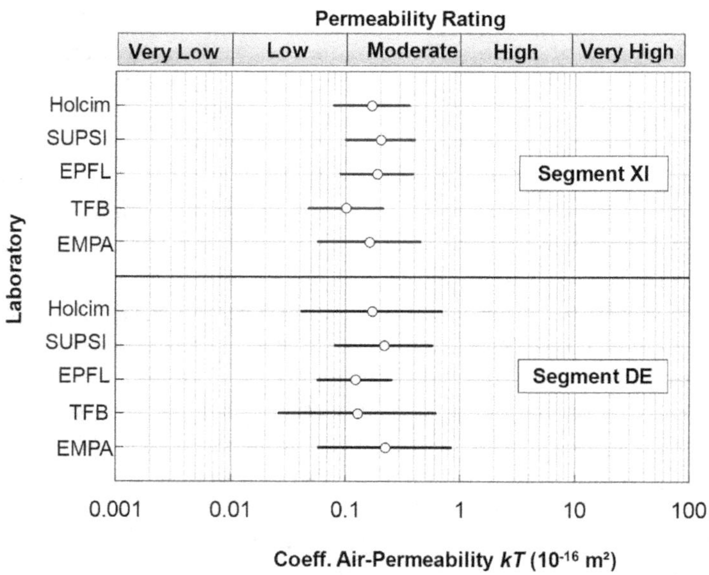

Figure 5.11 Values of $kT_{gm} \pm s_{LOG}$ obtained by five Swiss laboratories on two segments of same bridge.

Neves performed Analysis of Variance (ANOVA) tests for both *kT* and log *kT* and both failed to reject the hypothesis that one of the sets (for each bridge segment) is different from the other four. Furthermore, assuming no particular statistical distribution of *kT* results, the Kruskal-Wallis test yielded the same result.

Another reproducibility trial ground was the Application Test, organized by RILEM TC 230-PSC (2015). In it, eight concrete panels, made with different binders and *w/c* ratios and cured outdoors under different protection conditions, were blind tested by different laboratories, applying various test methods. Among the latter were Univ. of Zagreb (UZ) from Croatia, TFB from Switzerland and Materials Advanced Services (MAS) from Argentina, each of them using different units of the *PermeaTORR* instrument. The panels were measured in two occasions:

- first round, 15–16 April 2012, with the panels exposed outdoors under unsuitable test conditions (Age: 14–21 days, T=1°C–6°C, surface moisture m%=4.9%–5.7%), not complying with several prescriptions of Swiss Standard (SIA 262/1-E, 2019) (see Section 5.7). UZ and MAS participated
- second round, 9 July 2012, with the panels stored indoors (Age: 101–108 days; T=17°C–21°C, surface moisture m%=4.4%–5.5%). The test conditions complied with Swiss Standard (SIA 262/1-E, 2019). TFB and MAS participated

Figure 5.12 presents the results of UZ and TFB as function of those of MAS that participated in both rounds. The results of UZ were obtained on the same spots tested by MAS. In the case of TFB, some results were obtained

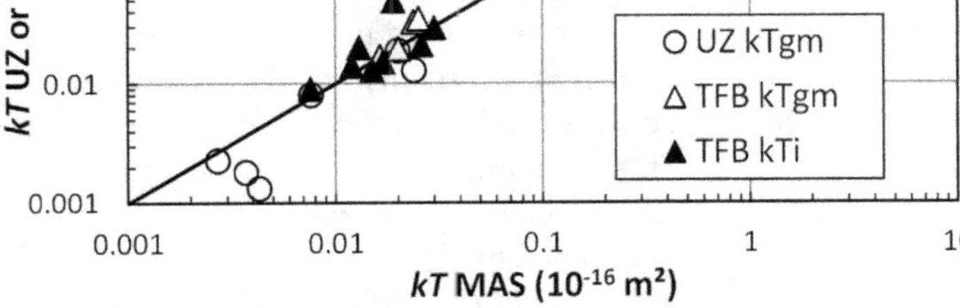

Figure 5.12 Reproducibility of *PermeaTORR* instruments on test panels (RILEM TC 230-PSC, 2015).

on the same spots (black triangles); the white triangles represent the geometric means of six tests performed by TFB and MAS on the same panels, but not on the same spots.

The reproducibility obtained in the second round (triangles) is remarkable; not so good, but acceptable, is that obtained in the first round (circles), under unfavourable conditions. It has to be mentioned that, due to the low ambient temperature, the instrument of UZ showed calibration problems; it was not possible to reach the requirements of SIA 262/1-E (2019), as reported in RILEM TC 230-PSC (2015). This explains the differences observed for concretes with very low permeability ($kT<0.01\times10^{-16}$ m^2), more affected by inaccurate calibration values, as a small difference in the calibration pressure has a strong impact on the small pressure rise recorded for low-permeability concretes.

5.6.4.2 Reproducibility for Different Brands

The *PermeaTORR* family of instruments introduced small changes in the test method that ended up in establishing slight but systematic differences in the measured *kT* value (Torrent, 2012). In particular, limiting the initial vacuum pressure to not less than 30 mbar (i.e. above water vapour pressure), avoided non-linear responses of the instrument in the plot ΔP_{eff} with $\sqrt{t}-\sqrt{t_0}$, that were attributed to water vaporization effects in the test chamber (Romer, 2005a), see Section 5.3.5.2.

Figure 5.13 presents results of measurements conducted with the *PermeaTORR* and the *TPT*, applied on the same samples by different researchers. Most data (Kurashige, Neves, Torrent) were obtained in the laboratory; other (Torrent+Jacobs Site) were obtained on the same spots of a Tunnel near Aarau, in Kanton Aargau, Switzerland (Jacobs et al., 2009); Jacobs with a *TPT*, Torrent with a *PermeaTORR*.

The results obtained by the *PermeaTORR* shall be considered as reference, as they are not affected by the pressure rise contribution of vaporized water in the vacuum chamber (Torrent, 2012). A difference between both instruments is appreciable only for low and very low permeability (Classes PK1 and PK2), attributable to the water vaporization effect discussed in Torrent (2012). Yet, a very good correlation exists between the results, through the conversion regression presented in Figure 5.13.

In 2016, a significant new concept was incorporated into the *PermeaTORR AC* (Active Cell) fourth-generation instrument (Torrent & Szychowski, 2016, 2017). Before and after launching the fourth-generation instrument, comparative tests with the third-generation *PermeaTORR* instrument were conducted in the laboratory (Torrent & Szychowski, 2017) and on site (Torrent et al., 2018). Two *PermeaTORR AC* (Serial Numbers #2 and #3) instruments and one *PermeaTORR* instrument (Serial Number #89) were used for the comparative test. Just three site tests were performed with both

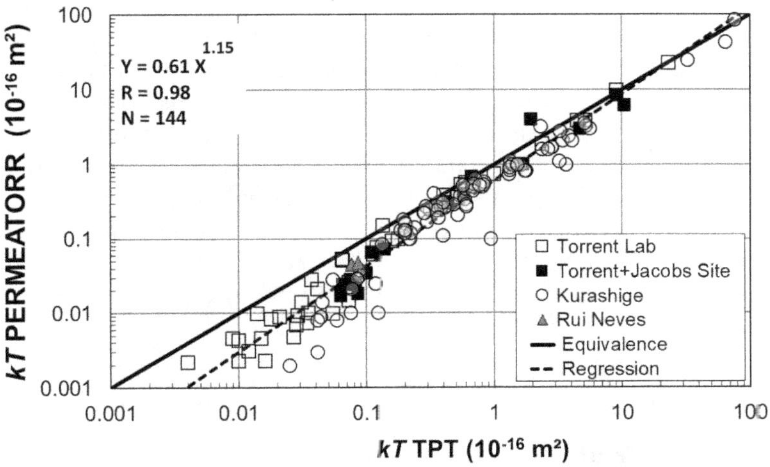

Figure 5.13 Correlation between *kT* results obtained with the *PermeaTORR* and *TPT* instruments.

instruments at a construction site in Como, Italy (Torrent et al., 2018). The results indicate that the Serial Numbers #2 and #3 of the fourth-generation *PermeaTORR AC* yielded slightly lower *kT* values than those of the third-generation *PermeaTORR* used for comparison. Both *PermeaTORR AC* units, tested at SUPSI Laboratory, yielded very similar results.

The conclusion is that both models: third-generation *PermeaTORR* and fourth-generation *PermeaTORR AC* can be used indistinctly, as both yield very similar results. In the case of the second-generation *TPT*, it yields higher results for concretes of low permeability (*kT* below $\approx 0.1 \times 10^{-16}$ m^2); in that case, a correction may be applicable, using the regression shown in Figure 5.13, where the permeabilities are expressed in 10^{-16} m^2.

5.7 EFFECTS AND INFLUENCES ON *kT*

For effects it is understood intrinsic properties or characteristics of concrete that have a direct impact on its permeability (e.g. *w/c* ratio, binder type, curing, cracks, etc.), which cannot be modified by the user of the test method. They are discussed in detail in Chapter 6, for air-permeability *kT* as well as for other permeability test methods.

For influences it is understood extrinsic factors, in general not related to the concrete quality, that have an impact on the air-permeability *kT* test results. Examples are the roughness, temperature and moisture of the tested surface. Contrary to the effects, the influences can be controlled to some extent by the user of the test method. The rest of this section is concerned with the impact of the influences on *kT*.

One special case is the age of concrete at the moment of test, which can be controlled, within certain limits, by the user of the test method. In this case, there is a dual impact; on the one hand, it is known that continued hydration will improve the tightness of concrete (Sections 3.1 and 3.2), so a concrete tested at a later age is expected to have less permeability than when tested earlier. On the other hand, particularly for site testing as is often the case for *kT*, the moment of test may determine the temperature and, especially, the moisture content of the concrete. These variables, especially moisture, have a strong effect on the measured value of *kT*, very specific to this particular test method. That is the reason why age is included within this section.

5.7.1 Influence of Temperature of Concrete Surface

5.7.1.1 Influence of Low Concrete Temperature

In principle, the temperature of the concrete surface should have a minor impact on the air-permeability *kT*, at least when above 0°C, because the viscosity of air in Eq. (5.13) does not change significantly with temperature (see Table 3.2). There is some influence on the instrument's response (possibly thermal dilation/contraction of components), reflected in systematic lower calibration values obtained at low temperatures (hence the need to repeat calibrations on site under changing temperature conditions – see Annex B). There was some debate within the Committee in charge of drafting ASTRA Recommendations for *in situ* testing of air-permeability (Jacobs et al., 2009) on the minimum acceptable temperature for running a test. Finally, a conservative approach prevailed, establishing a lower limit of 10°C, value that was included in the 2013 version of Swiss Standard (SIA 262/1-E, 2019). That limit was lowered to 5°C in the 2019 revision of the standard.

Some experiments were conducted by R. Torrent to check the possibility of lowering the minimum limit below 10°C (Torrent & Szychowski, 2016). In one of them, three concrete samples of different Permeability Classes (PK2, PK3 and PK4, see Table 5.2) were tested at the offices of Materials Advanced Services in Buenos Aires, Argentina, first indoors ($T \approx 22$°C) and afterwards outdoors in winter time ($T \approx 5.5$°C), using a third-generation instrument *PermeaTORR* (PT) and the first unit produced of the fourth-generation instrument *PermeaTORR AC* (PTAC#0). It was shown that both instruments measured very similar values at both temperatures on the three samples.

A similar experiment was conducted by R. Torrent in Switzerland with a *PermeaTORR AC* instrument (PTAC #2), on 17 (≈4-years old) concrete samples (150 mm cubes) covering a wider range of permeabilities. The samples were first measured (24 and 25 January 2017) in a laboratory room (SUPSI Univ., Lugano), where the temperature sensor of the instrument

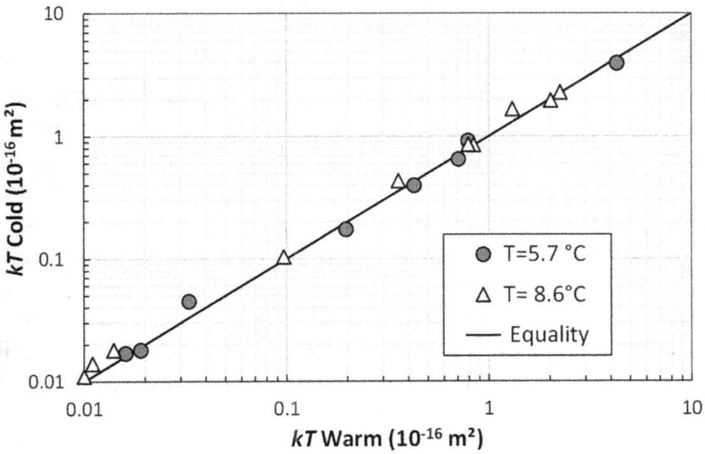

Figure 5.14 Effect of temperature on *kT* measured with *PermeaTORR AC.*

reported values between 20°C and 25°C, with an average of 23.4°C. The same samples were stored outdoors (protected with plastic sheets to avoid changes in moisture) at the offices of Materials Advanced Services in Coldrerio, Switzerland, where they were measured again in January 29 and 31, 2017, with reported temperatures of 5.7°C and 8.6°C, respectively. Figure 5.14 shows the results obtained on the samples under the different temperature conditions.

The results in Figure 5.14 support the proposed change in the Swiss Standard, allowing the measurement of *kT* for temperatures down to 5°C, instead of just 10°C.

5.7.1.2 *Influence of High Air Temperature and Solar Radiation*

During the interlaboratory test described in Section 5.6.4.2, unacceptably high calibration values (≈ 8 mbar) were observed when the *TPT* was calibrated at noon. This was attributed to the influence of solar radiation; exposed to the sunlight, the metal pressure regulator became very hot (≈43°C); on the contrary, the calibration values in the morning ranged between 2 and 4 mbar (Jacobs et al., 2009). Therefore, it is important to protect the instrument, whatever the brand and model, from direct exposure to solar radiation, by means of umbrellas or canopies.

5.7.2 Influence of Moisture of Concrete Surface

The strong influence the moisture content has on the gas-permeability of concrete is discussed in general terms in Section 6.9.

From the very beginning this influence was recognized as a formidable hindrance for the measurement of kT on site, due to the virtual impossibility of conditioning the test area prior to the measurement. The same problem was faced by other developers of site methods to test the air-permeability of concrete. For instance, Parrott (1994) considered a conversion factor between the air-permeability k_r measured with Hong-Parrott method (see Section 4.3.2.2) at a certain relative humidity r and at a reference relative humidity of 60% (k_{60}). Similarly, in the case of the Autoclam System (see Section 4.3.2.7), its Operating Manual states "...it is recommended that tests are carried out when the concrete is relatively dry (i.e. when the internal relative humidity of the cover concrete up to a depth of 10 mm is less than 80%)". The relative humidity of the cover concrete is typically measured with a probe inserted in a sealed cavity (10 mm deep) inside a drilled hole (Torrent, 2005).

The importance of this matter became obvious as soon as the newly developed "Torrent Method" was intended for site application. Since the test method is entirely non-destructive, the influence of surface moisture had to be dealt with also in an entirely non-destructive manner.

The first attempt was to use a surface moisture device called "H_2O meter", offered in the early 90s by James Instruments, that operated by pressing a sensor head onto the concrete surface. After trying it on specimens that had different moisture contents, the instrument was discarded.

The alternative was to combine the kT measurement with a complementary test, in order to compensate the influence of moisture. A suitable complementary property should show a higher "penetrability" for lower quality concrete (same as kT), and show a higher "penetrability" for higher moisture content (opposite to kT). A property that fulfils these conditions is electrical conductivity, with the advantage that it can be measured non-destructively by the Wenner method (in fact its reciprocal, the electrical resistivity ρ).

After some experimental work, reported in Torrent and Ebensperger (1993), a criterion was established by which a compensation of the measured value of kT (kT_m) was introduced via a parallel measurement of ρ. This compensation was of the form (Torrent & Frenzer, 1995):

$$kT' = \text{Max}\left[kT_m; \frac{3.5\ kT_m}{\rho}\right] \tag{5.34}$$

where

kT'=air-permeability compensated by moisture (10^{-16} m^2)
kT_m=measured air-permeability (10^{-16} m^2)
ρ=measured electrical resistivity by the Wenner method (kΩ.cm)

Equation (5.35) gives the limiting ρ values, as function of kT, below which the correction was necessary.

$$\rho = 3.5 \cdot kT_m^{-0.43} \quad (5.35)$$

It has to be mentioned that, at the time of developing the concept, virtually all concretes in Switzerland were made with OPC. With the advent of composite binders, containing PFA, SF or POZ, the correction ceased to be useful, since the inclusion of active additions can increase the ρ values by a factor of 5–10 (RILEM TC 154-EMC, 2000) and a specific compensation should be built for each binder type.

Other factors that were decisive in abandoning the combined kT–ρ approach were:

- the dependence of ρ with the temperature and the vicinity of steel bars (Jacobs et al., 2009, p. 641)
- the fact that, very often, the surface moisture of the site concrete is low enough to break the continuity of the electrolyte (the pore solution) making it impossible to get a reading of ρ (Jacobs et al., 2009 – Table C-2)

The combination of Torrent air-permeability and Wenner electrical resistivity has recently been revisited by Bonnet and Balayssac (2018), who concluded that "the results show that resistivity and Torrent permeability can be used for the combined assessment of carbonation depth and saturation degree in laboratory conditions". Based on data obtained at the Hong Kong-Zhuhai-Macao sealink, a similar approach has been proposed by Li et al. (2019). In addition, the advantages of combining different NDT techniques for assessing the service life of reinforced concrete structures has been advocated, based on intensive experimental work, by Sofi et al. (2019).

A research conducted at EMPA in Switzerland (Romer, 2005a), demonstrated the suitability of an impedance-based moisture meter ("*Concrete Encounter Moisture Meter*", from Tramex) to monitor the drying process of concrete samples, corresponding to four different mixes, stored in rooms at 35% and 70% RH, see Figure 5.15.

This research opened the way to an entirely and successful new approach, adopted by Swiss Standard (SIA 262/1-E, 2019), by which the air-permeability can be measured on site, only if the surface moisture of concrete, measured by an impedance-based instrument, does not exceed 5.5%. This limit derives from the work of a team of Swiss experts (Jacobs et al., 2009); according to Paulini (2014), this limit is too high.

The electrical-impedance method to assess the surface moisture of concrete has been standardized in USA for application on concrete floors (ASTM F2659, 2015).

The next sections investigate the matter in more detail.

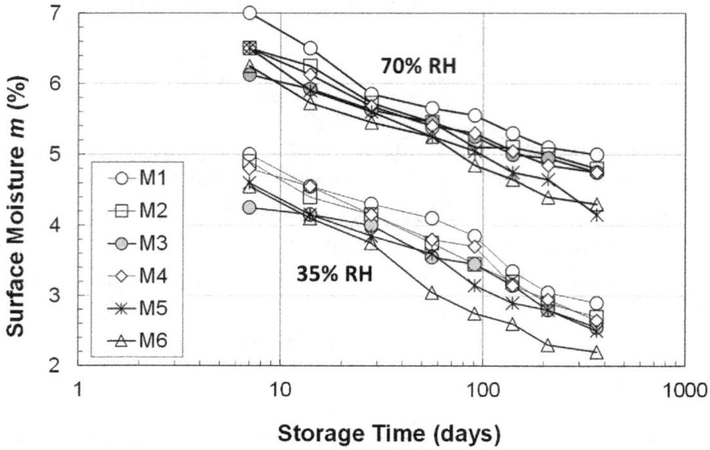

Figure 5.15 Monitoring the drying process of samples stored at 35% and 70% RH, with Tramex instrument, data from Romer (2005a).

5.7.2.1 Influence of Natural and Oven Drying on kT

A comprehensive investigation was conducted by Torrent et al. (2014, 2019) to explore in more detail the dependence of *kT* with the surface moisture *m*, measured by an electrical impedance-based instrument.

For this, 150 mm cubes were cast with two concrete mixes ($w/c = 0.40$ and 0.65) made with 9 widely different binders (see Table 5.4), totalling 17 concrete mixes.

The first letter of the code indicates the clinker used to produce the cements (H: Höver, Germany; M: Merone, Italy). The values in parentheses indicate the content and type of mineral additions originally included in the cement (MIC). When a mineral addition was added separately as

Table 5.4 Binders used in the concrete mixes

Code	*Brand (MIC)+SCM*	*EN 197 class*
H0	Holcim Pur-5N	CEM I 52.5N
H8M	92 % Holcim Pur-5N+*8% Silica Fume*	CEM II/A-D
H22S	Holcim Ferro 4 (22% GBFS)	CEM II/B-S 42.5R
H41S	Holcim Duo 4 (41% GBFS)	CEM III/A 42.5N
H68S	Holcim Aqua 4 (68% GBFS)	CEM III/B 42.5 L-LH/HS/NA
M0	I 52,5 R	CEM I 52.5 R
M26L	II/B-LL 32.5R (26% Limestone Filler)	CEM II/B-LL 32.5 R
M31FL	IV/A 32,5 R (27% PFA+4% LF)	CEM IV/A 32.5 R
M40FL	87% IV/A 32,5 R +*13% PFA*	CEM IV/B 32.5

supplementary cementitious material (SCM) into the concrete mix, the content and type are indicated in italics.

All 34 cubes (two per mix) were cured under water at 20°C for around 3 months, to ensure a high degree of hydration, so as to minimize further hydration during the drying period.

After the moist curing period, one cube of each mix was exposed to natural laboratory drying (stored in a room with still air at 18°C–23°C and 50%–65% RH). The companion cube was placed in a ventilated oven at 50°C (tests of the oven-dried cubes were performed after 24±2 hours cooling in the laboratory dry room). After the tests were completed (over 3 years for the lab-stored and over 4 months for the oven-dried specimens), the cubes were immediately returned to the dry room or oven, respectively.

At intervals, the following NDT and instruments were applied on the cubes:

a. Mass M (10 kg/0.1 g Kern scale)
b. Moisture by impedance method m (Tramex CMEX II), covering a range of 0.0%–6.9%
c. Wenner electrical resistivity ρ (RESI)
d. Air-permeability kT (*PermeaTORR*). The instrument measures above $kT=0.001\times10^{-16}$ m²

Tests b and d (dry tests) were applied on two opposite lateral faces of each cube. Two readings of m were made on each face, one in a horizontal position (m_h) and the other in a vertical position (m_v). The reported value m is the average of the four readings on each cube. One reading of kT was performed on two opposite faces of each cube, the reported value being the geometric mean of the resulting two values (assuming that kT follows a log-normal distribution, see Section 5.8). The results of electrical resistivity ρ are not discussed here because, after relatively short drying periods, it was impossible to get valid readings (Torrent et al., 2014).

When the 50°C oven-dried specimens had completed at least 100 days of drying, they were weighed and dried at 105°C till constant weight, recording the mass M_{105}, and measuring kT and m. Then, the samples were submerged under water at 20°C until constant weight, recording the saturated mass M_s. This allowed converting the values of mass M of all the cubes into degrees of saturation S_d (-) using Eq. (5.36).

$$S_d = (M - M_{105})/(M_s - M_{105}) \tag{5.36}$$

The change in m with the change in S_d is presented in Figure 5.16 for concretes made with Höver clinker. All specimens' results are plotted in the charts, including those under laboratory and oven drying, the latter identified by the framed symbols. It is worth mentioning that all the cubes showed that

- after drying to constant mass at 105°C (S_d=0.0) → m = 0.0% (bottom of instrument's scale)
- after saturation to constant mass (S_d=1.0) → m = 6.9% (top of instrument's scale)

This situation is indicated by the solid black line (indicative only) joining the extreme black squares in Figure 5.16, which presents the results obtained on the concretes based on Höver clinker, very similar to those obtained with Merone clinker.

Figure 5.16 shows the monotonic relation existing between S_d and m along the whole drying process, with the oven-dried specimens showing a reasonable continuity with those dried in the lab room.

It was observed that the binder type exerts a significant influence on the m vs S_d relationship. It can be seen from Figure 5.16 that, for the same m, concretes made with H0 show lower values of S_d than those containing hydraulic MIC and/or SCM.

In Figure 5.16, a segment with m=5.5% (maximum value admitted by Swiss Standard (SIA 262/1-E, 2019) for performing air-permeability test kT on site) has been drawn, indicating corresponding S_d values within 0.75–0.90 for Höver samples; the range for Merone clinker concretes was 0.60–0.90.

It must be borne in mind that the saturation degree S_d corresponds to the bulk of the cube, whilst the surface moisture m corresponds to a layer about 15–20 mm thick. The degree of saturation of that layer must certainly be lower than the bulk value, due to the moisture gradient established during drying.

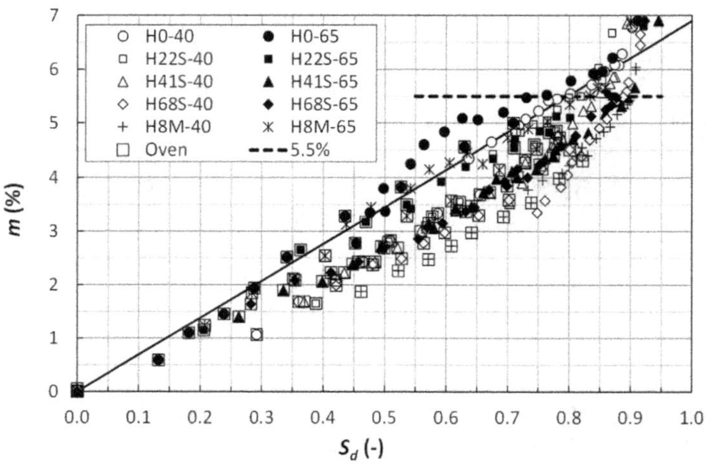

Figure 5.16 Relation between saturation degree S_d and surface moisture m (Höver clinker).

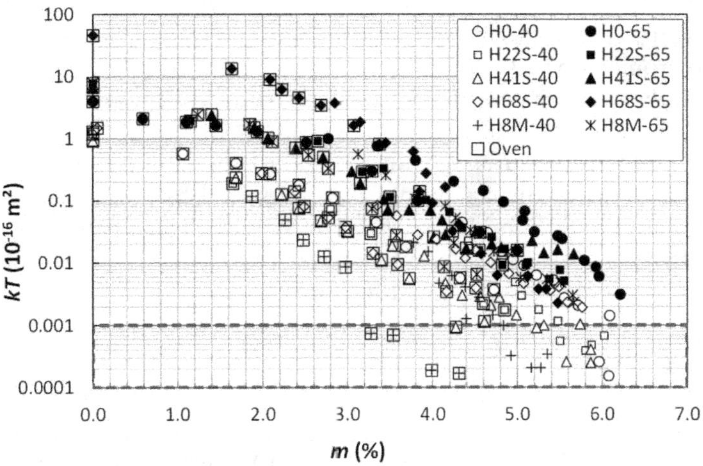

Figure 5.17 Relation *m* vs. *kT* (Höver clinker).

Figure 5.17 presents the relation found between *kT* and *m*, along the whole drying process, with the first reading taken after 1 day of drying, for concretes made with Höver clinker (a similar picture for Merone clinker can be found as Figure 6 of Bueno et al. (2021)). The values inside the broken-line box are not sufficiently accurate as they lie below the bottom limit of measurable *kT* (0.001×10^{-15} m^2).

Figure 5.17 shows a monotonic increase in *kT* as concrete gets drier, with the highest *kT* value corresponding to the cubes subjected to the drastic final drying at 105°C (m=0%). At first sight, the relation ln(*kT*) vs. *m* presented in Figure 5.17 shows some linearity and parallelism for the different mixes, particularly for values of *m* within the range 1.0%–6.0%. This opens the way for attempting a compensation of *kT* when the *m* values are too low (e.g. indoors elements), as discussed in the next section.

A similar picture was obtained by Kato (2013) who dried specimens of different configurations, intensively hydrated (28 days in water at 60°C), for 91 days at 20°C/60% RH. For w/c=0.50 and 0.70 he found an excellent linear correlation between log *kT* and the water loss of the specimens, with virtually the same slope.

5.7.2.2 Compensation of kT for Surface Moisture

An approach to compensate the *kT* values for the moisture content of concrete was developed by Misák et al. (2010). They prepared concrete slabs (300×300×100 mm) of an EN 206 C20/25 class concrete that were moist cured for 28+2 days and then stored in a dry room (23°C/48% RH), monitoring the influence of drying on *kT*. At a certain point, the samples were oven-dried at 50°C and finally at 105°C. The surface moisture content *w* was measured with a KAKASO capacitive humidity meter that was calibrated against bulk gravimetric moisture content. The KAKASO moisture meter hygrometer uses the property that water in the capillary pore environment greatly affects its permittivity; the instrument explores the 15–30 mm surface layer (Misák, 2018).

Results of 153 measurements of kT against the humidity obtained from the capacitive instrument readings were reported by Misák et al. (2010), converted into bulk humidity w through the established calibration curve, which differs slightly for different materials. Based on these results, the following relation between kT and w was proposed:

$$kT = kT_0 \cdot e^{-\alpha \cdot w} \tag{5.37}$$

For the particular mix investigated and their own moisture determination, the regression analysis yielded $kT_0=5.25\times10^{-16}$ m^2 and $\alpha=0.8623$; where kT_0 is the kT value for $w=0$ and α the slope of the line $\ln(kT) - w$.

There is a strong similarity between the data from Misák et al. (2010) and those presented in Figure 5.17. Therefore, a similar approach was pursued, but taking m (impedance-based) as the direct indicator of the surface moisture content of the concrete.

An initial analysis was made on the basis of the data obtained by Torrent et al. (2014, 2019), part of which are shown in Figure 5.17, that showed quasi-linear relations between $\log(kT)$ and m, to which regressions of the form indicated by Eq. (5.38) were fitted:

$$kT = kT_0 \cdot e^{-\delta \cdot m} \quad m \text{ expressed in } \% \tag{5.38}$$

Only the data with $1.0\% \le m \le 6.0\%$ and with $kT \ge 0.001\times10^{-16}$ m^2 were considered for the regression analysis. The mean correlation coefficient was $R=0.93$, with extreme values of 0.65 (Mix H8M-40) and 0.99 (Mix M26L-40).

The regression analyses showed that, despite the wide range of binders (Table 5.4) and w/c ratios used to prepare the concrete mixes, exponent δ remains within a limited range of 1.00–1.65 for 16 out of the 17 mixes investigated.

Relying on a single source of data for the analysis was deemed insufficient, reason why a more comprehensive cooperative investigation on the subject was organized, based on available data from other sources (Romer, 2005a, b; Nsama et al., 2018; Sandra et al., 2019; Nguyen et al., 2019, 2020). The main results of it were presented in Bueno et al. (2021) and are summarized in Figure 5.18, which shows the histogram of δ values obtained by regression of Eq. (5.38) on 50 series of test data, coming from the abovementioned sources, including the 17 obtained from Torrent et al. (2014, 2019) data.

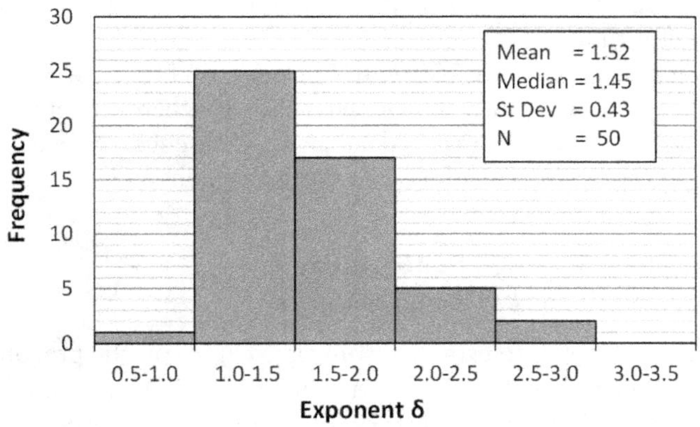

Figure 5.18 Histogram of exponent δ values obtained on 50 series of test data.

Figure 5.18 shows that the statistical distribution is positively skewed with a central value that can be assumed equal to the median $\delta=1.45$; moreover, 84% of the δ values fall within the range of 1.0–2.0. The average value of the correlation coefficient R for the 50 reported cases is $R=0.95$.

Based on the results presented in Figure 5.18, a single compensation formula is proposed in Eq. (5.39), using the median δ value of 1.45, where kT_m is the air-permeability value measured at surface moisture m:

$$kT_m = kT_0 \cdot e^{-1.45m} \tag{5.39}$$

The value kT_0 corresponds to the extrapolation of Eq. (5.39), valid for $1.0\% \le m \le 6.0\%$, to $m=0\%$, giving an unrepresentative high reference value. Therefore, it is proposed to take as reference the value kT_5, corresponding to a moisture $m=5.0\%$. From Eq. (5.39), we can write:

$$kT_5 = kT_0 \cdot e^{-1.45 \cdot 5.0} \tag{5.40}$$

Dividing Eq. (5.40) by Eq. (5.39) and introducing a compensation factor F_5 to the value kT_m measured at moisture m:

$$kT_5 = F_5 \cdot kT_m \tag{5.41}$$

with

$$F_5 = e^{1.45(m-5.0)} \quad \text{valid for } 1.0\% \le m \le 6.0\% \tag{5.42}$$

For m values between 4.5% and 5.5%, the compensation is not truly necessary, because the corresponding kT values differ from kT_5 by a factor of

2 and 0.5, respectively, which is acceptable given that kT varies over six orders of magnitude.

An interesting real case is presented in Section 11.4.4, where measurements conducted in the Atacama Desert in Chile, where extremely dry concrete was tested ($m<2\%$), resulted in large differences between kT_m and kT_5, showing the usefulness of the moisture compensation.

5.7.2.3 Pre-conditioning of Laboratory Specimens for kT Measurements

It is of practical interest to establish some guidelines on the preconditioning of laboratory specimens for kT measurements. In Annex B, guidelines are given for testing kT in the laboratory, case not covered by Swiss Standard (SIA 262/1-E, 2019) which refers exclusively to site testing. Regarding preconditioning of the samples, the guidelines specify the following procedure:

> At the age of 28 days, the specimens shall be dried in the ventilated oven at a temperature of 50±2°C, leaving a free distance of at least 20mm between the specimens and with the walls of the oven. The drying will continue until the moisture meter indicates a surface moisture of the specimens within the range 4.0%–5.5% (normally this is achieved within 4±2 days of drying of specimens at or near saturation).

This criterion has been applied to the data recorded in the investigation described in Section 5.7.2.1, with the result shown in Figure 5.19 (Torrent et al., 2019). Figure 5.19 shows the kT values obtained for the 17 mixes investigated, in abscissas for laboratory drying until $m \leq 5.5\%$ (limiting value prescribed by SIA 262/1-E (2019)) and in ordinates after 3 days in the oven at 50°C to reach a surface moisture $4.0\% \leq m \leq 5.5\%$; if not achieved,

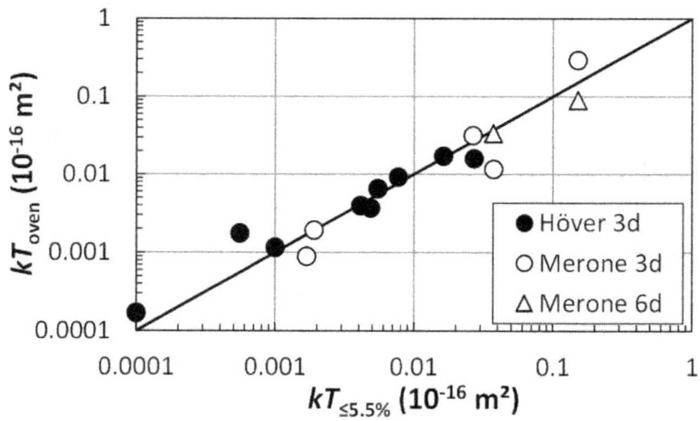

Figure 5.19 Relation between 50°C oven-dried (3 or 6 d.) and lab-dried (till $m \leq 5.5\%$) kT values.

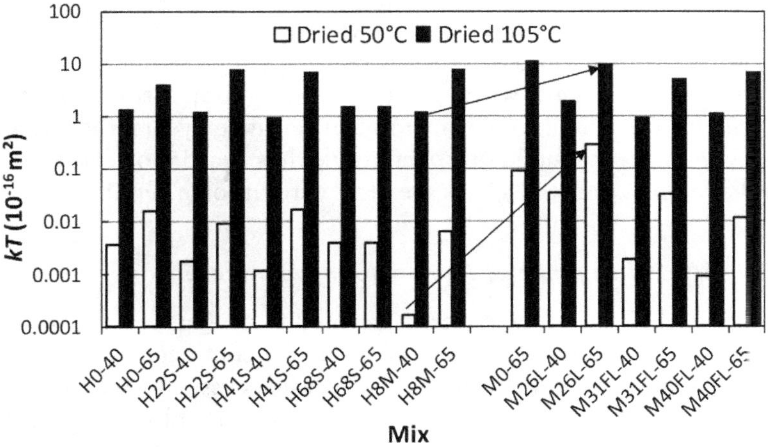

Figure 5.20 Effect of oven-drying conditions on the measured *kT* values.

the value obtained after 6 days in the oven was taken. It has to be mentioned that, in the case of Mix H0–40, after 3 days of oven-drying, the moisture dropped already to 3.8%. Only for mixes M0–65 and M26L–40, was it necessary to dry the specimens 6 days to reach $m \leq 5.5\%$, indicated by triangles in Figure 5.19.

Figure 5.19 shows that the criterion works well in that the *kT* values obtained according to the oven-drying procedure prescribed in Annex B leads to very similar values to those obtained under laboratory drying until reaching $m \leq 5.5\%$.

Figure 5.20 shows the *kT* values for each mix, measured under different oven-drying conditions: the white bars correspond to oven-drying 3–6 days at 50°C, whilst the dark bars to oven-drying at 105°C to constant mass.

It can be seen that, for 50°C oven drying, the range between the *kT* value for H8M-40 mix (lowest) and for M26L-65 mix (highest) covers three orders of magnitude, whilst for 105°C oven-drying the range is reduced to just one order of magnitude (see arrows in Figure 5.20). This confirms the well accepted practice of drying specimens for gas-permeability tests not above 50°C (see Section 4.3.1.1).

5.7.3 Effect/Influence of Age on *kT*

Originally, the Torrent test method to measure *kT* was designed for testing concretes at a mature stage, but not too old. Indeed, Swiss Standard (SIA 262/1, 2019) prescribes that the quality of the cover concrete has to be measured at ages between 28 and 120 days.

Nevertheless, there is some interest in measuring *kT* at earlier ages, say 7 days, so as to know well in advance whether the quality achieved on site matches the specifications or expectations of the stakeholders.

On the other extreme, when dealing with condition assessment of existing structures, exposed to the environment for many years, there is a need to assess *kT* data measured on concrete which is decades old.

In the following, both aspects are dealt with in some detail.

5.7.3.1 Effect/Influence of Age on Young Concrete

When testing *kT* on young concrete exposed to drying, two superimposed phenomena happen: a reduction in *kT* due to the continued hydration of the binder and an increase of *kT* due to drying as discussed in Section 5.7.2.

Figure 5.21 presents data from EPFL in Switzerland (Brühwiler et al., 2005) obtained on 800×800×200 mm reinforced concrete panels made with OPC mixes of w/c=0.43 and w/c=0.52. Face A of the panels was cast against the form, whilst face B was cast against a permeable formwork liner ("Zemdrain"), described in Section 7.2.3. The lateral faces (800×200 mm) of the panels were sealed with epoxy resin. The panels were moist cured for 7 days and thereafter kept in a dry room (20°C/60% RH). The coefficient of air-permeability *kT* of the four faces was measured at intervals between 4 and 389 days of exposure to the dry environment, with the results shown in Figure 5.21.

A consistent trend can be observed in Figure 5.21 for the measurements on the four surfaces in that, up to 11–21 days of exposure, *kT* decreases significantly, moment at which the trend is reversed and a continuous increase in *kT* is observed until the end of the experiment (389 days). At an exposure time around 100–200 days, the value of *kT* returns to the one measured after 4 days of drying.

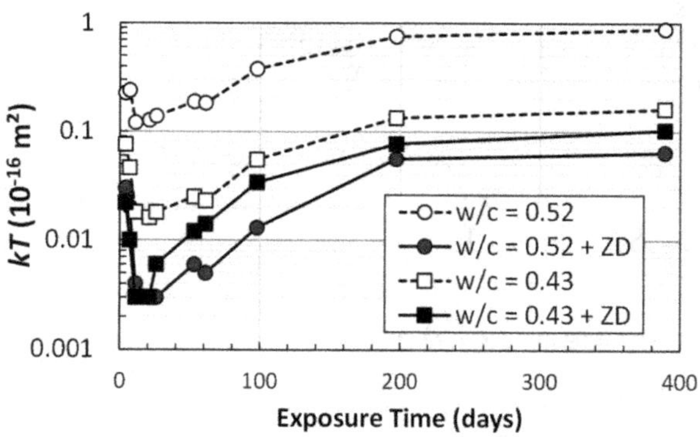

Figure 5.21 Evolution of air-permeability *kT* with exposure time, data from E. Denarié (Brühwiler et al., 2005).

The initial reduction in *kT* is attributed to the changes in microstructure of the young concrete, due to continued hydration. After 11–21 days exposure, the rate of hydration diminishes and the influence of drying prevails, by which more and more pores are freed, facilitating the flow of air, resulting in a gradual increase in *kT*. Interesting to see is that, based on either the 4 or 389 days *kT* data, the best quality surfaces are those treated with "Zemdrain" (black symbols +ZD), followed by the one with $w/c = 0.43$, the worst being the one with $w/c=0.52$ (both without "Zemdrain").

In the Application Test, organized by RILEM TC 230-PSC (2015), described in Section 5.6.4.2, tests were performed on eight concrete panels, made with different binders and *w/c* ratios. The panels were measured for *kT* at "young" 14–21 days of age (maturity age around 10 days) and, again, at "mature" 101–108 days of age. The ambient conditions were quite different during both tests, as described in Section 5.6.4.2. Figure 11.8 of RILEM TC 230-PSC (2015) presents the geometric means of *kT* results obtained with the same instrument and operator on the eight panels at both ages. It shows that the original *kT* values measured on the panels at "young" age are about one order of magnitude smaller than those obtained at a "mature" age. Interesting to note, though, is that there is a certain correlation between both sets of values. If this correlation could be established in advance, it would be possible to predict the *kT* values at "mature" ages from *kT* values measured on "young" concrete.

5.7.3.2 *Effect/Influence of Age on Mature Concrete*

By repeating field tests on old bridges located outdoors, unfortunately not always on the same spots, Adey et al. (1998) concluded: "Permeability measurements are repeatable, on existing concrete bridges, when taken in exactly the same location and there has been no change in internal water content" and stated: "The presence of water in the concrete may cause differences in the permeability that cannot be adjusted using the resistivity measurements". The latter statement already objected to the use of electrical resistivity to compensate for moisture content in the concrete, applied at the time of the paper, later substituted by the use of impedance-based moisture meters (see Section 5.7.2).

A similar experiment was conducted by Imamoto et al. (2012), reported in slide 14 of the paper's presentation, reproduced as Figure 5.22. They repeated *kT* measurements on two locations (black and white bars) of an external beam of 50 years old Tokyo's National Museum of Western Art (investigation described in detail in Section 11.4.1.1), at intervals during a period of about 3 months. Figure 5.22 indicates a very good stability of the results on both locations, even when measuring 1–2 days after a heavy rain.

In order to simulate the influence of the environment on the long-term values of *kT* of 0.9×0.9 m cross section columns, Kato (2013) prepared 450×450×400 mm specimens, sealing them as shown in Figure 5.23 (l.).

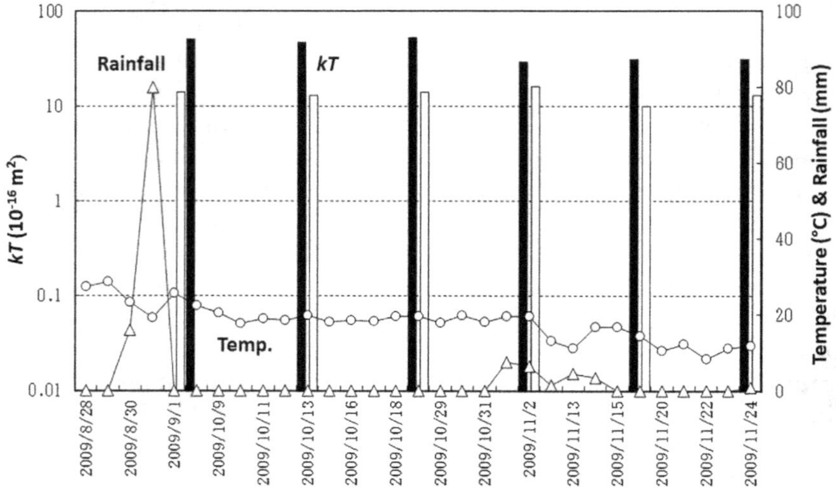

Figure 5.22 Stability of *kT* readings at two locations on an external beam of Tokyo's NMWA (Imamoto et al., 2012).

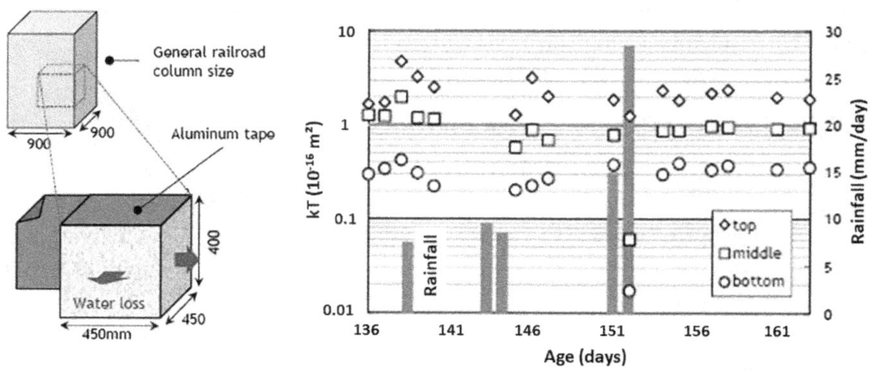

Figure 5.23 Simulation of 0.9×0.9 m column (l) and effect of exposure on *kT* (r) (Kato, 2013).

After an intensive curing (28 days under water at 60°C) and 5 days sealing, the sealed specimens were left drying for 91 days at 20°C/60% RH, before being exposed unprotected to the natural environment. The evolution of *kT* of the sealed specimens placed outdoors is presented in Figure 5.23 (r.) (dots). In some cases of rain (vertical bars), the water was wiped off the surfaces and *kT* measured on the moist surface. Figure 5.23 (r.) shows that the *kT* values remained reasonably stable, except at 152 days of age, when measured in correspondence with a heavy rainfall, where the values in the center and bottom part of the specimens dropped by more than one order of magnitude. Yet, as remarked by Kato (2013), "just two days later kT had almost returned to the state before the rain".

In a research by Yokoyama et al. (2017), full-scale concrete elements were cast with four different mixes, made with OPC and GBFS cements, subjected to different curing conditions and exposed to the environment during 6–7 years, monitoring the resulting changes in *kT* with time. After a significant increase in *kT* between 3 and 6 months, the values stabilize until around 45 months, after which some further increase was recorded. The main conclusion of the research by Yokoyama et al. (2017) is

> "Results showed that the difference in the air-permeability owing to curing conditions decreased with age. About one year later, the coefficient of air-permeability of specimens exposed to rainfall was approximately the same regardless of the curing method. The effect of curing was present even after one year for specimens under the roof-not exposed to rain, but the difference in the coefficient of air-permeability continued to decrease. On the other hand, the difference in surface air-permeability between concretes with different water to cement ratios remained even at the age of five years".

This section can be concluded with a practical recommendation of not measuring *kT* during a rainfall and to wait at least 2 days after the rain to carry out site measurements of exposed structures. In case of monitoring changes in *kT* with age, it is essential to mark the exact location of the test, to make repetitions precisely on the same spot.

5.7.4 Influence of Vicinity of Steel Bars

The presence of steel bars may have two effects on the measured *kT* values. On the one hand, the presence of the rigid steel bars may produce some aggregates particles segregation, may affect the compaction of the cover concrete and also create a sort of ITZ around them, that may affect the true "penetrability" of the *Covercrete*. On the other hand, the presence of the bars may obstruct the flow of air towards the central chamber and, thus, affect the measured *kT* value.

Within the comprehensive research sponsored by the Swiss Federal Highway Administration (Torrent & Ebensperger, 1993), a special investigation on the influence of the vicinity of steel bars was conducted.

For that, two sets of four concrete slabs (120×250×360 mm) were cast with a mix of the following characteristics:

- OPC content: 325 kg/m^3
- w/c=0.46
- 32 mm maximum size of aggregate
- 75 mm slump

- 4.8% entrained air
- Standard cube strength at 28 days=43.0 MPa

The slabs were cast in two layers, each one of them compacted on a vibrating table and carefully finished with a thick metal ruler. From each set of four slabs, one was unreinforced (used as Control) and the other three contained Ø18 mm bars, according to the pattern shown in Figure 5.24 (l.). The steel bars of the three reinforced slabs had a variable cover thickness of 11, 26 and 41 mm, respectively.

In one set (AO) the bars were placed close to the upper (finished) surface as cast; the four slabs of this test did not receive any moist curing. In the other set (BU), the bars were placed close to the bottom (moulded) surface as cast; the four slabs of this set had an initial curing of 7 days in the moist room (20°C/RH>95%). The *kT* tests were performed at the age of 90 days in a dry room (20°C/50% RH) where all slabs were stored after the moist curing (0 and 7 days).

Six *kT* readings were performed on each slab; in the case of the reinforced slabs, the central chamber of the vacuum cell was placed above each of the 3 bars (see Figure 5.24 (l.)). Afterwards, one Ø150 mm core, per slab, was drilled (across the steel bars, see Figure 5.24 (l.)) and cut to 50 mm thickness, to measure the oxygen-permeability *kO* (Cembureau method, described in Section 4.3.1.2). The Ø150×50 mm discs were oven-dried at 50°C for 6 days and cooled down 1 day in a desiccator before performing the *kO* test.

Figure 5.24 (r.) shows the results of the measurements for both slab sets (AO and BU), with *kT* being the median value of the six test results of air-permeability on each slab (corrected by factor 1.846 as described at the end of Section 5.3.4) and *kO* the single value of oxygen-permeability. The plain "Control" slab is represented by a "cover thickness" = 120 mm (thickness of the slab); data from Table 4-II of Torrent and Ebensperger (1993).

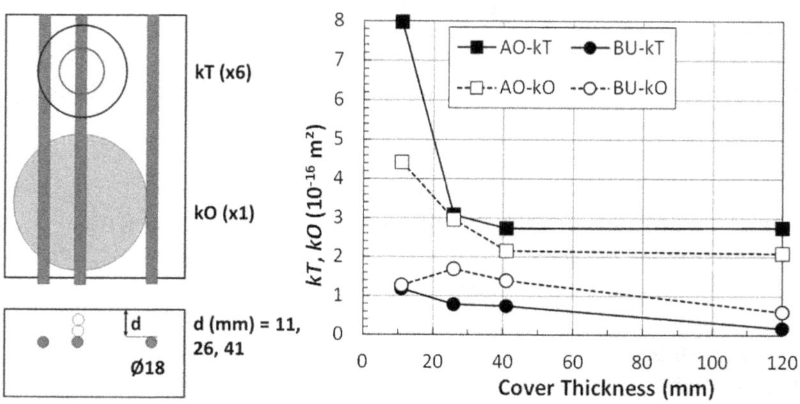

Figure 5.24 Sketch of "reinforced" slabs (l) and effect of steel bars on *kT* and *kO* (r).

Figure 5.24 (r.) indicates that the presence of the steel bars increases the permeability of the *Covercrete*, compared with that of plain concrete. The oxygen-permeability *kO* is raised by a factor of 2–3, whilst the air-permeability *kT* by a factor of 3–5. It seems that *kT* experiences a "jump" for the shallowest cover of 11 mm, which may indicate some effect of the vicinity of the bar on the quality of the *Covercrete*; the expected influence on gas flow into the central chamber would have been a decrease of *kT* (obstructing air-flow), that was not observed. The influence in any case should depend on the penetration of the test *L* (in turn, function of *kT*, see Figure 5.7). This influence could be investigated by modelling.

In a comprehensive research, Eddy et al. (2018) investigated the effect of shallow covers on the *Covercrete* quality. For that purpose, reinforced concrete specimens were cast with Ø19 mm steel bars embedded near the bottom-as-cast surface, with shallow cover thicknesses of 5, 15 and 30 mm, insufficient according to Japanese standards for durability (minimum cover=40 mm). Two concrete qualities were investigated, namely $w/c=0.45$; OPC=378 kg/m^3; 91 days $f'c_{cyl}=49.5$ MPa and $w/c=0.60$; OPC=283 kg/m^3; 91 days $f'c_{cyl}=38.6$ MPa. The specimens were sealed-cured for 7 days and thereafter kept indoors or exposed unprotected to Tokyo's summer-autumn environment until 91 days of age. The pore structure of small mortar samples taken at different depths, including the cover in contact with the steel was investigated by MIP. Also, measurements of air-permeability *kT*, surface moisture content (impedance-based method) and Wenner electrical resistivity were conducted. The results of air-permeability tests indicated that the *kT* measured on the 5 mm *Covercrete* was significantly higher than that on the 30 mm *Covercrete* or on plain concrete samples, whilst for 15 mm *Covercrete* the difference in *kT* was smaller. Since the MIP results also showed a more open pore structure for shallow covers it was concluded that the differences in *kT* are due to differences in quality of the *Covercrete* rather than to an interference of the bar with the air flow.

Based on the results of Figure 5.24, Swiss Standard (SIA 262/1-E, 2019) prescribes a minimum cover thickness of 20 mm for the measurement of *kT* in coincidence with steel bars.

5.7.5 Influence of the Conditions of the Surface Tested

5.7.5.1 *Influence of Specimen Geometry and Surface*

The measurement of *kT* in the laboratory poses the question about suitable shape and size of specimens and surfaces to be investigated.

Successful tests made on 150 and 200 mm cubes, Ø150 mm cylinders and disks and slabs (e.g. 250×360×120 mm) have been reported. Since the external diameter of the cell is 110 mm, this is a practical limit to the size of the surface to be tested. A reasonable limit to the minimum size of a surface to be tested is a diameter or side of at least 150 mm, whilst the depth should not be less than 50 mm, capable of accommodating the vacuum front penetration up to *kT* values of 2×10^{-16} m² (see Figure 5.7) without the need of correcting the measured value for depth.

An investigation dealing with the influence of the surface and geometry of the specimens on the measured *kT* values was conducted by Neves et al. (2015). With that aim, a series of 12 concrete mixes, made with different binders and *w/c* ranging between 0.43 and 0.61, were prepared, with which Ø150×300 mm cylinders, 200 mm cubes and 250×360×120 mm slabs were cast. From some cylinders, five slices 50 mm thick were obtained by saw-cutting. All specimens were moist cured for 7 days to be later kept in a dry room (20°C/60% RH) until the moment of test at 28 days of age. Not all specimens were prepared from each mix. A non-parametric statistical analysis of all the test results allowed (Neves et al., 2015) to conclude:

> "The investigation on the influence of specimens' geometry and type of surface on concrete air-permeability leads to conclude that there are no differences, with statistical significance, either when air-permeability is assessed in any of the three geometric shapes of the specimens used in this work, or when it is assessed in moulded or sawn surfaces. As no significant differences between tested geometries and surfaces of specimens were found, the selection of the specimen geometry and testing surface may vary according to the specific requirements of each use."

In general, it is a good practice to measure *kT* on moulded surfaces of specimens, as they are less affected by preparation, unless the effect of finishing techniques is to be assessed. Also, if unduly extensive vibration is applied to the specimens (even if shallow), segregation and excessive bleeding can occur, affecting the trowelled surface (see Section 6.6). Moreover, the curing of the trowelled surface may be different to that of surfaces protected by the form until stripping. When testing saw-cut surfaces, the flow of air from the exposed aggregates goes straight into the vacuum cell, whilst in reality, it always flows through a layer of h.c.p.; this is even more important when dealing with highly porous aggregates.

To conclude, the best specimens for conducting *kT* tests in the laboratory are cubes, at least 150 mm side length; for countries following ASTM standards, based on cylinder strength, the moulds for 150×150 mm cross-section beams, specified for flexural tests, can be used with metal separators to produce 150 mm cube-like specimens. The cube offers four lateral faces that can be considered identical for measuring *kT*.

5.7.5.2 Influence of Curvature

The vacuum cell of the different instruments measuring *kT* according to the Torrent method has been designed for application on surfaces that are relatively flat.

Practice indicates that, due to the soft rubber rings with which the cell is equipped, a certain curvature of the surface can be accommodated. For instance, Figures 11.12 and 11.48 show direct applications on curved surfaces, whilst Figure 5.25 shows two solutions developed by K. Imamoto to measure *kT* on cylindrical pillars. Figure 5.25 (l.) shows a flexible adaptor applied in Tokyo's Museum of Western Art (Imamoto, 2012; Imamoto et al., 2012) and Figure 5.25 (r.) a rigid one applied in a building in Locarno (Switzerland).

It is important to check that the accessory used, such as the ones in Figure 5.25, does not change the constant of the instrument V_c/A (see Eq. 5.13); otherwise, a correction of the measured value is necessary.

5.7.5.3 Influence of Roughness

In an investigation by Brühwiler et al. (2005), conducted on site on a road overpassing the railway, the concrete pavement to be tested showed a surface with irregularities, cracks and cement laitance due to bleeding. It was decided to conduct comparative measurements on the pavement natural surface and also after the application of sand-blasting to expose the aggregates. Three *kT* measurements were conducted on each surface – original and sand-blasted – with geometric means of 0.045 and 0.055×10^{-16} m², respectively, showing an insignificant effect of the sand-blasting on *kT*.

Regarding roughness, Figure 5.26 (l.) shows the application of the *PermeaTORR* on the rough surface of a concrete pavement, during an investigation conducted by TFB in Switzerland.

Sometimes, the irregularities are such (e.g. shotcrete or grooved pavements) that the surface must be treated before application of the test method. A real case happened in the city of Buenos Aires, Argentina, where the air-permeability of trench walls was measured on the excavated hidden side, due to uncertainties on the quality of the job done by the contractor. The surface was extremely rough, with a thick layer heavily contaminated with

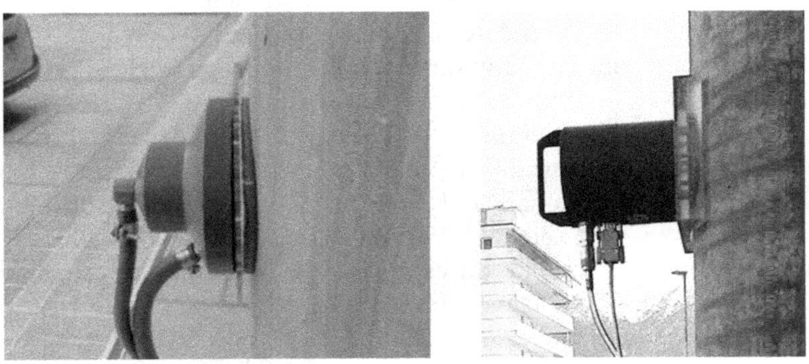

Figure 5.25 Flexible (l.) and rigid (r.) adaptors for curved surfaces.

Figure 5.26 *kT* test on concrete pavement surface (l., courtesy F. Jacobs, TFB) and on polished trench wall (r.)

bentonite, requiring the use of a polishing machine (Bosch GWS 6–115 Professional) to prepare the test areas.

Figure 5.26 (r.) shows a test being performed on a polished area. Despite the awful aspect of the raw surface, the tests yielded good results in terms of Permeability Classes, as defined in Table 5.2: 15% in PK0 (Negligible); 42% in PK1 (Very Low); 28% in PK2 (Low) and 15% in PK3 (Moderate). So, the aspect of the concrete *per se* is not a good indicator of its quality.

5.7.5.4 Effect/Influence of Surface Air-Bubbles

Large air bubbles (bug-holes) tend to appear on concrete surfaces cast against metal forms, especially when inclined, as they find it difficult to escape even under heavy vibration.

The appearance of bug-holes often attracts the attention of the owners, concerned on how they can affect the durability of the elements. To start with, in reinforced concrete structures, bug-holes may reduce the cover thickness, thus promoting an early localized corrosion.

An interesting real case is described in Section 11.3.4, in which comparative *kT* tests performed exactly on bug-holes (see Figure 11.23) and away from them showed no significant difference.

A dedicated investigation, in the laboratory and in the field, on the effect of bug-holes on the permeability of tunnel liners was reported by Hirano et al. (2014) and Maeda et al. (2014). In the laboratory, 300×300 mm prismatic specimens with a height of 750 mm were cast using a metal form that could be rotated to simulate different inclinations. They investigated the effect of the demoulding agent, of the vibration and of surface treatments (ceramic coating or permeable formwork liners) on the appearance of air bubbles and on the resulting *kT* measurements. The investigation included site measurements on treated and untreated surfaces of two tunnels' liners, the concentration of air bubbles (m^2/m^2), called "bubbles ratio", being computed from image analysis of pictures taken from the resulting surfaces. The air-permeability *kT* was measured on the same surfaces.

Figure 5.27 shows the influence of the inclination angle (from vertical position) on the "bubbles ratio", without a clear positive influence of the ceramic coating. The research showed a strong effect for air bubbles ratios on *kT*, especially between 0% and 0.5%, the effect becoming less marked for higher bubbles ratios.

The action and positive effect of permeable formwork liners in reducing the permeability of the concrete surface is discussed in Section 7.2.3. Figure 5.28 shows how the use of one of these liners ("Zemdrain") has eliminated the bug-holes by allowing the air bubbles to escape through the fabric. The left-hand zone of the element (untreated) shows plenty of bug-holes, which disappeared from the right-hand zone of the element, treated with "Zemdrain".

The laboratory results reported by Hirano et al. (2014) and Maeda et al. (2014) indicate that the presence of air bubbles increases significantly the air-permeability *kT* (unfortunately, the results of site tests on the tunnels' walls were not reported nor commented), which is in contradiction with what is presented in Section 11.3.4.4. More research is needed to elucidate this matter.

Nevertheless, when testing concrete surfaces showing high concentration of relatively large bug-holes, care has to be taken to avoid that the air bubbles "shortcut" the borders of the concentric chambers of the vacuum cell.

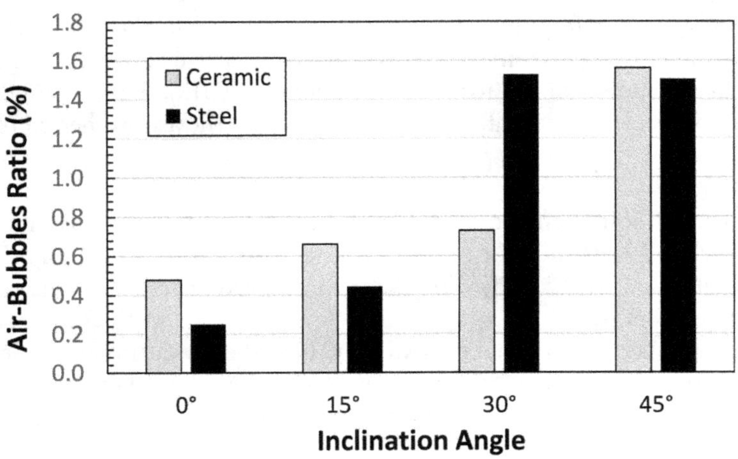

Figure 5.27 Effect of form inclination on bubbles ratio (Hirano et al., 2014).

Figure 5.28 Visible effect of "Zemdrain" (right-hand zone of element) in eliminating bug-holes.

5.7.6 Influence of Initial Pressure P_0

Occasionally, the initial pressure of the test (P_0) at time t_0 may be abnormally high, say above 100 mbar. This may be due to a number of factors: rough surface that impedes a tight sealing of the cells, high porosity of the substrate under test, low vacuum capacity of the pump, damaged O-rings, etc.

In principle, P_0 and, in general the pressure of the inner chamber should not be too high for two reasons:

- the cell can be detached from the surface, fall and get damaged (an emergency safety level of 200 mbar is ensured in the *PermeaTORR AC* and *AC+* instruments to prevent its expensive active cell of falling)
- the derivation of the formula to calculate kT (Eq. 5.13) assumes that P in the central chamber is much lower than the atmospheric pressure P_a (Eq. 5.8)

Therefore, although an occasional high P_0 value does not necessarily mean that a wrong kT result will be obtained, the recurrent testing at high P_0 values should be avoided and, when happening permanently, the instrument should be checked.

On the other extreme, very low P_0 values (below 30 mbar), especially when testing low-permeability concretes with the *TPT*, may lead to artificially high kT values, as discussed in Sections 5.3.5.2 and 5.6.4.2.

5.7.7 Influence of Porosity on the Recorded *kT* Value

The formula used to compute kT from the recorded rate of pressure increase in the inner chamber, Eq. (5.13), contains explicitly the porosity ε of the concrete tested. This value is usually not known, especially for site testing, so a default value ε=0.15 is assumed for the calculation. If ε is known, a corrected value kT_ε can be computed as

$$kT_\varepsilon = \frac{kT \cdot 0.15}{\varepsilon} \quad (5.43)$$

where kT is the value reported by the instrument assuming ε=0.15 and ε is the true porosity of the concrete.

The porosity of conventional concretes, including air-entrained and high-performance concretes, ranges typically between 0.05 and 0.25. Therefore, from Eq. (5.43), we can see that kT underestimates the permeability of low porosity concretes and overestimates that of high porosity. Since low porosity concretes tend to present low kT values and high porosity concretes to present high kT values (Eq. 5.31), the end result is that, by using a default value of ε=0.15, the range of permeabilities measured is artificially expanded. The "error" factor is between 0.6 and 3; if we take into consideration (Table 5.2) that kT varies across six orders of magnitude, this error becomes of little significance and the use of the default value is recommended.

However, if a more accurate calculation is required, then Eq. (5.43) should be applied, entering the true porosity ε, either measured or estimated through Eq. (5.31).

In the case of measuring materials other than concrete (e.g. mortar, paste, non-cementitious materials), a correction is needed, as mortar and paste present much higher porosities than concrete. If we assume a concrete containing, by volume, 28% cement paste, 32% fine aggregate and 40% coarse aggregate, the default porosities would become ε=0.15/0.60=0.25 for mortar and ε=0.15/0.28=0.54 for paste (assuming that the aggregates do not contribute porosity). These values should be used in Eq. (5.43) to obtain the coefficient of air-permeability corrected for porosity.

5.8 STATISTICAL EVALUATION OF *kT* TEST RESULTS

5.8.1 Statistical Distribution of *kT* Results

There is general consensus on that the kT values obtained in a set of measurements on nominally the same concrete are not normally distributed, with several cases reported where the log-normal distribution provides a suitable representation of the statistical distribution of kT results (Torrent, 2001; Brühwiler et al., 2005; Conciatori, 2005; Denarié et al., 2005; Jacobs & Hunkeler, 2006; Misák et al., 2008, 2017).

Indeed, as shown in Table 5.3, for large elements cast under laboratory conditions, the CoV of the test results range between 35% and 110%. It has been shown that, for CoV above 25%, properties that cannot accept negative values are not well represented by a normal distribution, with the

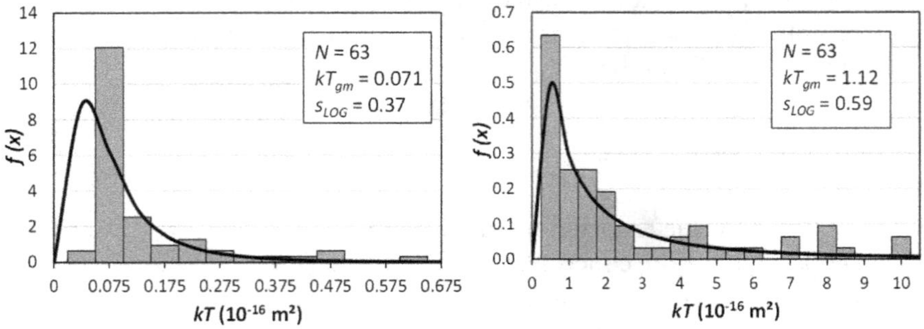

Figure 5.29 Log-normal distributions fit to *kT* results obtained on two different concretes, data from Conciatori (2005).

log-normal distribution being more appropriate for these cases (Torrent, 1978).

Figure 5.29 shows the statistical distribution of 63 *kT* results obtained by Conciatori (2005) on concretes of different *w/c* ratios: 0.52 (l.) and 0.73 (r.), with the lognormal function fitted to them. In ordinates, the probability density function *f(x)* is represented, so that the area under the histogram and curve is=1, see Torrent (1978).

Other examples of non-Gaussian distribution can be seen for Port of Miami Tunnel real case in Section 11.3.1.

Nevertheless, it has also been reported that some sets of *kT* results do not follow a log-normal or any other known statistical distribution (Jacobs & Hunkeler, 2006; Neves et al., 2012). Therefore, two different approaches for the statistical evaluation of *kT* results are presented: one parametric and the other non-parametric. The parametric approach for the analysis assumes that *kT* results follow a log-normal distribution and the non-parametric approach assumes that *kT* results do not follow any particular statistical distribution and will be evaluated through an Exploratory Data Analysis.

As for any other test method, the collected *kT* results shall be evaluated applying statistical analysis tools. The above-mentioned intrinsic high scatter of any gas-permeability assessment (Coutinho & Gonçalves, 1993; Andrade et al., 2000) presents some challenges regarding its statistical evaluation that will be addressed in the following, for the particular case of *kT* results.

5.8.2 Central Value and Scatter Statistical Parameters

5.8.2.1 Parametric Analysis

A suitable parameter for the central value of a data set or a population of *kT* is the geometric mean, given by:

$$kT_{\mathrm{gm}} = \left(\prod_{i=1}^{n} kT_i \right)^{\frac{1}{n}} \tag{5.44}$$

Equation (5.44) represents the *n*th root of the product of *n* individual *kT* results (kT_i).

The geometric mean can also be calculated taking advantage of the logarithm function properties:

$$kT_{\mathrm{gm}} = b^{\frac{\sum_{i=1}^{n} \log_b (kT_i)}{n}} \tag{5.45}$$

In this case, the geometric mean is calculated as the antilogarithm of the mean value of *n* logarithms of individual *kT* results $\log_b$ (kT_i). It shall be noticed that Eq. (5.45) is valid for any positive *b* not equal to 1, being common to find $b=10$ (LOG or $\log_{10}$) in statistical analysis of *kT* results. This second option to compute the geometric mean, although not so intuitive as the first, is less prone to underflow or overflow errors.

As a measure of scatter, given the (usually positive) skewness of the data, there is not an obvious indicator. The antilogarithm of the standard deviation of $\log_b$ (kT) cannot be considered as appropriate, in opposition to what happens with the mean for central value. Nevertheless, the standard deviation of $\log_{10}$ (kT), denoted as s_{LOG}, has been used to evaluate the scatter of a *kT* results set (see Section 5.6.2). A fair indication of s_{LOG} range can be found in Figure 5.9, whereas for an *a priori* estimation, $s_{\mathrm{LOG}}=0.40$ is a suitable likely value for site tests (Jacobs, 2006).

Actually, s_{LOG} is the shape parameter of the log-normal distribution of a *kT* population represented by the corresponding set of *kT* results. Therefore, it can be used to evaluate the scatter of a data set.

5.8.2.2 Non-Parametric Analysis

In non-parametric analysis, the median is considered a suitable statistic parameter to represent the central value. It corresponds to the 50th percentile of a set or it is the value that stands in the middle of a ranked set.

For the evaluation of scatter, two statistics can be used. One for the absolute scatter and one for the relative scatter. The interquartile range is a measure of absolute scatter and is calculated as the difference between the third and the first quartiles, Q_3 and Q_1, respectively, i.e. the difference between the 75th percentile and the 25th percentile. Given the wide range of measurable *kT* values, an interquartile range of $0.1\times10^{-16}\mathrm{m}^2$ can correspond to a low scatter for a PK4 concrete (see Table 5.2), while it will identify a large scatter for a PK1 concrete. Thus, it is important to have a relative indicator of scatter, like the quartile coefficient of dispersion (QCD):

$$QCD = \frac{Q_3 - Q_1}{Q_3 + Q_1} \tag{5.46}$$

According to our experience, QCD values of 0.23 may be expected for laboratory concrete, whereas for site concrete QCD of 0.50 may be a fair forecast.

5.8.3 Interpretation and Presentation of Results

Regardless of the intended type of analysis (parametric or non-parametric), it is recommended to start with a histogram, as exemplified in Figure 5.29. Although this representation of data is very useful for the analysis, it shall be kept in mind that a considerable number (at least 30) of results is required to have a meaningful histogram. Usually, the number of kT determinations within an assessment is under 20; therefore, other tools for the analysis of the data shall be considered.

Two major alternatives to the histogram are suggested: presentation of geometric mean with error bars and/or boxplot. The length of error bars is often taken as the standard deviation. Then, considering the information provided in 5.8.2.1, the lower and upper limits of error bars will be $kT_{gm} \div 10^{s\,\log}$ and $kT_{gm} \times 10^{s\,\log}$, respectively. Figure 5.11 is an example of this kind of plot.

The boxplot is a graphic representation of several descriptive statistics: minimum, maximum, quartiles and, eventually, outliers. The concept of outlier is an observation/result that is significantly distant from the rest. According to Tukey (1977), significantly distant refers to an observation that is more than one and a half times the interquartile range away from the first or third quartile. Outliers are often found in gas-permeability assessments, as a single void or a micro-crack may increase gas-permeability over ten-fold (Neves et al., 2012). The identification of the ranges of the different Permeability Classes in the charts is also recommended.

To exemplify, let us consider two different sets of results, one from a prefabricated slab in Portugal (Neves et al., 2012) and other from a prefabricated bridge segment in Switzerland (Jacobs et al., 2009), hereon identified as "Portugal" and "Switzerland". The values are ordered from smallest to highest.

"Portugal" = {0.012; 0.012; 0.014; 0.014; 0.017; 0.018; 0.020; 0.021; 0.027; 0.029; 0.040; 0.045; 0.047; 0.049; 0.061; 0.104; 0.105; 0.144; 0.158; 0.247} ($10^{-16}m^2$)

"Switzerland" = {0.057; 0.083; 0.102; 0.103; 0.107; 0.113; 0.115; 0.118; 0.122; 0.167; 0.577; 1.474; 1.507; 1.814; 2.695} ($10^{-16}m^2$)

Some descriptive statistics of these sets are presented in Table 5.5.

To build the suggested chart, the computation of lower and upper limits of the error bars is still required. The lower limits will be

Table 5.5 Descriptive statistics of example sets of *kT* assessment

Set	kT_{gm} ($10^{-16}m^2$)	s_{LOG} (-)	*Minimum* ($10^{-16}m^2$)	Q_1 ($10^{-16}m^2$)	*Median* (Q_2) ($10^{-16}m^2$)	Q_3 ($10^{-16}m^2$)	*Maximum* ($10^{-16}m^2$)	*QCD* (-)
"Portugal"	0.038	0.41	0.012	0.018	0.035	0.072	0.158[a]	0.60
"Switzerland"	0.222	0.58	0.057	0.103	0.118	1.474	2.695	0.87

[a] The largest value in the set (0.247) is not taken as the maximum, as it is considered an outlier: $0.247 > Q_3 + 1.5 \times (Q_3 - Q_1)$.

$0.038 \div 10^{0.41} = 0.015 \times 10^{-16} m^2$ and $0.222 \div 10^{0.58} = 0.059 \times 10^{-16} m^2$ for "Portugal" and "Switzerland", respectively, while the upper limits will be $0.038 \times 10^{0.41} = 0.097 \times 10^{-16} m^2$ and $0.222 \times 10^{0.58} = 0.840 \times 10^{-16} m^2$, for "Portugal" and "Switzerland", respectively. Based on the previous information, a chart like the one presented in Figure 5.30 can be built.

At first glance in Figure 5.30, it is possible to conclude that "Portugal" set represents a low-permeability concrete, whereas "Switzerland" set corresponds to a moderate permeability concrete, as defined in Table 5.2. A deeper analysis reveals that there is a large scatter in "Switzerland" set, with individual results in three different classes, whilst "Portugal" set has

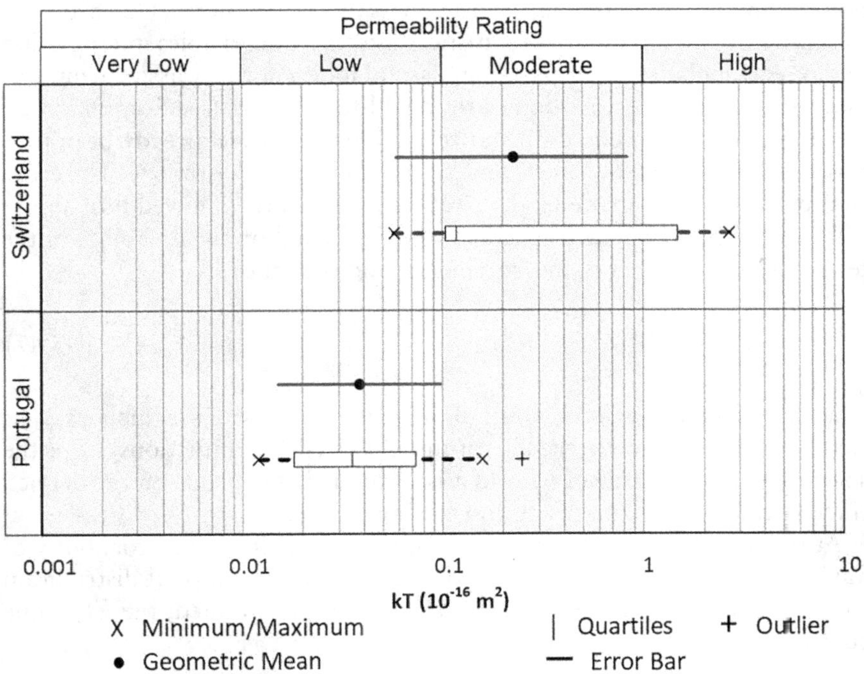

Figure 5.30 Two representation plots of both *kT* sets ("Portugal" and "Switzerland").

individual results in two classes. Furthermore, in "Switzerland" set, more than 25% are in PK4 (high permeability). Hence, the non-parametric box-plot representation, although visually more complex, provides more information on the statistical distribution of the measured values.

Beyond the specific findings for these two sets, there are also other conclusions from the presented information that can be extended to other sets. Taking the geometric mean as the representative value of a set is conservative, in comparison with the median, i.e. the geometric mean is higher than the median, unless there is a negative skewness (which is rare). The length of the error bars is longer than the interquartile range, as the first is limited by the 16%–84% fractiles (see below) and the second is limited by the 25–75 percentiles. If the median is equidistant from the first and third quartiles, and furthermore both extremes are also equidistant from the median – all in logarithmic scale – it is most likely that the results are log-normally distributed. Then, the geometric mean and the median assume similar values. This is the case for "Portugal" set, where the minimum, the maximum, Q_1 and Q_3 have an almost symmetric distribution around the median and the geometric mean is close to the median. In opposition, the "Switzerland" set exhibits the median much closer to Q_1 than to Q_3, and the geometric mean is fairly away from the median.

When the kT values obtained in a set of measurements follow a log-normal distribution, which happens in most cases, the parametric and non-parametric analysis will lead to the same conclusions. However, the parametric analysis comprises statistical inference, i.e. enables interpolation and/or extrapolation of results, being useful for conformity assessment, service life design and estimation of residual life purposes (see Section 9.5.2). Conformity criteria, based on statistical inference, have already been proposed by Denarié et al. (2005) and Jacobs and Hunkeler (2007). The criterion proposed by Denarié et al. (2005) is based on the allowed probability of having kT higher than a specified value (kT_s) and on the reliability of the tested sample (number of measurements) and is defined by

$$kT_{\text{gm}} \times 10^{(\lambda \cdot s_{\text{LOG}})} \leq kT_s \tag{5.47}$$

where kT_{gm} is the geometric mean of the kT determinations (sample), s_{LOG} is the standard deviation of the logarithms of the kT determinations, kT_s is the specified/required value of kT and λ is a factor that depends on the defined fractile, the level of confidence and the sample size (n).

At this point, it is important to recall that a fractile is the probability of having a result lower than a defined value. As the log-normal distribution assumes that the logarithms of kT are normally distributed, the kT_p value corresponding to a given fractile p (%) is calculated as

$$kT_p = 10^{\log kT_{\text{gm}} + z \cdot s_{\text{LOG}}} = kT_{\text{gm}} \times 10^{(z \cdot s_{\text{LOG}})} \tag{5.48}$$

where z is the argument of the standard Gauss distribution that delivers the probability p, e.g. for fractile $1-p=95\%$ is $z=1.65$; for fractile $1-p=10\%$ is $z=-1.28$, as detailed in Eqs. (5.49) and (5.50).

$$\Phi(z)=0.95 \Leftrightarrow z=\Phi^{-1}(0.95) \Leftrightarrow z=1.65 \tag{5.49}$$

$$\Phi(z)=0.10 \Leftrightarrow z=\Phi^{-1}(0.10) \Leftrightarrow z=-1.28 \tag{5.50}$$

The criterion proposed by Jacobs and Hunkeler (2007) is similar to the criterion proposed by Denarié et al. (2005), with $\lambda=1$, regardless of the size sample:

$$kT_{gm} \times 10^{s_{LOG}} \leq kT_s \tag{5.51}$$

Please notice that Eq. (5.51) is equivalent to

$$\log_{10}\left(kT_{gm} \times 10^{s_{LOG}}\right) \leq \log_{10}\left(kT_s\right) \Leftrightarrow \log_{10}\left(kT_{gm}\right) + s_{LOG} \leq \log_{10}\left(kT_s\right) \Leftrightarrow z \geq 1 \tag{5.52}$$

with

$$z = \frac{\log_{10}\left(kT_s\right) - \log_{10}\left(kT_{gm}\right)}{s_{LOG}} \tag{5.53}$$

This corresponds to the fractile $1-p=84\%$, as Φ (1)=0.84. In practice, this means allowing a maximum probability $p=0.16$ of having kT results higher than the specified value (kT_s).

Although non-parametric analysis does not allow statistical inference, still it can be used in conformity assessment. The specified value of kT_s can be compared with a defined percentile of a kT sample, and if the latest is lower, the assessed concrete surface can be considered as conform to the specification. If the defined percentile is the 50th, then the median will be the characteristic value of kT. The median has been adopted as characteristic value in the assessment of several concrete properties, such as resistivity (Polder et al., 2000), surface hardness (CEN, 2001) and carbonation depth (McGrath, 2005). However, defining just a percentile is arguable. The sample size shall be considered and, like in parameter λ of Eq. (5.47), the percentile may be defined as function of sample size. Another option for non-parametric analysis is the inspection by attributes. For this approach there are already defined criteria, such as that adopted by Swiss Standard SIA 262/1: "Concrete Construction – Supplementary Specifications" (SIA 262/1, 2019), described in Annex B. A statistical analysis of that conformity criterion can be found in Section D-2 of Jacobs et al. (2009); warning, Figure D-5 of that document is wrong, Figure 8.20 of this book is the correct one.

5.9 TESTING PROCEDURES FOR MEASURING *kT* IN THE LABORATORY AND ON SITE

As with all test methods, when conducting measurements of air-permeability *kT*, it is important to observe the same or, at least, similar testing procedures in order to get meaningful and comparable results.

Regarding measuring *kT*, two well-differentiated cases have to be considered: laboratory testing and site testing. Different to laboratory testing, where typically plain concrete specimens are cast and preconditioned to be tested in rooms under rather stable and known ambient conditions, site testing is faced with many uncontrolled variables, typically: rough or irregular surfaces, extreme temperatures, variable humidity conditions of the concrete surface, the eventual presence of steel reinforcement too close to the surface, cracks, etc.

In Annex B, procedures for measuring *kT* in the laboratory and on site are given, in a pre-Standard format, as a basis for elaborating laboratory protocols or even local or regional standards.

Annex B refers primarily to site testing of relatively young structures, say up to 1 year of age. Sometimes, *kT* is used as indicator of the durability of old concrete structures, becoming a tool for their condition assessment (JCI, 2014). A good example of such application can be found in Imamoto (2012), Imamoto et al. (2012) and Neves et al. (2018). This kind of application requires a special planning.

REFERENCES

Adey, B., Roelfstra, G., Hajdin, R. and Brühwiler, E. (1998). "Permeability of existing concrete bridges". *2nd International PhD Symposium on Civil Engineering*, Budapest, 8 p.

Andrade, C., González Gasca, C. and Torrent, R. (2000). "The suitability of the 'TPT' to measure the air-permeability of the covercrete". ACI SP-192, 301–318.

ASTM F2659 (2015). "Standard guide for preliminary evaluation of comparative moisture condition of concrete, gypsum cement and other floor slabs and screeds using a non-destructive electronic moisture meter", 6 p.

Bonnet, S. and Balayssac, J.P. (2018). "Combination of the Wenner resistivimeter and Torrent permeameter methods for assessing carbonation depth and saturation level of concrete". *Constr. & Building. Mater.*, v188, 1149–1156.

Brühwiler, E., Denarié, E., Wälchli, Th., Maître, M. and et Conciatori, D. (2005). "Applicabilité de la mesure de perméabilité selon Torrent pour le contrôle de qualité du béton d'enrobage". Office Fédéral Suisse des Routes, Rapport n. 587, Avril, Bern, Suisse, 1–48.

Bueno, V., Nakarai, K., Nguyen, M.H., Torrent, R.J. and Ujike, I. (2021). "Effect of surface moisture on air-permeability *kT* and its correction". *Mater. & Struct.*, v54, 89, 12 p.

CEN (2001). "Testing concrete in structures. Non-destructive testing. Determination of rebound number". Comité Européen de Normalisation, European Standard EN 12504-2, 1–8.

Conciatori, D. (2005). "Effet du microclimat sur l'initiation de la corrosion des aciers d'armature dans les ouvrages en béton armé". Ph.D. Thesis N° 3408, EPFL, Lausanne, 264 p.

Coutinho, A. de S. and Gonçalves, A. (1993). *Fabrico e Propriedades do Betão.* V.III, 2nd ed., LNEC, Lisboa, Portugal.

Denarié, E., Conciatori, D., Maître, M. and Brühwiler, E. (2005). "Air-permeability measurements for the assessment of the in situ permeability of cover concrete". ICCRRR, Cape Town, November 21–23.

Dobel, T., Balzamo, H. and Fernández Luco, L. (2010). "Aplicaciones de la Medida de Permeabilidad al Aire del Hormigón de Recubrimiento en Situaciones de Heterogeneidad del Sustrato". *18°. R.T AATH*, Mar del Plata, Argentina, Noviembre 8–10, 8 p.

Dobel, T. and Fernández Luco, L. (2012). "Coefficient of air permeability of non-homogeneous substrates and drying concrete". *Microdurability 2012*, Amsterdam, April 11–13, Paper 191.

Eddy, L., Matsumoto, K., Nagal, K., Chaemchuen, P., Henry, M. and Horiuchi, K. (2018). "Investigation on quality of thin concrete cover using mercury intrusion porosimetry and non-destructive tests". *J. Asian Concr. Federation*, v4, n1, 47–66.

Hirano, M., Yoshitake, I., Hiraoka, A. and Inagawa, Y. (2014). "Evaluation of air-bubbles distributed on concrete surface of side wall of tunnel lining". *Cem. Sci & Concr. Technol.*, v67, n1, 252–258. In Japanese.

Imamoto, K. (2012). "Non-destructive assessment of concrete durability of the National Museum of Western Art in Japan". *Microdurability 2012*, Amsterdam, April 11–13, 22 slides.

Imamoto, K., Tanaka, A. and Kanematsu, M. (2012). "Non-destructive assessment of concrete durability of the National Museum of Western Art in Japan". Paper 180, *Microdurability 2012*, Amsterdam, April 11–13.

Jacobs, F. (2006). "Luftpermeabilität als Kenngrösse für die Qualität des Überdeckungsbetons von Betonbauwerken". Office Fédéral des Routes, VSS Report 604, Bern, Switzerland, 85 p.

Jacobs, F., Denarié, E., Leemann, A. and Teruzzi, T. (2009). "Empfehlungen zur Qualitätskontrolle von Beton mit Luftpermeabilitätsmessungen". Office Fédéral des Routes, VSS Report 641, December, Bern, Switzerland, 53 p.

Jacobs, F. and Hunkeler, F. (2006). "Non destructive testing of the concrete cover – Evaluation of permeability test data". *International RILEM Workshop on* Performance Based Evaluation and Indicators for Durability, Madrid, Spain, March 19–21, 207–214.

Jacobs, F. and Hunkeler, F. (2007). "Air-permeability as a characteristic parameter for the quality of cover concrete". Concrete Platform, 173–182.

JCI (2014). "Guidance for Assessment of existing concrete structures". Japan Concrete Institute, 49.

Kato, Y. (2013). "Characteristics of the surface air-permeability test and the evaluation of quality variation in cover concrete due to segregation of concrete". *J. Adv. Concr. Technol.*, v11, 322–332.

Kreijger, P.C. (1984). "The skin of concrete: Composition and properties". *Mater. & Struct.*, v17, n100, 275–283.

Li, K., Zhang, D., Li, Q. and Fan, Z. (2019). "Durability for concrete structures in marine environments of HZM project: Design, assessment and beyond". Cem. & Concr. Res., v115, 545–558.

M-A-S (2019). PermeaTORR AC (Active Cell). http://m-a-s.com.ar/eng/product.php.

Maeda, T., Honma, H., Hirano, M. and Yoshitake, I. (2014). "Permeability of tunnel lining with air/water bubbles on concrete surface". Sustainable Solutions in *Structural Engineering and Construction*, ISEC Press, 321–325.

Mayer, A. (1987). "The importance of the surface layer for the durability of concrete structures". ACI SP-100, v1, 49–61.

McGrath, P.F. (2005). "A simple chamber for accelerated carbonation testing of concrete". ConMat, Vancouver, University of British Columbia, Canada.

Misák, P. (2018). Private e-mail communication, 09.01.2018.

Misák, P., Kucharczyová, B. and Vymazal, T. (2008). "Evaluation of permeability of concrete by using instrument Torrent" (in Czech), *JUNIORSTAV 2008*, 2.5 Stavebni zkusebnictví, Brno, January 23, 3 p.

Misák, P., Kucharczyková, B., Vymazal, T., Daněk, P. and Schmid, P. (2010). "Determination of the quality of the surface layer of concrete using the TPT method and specification of the impact of humidity on the value of the air-permeability coefficient". *Ceramics – Silikáty*, v54, n3, 290–294.

Misák, P., Stavař, T., Rozsypalová, I., Kocáb, D. and Põssl, P. (2017). "Statistical view of evaluating concrete-surface-layer permeability tests in connection with changes in concrete formula". *Materiali in tehnologije/Mater. Technol.*, v51, n3, 379–385.

Neves, R., Branco, F. and de Brito, J. (2012). "About the statistical interpretation of air-permeability assessment results". *Mater. Struct.*, v45, n4, 529–539.

Neves, R., Branco F. and de Brito, J. (2015). "Study on the influence of surface and geometric factors on the results of a nondestructive onsite method to assess air-permeability". *Exp. Technique.*, v40, n3, August, 1–8.

Neves, R., Torrent, R. and Imamoto, K. (2018). "Residual service life of carbonated structures based on site non-destructive tests". *Cem. Concr. Res.*, v109, 10–18.

Newman, K. (1987). "Labcrete, realcrete, and hypocrete. Where we can expect the next major durability problems". ACI SP-100, v2, 1259–1283.

Nguyen, M.H., Nakarai, K., Kai, Y. and Nishio, S. (2020). "Early evaluation of cover concrete quality utilizing water intentional spray tests". *Constr. & Build. Mater.*, v231: 117144.

Nguyen, M.H., Nakarai, K. and Nishio, S. (2019). "Durability index for quality classification of cover concrete based on water intentional spraying tests". *Cem. & Concr. Composites*, v104: 103355.

Nsama, W., Kawaai, K. and Ujike, I. (2018). "Influence of bleeding on modification of pore structure and carbonation-induced corrosion formation". *SLD4*, Delft, Netherlands, 674–685.

Parrott, L. (1994). "Design for avoiding damage due to carbonation-induced corrosion". ACI SP-145, 283–298.

Paulini, P. (2014). "Empfehlungen zur Bestimmung der Gaspermeabilität von Beton". Contribution to RILEM TC 230-PSC, June, 7 p.

Polder, R., Andrade, C., Elsener, B., Vennesland, Ø., Gulikers, J., Weidert, R. and Raupach, M. (2000). "Test methods for on site measurement of resistivity of concrete". *Mater. Struct.*, v33, n10, 603–611.

Proceq (2019). "Torrent Permeability Tester". https://www.proceq.com/uploads/tx_proceqproductcms/import_data/files/Concrete%20Testing%20Products_Sales%20Flyer_English_high.pdf.

Proceq (not dated). "Operating instructions – Permeability tester TORRENT". Proceq S.A., 11 p.

RILEM TC 154-EMC (2000). "Electrochemical techniques for measuring metallic corrosion". *Mater. Struct.*, v33, December, 603–611.

RILEM TC 230-PSC (2015). "Performance-based specifications and control of concrete durability". Beushausen, H. and Fernández Luco, L. (Eds.), RILEM Report V18, 373 p.

Romer, M. (2005a). "Effect of moisture and concrete composition on the Torrent permeability measurement". *Mater. Struct.*, v38, July, 541–547.

Romer, M. (2005b). Personal communication and supply of test data.

Romer, M. (2005c). "Multiscale durability aspects of concrete structures exposed to ground water". *COE Workshop on Material Science in 21st Century for the Constr. Ind.*, Hokkaido Univ., Sapporo, Japan, August 11, 24 slides.

Sakai, Y. (2019). "Correlations between air-permeability coefficients and pore structure indicators of cementitious materials". *Constr. Build. Mater.*, v209, 541–547.

Sakai, Y., Nakamura, C. and Kishi, T. (2013). "Correlation between permeability of concrete and threshold pore size obtained with epoxy-coated sample". *J. Adv. Concr. Technol.*, v11, August, 189–195.

Sakai, Y., Nakamura, C. and Kishi, T. (2014). "Evaluation of mass transfer resistance of concrete based on representative pore size of permeation resistance". *Constr. Building Mater.*, v51, 40–46.

Sandra, N., Kawaai, K., Ujike, I., Nakai, I. and Nsama, W. (2019). "Effects of bleeding on corrosion of horizontal steel bars in reinforced concrete column specimen". *Mater. Sci. Eng.*, v602: 012058.

SIA 162/1 (1989). Test No. 7 'Porosity'. EMPA Guidelines for Testing, Dübendorf.

SIA 262/1 (2019). Swiss Standard SIA 262/1:2019, "Construction en béton. Spécifications complémentaires". Norme Suisse, March 1, 60 p. (in French and German).

SIA 262/1-E (2019). Swiss Standard SIA 262/1:2019, "Construction en béton. Spécifications complémentaires". Annex E: 'Perméabilité à l'air dans les structures.

SIA 262/1-K (2019). "Concrete construction – Complementary specifications". Swiss Society of Engineers and Architects. Annex K: 'Characteristics of the pores'.

Sofi, M., Oktavianus, Y., Lumantarna, E., Rajabifard, A., Colin Duffield, C. and Mendis, P. (2019). "Condition assessment of concrete by hybrid non-destructive tests". J. Civil Struct. Health Monitoring, v9, 339–351.

Szychowski, J. (2010). "PermeaTORR: Relación entre profundidad de penetración y parámetros constructivos". Informe de DISTEK S.R.L., Buenos Aires, Argentina.

Torrent, R.J. (1978). "The log-normal distribution: a better fitness for the results of mechanical testing of materials". *Mater. Struct.*, v11, n64, July–August, 235–245.

Torrent, R. (1991). "Un nuevo método no destructivo para medir la permeabilidad al aire del recubrimiento de hormigón". 10a. Reunión Técnica de la AATH, Olavarría, Argentina, Octubre, vI, 307–323.

Torrent, R.J. (1992). "A two-chamber vacuum cell for measuring the coefficient of permeability to air of the concrete cover on site". *Mater. Struct.*, v25, n150, July, 358–365.

Torrent, R. (1997). "Your enquiry about the Permeability Tester". Telefax to Carola Edwardsen (COWIConsult), Holderbank, Switzerland, February 4, 7 p.

Torrent, R. (2001). "Diseño por Durabilidad - Técnicas de Ensayo y su Aplicación". CENCO Seminar on Durability of Concrete and Evaluation of Corroded Structures, Instituto Eduardo Torroja, Madrid, April 17–19.

Torrent, R. (2005). "Update of comparative test – Part I – Comparative test of penetrability methods". *Mater. Struct.*, v38, December, 895–906.

Torrent, R. (2012). "Non-destructive air-permeability measurement: From gas-flow modelling to improved testing". Paper 151, *Microdurability 2012*, Amsterdam, April 11–13.

Torrent, R., Bueno, V., Moro, F. and Jornet, A. (2019). "Suitability of impedance surface moisture meter to complement air-permeability tests". RILEM PRO 128, Durability, Monitoring and Repair of Structures, March, 56–63.

Torrent, R., di Prisco, M., Bueno, V. and Sibaud, F. (2018). "Site air-permeability of HPSFR and conventional concretes". *ACI SP-326*, Paper 84.

Torrent, R. and Ebensperger, L. (1993). "Methoden zur Messung und Beurteilung der Kennwerte des Überdeckungsbetons auf der Baustelle". Office Fédéral des Routes, Rapport No. 506, Bern, Switzerland, Januar, 119 p.

Torrent, R. and Frenzer, G. (1995). "Methoden zur Messung und Beurteilung der Kennwerte des Ueberdeckungsbetons auf der Baustelle -Teil II". Office Fédéral des Routes, Rapport No. 516, Bern, Suisse, October, 106 p.

Torrent, R. and Jornet, A. (1990). "Covercrete study – Part I: Scope of the research, test methods, characteristics of raw materials and concrete mixes, test results". HMC Report MA 90/3815/E, September, 48 p.

Torrent, R., Moro, F. and Jornet, A. (2014). "Coping with the effect of moisture on air-permeability measurements". Inter*national* Workshop on Performance-based Specification and Control of Concr*ete* Durability, Zagreb, Croatia, June 11–13, 489–498.

Torrent, R.J. and Szychowski, J. (2016). "Medición no destructiva de la permeabilidad al aire: evolución e innovación". Revista Hormigón n54, AATH, Buenos Aires, Argentina, Ene-Jun.

Torrent, R. and Szychowski, J. (2017). "Innovation in air-permeability NDT: Concept and performance". XIV DBMC, Ghent, May 29–31, Proceedings, paper 313.

Tukey, J.W. (1977). *Exploratory Data Analysis*. Addison-Wesley Publishing Company, Boston, MA.

Urdan, T.C. (2011). *Statistics in Plain English*. Routledge, Taylor & Francis Group, New York.

Yokoyama, Y., Sakai, Y., Nakarai, K. and Kishi, T. (2017). "Change in surface air-permeability of concrete with different mix designs and curing". *Cem. Sci. & Concr. Technol.*, v71, n1, 410–417 (in Japanese).

Chapter 6

Effect of key technological factors on concrete permeability

6.1 INTRODUCTION

As discussed in Chapter 3, the permeability of concrete depends on the microstructure of the material, with the flow of matter taking place predominantly through the pores in the cement paste and in the ITZ, but also along eventual cracks. Therefore, all technological factors that affect the volume, size, tortuosity and connectivity of pores and voids will exert an influence on the permeability of concrete.

The steady flow of liquids under pressure through a concrete sample or element requires that the pores are saturated with the fluid; otherwise, part of the inflow of liquid will end up filling the empty voids. On the contrary, the gas flow under pressure requires that the pores are sufficiently dry so as to leave empty paths for the gas molecules to be transported without hindrance through the material. Hence, it is expected that the degree of saturation or moisture content of the concrete exerts an important influence on the permeability of the material to liquids and gases.

This chapter is devoted to present and discuss experimental data, obtained both in the laboratory and on site, on the influence of key technological parameters of concrete (and of its temperature and moisture) on the permeability of the material to liquids and gases, measured using some of the test methods described in Chapters 4 and 5. This topic was dealt with in depth in Chapter 4 of "Performance criteria for concrete durability" (RILEM TC 116-PCD, 1995) to which the reader is referred. Here, new data on permeability of concrete to water and gases are contributed, obtained from tests covered by standards, namely:

W_p: Water penetration under pressure (EN 12390-8, see Section 4.1.1.2)

a, A: Water sorptivity (SIA 262/1-A, ASTM C1585, see Section 4 2.1)

kO, K_g, K_{int}: Cembureau gas-permeability (UNE 83981, LNEC E 392, see Section 4.3.1.2)

OPI, k_{OPI}: Oxygen-Permeability Index (SANS-3001-CO3-2, see Section 4.3.1.3)

kT: Torrent air-permeability (SIA 262/1, IRAM 1892, see Chapter 5)

When other methods are applied, they are briefly described in the text.

DOI: 10.1201/9780429505652-6

It is expected that this chapter will be useful to the reader in designing concrete mixes, not just for strength but for permeability as well, considering the effect of the key technological factors involved. It will also help contractors and inspectors to identify key aspects of jobsite concrete practices affecting its permeability.

6.2 EFFECT OF *W/C* RATIO AND COMPRESSIVE STRENGTH ON CONCRETE PERMEABILITY

The water/cement ratio (*w/c*) is a key factor in the design of concrete mixes because, as explained in Section 3.2.2, it determines to a large extent the volume and size of the capillary pores in the hydrated cement paste and, hence, the transport properties of concrete. The relation between compressive strength and *w/c* ratio is at the root of most mix design methods, because compressive strength is strongly affected by the pore structure. Therefore, it is to be expected that significant correlations of the permeability of concrete to gases and water with *w/c* ratio and compressive strength exist. The following sections confirm that expectation, based on abundant experimental evidence.

6.2.1 Data Sources

The data presented and discussed in Section 6.2 were collected from comprehensive investigations made (some by the authors of this book) predominantly in Switzerland, Japan and South Africa. They are summarized in the following sections.

6.2.1.1 HMC Laboratories

During several years starting 1990, "Holderbank" Management & Consulting Ltd[2] (HMC) conducted, in its laboratories located in Holderbank (Canton Aargau, Switzerland), several research projects, some with own funding, some sponsored by the Swiss Federal Highways Administration (ASTRA). Concretes of different compositions (binder types and proportions) were prepared, to a large extent with the same aggregates used as reference in HMC Laboratory, and tested following strictly the same procedure. Regarding permeability tests, 250×360×120 mm slabs were cast in two layers, each layer compacted on a vibrating table, undergoing three curing regimes:

A (dry): immediately after casting, the slabs were kept (stripped at 24 hours) in a dry room at 20°C/50% RH until the age of 28 days.

[2] Later branded as Holcim Technology Ltd.

B (reference): immediately after casting, the slabs were kept (and stripped) in a moist room (20°C/>95% RH) for 7 days and later moved to a dry room at 20°C/50% RH until the age of 28 days.

C (standard): immediately after casting, the slabs were kept (and stripped) in a moist room (20°C/>95% RH) until the age of 28 days.

In all cases, 120×120×360 companion prisms were cast, cured after procedure C, and tested at 28 days, first in three-point bending and the resulting halves in compression as equivalent 120 mm cubes.

Several non- or slightly destructive test methods were applied on the slabs, after 21 days storage in the dry room, in particular the here reported air-permeability *kT*. Immediately afterwards, Ø150×120 mm cores were drilled from the slabs and cut to 50 mm depth, from both the top and bottom (as cast) surfaces. The resulting Ø150×50 mm discs were oven-dried at 50°C during 6 days, followed by 1-day cooling in a desiccator inside the testing room at 20°C. Then, the discs were placed in the testing cell and the coefficient of O_2-permeability *kO* was measured, averaging the results obtained at relative gas pressures of 0.1 and 2.5 MPa (after waiting 30 minutes to achieve steady-state conditions). After the *kO* test was completed, the discs were weighed and placed (without any further treatment) with the external surface in contact with 3 mm of water in a closed container. At intervals, the mass gain was measured and the value after 24 hours of contact with water was recorded and the test result a_{24} (g/m²/s^½^) reported as the mass gain, divided by the surface area and the square root of time (24 hours).

The reference condition for performance comparison, based on *kT*, *kO* and a_{24}, was the bottom (as cast) surfaces of the samples subjected to curing regimen B.

The results were reported in Torrent and Jornet (1990, 1991), Torrent and Gebauer (1992a, b), Torrent and Ebensperger (1993) and Torrent and Frenzer (1995).

6.2.1.2 ETHZ Cubes

This was an exercise, conducted within a project financed by ASTRA, aimed at checking the suitability of several test methods to discriminate the permeability of concretes, prepared by the Swiss Federal Institute of Technology in Zürich (ETHZ) and blind tested at ETHZ laboratory by HMC personnel (Torrent & Frenzer, 1995). The permeability tests discussed here are *kT*, *kO* and a_{24}; the geometric mean of *kT* and the arithmetic means of *kO* and a_{24} were reported in Table 5.3. Details of the composition of the cubes can be found in Section 5.6.2 and in Torrent and Frenzer (1995). Suffice it to say that, from each of the two opposite faces where five measurements of *kT* were conducted, two Ø150×50 mm discs were drilled and cut, dried and tested for *kO* and a_{24}, exactly in the same way as described previously in Section 6.2.1.1.

6.2.1.3 General Building Research Corporation of Japan

In a laboratory research, a series of 16 OPC concrete mixes was prepared with the same constituents, varying just the proportions to achieve *w/c* ratios within the range 0.30–1.00 (cement contents between 200 and 580 kg/m^3), resulting in cylinder compressive strengths within the range 9–71 MPa. Data of the mixes can be found in Table 2 of Imamoto et al. (2009). With each mix (except No. 9), Ø150×50 mm discs were cast and moist cured during 1 month, followed by 1-month storage in a dry room at 20°C/60% RH, moment at which they were tested for air-permeability *kT* and O_2-permeability *kO*.

6.2.1.4 University of Cape Town

In Starck (2013) and Starck et al. (2017), a comprehensive investigation is reported in which a series of 150 mm concrete cubes, made with two binders and three different *w/b* ratios, were cured under conditions representing winter and summer conditions in Cape Town, South Africa. The binders were an OPC and a 50%+50% blend of OPC and slag (GBFS) and the *w/b* were 0.50, 0.65 and 0.80. The (favourable) winter curing was simulated by exposing the cubes protected outdoors and soaked daily during 35 days. Then, the cubes were stored for 5 days in a dry room (20°C/53% RH). The summer conditions were simulated by keeping the cubes up to 35 and 90 days in the dry room immediately after stripping. After measuring *kT*, Ø70×30 mm discs were drilled and saw-cut from the cubes, oven dried at 50°C for 7 days and tested for O_2-permeability *OPI*.

6.2.1.5 KEMA

A comprehensive research (Van Eijk, 2009) was conducted in the Netherlands to check the suitability of two permeability test methods[2] to assess the quality of concrete surfaces, namely: maximum water penetration under pressure WP_{max} and air-permeability *kT*. For that purpose, four (1.2×1.2×0.3 m) walls were cast in the city of Utrecht on 2 March 2009, two with OPC concrete (*w/c*=0.40 and 0.57) and the other two with same *w/c* ratios but using GBFS cement as binder. The form of the *N* side of each wall was kept 7 days in place, whilst that of the *S* side was kept just 1 day in place, giving eight walls' surface qualities; later the walls were exposed to the natural outdoors' environment until the age of 28 days.

At that age, *kT* was measured non-destructively at three positions on each of the eight surfaces, followed by Ø150×150 mm cores drilling from the same spots. The maximum water penetration *WP* through the exposed surface of the cores was measured.

[2] The Wenner electrical resistivity was also measured but the results were not fully reported

6.2.1.6 Other

Results from other sources are described in the corresponding sections.

6.2.2 Effect of *w/c* Ratio and Strength on Gas-Permeability

6.2.2.1 Cembureau Test Method

Figure 6.1 has been built with gas-permeability test results obtained in the investigations described in Sections 6.2.1.1 (legend Torrent et al.), 6.2.1.3 (Imamoto et al.) and 6.1.2.2 (ETHZ cubes), plotted against the *w/c* ratio of the OPC concretes tested. The permeability tests reported correspond to specimens moist cured for at least 7 days.

Figure 6.2 presents the same gas-permeability data but plotted against the standard compressive strength measured on cubes, moist cured during 28 days (the strength data from Imamoto et al., measured on cylinders fc_{cyl}, were converted into cube strengths fc_{cube} applying conversion Eq. (6.1), that gives a very good approximation to the relation shown in Tables 5.1–5.3 of Model Code 2010 (*fib*, 2012)):

$$fc_{cyl} = fc_{cube}\left[0.76 + 0.2 \cdot \log_{10}\left(fc_{cube}/19.6\right)\right] \quad fc \text{ in MPa} \tag{6.1}$$

Despite the different experimental conditions, the agreement between the three sets of data is remarkable.

The line shown in Figure 6.1 corresponds to the prediction of gas-permeability as function of the *w/c* ratio, by the formula proposed in Eq. (2.1-107) of CEB-FIP (1991), valid in the range $0.4 < w/c < 0.7$, extrapolated as dotted line:

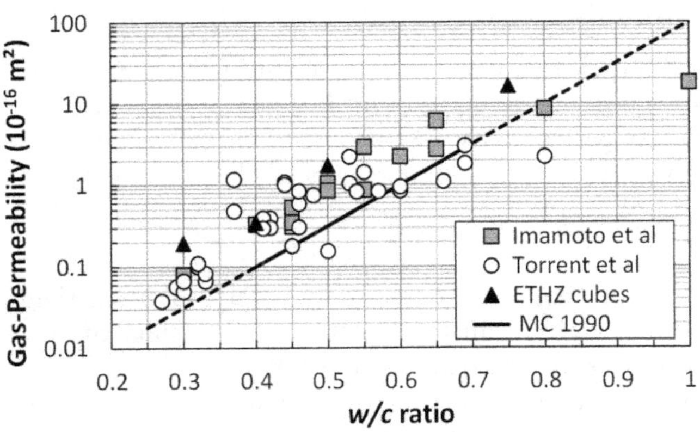

Figure 6.1 Effect of *w/c* ratio on Cembureau gas-permeability of OPC concretes.

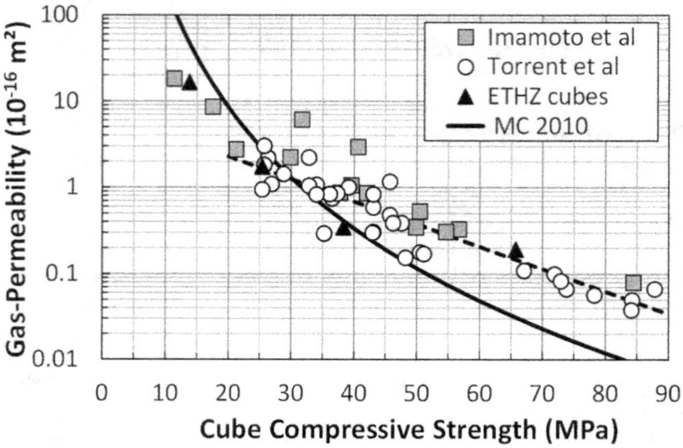

Figure 6.2 Effect of compressive strength on Cembureau gas-permeability of OPC concretes.

$$\log_{10} K_g = -\left(19 - 5 \cdot w/c\right) \tag{6.2}$$

where

K_g=gas-permeability (m²)

Equation (6.2) seems to reflect quite well the trend (slope) of $\log_{10} K_g$ vs *w/c* but underestimates the gas-permeability values measured by the Cembureau test method, under the applied test conditions.

The full line in Figure 6.2 corresponds to the prediction of gas-permeability as function of the compressive strength, by the formula proposed in Eq. (5.1–123) of Model Code 2010 (*fib*, 2012) relating K_g (m²) and the mean cylinder compressive strength f_{cm} (MPa):

$$K_g = 2 \times 10^{-10} / f_{cm}^{4.5} \tag{6.3}$$

The full line curve in Figure 6.2 was plotted, converting the cylinder strength of Eq. (6.3) into cube compressive strength by means of Eq. (6.1).

It can be seen in Figure 6.2 that *fib* Eq. (6.3) grossly underestimates the Cembureau gas-permeability test results. The dotted line represents the regression line fitted to the 34 test results labelled "Torrent et al.":

$$kO = 7.6\, e^{-0.06\, f_c} \qquad R = 0.92 \tag{6.4}$$

where f_c is the cube compressive strength (MPa) and *kO* is the Cembureau coefficient of O_2-permeability (10^{-16} m²). The results of the other two sources fit quite well to the regression of Eq. (6.4).

For the sake of completeness, the relation by Eq. (5.1–122) of Model Code 2010 (*fib*, 2012) between water-permeability coefficient K_w (m/s) and the mean cylinder compressive strength f_{cm} (MPa) is

$$K_w = \frac{4 \times 10^{-3}}{f_{cm}^{6}} \tag{6.5}$$

6.2.2.2 OPI Test Method

Figure 6.3 presents data of the coefficient of O_2-permeability (K_{OPI}), coming from different sources, two from South Africa (Starck, 2013; Gopinath, 2020) and one from Switzerland (Romer & Leemann, 2005; Romer, 2005); in all cases the specimens tested were cubes. Some of the mixes tested were prepared with OPC (circles), some with 50% GGBS (triangles) and others with 30% PFA (diamonds) in the binder. The data from Gopinath (2020), black symbols, were obtained on cubes after 28 days of moist curing: those from Starck (2013), white symbols, are average values of K_{OPI} obtained after 35 days of intermittent soaking, see Section 6.2.1.4. The data from Romer and Leemann (2005), grey symbols, correspond to cubes tested after 1 year storage at 20°C/90% RH. In all cases, cores were drilled to perform the OPI test, as described in Section 4.3.1.3.

Figure 6.3a shows the K_{OPI} results as function of the *w/c* ratio of the mixes and Figure 6.3b as function of the cube compressive strength. The plots K_{OPI} vs. *w/c* obtained on OPC mixes by Starck (2013), white circles, and (Romer & Leemann, 2005), grey circles merge quite nicely (Figure 6.3a). The K_{OPI} results of Gopinath (2020), black symbols, are lower, for the same *w/c* (for both OPC and GGBS binders), than those reported by the other two sources. It is interesting to observe that, for Gopinath (2020), GGBS concretes perform better than OPC concretes, whilst the opposite happens with (Starck, 2013) results, a fact attributable to the different curing conditions applied to the cubes.

Figure 6.3b shows that the relation K_{OPI} vs. cube strength is less dependent on the curing regime (and age) and also on the cement type, possibly

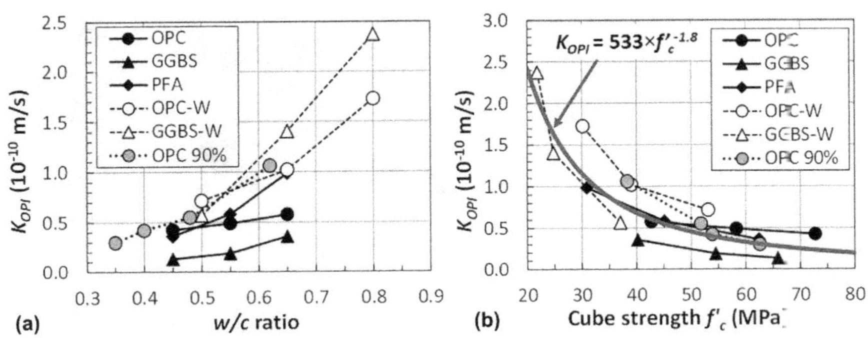

Figure 6.3 Effect of (a) *w/c* ratio and (b) cube strength on K_{OPI}; data from Romer (2005), Romer and Leemann (2005), Starck (2013) and Gopinath (2020).

because both properties are affected in a similar sense by those factors. Still, for the same strength, the OPC mixes tend to show higher K_{OPI} values. Worth noticing is the nice continuity between the results obtained in both South African investigations with GGBS binder (white and black triangles). A regression line was fitted to all test results, shown and plotted as a continuous curve in Figure 6.3b, with a correlation coefficient $R=0.83$.

6.2.2.3 Torrent kT Test Method

To investigate the effect of the *w/c* ratio on *kT*, it is important to count with data in which just the *w/c* ratio is varied, whilst all the other variables are kept constant (concrete constituents, surface tested, curing and storing conditions, age, etc.). An ideal set of results is, then, that obtained in the blind test of 0.5 m ETHZ cubes (Torrent & Frenzer, 1995), presented in Table 5.3 and described in Section 6.2.1.2, where four OPC concrete mixes with different *w/c* ratios were investigated.

The geometric mean of the ten individual *kT* tests conducted on each cube is plotted in Figure 6.4a, as function of the *w/c* ratio of each of the four mixes investigated, differentiating the cubes that were moist cured for 7 days from those without moist curing. A linear relationship between the logarithm of *kT* and the *w/c* ratio is observed, in very good agreement with Eq. (6.2), taken from Model Code 1990 (CEB-FIP, 1991).

In Figure 6.4b, results reported by Torrent and Ebensperger (1993) are plotted (7 d curing). These data correspond to laboratory concrete mixes made with another OPC, changing the *w/c* ratio and the aggregate type (three types from N, W and S Switzerland, same 32 mm Fuller grading); all mixes but one contained entrained air. Results of two mixes, where 8% silica fume (SF) was added to the OPC, are also shown (no air-entrainment); SF was added to OPC content to compute *w/c*. The tests were conducted on the bottom side of 250×360×120 mm concrete slabs, moist cured during 7 days and later kept 21 days in a dry room (20°C/50% RH).

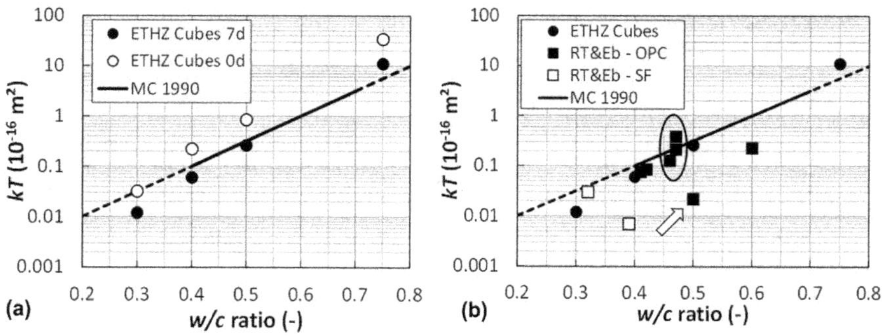

Figure 6.4 (a) Effect of *w/c* and curing on *kT* (OPC); (b) effect of *w/c*, aggregate and SF on *kT*.

The results in Figure 6.4b, enclosed by an oval, correspond basically to the same mix, but made with the three different aggregates; the mix pointed by an arrow corresponds to the OPC mix without air-entrainment.

Figure 6.4b shows that, whilst the same trend found in Figure 6.4a still holds valid, the changes in binder, aggregates and air content have increased the scatter of the results.

Figure 6.5 presents the relation between *kT* and *w/c* ratio, compiled from several sources by Jacobs et al. (2009), including laboratory and site tests (Torrent & Ebensperger, 1993; Romer & Leemann, 2005; Conciatori, 2005; Jacobs, 2006; RILEM TC 189 NEC, 2007). The white circles correspond to site tests conducted on many Swiss structures by Jacobs (2006) and the line corresponds to Eq. (6.2).

Although the expected general trend of higher *kT* for higher *w/c* can still be observed in Figure 6.5 (in harmony with that predicted by Eq. (6.2), a large scatter is evident. This scatter is attributable to the different raw materials used for the preparation of the concretes, but also to the different exposure and experimental conditions under which the measurements were performed, especially those conducted on site. The effect of the binder type on *kT* is discussed separately in Section 6.3.

Figure 6.5 is a confirmation of the unsuitability of *w/c* ratio as durability indicator, as discussed in detail in Section 1.6.2.

Figure 6.6 shows the same *kT* data of Figure 6.4, but here against the cylinder compressive strength at 28 days instead of the *w/c* ratio (the cube strength values from Torrent and Ebensperger (1993) were converted into cylinders using Eq. (6.1).

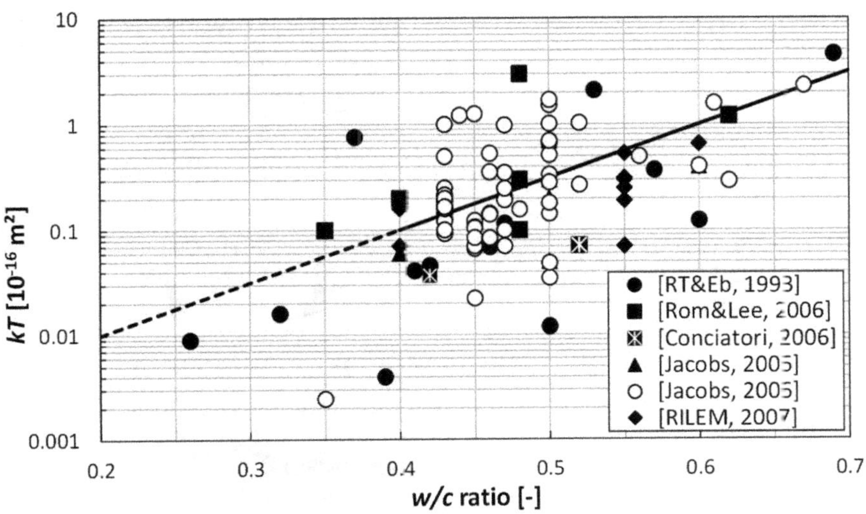

Figure 6.5 Relation between *kT* and *w/c* ratio; data from Jacobs et al. (2009).

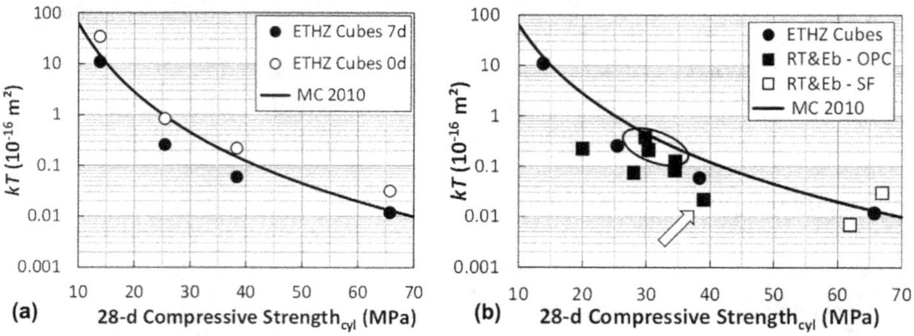

Figure 6.6 (a) Effect of compressive strength and curing on *kT*; (b) effect of strength, aggregate and SF on *kT*.

The full line in Figure 6.6 represents Eq. (6.3), from Model Code 2010, that fits quite well the test results.

A special investigation worth citing is that reported in Mohr et al. (2000), in which the relation between compressive strength and different "penetrability" tests was studied, based on cores drilled from 15 pavements, aged between 11 and 51 years, across nine States of USA. The "penetrability" tests were: Rapid Chloride Permeability Test (RCPT) (see Section A.2.1.1), Field Permeability Test (Section 4.1.2.3) and air-permeability *kT*. In the case of *kT* and RCPT, the measurements were conducted on saw-cut discs of Ø150 and Ø100, respectively, corresponding to top, middle and bottom parts of the cores. The compressive strength was measured on Ø150×300 mm saw-cut cores (corrected by slenderness).

Figure 6.7 shows the relation between *kT* and compressive strength. Two aspects are evident from Figure 6.7: one of them is that the air-permeability

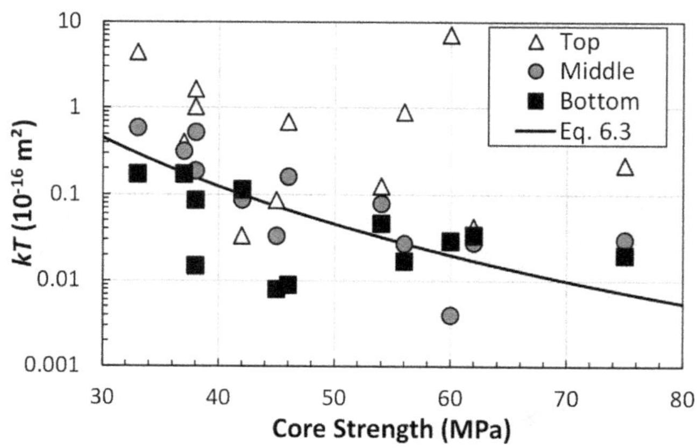

Figure 6.7 Relation *kT* vs compressive strength of cores drilled from several pavements across USA; data from Mohr et al. (2000).

of the cores decreases with depth, in particular showing high kT values for the top discs, which reflects the deleterious effect of weathering and traffic loads on the microstructure (an effect observed also for the RCPT). The second aspect is that Eq. (6.3) seems to reflect reasonably well the relation between kT (measured on the middle and bottom discs) and compressive strength of aged concrete.

6.2.3 Effect of *w/c* Ratio on Water-Permeability

6.2.3.1 *Water Penetration under Pressure*

An example of the effect of *w/c* ratio on water penetration of concrete under pressure W_p was found in the research reported by Sezer and Gülderen (2015), in which the replacement of a reference limestone aggregate (coarse and fine) by steel slag was investigated in concretes of w/c=0.40, 0.55 and 0.70. In one set of mixes, just the coarse limestone was replaced by steel slag, and in another set, just the fine limestone was replaced. An attempt to replace both aggregate fractions by slag ended in unworkable mixes that could not be tested. The tests applied to the OPC concretes were: cube compressive strength, W_p, and freeze-thaw resistance (not discussed here).

Figure 6.8 presents the effect of the *w/c* ratio on the water penetration (not disclosed whether mean or maximum) of concretes made with limestone and with replacement of coarse and fine fractions by steel slag. It can be seen that the penetration of water increases with the *w/c* ratio for all sets of mixes. Coarse steel slag shows no negative influence, but fine steel slag increased significantly the water penetration and reduced the strength, a

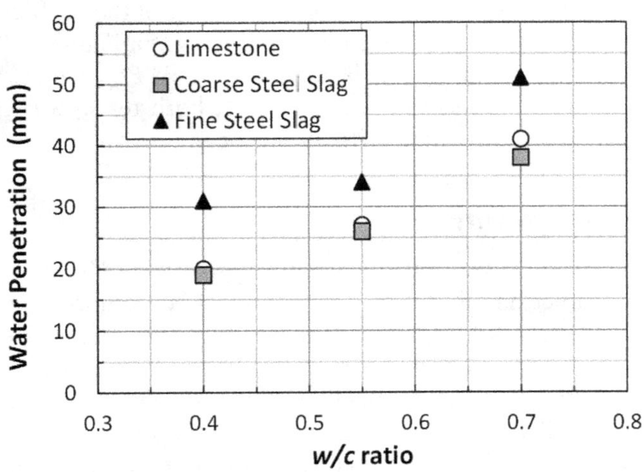

Figure 6.8 Effect of *w/c* and aggregate type on water penetration W_p; data from Sezer and Gülderen (2015).

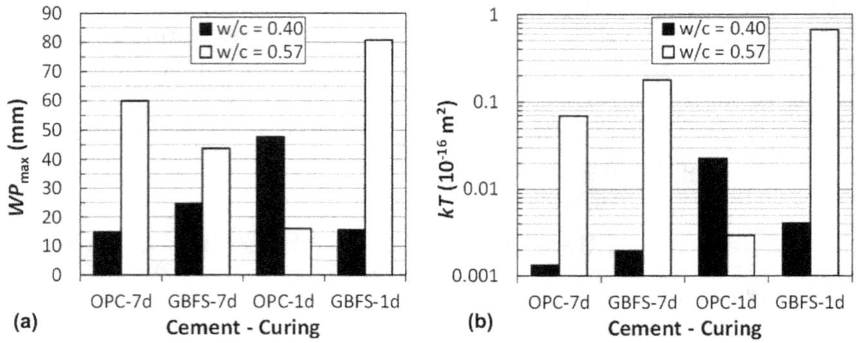

Figure 6.9 Effect of *w/c*, cement and curing on (a) water penetration WP_{max} and (b) air-permeability *kT* of walls; data from Van Eijk (2009).

fact attributed by Sezer and Gülderen (2015) to workability issues caused by the higher density, angularity, roughness and fineness of the slag fine aggregate compared to the limestone sand.

Figure 6.9a shows the water penetration test results for the eight wall surfaces tested in The Netherlands (Van Eijk, 2009), in the research described in Section 6.2.1.5. We can see that, in 3 out of the 4 cases, the water penetration increases with the *w/c* ratio, as expected. Yet, for the S side of OPC wall (OPC-1d), the opposite happened. It is difficult to justify this abnormal behaviour, which could be due to some batching mistake or by accidental extra curing of the N side of the OPC wall made with *w/c*=0.57, without ruling out a possible experimental error. The results of the coefficient of air-permeability *kT* measured at the same spots, shown in Figure 6.9b, display the same pattern of results, which confirms the WP_{max} values presented in Figure 6.9a. As discussed in Section 8.3.3, both tests, WP_{max} and *kT*, showed an excellent correlation with each other.

6.2.3.2 Water Sorptivity

Figure 6.10a and b present data of water sorptivity at 24 hours a_{24} of OPC concretes, as function of the *w/c* ratio and cube compressive strength, respectively. The circles in Figure 6.10 represent 20 values obtained on concretes prepared and tested at HMC Laboratory (see Section 6.2.1.1). In addition, the values obtained on the ETHZ cubes (Section 6.2.1.2) are plotted as triangles. All a_{24} values plotted correspond to specimens under 7 days moist curing. Figure 6.10 shows that a_{24} relates linearly with both *w/c* and compressive strength.

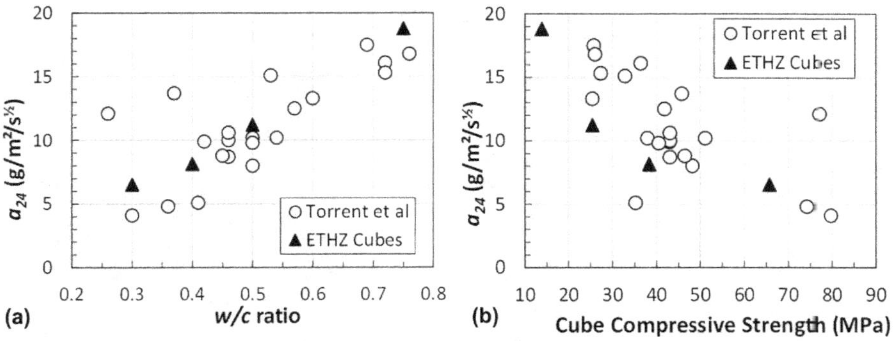

Figure 6.10 Effect of (a) *w/c* and (b) cube strength on water sorptivity at 24 hours.

6.3 EFFECT OF BINDER ON CONCRETE PERMEABILITY

6.3.1 Effect of OPC Strength on Permeability

An experimental research was conducted at HMC laboratories in Holderbank, with the aim of assessing the effect of OPC characteristics (chiefly mortar strength) on the permeability of concretes of same strength (instead of the same *w/c* ratio) (Torrent & Jornet, 1991). The lack of durability of concrete structures built since the 1970s was attributed to the improvement in strength of modern cements (demanded by the market), which allowed reaching the required concrete strength with higher *w/c* ratios (and less cement and associated costs) but, predictably, leading to more "penetrable" microstructures. In addition, modern cements, more finely ground, would develop less hydration after 28 days, with less "safety" reserve. This line of thinking is superbly presented in Section 3.1 of Neville (2003).

For that purpose, three industrial OPCs were chosen (GA, U2 and U5) that were representative of "modern" cements and one (RO) that had composition and fineness typical of "old" cements. The composition and main characteristics of the four OPCs selected are presented in Table 6.1. Cement U5 was made with similar constituents than U2, but ground in an open circuit grinding system (hence its lower RR Slope *n*).

It can be seen that cement RO, purposely manufactured to simulate an "old" cement, presents a relatively low C_3S content ("belite cement") and has been relatively coarsely ground (see low Blaine and high *d*' values) and presents a relatively low early strength. Its high water-demand on paste, accompanied by low mortar flow, is to be noticed.

On the other extreme is GA cement, with high C_3S content ("alite cement"), finely ground and showing the highest strength at all ages.

Table 6.1 Composition and characteristics of cements used in HMC investigation

Cement	*RO*	*U2*	*U5*	*GA*
Composition				
C_3S (%)	38	46	48	57
C_2S (%)	36	28	25	18
C_3A (%)	11	9	8	6
C_4AF (%)	5	8	8	11
Alkalis (% Na_2O_{equiv})	0.96	0.97	0.93	0.52
Physical properties				
Density (g/cm^3)	3.09	3.16	3.14	3.12
Blaine (m^2/kg)	247	290	306	335
RR slope *n*	0.930	0.963	0.875	0.945
RR diameter *d′* (μm)	31.4	23.2	28.1	18.4
Water demand on paste (%)	30.0	26.2	24.5	25.7
Flow of ISO mortar (%)	100	140	140	141
Mechanical properties: compressive strength ISO mortar (MPa)				
2 days	13.3	19.7	20.1	23.0
7 days	29.8	33.0	31.2	41.6
28 days	44.6	43.8	41.5	58.0
90 days	51.6	59.5	53.1	68.0
365 days	53.0	67.4	62.0	69.8

Note: *n* and *d'* are the "uniformity" factor (higher *n* means more uniform the particles' size) and "characteristic size (36.8% residue)" of the Rosin-Rammler particle size distribution, often used in the cement industry.

U2 is an average cement produced in a closed-circuit grinding system with high-efficiency separator (hence its higher RR Slope *n*). To simplify, cement U5 will not be considered in the following discussion.

Air-entrained (4.7%–6.0% air) concrete mixes were prepared without water reducers, aimed at reaching compressive strengths of 25, 35 and 45 MPa, using the RO, U2 and GA OPCs described in Table 6.1. On each mix, the coefficient of oxygen-permeability *kO* was measured on cores taken from the bottom (as cast) surfaces of samples moist-cured for 7 days. In parallel, the equivalent cube compressive strength was measured on specimens moist cured for 28 days. The testing procedure was that described in Section 6.2.1.1.

Figure 6.11a shows the relation cube strength vs. *w/c* ratio for the three OPCs. As expected, the cement with higher mortar strength (GA) shows the best strengths for the same *w/c* ratio, whilst the coarsely ground cement (RO) showed the worse performance (despite its mortar being slightly stronger than U2).

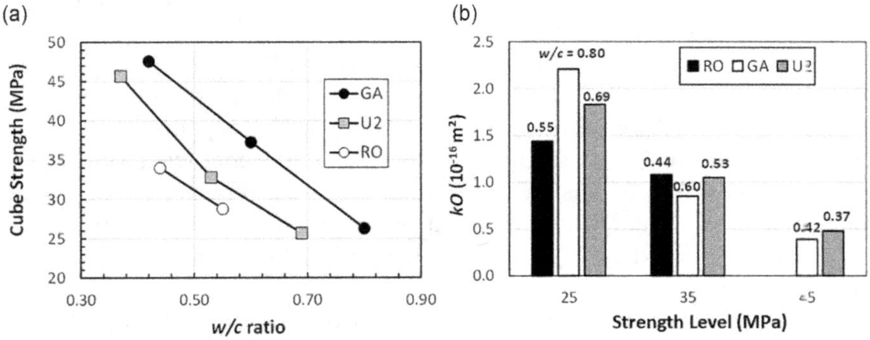

Figure 6.11 (a) Relation strength vs *w/c* ratio for three OPCs; (b) effect of strength level and OPC on *kO*; data from Torrent and Jornet (1991).

Figure 6.11b presents the results of oxygen-permeability *kO* for concretes made with the three cements at each compressive strength level; the *w/c* of the mix is indicated above each bar. It can be seen that at the lowest strength level (25 MPa), the strongest cement GA shows the highest permeability, due to the high *w/c* (0.80) required to achieve that low strength. The concrete made with the "old" cement RO shows the lowest permeability, as a *w/c*=0.55 is required to achieve the 25 MPa strength level. For concretes of strength levels 35 and 45 MPa, the strongest cement GA develops the lowest oxygen-permeability. From these results it seems that the lack of durability of "modern" cements should be attributed primarily to the use of concretes of low strengths (high *w/c* ratios) rather than to the high strength of the OPC.

6.3.2 Effect of Binder Type on Permeability

For the development of this topic, binders are divided into two categories: "Conventional" and "Unconventional". The former includes cements made with Portland cement clinker alone or combined with "conventional" mineral additions, i.e. those accepted by most national and international standards: SF, PFA, GBFS, natural or artificial (e.g. calcined clays) POZ, LF, etc. This includes the incorporation of the mineral additions at the cement plant (MIC: mineral components) or batched separately at the concrete plant (SCM: supplementary cementitious materials).

Under "unconventional" binders we include those made with Portland cement clinker and "unconventional" mineral additions as well as those made without Portland cement clinker.

6.3.2.1 *"Conventional" Binders*

In the research described in Section 6.3., the influence of adding SCMs to OPC U2 was also investigated (Torrent & Jornet, 1991). Four blended binders were manufactured in the laboratory, with the following compositions:

FA composed of 80% U2+20% of a commercial low-calcium PFA
SL composed of 65% U2+35% of an industrial GGBFS
LF composed of 85% U2+15% of an industrial LF
SF composed of 92% U2+8% of a commercial densified condensed SF

The blends were obtained in the laboratory by homogenizing the powders in a "Turbula" powder mixer. On each mix, the coefficient of oxygen-permeability *kO* as well as the initial water absorption rate a_3 (after 3 hours of contact with water) was measured on the bottom (as cast) surfaces of samples moist-cured during 28 days (other curing durations were also investigated, discussed in Section 6.7.2.1).

Figure 6.12a and b presents the results of oxygen-permeability *kO* and initial water absorption rate at 3 hours a_3, respectively, of mixes of similar cube strength (≈ 35 MPa), made with the OPC U2 and with the four blends described above. Both charts show a reduction in gas and water-permeability when PFA and SF and, to a lesser extent GGBFS, are added to OPC U2, whilst the effect of adding LF reduces the O_2-permeability but increases the initial sorptivity a_3.

In 2012–2013, Holcim Technology Ltd. conducted a comprehensive experimental research project in its own laboratories in Holderbank, Switzerland, complemented by tests performed at EMPA (Dübendorf, Switzerland) and SUPSI (Lugano, Switzerland). The aim of the research project was to study the influence of *w/b* ratio and binder type on a large variety of durability indicators (Moro & Torrent, 2016). Here we will focus just on the permeability tests *kO*, *kT* and WP_{max}.

A total of 18 concrete mixes, 9 with *w/b*=0.40 and 380 kg/m³ of binder and 9 with *w/b*=0.65 and 280 kg/m³ of binder, were prepared at EMPA with same 22 mm aggregates and admixtures, predominantly reaching "plastic" consistencies. The nine binders used are described in Table 5.4; more details on the composition of the binders can be found in Leemann and Moro (2017).

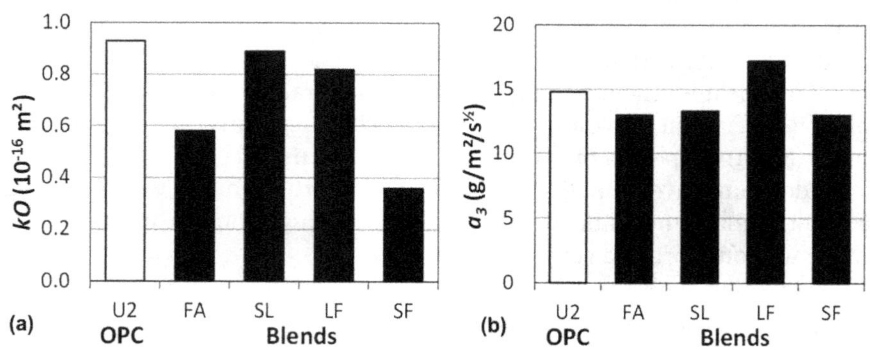

Figure 6.12 Effect of binder type on (a) O_2-permeability *kO*, (b) initial sorptivity a_3 data from Torrent and Jornet (1991).

The first letter of the code in Table 5.4 indicates the clinker used to produce the cements (H: Höver, Germany; M: Merone, Italy). The values in brackets indicate the content and type of mineral additions originally included in the cement (MIC). When a mineral addition was added separately as SCM into the concrete mix, the content and type is indicated in *italics*. All cements and SCMs are industrial products sold on the market.

The specimens for measuring *kO* and *kT* were three Ø150×50 mm discs, saw-cut from the bottom of cast Ø150×300 mm cylinders that were moist cured (moist room 20°C/RH>95%) for 28 days. Prior to testing, the discs were oven-dried at 50°C for 6 days. The specimens for measuring WP_{max} were 150 mm cubes moist cured for 28 days. For the three tests, the bottom-as-cast faces of the specimens were investigated.

Figures 6.13 a and b present the coefficients of permeability to oxygen (*kO*) and to air (*kT*), respectively, of the mixes investigated. Figure 6.13 shows the wide range of *kO* and, especially, of *kT* values obtained for each *w/b* ratio (binder=cement+SCM), for the nine different binders investigated. Figure 6.13b shows that the values of *kT*, for the same *w/b*=0.40, span two orders of magnitude for the different binders.

The charts in Figure 6.14 present the same results of *kO* and *kT* of Figure 6.13 but now plotted as function of the cube compressive strength of the mixes at 28 days.

Figures 6.13 and 6.14 show that the higher *kT* values, for same *w/b* or strength, correspond to the concretes made with OPC (H0 and M0) and with 26% of LF (M26L). The beneficial effect of hydraulic MICs or SCMs (SF, GBFS and PFA) on *kT* can be appreciated. Binders containing 8% of SF (H8M) and 68% of GBFS (H68S) produced the less permeable concretes. This is not so obvious from the *kO* results that show high values for the mix containing 68% of GBFS (H68S).

The reduction in *kT* by the addition of SF was also reported by Mittal et al. (2006).

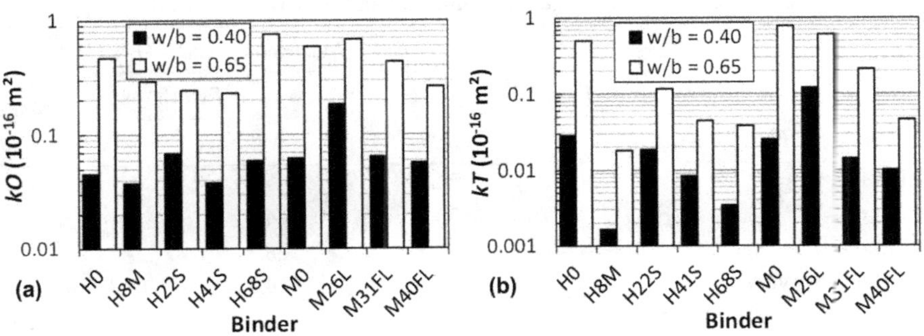

Figure 6.13 Effect of binder type on (a) *kO* and (b) *kT*, measured on concretes of two *w/b* ratios; data from Moro and Torrent (2016).

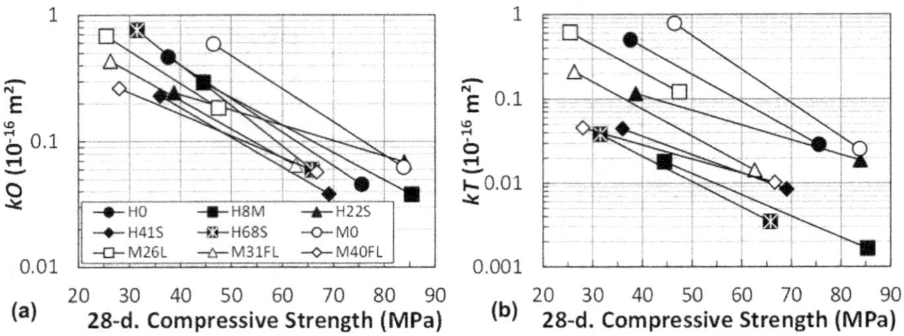

Figure 6.14 Effect of binder type on the relation of (a) *kO* and (b) *kT* with compressive strength; data from Moro and Torrent (2016).

The beneficial effect on *kT* of adding GBFS was experimentally confirmed by Panesar and Churchill (2010), who found that replacing OPC with GBFS in a $w/b=0.38$ concrete mix reduced *kT* (10^{-16} m^2) from 0.077 (OPC mix) to 0.028 and 0.018 for 25% and 50% replacement levels, respectively. These authors conducted a comparative service life and life cycle analysis of culverts on the basis of those results.

Glinicki and Nowowiejski (2013) investigated the effect of incorporating different levels of a Polish high calcium fly ash (W) on the mechanical and durability performance of concrete mixes made with *w/c* ratios 0.45 and 0.55. The composition of the cements used can be seen in Table 6.2 and includes binary and ternary blends by selective addition of a siliceous fly ash (V) and also GBFS (S). The cements were obtained by inter-grinding the constituents to a Blaine fineness of ≈ 380 m^2/kg.

Slabs (500×500×100 mm) were cast with each concrete mix, cured 21 days at 20°C/95% RH and then stored for 7 days at 20°C–22°C, 50%–60% RH.

Table 6.2 Types of cements investigated by Glinicki and Nowowiejski (2013)

	Main constituents (%)			
		Fly ash		*GBFS*
Cement type (EN 197-1 classification)	*Clinker*	*W*	*V*	*S*
CEM I	94.5	-	-	-
CEM II/A-W	80.9	14.3	-	-
CEM II/B-W	67.4	28.9	-	-
CEM II/B-M (V-W)	66.6	14.3	14.3	-
CEM II/B-M (S-W)	66.6	14.3	-	14.3
CEM V/A (S-W)[a]	47.9	23.9	-	23.9

[a] Not defined in EN Standard 197-1.

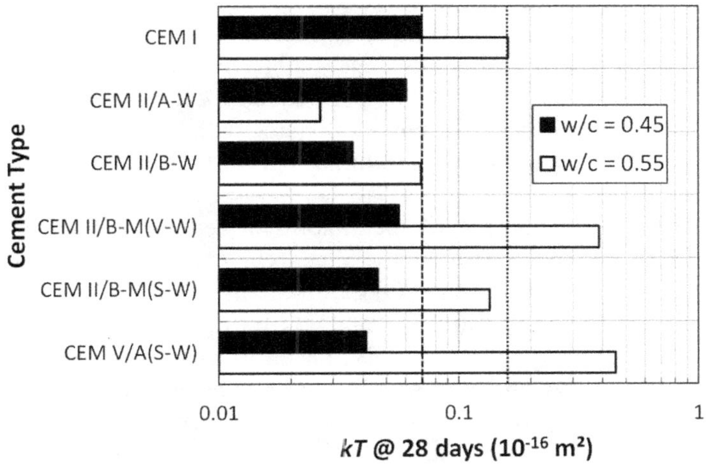

Figure 6.15 *kT* values of concrete mixes of *w/c*=0.45 and 0.55 made with different cement types; data from Glinicki and Nowowiejski (2013).

Figure 6.15 shows the *kT* values measured by Glinicki and Nowowiejski (2013) at 28 days of age for both *w/c* ratios. It can be seen that, for *w/c*=0.45, all the concretes containing active additions presented lower *kT* values than the reference CEM I (broken line) cement. For *w/c*=0.55 the measured performance of the concretes was less consistent, with two ternary cements showing higher *kT* than the reference CEM I (dotted line); the same effect was also observed applying the water penetration under pressure test.

An investigation was carried out in Turkey (Beglarigale et al., 2014) to study the effect of replacing 15% and 30% of OPC by a high calcium fly ash on the durability of concretes of 30 and 50 MPa cube strength. They also found a reduction in both air-permeability *kT* and water sorptivity at 24 hours with increasing replacement levels of OPC by fly ash, accompanied by a decrease in the charge passed in the RCPT and an increase in Wenner electrical resistivity (see A.2.1.1. and A.2.2.2).

Mathur et al. (2005) conducted a research aimed at studying the performance of High-Volume PFA concretes. They tested Reference and High-Volume PFA concretes (30%–50% of PFA in the binder) of three *w/b* ratios (between 0.33 and 0.65) for durability, applying the RCPT (ASTM C1202) and air-permeability *kT* test methods. They found that the addition of PFA in such volumes reduced both the "chloride"- and gas-permeability, reaching extremely low values of both properties: <1,000 Coulombs and $<0.005\times10^{-16}$ m², respectively.

Paz Montes de Oca (2016) made a comparative study on the air-permeability of concretes made with OPC (P-35) from Carlos Marx plant and a ternary blend Limestone Calcined-Clay Cement (LC^3) from Siguaney plant. The LC^3 cement contained 50% clinker and 41% of a blend of calcined clay and limestone, with an estimated reduction of CO_2 emissions of 25% with respect to OPC (Vizcaino et al., 2015). Specimens were cast with concrete of *w/c*=0.42, made with both cements, subjected to moist curing during 0, 3 and 28 days and tested for air-permeability *kT*. The results obtained showed lower *kT* values for the concretes made with LC^3 cement for all curing conditions (it has to be borne in mind that both cements were produced in different plants).

The positive contribution of metakaolin in reducing gas- and water-permeability was demonstrated by Shekarchi et al. (2010). A concrete mix with *w/c*=0.32 (400 kg/m^3 OPC), taken as reference, was replicated but replacing 5%, 10% and 15% of OPC by metakaolin. Among other tests, the permeability to gas K_N (N_2-permeability after Cembureau test method) and to water under pressure W_p were investigated, with the results shown in Figure 6.16. A reduction in both "permeabilities" with increasing metakaolin content is observed. The absolute values for 10%–15% replacement with metakaolin were similar to those recorded for a mix with 10% replacement by SF. The addition of metakaolin also increased significantly the electrical resistivity of the mixes.

In Perú, it was found that 5%–10% replacement of cement by diatomaceous earth produced high-performance concretes (slumps around 250 mm) of excellent behaviour (including air-permeability *kT*), comparable to that achieved with 10% SF, but at a significantly lower cost (Sánchez Stasiw, 2008).

The data presented in this section provide quantitative confirmation of the fact that binders containing hydraulic MIC or SCM (GBFS, PFA, SF) tend to show, for the same *w/b* ratio, lower permeability than those made with OPC or LF cement. This positive effect seems to be more marked for *w/b* ratios below 0.50 and prolonged moist curing.

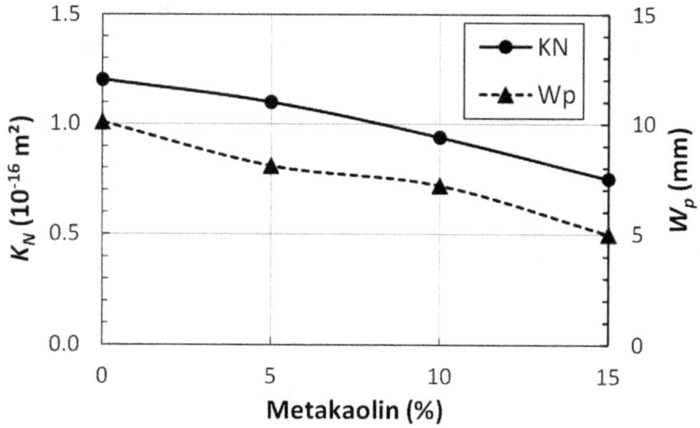

Figure 6.16 Effect of replacement of OPC by metakaolin on gas (K_N) and water (W_p) permeability; data from Shekarchi et al. (2010).

6.3.2.2 "Unconventional" Binders

A thorough investigation was conducted by Bahurudeen and Santhanam (2014) about the effect of replacing OPC by up to 20% of processed sugar cane bagasse ash (SCBA) on the durability performance of concrete. Four concrete mixes, with binder content of 360 kg/m³ and $w/b=0.45$, were prepared for the durability performance evaluation, with replacement of OPC by SCBA of 0%, 10%, 15% and 20%. Some specimens received 28 days of moist curing, others 56 days. Among the durability properties investigated were permeability to gases (*OPI* and *kT*) and to water (W_p). Figure 6.17 shows the beneficial effect of SCBA on the gas- and water- permeability measured after 28 days of moist curing.

In addition, the RCPT (Section A.2.1.1) and the CCI test (Section A.2.2.3), not discussed here, showed very similar improvement trends.

The effect on concrete performance of replacing 10% and 20% of OPC by two rice-husk incineration ashes (RHA) was investigated by Rodríguez de Sensale (2006). The origin of one ash was a rice paddy milling industry in Uruguay (UY RHA), whilst the other was a homogeneous ash produced by controlled incineration from the United States (US RHA). The ashes were dry-milled so as to achieve similar particle size characteristics: 28,800 and 24,300 m²/kg N_2 absorption for the UY RHA and US RHA, respectively, resulting in high activity indices (ASTM C311) of 92.9% and 92.4%, respectively. Two sets of concrete mixes were tested for air-permeability *kT*: one of $w/b=0.32$ and 534 kg/m³ of binder and the other with $w/b=0.50$ and 408 kg/m³ of binder. In each set, 10% and 20% of the OPC was replaced by each RHA. Figure 6.18 shows the effect of RHA replacement on air-permeability *kT*, for concretes with $w/b=0.32$ and 0.50.

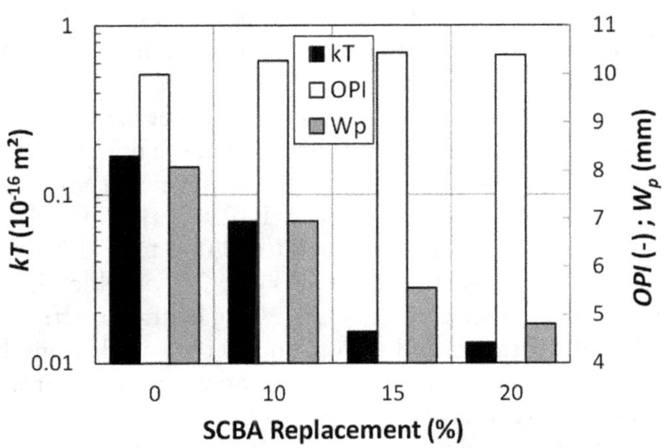

Figure 6.17 Effect of SCBA replacement on *kT*, *OPI* and W_p; data from Bahurudeen and Santhanam (2014).

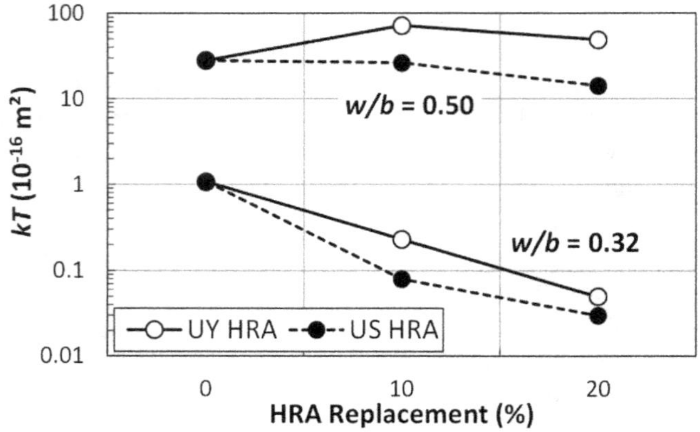

Figure 6.18 Effect of RHA replacement on *kT*, for concretes with *w/b*=0.32 and 0.50; data from Rodríguez de Sensale (2006).

What Figure 6.18 shows is that the incorporation of RHA into the concrete mix has a very positive effect in reducing air-permeability *kT* for *w/b*=0.32, whilst it is very small, if any, for *w/b*=0.50. Systematically, US RHA performs slightly better than UY RHA. Rodríguez de Sensale (2006) discusses the dual effect of RHA on concrete performance: pozzolanic and micro-filler effects. The results in Figure 6.18 may suggest a predominant role of the micro-filler effect, that is more marked in a more compact cement paste (*w/b*=0.32) than in a more dispersed paste (*w/b*=0.50). Figure 6.18 indicates that the inclusion of both RHAs is a good measure to achieve high performance concretes (cylinder compressive strength between 50 and 60 MPa and low permeability $kT<0.10\times10^{-16}$ m^2).

Two separate researches (Neves et al., 2015) and (Gurdián Curran, 2016) studied the suitability of spent fluid cracking catalysts (SFCC) to be used as hydraulic addition. The petrochemical industry uses, in its fluid catalytic cracking units, zeolites as catalysts. The zeolites promote breaking and rearrangement of the hydrocarbon molecules in order to generate new products. After several cycles of use and regeneration, the fluid cracking catalyst becomes spent, due to the accumulation of heavy metals and carbon from the hydrocarbon stream on their surface. In 2013, the annual quantity of SFCC generated worldwide is estimated at 180,000 tons. SFCC is a porous material of high specific surface (250 m^2/kg), 90% of which is constituted by silica and alumina. SFCC presents interesting hydraulic activity, reflected in acceleration of hydration and setting and in increased strength development, but higher carbonation susceptibility. Sometimes it is ground and/or thermally treated to further increase its reactivity.

In the research described in Neves et al. (2015), a concrete mix made with 280 kg/m^3 of OPC and *w/c*=0.69 was used as control, the performance of which was compared with a mix of same composition, except that the binder was composed of 238 kg/m³ of OPC and 42 kg/m³ of SFCC (15% substitution). Other mixes tested, not discussed here, contained a corrosion inhibitor. The mix containing SFCC showed an 8% reduction in 28 days compressive strength. The permeability of the concretes was measured by *kT* and by the water absorption coefficient at 4 hours a_4. The results, obtained for both properties with both binders, indicated almost the same *kT* value and a decrease of water sorptivity a_4 for the mix containing SFCC.

In the case of the research reported by Gurdián Curran (2016), the 100% OPC control mix was compared with one containing a ternary blend of 50% OPC+35% PFA+15% SFCC, which makes more difficult to assess the individual effect of SFCC. The performance of the concrete made with the ternary blend was significantly inferior to that of the control in terms of 28-day strength (–20%). The durability performance was measured in terms of air-permeability *kT*, of water sorptivity *a* and of water penetration under pressure W_p, all measured after 28 days of moist curing for both concretes. Both *kT* and *a* were smaller for the ternary blend, whilst W_p was higher.

It is worth mentioning that both investigations included, besides permeability, other tests such as carbonation, chloride penetration and corrosion tests.

Although not entirely conclusive, it seems that SFCC presents an interesting potential as pozzolanic material that deserves being further investigated.

Another experience worth citing is that reported by Kubissa (2020), describing the development of a concrete made with a binder composed of both OPC and CSA (calcium sulphoaluminate) cements. The concrete, aimed for repair work of concrete pavements, was expected to develop strength quite rapidly but also ensure durability by means of low permeability and high freeze-thaw-salts resistance. A series of nine concrete mixes containing 440 kg/m³ of binder and a *w/b* ratio of 0.40 were prepared. The variables studied were the amount of CSA cement in the binder (0%, 15% and 30% of the total binder) and the amount of air-entraining agent used (0% and ≈0.14% and 0.28% of binder content), which led to air contents of ≈1.3%, 4.5% and 7.5%, respectively. The slump of the mixes ranged between 160 and 200 mm and citric acid, in a proportion of 0.2% of the total binder content was added as setting retarder. The inclusion of CSA cement in the binder developed high early cube compressive strength of between 8 and 20 MPa at 2 days, compared with less than 1 MPa for the 100% OPC mix; at 28 days, all the nine mixes developed compressive strengths between 40 and 58 MPa. The permeability of the mixes was measured through the coefficient of water absorption at 6 hours a_6 at 28 days (samples dried at 105°C) and the coefficient of air-permeability *kT* at 56 days (on samples kept 7 days in dry air, which yielded suitable surface moistures for the test). In addition, a freeze-thaw-salts scaling test was performed, reporting the mass loss δ_m after 56 freeze-thaw cycles. An increase in *kT* with the CSA content was observed, especially between 0% and 15%, as well as an increase of *kT* with the air-entrainment level. The results of a_6 were not very consistent, showing lower values for the mixes containing CSA cement. It is worth mentioning that the scaling test, besides confirming the well-known beneficial effect of air-entrainment on frost resistance showed a systematic poor performance of the concretes containing 30% of CSA cement.

Within the context of the research described in Section 6.3.2.1 (Moro & Torrent, 2016), on top of the mixes with the nine binders described in Table 5.4, two more mixes of w/c=0.40 were investigated (Moro, 2013). One was made with an alkali-activated slag cement (AAS) and the other with a supersulphated cement (SSC). Following the same procedures as described in Section 6.3.2.1, the permeability of the concretes was tested after 28 days curing, measuring: O_2-permeability kO; air-permeability kT, coefficient of water absorption a_{24} and maximum water penetration under pressure W_p. Figure 6.19a shows the results of the gas-permeability of the AAS and SSC concretes, compared with those of OPCs H0 and M0 (see Table 5.4). Figure 6.19b presents the water-permeability test results.

Figure 6.19a shows that the O_2-permeability kO of both AAS and SSC concretes is significantly higher than for the OPC concretes; in the case of air-permeability kT the high permeability was only observed for the AAS concrete. The results of water sorptivity follow a similar trend, see Figure 6.19b, with the a_{24} value of SSS and AAS concretes being moderately and significantly higher, respectively, than for both OPC concretes. On the contrary, the results of water penetration W_p, Figure 6.19b, show the opposite trend, with both SSC and AAS concretes performing better than the OPC concretes. This abnormal behaviour may be explained by an alleged sensitivity of concretes made with these special binders to drying (even at 50°C) and, consequently, to the possible generation of internal micro-cracks. Thus, the results of W_p (the only test that does not require pre-drying) of the AAS and SSC concretes are better. This thermal sensitivity needs to be clarified as it may impair the field performance of these systems. It is worth mentioning that the response of the SSC and AAS concretes to the migration

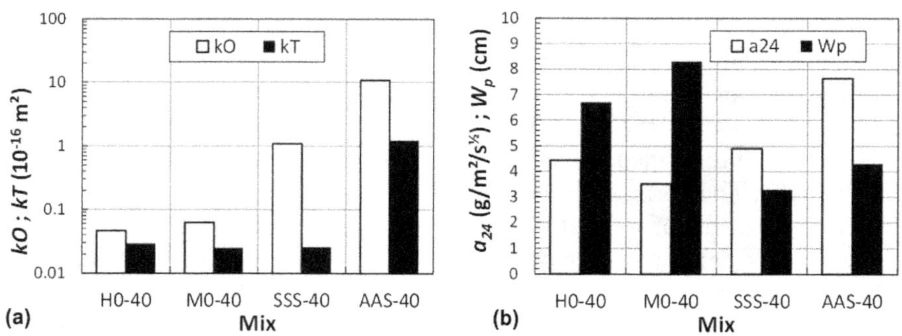

Figure 6.19 (a) Gas-permeability and (b) water-permeability of SSC and AAS concrete, compared with two OPCs (w/c = 0.40) (Moro, 2013).

tests (See Annex A) was superlative, with Q values below 500 coulombs ("RCPT", ASTM C1202) and migration coefficient M (Tang & Nilsson method, Section A.2.1.2) below 0.05×10^{-12} m²/s. On the other hand, their carbonation rate was 4–7 times higher that of the OPC concretes.

A field investigation was conducted on a 4.8×7.1 m slab, 0.6 m thick, built in Campbellfield, Australia, a suburb of Melbourne. The slab was cast with a geopolymer concrete mix (w/b=0.30), using as binder a 75%–25% blend of PFA and GGBFS, activated by a Na+K alkali hydroxide complemented by Na_2SiO_3 solution (Pasupathya et al., 2016). After eight years of exposure to the relatively mild environment of Campbellfield (summer averages around 26°C and winter averages around 6°C) a field investigation was conducted involving electrical resistivity, corrosion potential and air-permeability kT. The values of kT were typically within the range 1.0–10 ($\times10^{-16}$ m²), corresponding to the "High" Permeability Class (see Table 5.2). This fact, accompanied by low values of electrical resistivity and corrosion potential, cast some doubts about the long-term durability of the slab.

In Section 9.8 of RILEM TC 224-AAM (2014), it is concluded that the transport-related durability properties of alkali-activated binders depend strongly on their pore structure, which is determined by the binder chemistry and maturity (curing). It seems that the durability performance of concretes made with these special binders does not fit well to the trends and relations observed for Portland cement clinker binders and that relying just on one or two test methods may lead to wrong conclusions.

6.4 EFFECT OF AGGREGATE ON CONCRETE PERMEABILITY

As discussed in Section 3.2.1, in conventional concrete the aggregate particles are isolated, coated by a layer of cement paste that separates one another. Therefore, in order to reach an aggregate particle, any substance penetrating concrete has to travel first across the hardened cement paste (h.c.p.). One can imagine, then, that a concrete made with permeable aggregate particles, but surrounded by a low permeability h.c.p. will still show a low permeability. This concept is discussed in Section 6.4.1.

However, as discussed in Section 3.2.3, the presence of the aggregate particles affects the layers of h.c.p. around them, leading to the concept of ITZ, with its influence on permeability being discussed in Section 6.4.2.

6.4.1 Effect of Bulk Aggregate on Concrete Permeability

6.4.1.1 Porous Aggregates

In México, there is an abundance of very porous aggregates which, for economic reasons, are used in making concrete. An investigation of the effect

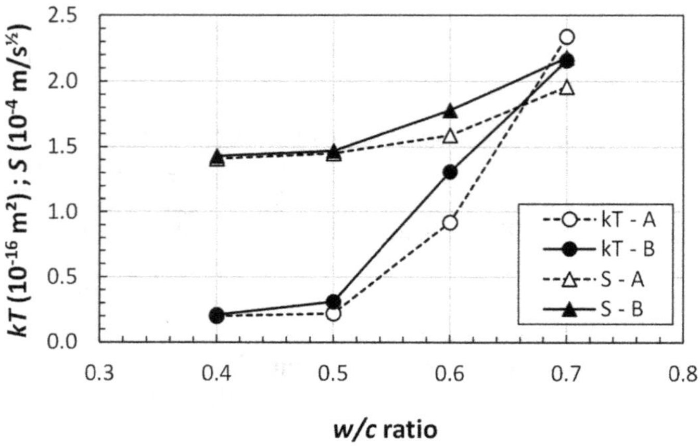

Figure 6.20 Effect of *w/c* ratio and aggregate type (B twice as porous as A) on air-permeability *kT* and water sorptivity *S* of concrete; data from Solís-Carcaño and Alcocer-Fraga (2019).

of using aggregates of different porosities on the air-permeability and sorptivity of concrete (among other properties) was conducted by Solís-Carcaño and Alcocer-Fraga (2019).

In particular, a coarse aggregate B showed almost twice the porosity (13.1%) than the already porous coarse aggregate A (6.7%). OPC mixes with four different *w/c* ratios (0.4, 0.5, 0.6 and 0.7) were prepared, one series with aggregate A and the other with aggregate B, moist cured 28 days and thereafter left drying under ambient conditions, prior to being tested for air-permeability *kT* and water sorptivity *S* (ASTM C 1585).

Figure 6.20 shows the coefficient of air-permeability *kT* and the water sorptivity *S* measured on the two series of concretes, differentiating those made with aggregate A (white symbols and dotted lines) from those made with aggregate B (black symbols and full lines).

It can be seen that at low *w/c* ratios the influence of the aggregate type is negligible on both *kT* and *S*, which could be explained by the dominant effect of the low permeability of the h.c.p. At higher *w/c* ratios, some differences, but not very significant, can be appreciated between the concretes made with both aggregates. In any case, the effect of the *w/c* is more significant than that attributable to the aggregate type.

In the case of *kT* test, for concretes of low permeability the penetration of the test may not be sufficient to reach or traverse the coarse aggregate particles, which also may explain the lack of influence of aggregate type on *kT*.

6.4.1.2 Recycled Aggregates

There is a trend promoted by some authorities to include aggregates obtained by recycling demolition waste into concrete mixes, instead of disposing the rubble in dumps. For instance, since 2005 it has become mandatory in the City of Zürich (largest in Switzerland with about 450,000 inhabitants) for all public buildings to be built with recycled aggregates (Zürich, 2019). For high concrete grade (RC-C), the aggregate shall contain at least 50% by mass of particles originated by the processing of waste concrete, concrete products, mortar or concrete blocks. For lower concrete grade (RC-M), the aggregate shall contain at least 25% of mixed demolition waste (clay bricks, roofing tiles, calcium silicate autoclaved bricks and non-floating aerated concrete). It is stated:

> "While RC-M contains a lower rate of recycled aggregate than RC-C, Zurich has chosen to focus on this, in order to incentivize the recovery of this otherwise more difficult to recover mixed demolition waste, which makes up around 60% of Switzerland's 10 million tonnes of mineral construction waste produced per year."

The range of materials that can be labelled as "recycled aggregates" is enormous, including even recycled aggregates obtained as "ashes" from urban waste incineration furnaces, which contain a myriad of different materials (including metals, ceramics, glass, etc.).

There is general agreement that well-processed recycled aggregates originated from concrete demolitions, used in a proportion of 10%–30%, do not negatively affect the performance of concrete, chiefly its workability and compressive strength (WBCSD, 2009). However, when including into the definition of "recycled aggregates" other wastes, the range of performance of concrete containing them becomes very large, which makes it impossible to generalize. Each recycled aggregate has to be judged on its own merits, by measured performance with meaningful tests of representative samples from representative concrete mixes.

Four examples will be presented here, three in which processed concrete demolition waste was used as recycled aggregate and the other involving processed clay bricks as recycled aggregates.

Regarding specifically the effect on concrete permeability of replacing natural coarse aggregate by recycled concrete aggregates, a research was conducted by Gurdián Curran (2016). In it, the natural coarse aggregate of a concrete mix made with 380 kg/m³ of OPC and w/c=0.45 was replaced by 20% and 100% of a recycled concrete aggregate provided by a local (Spanish) industrial supplier. The recycled aggregate used was free from "impurities" such as ceramics, asphalt, light particles, etc. For general information, the density of the concretes was reduced by 1% and 8% with the incorporation of 20% and 100% of recycled aggregate whilst the compressive strength was reduced by 3% and 16%, respectively. The permeability performance was measured by various techniques: coefficient of air-permeability kT, coefficient of water absorption a and depth of water penetration under pressure W_p.

The second case corresponds to a research reported by Yamamoto (2016), who investigated the effect of replacing 20% and 50% of the natural coarse aggregate by recycled concrete particles of almost the same grading. She investigated two concrete mixes, with *w/c* ratios 0.57 and 0.43. The composition of each concrete type was essentially the same, only the dosage of superplasticizer was adjusted to get a slump of 100 mm (the increase in dosage was significant only for the mixes containing 50% of recycled coarse aggregate). The recycled concrete aggregate was obtained from a local (Argentine) commercial supplier, presumably originated from rubble of widely different concrete qualities; the particles were partially covered with adhered mortar coatings, showing an absorption after 24 hours immersion of 5%–6%, compared with 0.3% for the natural coarse aggregate. Before preparing the mixes, the recycled coarse aggregate was pre-mixed and kept in contact during 30 minutes with the amount of water absorbed in that time interval. The permeability performance of the mixes was measured through the coefficient of air-permeability *kT* and the coefficient of water absorption at 24 hours a_{24} (the chloride migration ASTM C1202 and Wenner electrical resistivity were also measured).

The third case corresponds to a research reported by Jiménez and Moreno (2015), in which they replaced the reference crushed limestone coarse aggregate by 25%, 50%, 75% and 100% of recycled coarse aggregate obtained from heterogeneous structural concrete rubble. They tested two sets of mixes: one with *w/c*=0.50; OPC=410 kg/m³ and the other with *w/c*=0.70; OPC=293 kg/m³, water-curing the specimens for 28 days. For measuring porosity, water sorptivity *S* and air-permeability *kT*, the specimens were kept, after moist curing, at 28°C/80% RH until the age of test (91 days). Also measured were accelerated carbonation (4% CO_2) and chloride-diffusion by immersion in synthetic seawater.

The results of the permeability tests from the three investigations described above are presented in Figure 6.21a–c, data from Gurdián Curran (2016), Yamamoto (2016) and Jiménez and Moreno (2015), respectively.

Figure 6.21a indicates a moderate increase in *kT* and virtually no change in *a* with increasing contents of recycled aggregate, with a significant increase in W_p only for 100% of recycled concrete as coarse aggregate. Figure 6.21b presents similar results, indicating only moderate increases in *kT* and a_{24} with increasing content of recycled concrete aggregates. Figure 6.21c presents also a similar pattern; it is interesting to see that the mix with *w/c*=0.50 and 100% of recycled aggregate performs better (*kT*) or equal (*S*) than the corresponding mix with *w/c*=0.70 without recycled aggregate. In this research, the results of carbonation and chloride-diffusion showed a similar pattern, correlating very well with both sorptivity *S* and air-permeability *kT* (Jiménez & Moreno, 2015).

The fourth example corresponds to research work by Zong et al. (2014), in which the permeability of concrete made with 414 kg/m³ of binder (OPC+low-Ca PFA), *w/b*=0.45 and 100% natural aggregates was adopted as Control. The focus of the investigation was placed on the impact on the performance of concrete of replacing 30%, 40% and 50% of the

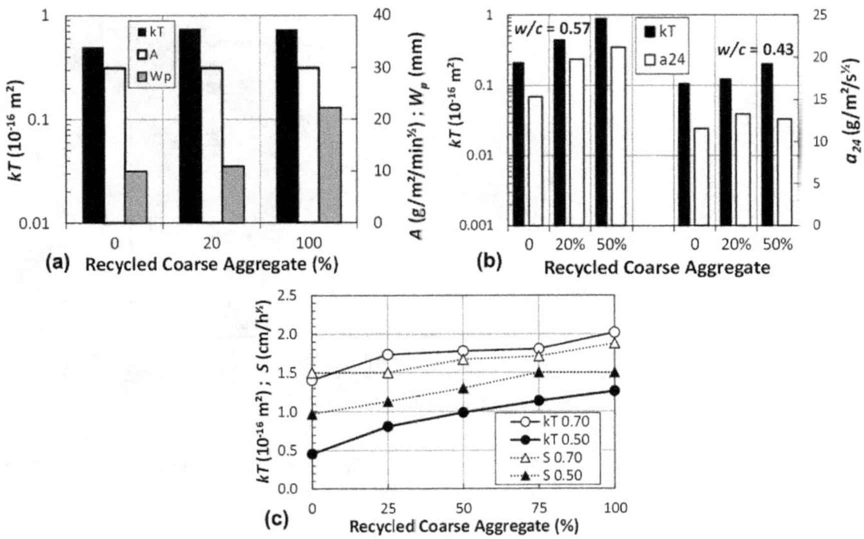

Figure 6.21 Permeability of concretes made with recycled concrete as coarse aggregate; data from (a) Gurdián Curran (2016), (b) Yamamoto (2016) and (c) Jiménez and Moreno (2015).

natural coarse aggregate of the mix by a recycled aggregate made of crushed and washed clay bricks obtained from a demolition site. A first negative effect of the recycled aggregate was that it greatly increased the water demand and thus, the resulting *w/b* ratio of the mixes became 0.48, 0.52 and 0.57 for substitution levels of 30%, 40% and 50%, respectively. There was a significant decrease of concrete density with increasing levels of replacement, reaching 12% for 50% replacement. In parallel, the compressive strength decreased by 16%, 27% and 44% for substitution levels of 30%, 40% and 50%, respectively. Regarding the permeability of the investigated concretes, the coefficients of air-permeability *kT* and of water-permeability K_w were reported (the method used for the latter is not indicated in the paper), as well as the capillary suction rate. Figure 6.22 shows the increase in air- and water-permeability of the concretes made with increasing content of recycled aggregate. The amount of water sucked by the concretes is represented by A_{24}, expressed in m^3 of water absorbed in 24 hours per m^3 of concrete.

Figure 6.22 shows clearly the negative impact of increasing replacement levels of coarse aggregate with recycled clay bricks on the gas- and water-permeability of concrete.

6.4.1.3 Spherical Steel Slag Aggregates

Electric Arc Furnace oxidizing slag (EAF slag) is a steel manufacturing by-product which, under a special production method, is obtained naturally

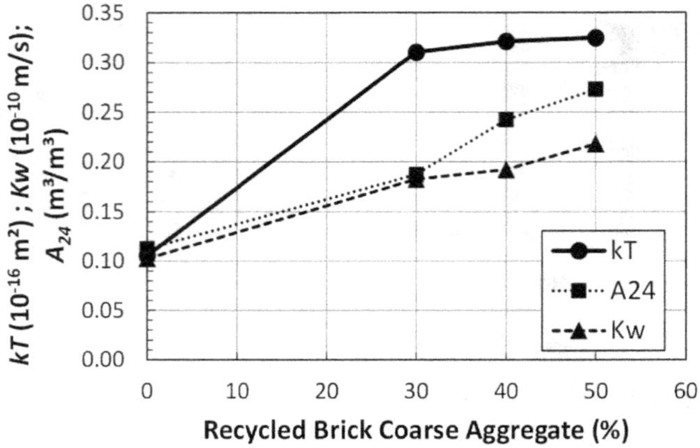

Figure 6.22 Permeability of concretes made with recycled crushed clay bricks as coarse aggregate; data from Zong et al. (2014).

in the form of fine spherical particles, in opposition to the angular-shaped particles obtained by crushing. An investigation was carried out by Roy et al. (2018) to evaluate the potential of these spherical slag particles to be used as partial or total replacement of natural sand in concrete-making, with a particular emphasis on concrete pavement application. The investigation showed that the inclusion of the spherical slag particles has a beneficial influence on the water demand, although accompanied by a slightly higher tendency for the concrete to segregate and bleed. Mechanical properties, such as compressive strength, E-modulus and flexural strength, showed improvement with the inclusion of the spherical slag particles, even to a 100% replacement ratio. The permeability of the concrete (*w/c* ratio = 0.40; 50 mm slump; 5% air content) was measured by the *kT* method as well as by a sort of Karsten tube water sorptivity test method (see 8.3.3.3). The *kT* results showed a moderate reduction in air-permeability for increasing spherical slag replacement ratios, falling within the "Low" Permeability Class, according to Table 5.2; the reduction of water sorptivity was more marked. For both tests, the concretes containing spherical slag outperformed those made with conventional angular steel slag.

6.4.2 Effect of ITZ on Concrete Permeability

The concept and relevance of the ITZ for the microstructure of concrete has been described in Sections 3.2.1 and 3.2.3. Suffice here to remind that ITZ (the 10 – 40 μm thick rim around the aggregate particles) presents a more open and porous microstructure than the bulk hydrated cement paste and constitutes a preferential path for the transport of matter.

Although it is not easy to measure the flow of matter through the ITZ, there is some experimental evidence that it strongly affects the transport

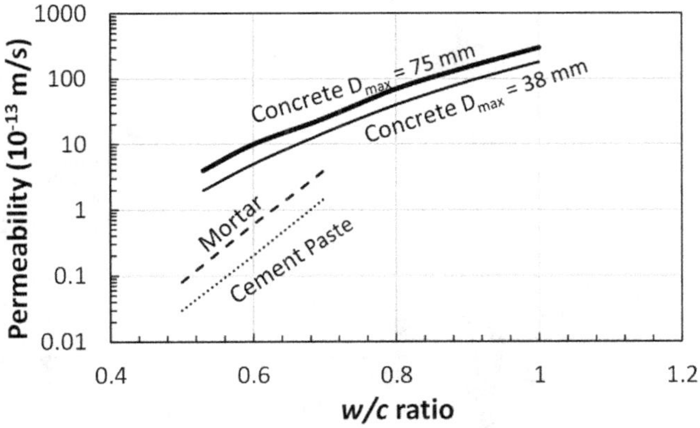

Figure 6.23 Water-permeability of paste, mortar and concretes as function of the *w/c* ratio (USBR, 1975) apud (Carcasses & Ollivier, 1999).

properties of concrete (Carcasses & Ollivier, 1999). Figure 6.23 shows the water-permeability measured on paste, mortar and concretes with two maximum aggregate sizes, as function of the *w/c* ratio (USBR, 1975), cited by Carcasses and Ollivier (1999).

It can be seen that for the same *w/c* ratio, the permeability of mortar is higher than that of paste which, assuming that the aggregate is less permeable than the paste, can be entirely attributed to the effect of the ITZ. The permeability of the concretes is higher than for mortar, with that containing 76 mm maximum size of aggregate being still more permeable than that made with 38 mm aggregate. It is difficult to assess the exact impact of the ITZ from these data, because the volume of ITZ is smaller and the tortuosity of the paths is higher in the 76 mm concrete, compared with the 38 mm concrete and yet, the permeability is higher. A possible explanation is that the ITZ around larger particles is more porous and permeable than that around smaller particles.

Watson and Oyeka (1982), cited by Carcasses and Ollivier (1999), measured the permeability of paste and concrete to crude oil, observing that the permeability of concrete specimens is about 100 times greater than that of paste specimens. In addition, by changing the volume of aggregate in the mixes they could study its impact on oil-permeability, showing that the permeability of concrete increases with the volume of aggregates in the mix, which is attributed to more oil flowing along the higher volume of ITZ.

Metha (1986) found that the permeability of concrete increased due to the presence of micro-cracks in the ITZ.

A detailed investigation on the ITZ around non-ferrous slag particles was reported by Harashina et al. (2018). Figure 6.24 shows the ITZ formed around a ferronickel slag aggregate particle. The main component of the ferronickel slag is MgO (see top right diffractogram). The intensity of the

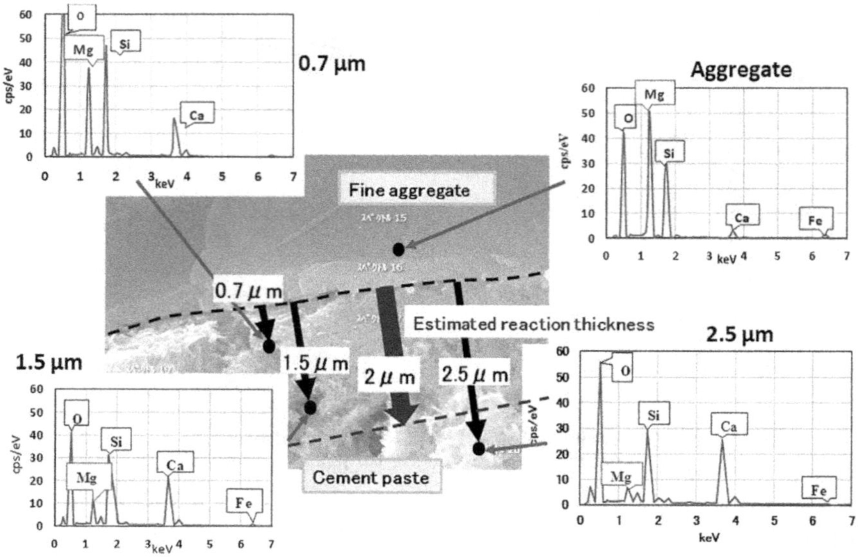

Figure 6.24 ITZ composition around ferronickel slag aggregate (Harashina et al., 2018).

Mg peak, observed in the paste at 0.7 µm from the surface of the aggregate (top left diffractogram), decreases with the distance from it (bottom diffractograms), suggesting that MgO in the paste was originated from the neighbouring slag particle (Harashina et al., 2018).

The thickness of ITZ can be assumed to be 2 µm as a midpoint of weak intensities at 1.5 and 2.5 µm (bottom left and right diffractograms, respectively). Consequently, the volume of reaction products (ITZ) can be calculated according to Eq. (6.6).

$$V_{tz} = \left[\left(\frac{r_2}{r_1} \right)^3 - 1 \right] . V_a \,.\, 100 \tag{6.6}$$

where

V_{tz} = volume of reaction product (ITZ) (%)
r_2 = average aggregate particle diameter + reaction thickness
r_1 = average particle diameter
V_a = volume ratio of slag aggregate particle

The volume of reaction products for different inclusions was calculated as shown in Figure 6.24, applying Eq. (6.6). The reference concrete contained a crushed coarse aggregate and natural sand (NS); part of the latter was replaced (30%, 50% and 100%) by different ferro-nickel slags (FNS) and copper slags (CUS).

Figure 6.25a shows the relation between the 28-day compressive strength of the concretes as function of the volume of reaction products in the ITZ. It can be seen that the compressive strength increases with an increase in volume of ITZ reaction products, indicating a densification of the ITZ. Figure 6.25b shows the relationship between the volume of ITZ reaction products and the coefficient of air-permeability *kT*, confirming the same effect.

A possible confirmation of the improved ITZ offered by CUS as aggregate is provided by the work of Kubissa and Jaskulski (2019). They studied the effect of replacing natural sand by a CUS-based waste (from surface blast-cleaning) on the performance of concretes made with different cements. The cements employed were (EN 197 classification in brackets): an OPC (CEM I 42.5R), a PFA Cement (CEM II/B-V 42.5N) and a GBFS Cement (CEM III/A 42.5N). Three mixes containing 360 kg/m³ and w/c=0.45 were prepared with each of the three cements; the control mix was made with natural sand 0–2 mm as fine aggregate. In the other two mixes, 66% of the natural sand was replaced by the (slightly finer) 0–2 mm CUS, one of similar consistency as the control (Mix 66) and the other with a more fluid consistency (Mix 66F), achieved with a higher dosage of superplasticizer. The permeability tests applied to the concrete specimens, moist cured during 28 days, were air-permeability *kT* (on samples dried at 20°C/55% RH until the age of 90 days) and water sorptivity at 6 hours a_6 of specimens oven-dried at 105°C until constant mass. These tests were complemented by chloride penetration depth (colorimetric test) of samples after 60 cycles of spraying (10% NaCl solution) and drying at 35°C and loss of mass and strength after an undisclosed number of freezing (–18°C) and thawing (+18°C) cycles, not discussed here. Figure 6.26a presents the data of *kT* and Figure 6.26b of a_6 for the three mixes prepared with the three cements described above.

Figure 6.26 shows the beneficial effect of replacing 66% of natural sand by CUS on the gas- and water-permeability of concrete, especially when

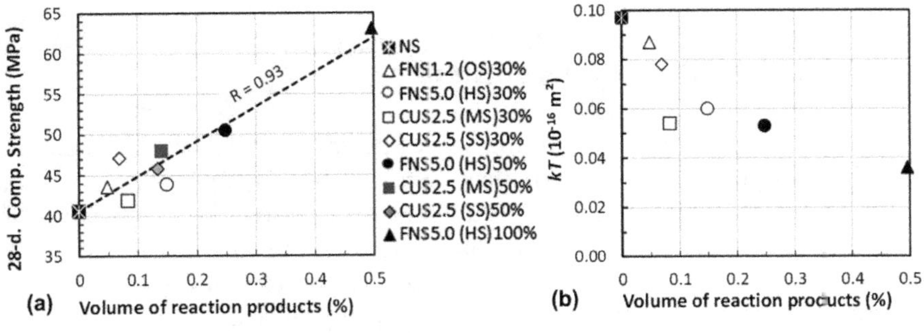

Figure 6.25 Effect of volume of reaction products in ITZ on (a) compressive strength, (b) air-permeability *kT*; data from Harashina et al. (2018).

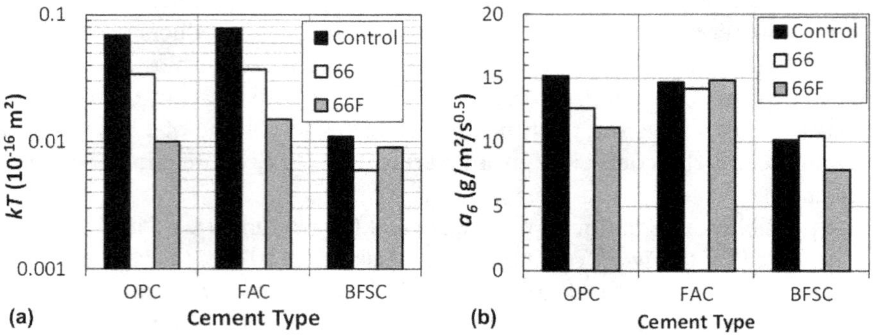

Figure 6.26 Effect of the addition of CUS sand on (a) *kT* and (b) a_6 of concretes made with different cements; data from Kubissa and Jaskulski (2019).

used in combination with an OPC. It is known that OPC, due to the preferential deposition of $Ca(OH)_2$ crystals near the aggregates, tends to produce weaker ITZs than cements containing active additions. Therefore, it appears logical to attribute the beneficial effect of CUS to an improved ITZ.

A contribution to the discussion on the role of ITZ and OPC vs Pozzolan Cements, widely used in Chile, on air-permeability and water sorptivity can be found in Bustos et al. (2015).

Although difficult to verify experimentally, there is enough evidence to claim that the ITZ constitutes a preferential path for the flow of matter through concrete and, thus, a weak link in the chain. Densifying the ITZ by choosing suitable aggregates and binders should be considered as a strategy to improve concrete durability.

The use of SCMs may improve particle packing and generate additional hydration products that refine the overall microstructure of concrete, in particular of the ITZ (Larbi, 1993; Ollivier et al., 1995; Scrivener et al., 2004), an effect that shall be confirmed by proper testing on a case-by-case basis.

6.5 EFFECT OF SPECIAL CONSTITUENTS ON CONCRETE PERMEABILITY

Exploring the possibility of adding some special constituents to concrete, i.e. those beyond the classical ones (cement, SCMs, aggregates, water, water-reducers or air-entrainers), is a fertile ground to enlarge the range of properties and applications of the material. Such special constituents may be: pigments to improve aesthetics, fibres to improve toughness, polymers to improve strength, corrosion inhibitors to improve protection of the embedded steel, expansive agents to reduce shrinkage-induced cracking, etc.

However, doubts often arise on the possible negative impact of such materials on the durability performance of the involved concrete structures, which are addressed by testing, permeability testing being one of the preferred tools.

In this section experimental results of permeability testing of concretes containing such special constituents are presented.

6.5.1 Pigments

Pigments are used to extend the aesthetic possibilities of concrete and are widely used, the addressed question being: do they affect permeability and, hence, durability?

A comprehensive investigation was conducted by Positieri et al. (2011), in which different powdered pigments – red (R), yellow (Y) and black (B) – were added to three grey concretes (G), used as control; the three concretes are identified as of low, medium and high *w/c* ratio (approximately 0.45, 0.55 and 0.61, respectively). To the three control G concretes, 26, 22 and 19 kg/m³ of R, Y and B pigments were added, respectively. After 28 days of curing, three different permeability tests were applied to specimens made with the resulting 12 concrete mixes: air-permeability *kT* and water-permeability (capillary suction *A* and maximum water penetration under pressure W_p). Figure 6.27a and b shows the results of the *kT* and *A* tests, respectively.

Figure 6.27 shows that the inclusion of the pigments is not only unharmful to the permeability of concrete but, even, in several cases it is beneficial. This positive effect of the pigments was attributed to their fine particles contributing to the reduction of the reported bleeding, especially for medium and high *w/c* ratios (Positieri et al., 2011).

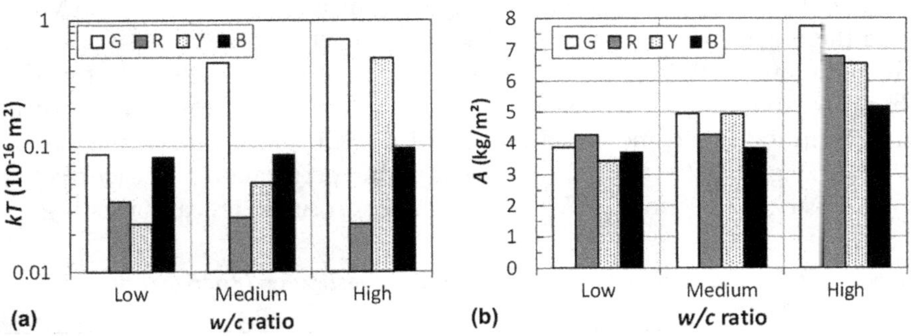

Figure 6.27 Effect of pigments and *w/c* on (a) *kT* and (b) capillary suction *A*; data from Positieri et al. (2011).

6.5.2 Fibres

An investigation was conducted to study the effect of incorporating synthetic and metal fibres into two concrete mixes, typical for precast elements construction, one conventional and the other self-compacting (Rodríguez de Sensale et al., 2018). The characteristics of the fibres used, as declared by the manufacturers, are shown in Table 6.3.

The synthetic fibres are polyolefin corrugated macrofibres, whilst the hooked steel fibres StF are made of cold-drawn low carbon steel. The fibres were added in two contents by mass: C1 (4 and 20 kg/m^3 for synthetic and StF, respectively) and C2 (8 and 30 kg/m^3 for synthetic and StF, respectively).

The fibres were added to two control mixes: conventional concrete CC35 (350 kg/m^3 OPC; w/c=0.46; cylinder $f'c_{28d}$=37 MPa) and self-compacting concrete SCC35 (390 kg/m^3 OPC; w/c=0.45; cylinder $f'c_{28d}$=48 MPa). Besides the mechanical performance in compression and bending, the permeability to air kT and the performance under the RCPT (ASTM C1202) were studied.

Figure 6.28a and b presents the effect of the inclusion of both types of fibres on the air-permeability kT of CC35 and SCC35 mixes, respectively; values represented are the geometric mean of the reported individual values.

Figure 6.28 shows that, except for the case of adding 8 kg/m^3 of synthetic fibres to the CC35, the incorporation of fibres (both synthetic and metallic) does not change significantly the permeability to air kT of conventional and self-compacting concrete. One explanation of the higher kT for high content of synthetic fibres may lay in clustering of the fibres or some negative effect on the consolidation of the conventional concrete.

The effect of including StF in loaded and unloaded concrete (Paulík & Hudoba, 2009) is discussed in Section 6.10.1.

Another investigation was conducted (Moreno et al., 2013) to study the effect of incorporating multifilament polypropylene fibres (PPF) on the performance of concretes of relatively high w/c ratios. Two control mixes were prepared, one with (w/c=0.62; 363 kg/m^3 cement; cylinder $f'c_{28d}$=25 MPa) and the other with (w/c=0.80; 281 kg/m^3 cement; cylinder $f'c_{28d}$=16 MPa). To these mixes, 0.9 kg/m^3 of PPF were added, evaluating the effect of the fibres on the mechanical performance of the concretes as well as on their air-permeability kT, measured on 150×150×600 mm beams at the age of 90 days. Figure 6.29 shows the results of the air-permeability tests (corrected by porosity, see Section 5.7.7, given the high porosity aggregates used).

Table 6.3 Characteristics of the fibres used in the investigation (Rodríguez de Sensale et al., 2018)

Fibre	*Diameter(mm)*	*Length(mm)*	*L/D ratio(-)*	*Tensile strength(MPa)*	*Density(kg/m^3)*
Synthetic	1.37	48	35	>550	920
Steel	1.00	50	50	>1,100	7,850

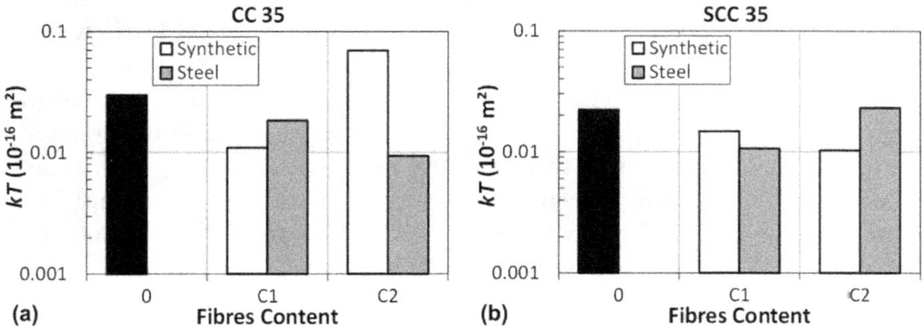

Figure 6.28 Effect of type and content of fibres on air-permeability *kT* of (a) conventional concrete and (b) SCC; data from Rodríguez de Sensale et al. (2018).

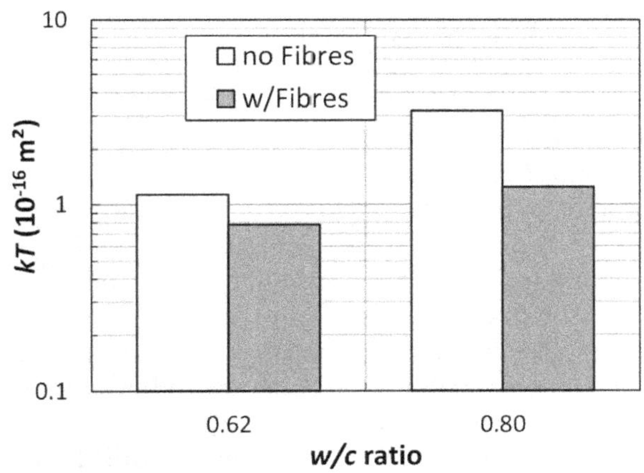

Figure 6.29 Effect of PPF on the air-permeability *kT* of two concretes; data from Moreno et al. (2013).

Figure 6.29 shows that the addition of PPF to concretes of relatively high *w/c* ratios, moderately reduces the air-permeability *kT*; the concrete of *w/c*=0.80 with fibres is still slightly more permeable than the one with *w/c*=0.62 without fibres.

A similar reduction in *kT* by the addition of PPF was reported by Adámek and Juránková (2009).

The possibilities of the combined use of locally available dune sand and alfa grass (*Stipa tenacissima* L.) to produce a repair mortar was investigated in Algeria by Krobba et al. (2018). Alfa grass is a tussock grass widely disseminated in semi-arid and arid regions, in North Africa and southern Spain (where it is called "Esparto"). For this investigation, Stipa fibres of 150–250 μm diameter were hand-cut to lengths between 3 and 5 mm. Dune sand mortars with and without the fibres (content within 0.1% and 1.25% by volume of mortar) of exactly the same composition were prepared (including 2% of superplasticizer); the mortars flow ranged between 110% for the control mortars and 108% for mortars with high fibre contents, indicating that the fibres do not affect sensibly the workability. Flexural and compressive strength, as well as drying shrinkage of the mortars, were investigated, as well as their permeability through capillary suction and N_2-permeability (Cembureau test method). Reported test results of the mortar with 0.75% vol. fibres showed moderately higher (16%) a_{24} values and identical intrinsic N_2-permeability than the control mix without fibres.

6.5.3 Polymers

In the attempt to develop high-performance concrete for industrial floors, the use of polymers was investigated by Holcim Brazil (Kattar et al., 1995). A concrete mix with w/c=0.46; 390 kg/m^3 of high-early strength OPC and cylinder $f'c_{28d}$=38.6 MPa was used as control, to which four different polymers were added: polyacrylate (ACR), styrene-butadiene (SB) and two brands of polyvinyl acetate (PVA1 and PVA2), in dosages indicated by the manufacturers or established by trial mixes. The control concrete specimens were moist cured for 28 days whilst those of the four polymer-modified concretes were cured 24 hours in the moulds at 23°C/95% RH and, thereafter in a room at 23°C/65% RH, until the age of test. Besides the mechanical performance, the permeability of the concretes was determined by three different methods: air-permeability kT, coefficient of capillary suction A and water penetration under pressure W_p. Figure 6.30 presents the permeability results obtained by the three test methods on the control mix and on the polymer-modified mixes.

Figure 6.30 shows that, in general, polymer-modified concretes have a lower permeability than the control mix, almost irrespective of the test method. In particular, the permeability performance of the concretes with ACR and SB polymers is superlative.

The good performance of SB polymer in concrete was confirmed by Chmielewska (2013), who investigated the effect of adding it at dosages of 5%, 10%, 15% and 20% by mass of cement to a control mix made with 356 kg/m^3 OPC and w/c=0.45. She measured the gas-permeability K_g applying the Cembureau method, using Helium as transport gas, finding a very strong reduction of several orders of magnitude in K_g as the dosage of polymer raised (Figure 6.31), to the extent that, for 20% SB polymer, no gas flow was detectable across the specimen.

Bhutta et al. (2009) developed a hardener-free epoxy-modified mortar aimed for repair work. A bisphenol A-type epoxy resin without any hardener was added in mass polymer/cement proportions (P/C), of 0%, 5%, 10% and 15%, to a 1:3 cement:sand mass ratio mortar. The compressive strength of the mortar increased with P/C up to 10%, decreasing for higher P/C ratios. In parallel, the air-permeability kT decreased monotonically

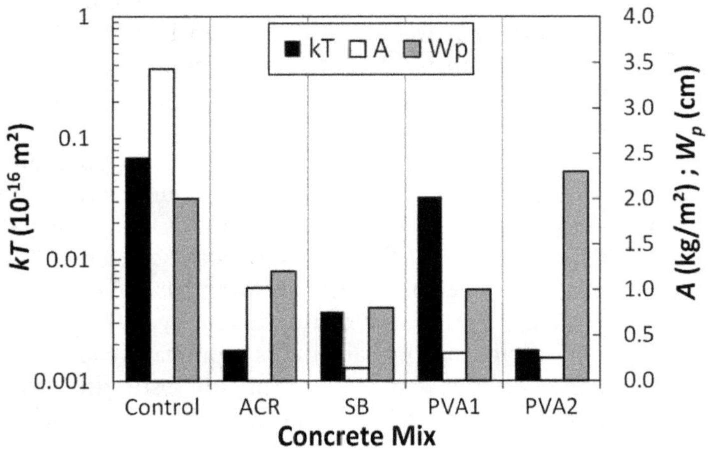

Figure 6.30 Permeability of polymer-modified concretes compared with a control mix; data from Kattar et al. (1995).

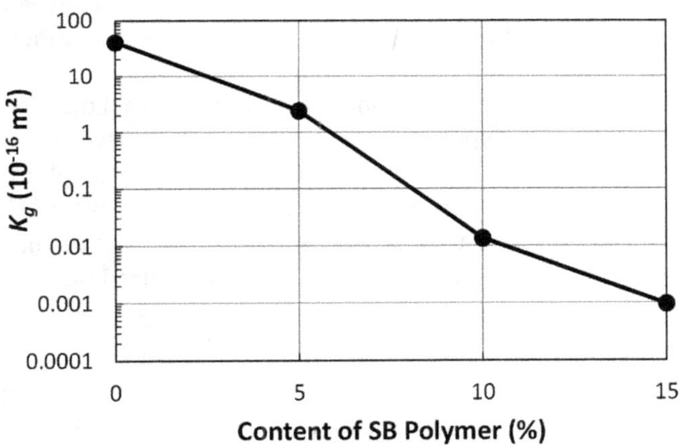

Figure 6.31 Helium-permeability of polymer-modified concretes as function of the SB polymer content; data from Chmielewska (2013).

with increasing P/C contents, becoming ten times lower for P/C contents of 10% or higher. Surface moisture measurements showed a marked increase for increasing P/C contents, which may have contributed to the lower *kT* measured values.

6.5.4 Expansive Agents

Expansive agents are used in concrete, typically to compensate for shrinkage deformations due to drying or cooling. CaO-, calcium sulphoaluminate- and MgO-based expansive agents are the most commonly used.

In Argentina, the use of shrinkage-compensated concrete for industrial floors became quite popular, after a pioneer start in 1999 (Fernández Luco et al., 2003), reason why an investigation was conducted on the performance of such concretes, summarized by Fernández Luco (2004). Among the properties investigated, air-permeability *kT* was included. Figure 6.32 shows the values of *kT* obtained on concretes made with different cement types and dosages (in percentage of the cement content) of the CaO-based product Onoda-Expan. The cements used were an OPC, a blast-furnace slag cement BFSC, a LF cement LFC and a composite cement CPC.

Figure 6.32 shows the beneficial effect of adding 10% of expansive agent (EA) to concrete mixes; irrespective of the cement type, the permeability falls into the "Low" Permeability Class and is significantly lower than that of the OPC concrete without expansive agent. Interesting to observe is a deliberate overdose (16%) of the EA on the OPC concrete which shows a huge increase in permeability, attributable to micro-cracking. The decrease in permeability at 10% dosage of EA can be attributed to the obturation of pores with hydration products of the EA; under real conditions, the consolidation of the microstructure due to the restrained expansion would probably decrease the permeability even further. Similarly, the large increase of permeability due to an overdose of EA may not take place under restrained conditions.

In this respect, a commercial concrete supplied by a Japanese producer, containing 20 kg/m^3 of expansive agent, was placed in two 0.6×0.6×0.9 m prismatic forms (Van et al., 2019). One of the prisms was plain whilst the other contained a cage of vertical Ø19 rebars and horizontal Ø16 stirrups. After 7d moist curing, the prisms were stored indoors until the ages of 28 and 56 days, when the air-permeability *kT* was measured on eight points of

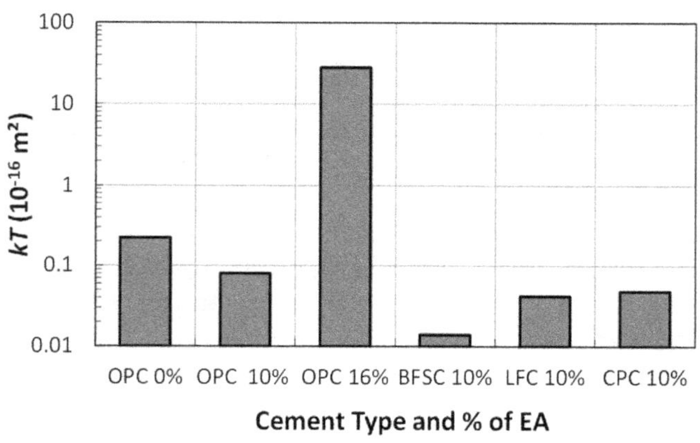

Figure 6.32 Effect of cement type and dosage of expansive agent on concrete air-permeability *kT*; data from Fernández Luco (2004).

each prism. At both ages, lower *kT* values were measured on the reinforced prism compared to the plain one (50% lower at 28 days and 75% lower at 56 days), which was attributed to the restraining effect of the bars to the expansion of the concrete, creating some "chemical precompression".

An investigation was conducted (Wang et al., 2011) to study the effect of a MgO expansive agent, alone and in combination with steel fibres (StF), on the performance of a concrete mix made with 340 kg/m³ OPC; $w/c=0.42$; $f'c_{28d}=57$ MPa. To this mix, 27.2 kg/m³ of EA or/and 78 kg/m³ of hooked-end StF, (Ø=0.4 mm and $L=25$ mm) were added; Figure 6.33 shows the air-permeability *kT* of the four mixes investigated. The MgO EA produced a slight reduction of *kT*, whilst the StF did not have any effect on *kT* (in line with what was discussed in Section 6.5.2). The combination of EA and StF, shows some synergy, reducing the air-permeability by 40%, possibly due to the restraining effect of the fibres to the concrete expansion leading to a certain pre-compression.

The performance of concretes made with and without a calcium sulphoaluminate-based expansive agent (used in conjunction with a Glycol-based shrinkage reducer) was investigated by Koh et al. (2006). They found that reducing the *w/c* ratio from 0.50 (normal concrete) to 0.30 (high-performance concretes) improved significantly the durability performance, measured through several tests, including chloride migration, carbonation and water- and air-permeability *kT*, performance that was further improved when the expansive agent was added at a dosage of 30 kg/m³ to the mix with $w/c=0.30$.

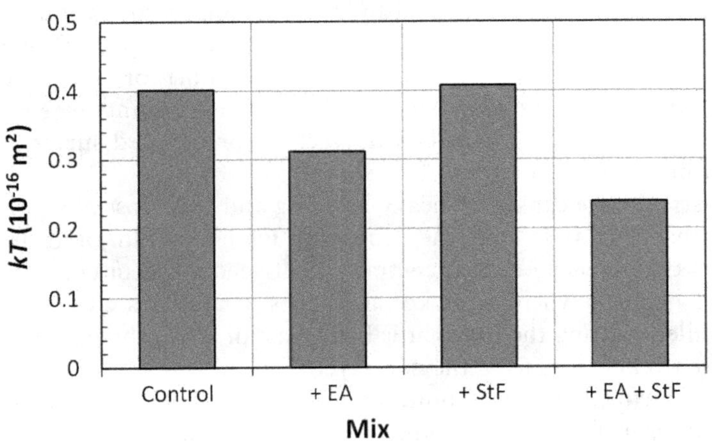

Figure 6.33 Effect of addition of MgO expansive agent (EA) and StF on concrete air-permeability *kT*; data from Wang et al. (2011).

6.6 EFFECT OF COMPACTION, SEGREGATION AND BLEEDING ON PERMEABILITY

Placing and compaction/consolidation are among the most important operations in concrete construction. According to Mehta and Monteiro (2006), compaction or consolidation is a process that envisages moulding concrete within the forms and around embedded elements and rebars and the elimination of voids pockets and entrapped air. Sometimes, to spare time and labour, concrete is poured from great heights and at considerable rates, which promotes the appearance of coarse aggregate nests (honeycombing) due to segregation. Low paste-aggregate ratios and low fine-to-coarse aggregate ratios also favour segregation.

Excessive bleeding in concreting of deep elements can be avoided if the poured concrete is gradually less plastic as the layers approach the top surface. Also, concrete shall be poured near its final position and in layers of thickness that enables a proper compaction. The previous layer shall be compacted before start pouring the next one, whilst the next layer shall be placed while the preceding one is still in a plastic state.

A proper compaction is of utmost importance to obtain a suitable finishing of moulded surfaces and to make the concrete as dense as possible, i.e. to minimize voids and to promote a uniform distribution of aggregates within the concrete mass. This is achieved by supplying vibration/tamping energy to overcome cohesion and friction forces between concrete particles.

Concrete consistency plays an important role in the result of concreting operations and should match the available tools for its placement and consolidation.

The main consequences of inadequate placing or poor compaction are: honeycombing, excessive voids, bug-holes, sand streaks and cold joints.

Honeycombing may be caused by poor placing, segregation or insufficient consolidation. Excessive voids are due to entrapped air that could not be released because vibration time was too short, or vibration power was insufficient, or the distance between vibration points was not close enough; they tend to appear behind vertical or inclined surfaces in the form of bug-holes (see Figure 11.23).

Sand streaks are caused by heavy bleeding and paste loss along the form surface that may arise from the deposition method. Cold joints are originated by the absence of interpenetration between adjacent concreting layers. This happens when the poker-head does not penetrate the underlying layer while vibrating the one that is being cast or when the preceding layer has stiffened before placing the new layer.

From a performance viewpoint, these problems affect aesthetics and produce heterogeneities, poor bonding between reinforcement and concrete, and loss of effective cross-section and cover depth. Indeed, the depth of honeycombs or bug-holes corresponds to an equal diminution of concrete cover.

All things considered, concrete permeability will be affected by placing and compaction. For instance, according to Mehta and Monteiro (2006), inadequate compaction is a typical cause of insufficient water-tightness. Analysing previously published data, Gonçalves (1999) concluded that compaction has a larger effect on resistance to chloride penetration than on compressive strength. Nevertheless, when trying to assess compaction effects on concrete permeability at a laboratory scale, the results are not conclusive because mimicking the lack of compaction in small – easy to compact – laboratory specimens is not easy. Although Gonen and Yazicioglu (2007) have found a consistent increase in sorptivity and porosity when lowering the compaction efficiency, it is quite common that researchers do not find conclusive variations of permeability with compaction degree. This happened, for instance, with Neves et al. (2011), Starck (2013) and Nishimura et al. (2015) for gas-permeability results and with Kumar and Bhattacharjee (2004) for surface water absorption.

Gonen and Yazicioglu (2007) assessed sorptivity and porosity under vacuum of a mix with 415 kg/m³ of OPC and a *w/c* ratio of 0.53. Their specimens were either vibrated, compacted by spading 25 or 15 times, or not compacted at all. Gonen and Yazicioglu (2007) concluded that compaction has a very important effect on concrete durability as they found consistent and relevant differences in concrete sorptivity and porosity with the compaction degree (Figure 6.34), while only minor differences in compressive strength were found, except for the non-compacted specimens.

Neves et al. (2011) carried out a research to investigate the effect of compaction on concrete carbonation resistance. Their study comprised four

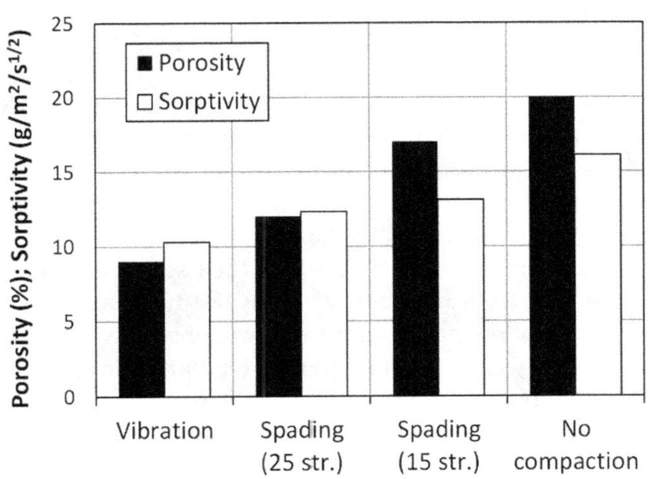

Figure 6.34 Sorptivity and porosity of concrete under various compaction procedures; data from Gonen and Yazicioglu (2007).

concrete mixes and oxygen-permeability *kO* was assessed on Ø150×50 mm discs, sliced from 150 mm diameter and 300 mm height cylinders. For each batch, a first set of specimens was compacted using a poke vibrator during a time judged by the operator as sufficient to achieve full compaction. A second set was compacted the same way but during half of the time. Finally, the last set was compacted by rodding. Although a good correlation between *kO* and carbonation depth was found, the effects of compaction on *kO* were not clear. Instead, the binder content and *w/b* ratio appeared as dominant factors defining the oxygen-permeability.

Within the scope of a study to investigate permeation quality of concrete, Kumar and Bhattacharjee (2004) applied different compaction methods (vibration and rodding) aimed at achieving concretes with different permeability. Then, they applied the Initial Surface Absorption Test (ISAT) method (Section 4.2.2.1) to evaluate water-permeability. It was observed that in four out of the five cases the ISA of the rodded samples exceeded that of the vibrated samples.

A successful experiment at relatively large scale was conducted by Liang et al. (2013) that will be described in some detail. In this research, a comprehensive theoretical and experimental analysis was made for optimizing the distance and time of vibration, using mock up elements at industrial scale. In particular, they prepared concrete blocks (500×500 and 800 mm high) in wooden moulds, cast in two layers 400 mm high each, by pouring concrete through an inverted street marking cone (bottom opening Ø100 mm) from the four angles of the mould. Each layer was compacted applying a poker vibrator (Ø50 mm) fixed in the centre of the section during predefined times.

Two concrete mixes were tested, both with *w/c*=0.50; one having a slump of 80 mm (OPC=300 kg/m^3) and the other of 150 mm (OPC=324 kg/m^3). Both mixes were air-entrained: 5.1% and 4.5%, respectively. The 28-day compressive strength of the mixes was about 40 MPa, based on drilled cores' testing. Two blocks were cast with the 80 mm slump mix, with vibration durations of 15 and 60 seconds, respectively, and another two with the 150 mm slump mix, with vibration durations of 15 and 30 seconds, respectively. Only the block of 80 mm slump, vibrated just 15 seconds showed surface anomalies in the form of air voids. The four blocks were kept in the forms during 5 days and then stored at 20°C/60% RH until the age of 28 days. At that age, 20 cores (Ø100×200 mm) were drilled from one surface, following a pattern of four cores in width per five cores in height. The cores were tested for strength (compressive and splitting) and for accelerated carbonation. Prior to that, air-permeability tests *kT* were conducted *in situ* on same places where the cores were later drilled. The *kT* results obtained on the four blocks are presented in Figure 6.35, showing the geometric mean $kT_{gm} \pm s_{LOG}$ (see Section 5.8.3). Unfortunately, the paper does not identify the results as function of the height, although the cores'

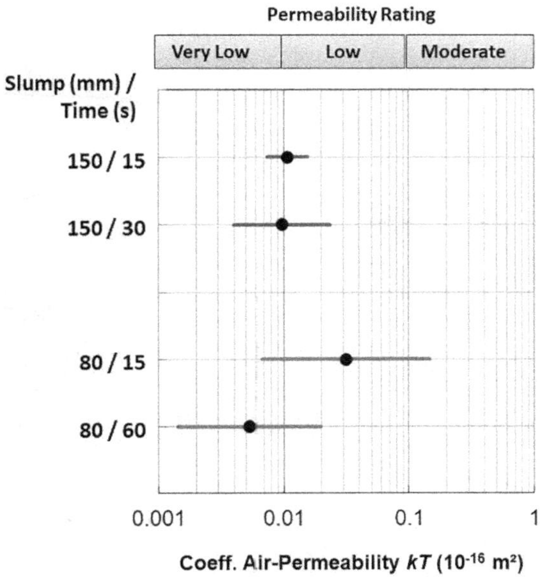

Figure 6.35 Air-permeability *kT* results obtained for different compaction practices; data from Liang et al. (2013).

tests did not indicate a significant effect of the position of the core on the strength results.

Figure 6.35 shows that the mix with 80 mm slump, thoroughly compacted (60 seconds vibration) had the best performance in terms of kT_{gm} air-permeability. On the opposite, the same mix but inadequately compacted (15 seconds vibration) showed the worst results. The mix with 150 mm slump shows an intermediate result, without a significant beneficial effect in extending the vibration time from 15 to 30 seconds. It is also interesting to note the smaller variability in *kT* of the mixes with 150 mm slump compared to those of the 80 mm slump, which may indicate a more homogeneous compaction of the softer mix. The accelerated carbonation test also showed 15% higher values for the 80 mm slump concrete vibrated for 15 seconds, compared to the other three blocks.

Within a comprehensive research conducted by Hayakawa and Kato (2012), Hayakawa et al. (2012) and Kato (2013), the effect of segregation and bleeding on the air-permeability *kT* was investigated. A set of 14 OPC concrete mixes, of widely different characteristics, were prepared in the laboratory; the composition and fresh state characteristics (slump, air content and bleeding) are described in Kato (2013).

The effect of bleeding on air-permeability was investigated by casting three 150 mm cubes with concrete mixes of same $w/c=0.55$ but different compositions: slump between 30 and 225 mm and bleeding between 0.0

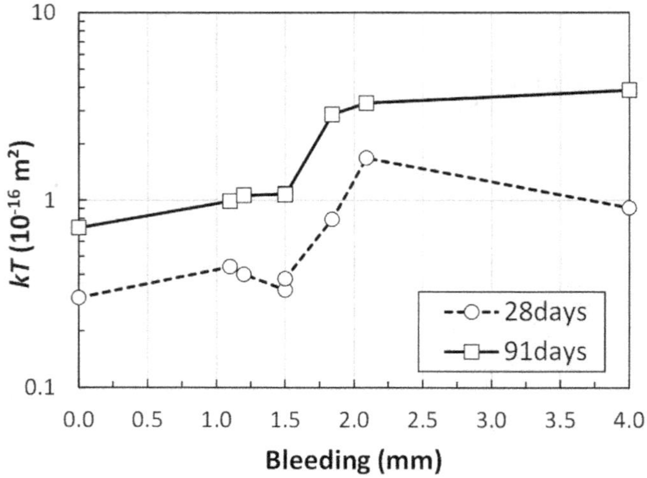

Figure 6.36 Effect of bleeding on air-permeability *kT* measured on lateral faces of 150 mm cubes; data from Kato (2013).

and 4.0 mm). The zero-bleeding mix was achieved by adding a viscosity-enhancing agent. After casting, the cubes were seal-cured for 7 days and thereafter exposed to a room with 20°C/60% RH until the time of test (28 and 91 days). The air-permeability *kT* was measured on the lateral sides of the cubes reporting the average of three cubes, shown in Figure 6.36. The *kT* results obtained at both 28 and 91 days show an increase in permeability with the amount of bled water. Kato expected a different result, since the amount of bleeding on the upper surface should have led to a lower *w/c* ratio in the rest of the cube and a possible reduction in *kT* with bleeding. He concludes that bleeding channels may increase the connectivity of the pore network in the specimen leading to an increase of *kT*. Unfortunately, *kT* values on the top surface were not measured or reported. A similar result was obtained by Positieri et al. (2011) on concretes with and without powdered pigments, already discussed in Section 6.5.1. They found an increase in *kT* with the amount of bleeding of the mixes; a similar effect was found on water sorptivity. Figure 6.36 suggests that, when bleeding exceeds 1.5 mm, the effect on *kT* becomes more significant (suggestion supported by the results of Positieri et al. (2011)).

Another aspect investigated by Kato (2013) was the effect of segregation, with and without the presence of reinforcement bars, on the air-permeability *kT*. The effect of segregation was studied on a block (500×500 and 800 mm high) with three of the sides reinforced as detailed in Figure 6.37. The figure also shows (dotted circles) the locations of the air-permeability measurements that allowed establishing profiles of *kT* values with height. The mix selected for the investigation had *w/c*=0.55, 125 mm slump, 5.0% air and 1.2 mm bleeding.

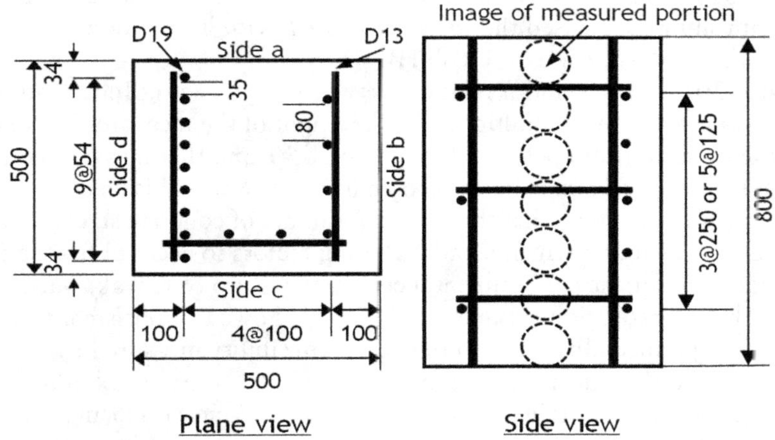

Figure 6.37 Details of the "segregation" block (Kato, 2013).

The kT profiles obtained along the four lateral faces of the block are presented in Figure 6.38 (the reinforcement pattern is expressed as separation of vertical bars – separation of stirrups in mm). The thick vertical line ($kT=0.61\times10^{-16}$ m^2) indicates the kT value measured on a companion 150 mm unreinforced cube. Figure 6.38 shows a general trend of increasing kT values with height, attributable to settlement and segregation of solids in the fresh concrete. The effect is more acute for the unreinforced side of the block (showing a ratio of 30 between the extreme top and bottom kT measurements), which suggests that the reinforcement, even with the relatively large cover thickness of the experiment, helps in preventing the settlement and segregation of the concrete. This was confirmed by Hayakawa and

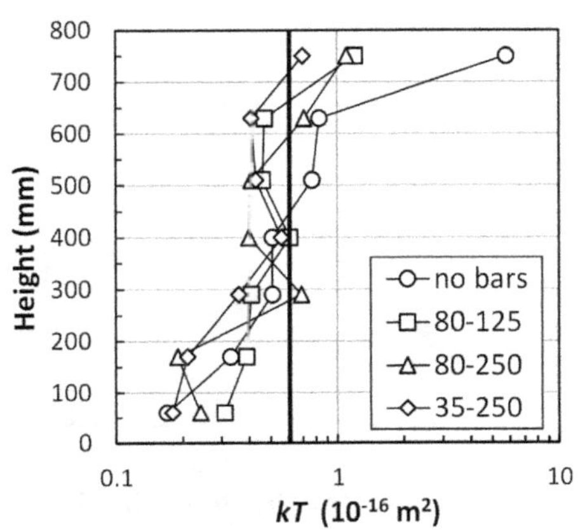

Figure 6.38 *kT* profiles with height; data from Kato (2013).

Kato (2010), showing also that excessive vibration of a relatively soft mix (180 mm slump) increased the variability of *kT* with height along relatively large prisms 0.9×0.9×1.2 m (L×W×H).

Kato (2013) also found a relation between the Bleeding Index (amount of bleeding water in cm multiplied by the height of the measurement in cm) and the *kT* ratio (ratio between the measured *kT* and that measured on the lateral face of a 150 mm cube), as shown in Figure 6.39. This bleeding index is used in a model to predict the carbonation rate of concrete structures, on the basis of *kT* measurements and including factors to account for the type of binder and for the environmental conditions (Kato & Hayakawa, 2013).

Another investigation (Nsama et al., 2018) showed a consistent trend of higher gas-permeability (*kT* and oxygen-permeability measured by an electrochemical technique) near the top of 300×300 ×1,500 mm columns, compared to the results obtained near the bottom. MIP measurements made on samples taken at different heights confirmed a larger porosity near the top than near the bottom of the columns. Nsama et al. (2018) concluded that

> "air-permeability coefficients tended to be higher in the upper parts of the concrete column specimens compared to those observed in the lower parts, due to segregation in form of bleeding. This can be better accounted for by the larger sizes of interconnected pores formed in those locations, especially in the OPC, as investigated via MIP. The variation observed in the rate of oxygen-permeability measured using cathodic polarization technique was like those observed in the results of air-permeability coefficients taking into consideration of moisture content. The observed results suggested that the upper parts of column specimens are more prone to corrosion, attributed to the adverse effects

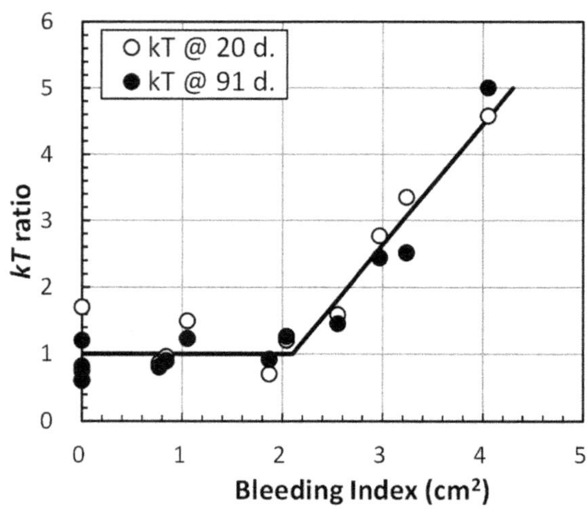

Figure 6.39 Relation between *kT* ratio and Bleeding Index; data from Kato (2013).

> of bleeding on the cover concrete and integrity of concrete and horizontal steel bars."

The findings of these investigations definitely provide ground to the importance of testing concrete permeability not just on laboratory specimens but also on site, as discussed in Chapter 7.

6.7 EFFECT OF CURING ON PERMEABILITY

6.7.1 Relevance of Curing for Concrete Quality

Curing is an essential step in concreting practice, very often overlooked; its importance for cement hydration was discussed in Section 3.1.2.

Curing can be defined as the measures to be undertaken to provide hydrothermal conditions to young concrete that are favourable for the material to fully develop its potential performance, which implies:

- advanced development of hydration reactions
- avoidance of plastic shrinkage cracking
- avoidance of drying shrinkage cracking
- avoidance of thermal cracking
- avoidance of frost damage by premature exposure to sub-zero temperatures

Regarding concreting practices in general and curing in particular, it is worth citing from Chapter 3 of Neville (2003):

> "In my experience, achieving durable concrete requires all the operations of concreting to be as nearly perfect as possible.
>
> I would like to move now to several more practical aspects of making good concrete. Because they are practical, some people, mainly academics, regard them to be unworthy of consideration and study; above all, these problems cannot be solved by an elaborate computer program.
>
> Curing concrete is the lowest of low-tech operations... it is seen by many as a silly operation, a non job...and [bad] curing does not show... If I emphasize ensuring curing, it is because curing can make all the difference between having good concrete and having good concrete ruined by the lack of a small effort."

Multiple references to curing effects on concrete properties can be found in the literature. Several authors (Powers et al., 1954; Page et al., 1981; Gräf & Grube, 1984; Rasheeduzzafar & Al-Saadoun, 1989; Balayssac et al., 1995; Meeks & Carino, 1999; Bai et al., 2002; Güneyisi et al., 2007) reported that early drying causes higher "penetrability" and that curing conditions, especially curing time, have a large effect on the durability of concrete. According to Hansen (1986), in ordinary mixes, poor curing can easily triplicate capillary porosity. Although curing has also an effect on concrete strength, it is acknowledged that it has more impact on durability, because it affects more strongly the surface layers, which are vital to prevent the penetration of aggressive agents (Ramezanianpour & Malhotra, 1995; Fernández Luco, 2010; Maslehuddin et al., 2013).

When moist curing is stopped before binder hydration is sufficiently developed (the most common situation), as the water loss is not uniform across concrete volume, there will be a zone, near the surface where the external environment has stronger effect on the local humidity regime, the so-called curing-affected-zone (CAZ) (Cather, 1994). This effect is reflected in different concrete properties between the CAZ and the remaining concrete volume; in fact, inadequate curing can result in a very weak and porous material near the concrete surface (Gowripalan et al., 1990; Ewertson & Petersson, 1993). This topic is dealt with in more detail in Chapter 7.

6.7.2 Effect of Curing on Permeability

6.7.2.1 Investigations in the Laboratory

The investigation described in Section 6.2.1.1 provided useful experimental information, not just on the type of binder but also on the effect of curing on several characteristics of concrete, including water- and gas-permeability. Figure 6.40a shows the effect of moist curing on the oxygen-permeability *kO* and Figure 6.40b on the initial water absorption rate a_3 (measured after 3 hours of contact with water). The data presented correspond to mixes prepared with various OPCs (RO, GA and U2, described in Table 6.1), exposed to Curing regimes A, B and C (1, 7 and 28 days moist curing), described in Section 6.2.1.1 and to cores drilled from the bottom side of the slabs as cast (hence why 1 day moist curing is attributed to the slabs that were kept in their moulds for 24 hours). The mixes are referenced by the OPC code (RO, GA, U2) followed by the concrete compressive strength level in MPa.

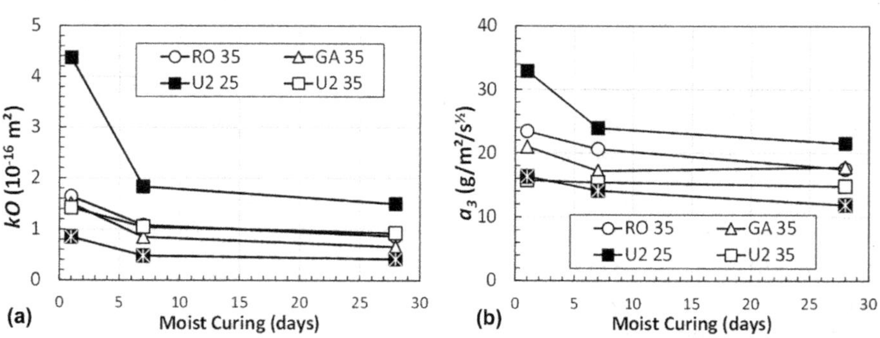

Figure 6.40 Effect of curing and concrete strength on (a) O_2-permeability *kO* and (b) water sorptivity a_3.

Both figures present basically the same pattern, which is very typical for most water- and gas-permeability tests on OPC concretes: moist curing in the laboratory is extremely beneficial in the first 7 days, but extending it beyond 7 days causes only marginal reduction in permeability. Notice also that the mix of lower strength (25 MPa) is more sensitive to the lack of curing than mixes of higher strength. This can be explained by the fact that poorer mixes have more and coarser pores that facilitate the evaporation of water, especially at early ages.

Figure 6.41 shows the ratio between the oxygen-permeability *kO* measured on concretes moist cured just 1 day and that measured after 7 and 28 days moist curing. The figure includes concretes made with OPCs and with the four blends of U2 with SCMs: 20% PFA (FA); 35% GGBFS (SL); 15% LF and 8% SF, described in Section 6.3.2. All mixes were of strength level 35 MPa, except the SF blend which was slightly stronger. The top of the black bar shows the ratio of *kO* between 1 and 7 days moist curing and the top of the white bar the ratio between 1 and 28 days moist curing.

What can be seen in Figure 6.41 is the already discussed effect that the first 7 days of curing are determinant in reducing gas-permeability, with a less relevant impact of extending the moist curing in the laboratory beyond 7 days. This is valid for all mixes, except for that made with OPC U5, which may be due to an experimental error. Now, if we compare the mix made with OPC U2 with those made with the four blends, the higher sensitivity of the latter to the lack of curing is evident, even – although to a lesser extent – for that of the blend containing LF.

Figure 6.42 presents results (Hooton, 2011) of water-permeability K_w (test method not disclosed) of concretes of $w/b=0.45$, made with OPC and with 25% and 50% replacement by blast-furnace slag. One set of specimens was moist cured until the age of testing (91 days) whilst the other set was

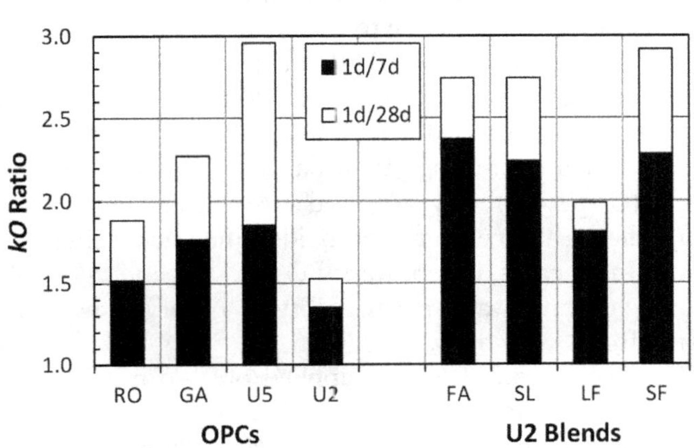

Figure 6.41 Effect of moist curing duration on O_2-permeability *kO* of OPC and blended cements concretes.

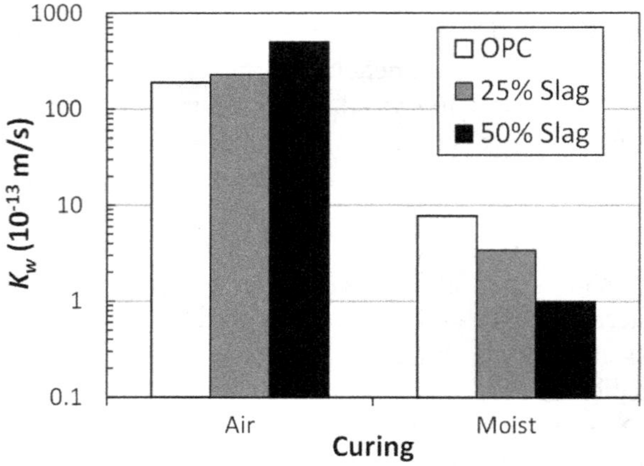

Figure 6.42 Effect of curing on water-permeability of OPC and slag concretes; data from Hooton (2011).

air-cured. Figure 6.42 shows the beneficial effect of incorporating slag, provided the concrete is thoroughly moist cured; on the contrary, if concrete is not well cured, the incorporation of slag has a negative impact on the water-permeability.

In another research by Alamri (1988), specimens from different concrete mixes (made with OPC and 30% and 60% replacement by PFA and GGBFS, respectively) were exposed to environmental conditions of 50°C/15% RH, 40°C/60% RH and 20°C/100% RH, until preconditioning them for air-permeability test at 28 days, using the Figg method (Section 4.3.2.1). The reported results show that harsh environmental conditions during binder hydration can lead to significant increases in air-permeability of the concrete surface and that both higher *w/c* ratios and the presence of SCM increase the sensitivity of concrete to the environmental conditions.

Concretes of *w/c*=0.40 and 0.70, prepared with different cement types, were tested by Gonçalves et al. (2000). Their test specimens were wrapped with a plastic sheet during 3 and 7 days and afterwards exposed to 20°C/75% RH until testing (at 28 days). The results are presented in Figure 6.43a (O_2-permeability *kO*) and 6.43b (initial water sorptivity at 1 hour a_{1h}), where each cement type is identified according to EN 197-1 classification. In all cases, except for OPC (CEM I) and *w/c*=0.40 a reduction in both gas- and water-permeability is observed by extending the sealed-curing from 3 to 7 days.

The WIST (Section 4.2.2.6) was applied on two mixes with the same binder content (340 kg/m^3) and *w/b* ratio (0.50), but different binder types (OPC and GBFS) (Nguyen et al., 2019a). Relatively large specimens, after being sealed-cured for different periods, were kept in an environment with average conditions T=16.4°C/RH=69.1% until the age of testing (90 days).

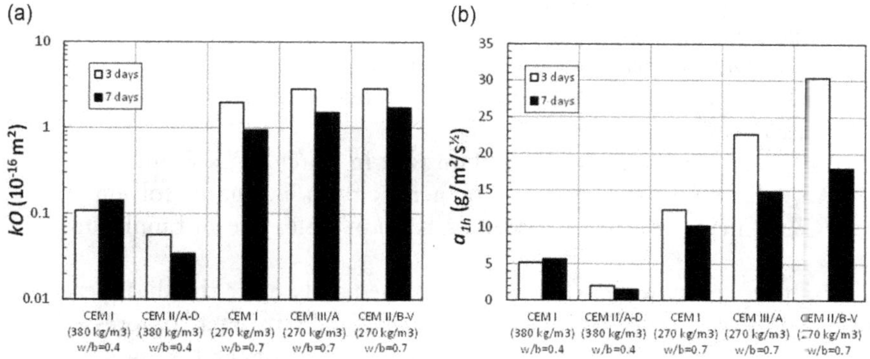

Figure 6.43 Effect of seal-curing duration on (a) O_2-permeability and (b) initial sorptivity of concretes; data from Gonçalves et al. (2000).

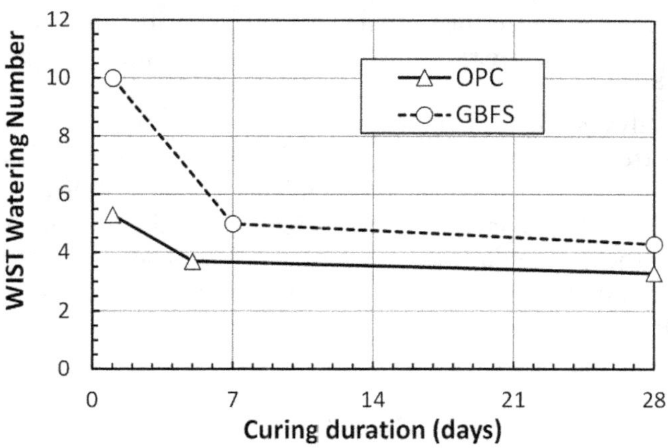

Figure 6.44 WIST results of concretes with different curing duration; data from Nguyen et al. (2019a).

The mean number of water-spraying repetitions until concrete surface saturation is shown in Figure 6.44 (the higher the number, the more absorptive the surface).

The results in Figures 6.41–6.44 confirm the fact that concretes containing MIC or SCM require better curing to develop their full durability potential.

6.7.2.2 *Investigations in the Field*

Here we refer to an investigation conducted by Surana et al. (2017), where five different curing regimes were applied to 1.2×1.2×0.2 m reinforced slabs, cast with an OPC concrete of *w/c* ratio 0.55 and 35 MPa compressive strength. After casting, the following curing regimes were applied to the slabs:

(7dH): The slab was cured using two layers of wet hessian-cloth until the age of 7 days, after which it was exposed to air. Hessian cloth was kept wet by sprinkling water intermittently
(Air): The slab was left exposed to air after casting without any deliberate curing measures
(WX): Curing compound based on wax in water (Wax Emulsion)
(AS): Curing compound based on acrylic resin in organic solvent
(RW): Curing compound based on resin in water (Resin Emulsion)

The curing compounds were applied immediately after the disappearance of bleed water sheen from the concrete surface by using a compressed air-assisted spraying gun, at a coverage rate of 5–6 m^2/L, taking care of a similar and proper coverage of the slab surface. After finishing each curing process, the slabs were placed unprotected outdoors, exposed to the weather conditions of Chennai, India. At 28 days of age, Ø70 mm cores were drilled from the upper side of the slabs and cut to 30 mm thickness, following the procedure for measuring *OPI*, described in Section 4.3.1.3 and *WSI* (Water Sorptivity Index). The results obtained are presented in Figure 6.45.

When analysing Figure 6.45 it has to be borne in mind that a higher *OPI* means a better concrete of lower permeability. Figure 6.45 shows that the worst performance in terms of gas- and water-permeability corresponds to the air-cured (air) concrete – something expected – and to the concrete cured with the wax emulsion compound (WX). The overall best performance corresponds to the concrete cured with the resin-based compounds (RW and RS).

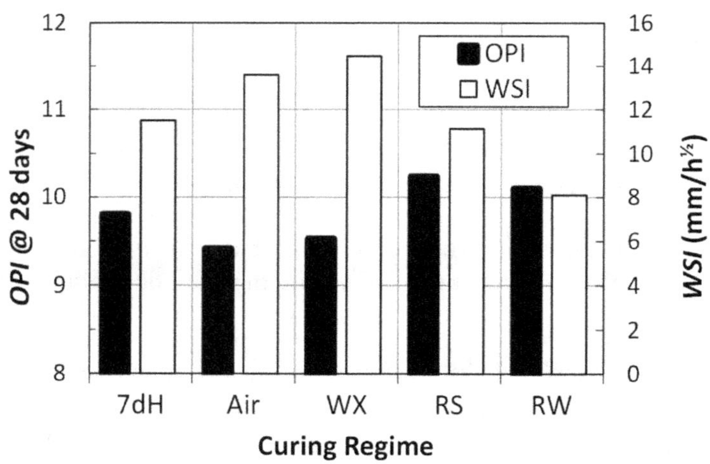

Figure 6.45 Effect of curing method on *OPI* and *WSI*; data from Surana et al. (2017).

6.7.3 Effect of Curing on Air-Permeability *kT*

6.7.3.1 Conventional Curing

Several laboratory investigations have been conducted to assess the effect of the curing conditions on the air-permeability *kT* of the *Covercrete* (Torrent & Ebensperger, 1993; Torrent & Frenzer, 1995; Kubens et al., 2003; Quoc & Kishi, 2008, 2009; Ichiro et al., 2009; Solís-Carcaño & Moreno, 2009; Fernández Luco, 2010; Kurashige & Hironaga, 2010; Song et al., 2014; Okasaki et al., 2015; Kawaai et al., 2015; Neves & Torrent, 2016; Ebensperger & Olivares, 2017; Yokoyama et al., 2017; Nakarai et al., 2019). These investigations consisted in casting specimens with concretes of different compositions, subjecting them to different curing conditions and in measuring the air-permeability *kT* of the resulting concrete qualities. In what follows some of them will be discussed in some detail.

In one research, Quoc and Kishi (2008, 2009) investigated the impact of the duration of water curing on air-permeability *kT*. For this, they prepared specimens with OPC concrete mixes of w/c = 0.30, 0.45 and 0.60 and cured them under water or sealed at 20°C during 0, 3, 7 and 28 days. Thereafter, the specimens were kept in a dry room (20°C/65% RH) until the age of 56 days when *kT* was measured. Figure 6.46 shows the beneficial effect that the duration of water curing has on *kT*, displaying the typical trend for laboratory testing, in that the first 7 days of moist curing are critical to achieve a tight concrete, whilst extending the curing beyond 7 days has only a small effect on *kT*. A similar effect can be found also for the

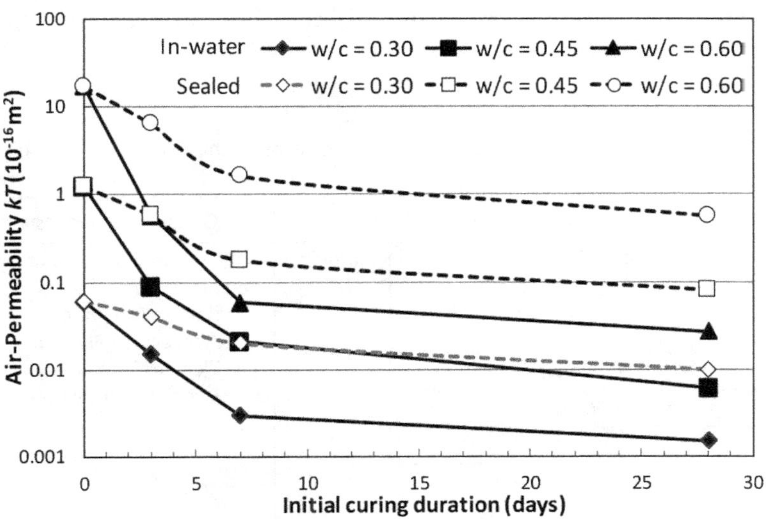

Figure 6.46 Effect of the length of water (a) or sealed (b) curing period on *kT*; data from Quoc and Kishi (2008, 2009).

sealed specimens. Comparing both curing methods, the higher impact of water curing over sealed curing becomes evident, with the former bringing *kT* about one order of magnitude lower compared with the same duration of sealing (sealing method not disclosed in the paper).

Similar to what was discussed in Section 6.2.2.3, Quoc and Kishi (2008) proposed a linear relation between log *kT* and *w/c* ratio, with the proportional factor (designated as curing factor) being a function of the duration of the moist or sealed curing period.

In a comprehensive investigation, several non-destructive tests were performed on 150×150×530 mm prisms, made with an air-entrained OPC concrete of *w/c*=0.50 (Kurashige & Hironaga, 2010). The prisms were subjected to different curing conditions, combining different durations in the moulds (1, 5 or 14 days), followed by exposure to different storage conditions, as described in Table 6.4. Companion 100×100×400 mm prisms were used to measure accelerated carbonation after 7, 28, 91 and 182 days exposure at (20°C/60% RH; 5% CO_2).

Figure 6.47 shows, in decreasing order, the results of *kT* measured at 91 days of age on prisms subjected to the curing and storage conditions described in Table 6.4. It can be seen that the prisms with worst curings 1–40 and 1–60 (1 day in the mould and permanently exposed to dry air afterwards) show the higher *kT* values. The 1–40H bar shows that a late exposure to humid air reduces *kT* but with poorer performance than the prisms subjected initially to moist curing (1–H and, especially, 1–W). Regarding permanence in the moulds as curing technique, Figure 6.47 shows the effectiveness of 5 days (prisms 5–40, 5–60 and 5–H), with little influence of the exposure after demoulding. Extending the permanence in the moulds to 14 days (prism 14-60) has little effect on *kT*, confirming the trend shown in Figure 6.46.

Table 6.4 Different curing, storage conditions and testing schedule, from Kurashige and Hironaga (2010)

Code	*Exposure environments (T=20°C)*					*Conditioning and testing*
1-40H		40% RH			Humid air	60% RH until 91 days, when tested for: • Rebound number • Air-permeability *kT* • Electrical resistivity • Accelerated carbonation
1-40		40% RH			60% RH	
1-60		60% RH				
1-H		Humid air				
1-W		In Water				
5-40			40% RH			
5-60			60% RH			
5-H			Humid air			
14-60	**In the moulds**			60% RH		
Age (days)	1	5	14	28	56	91

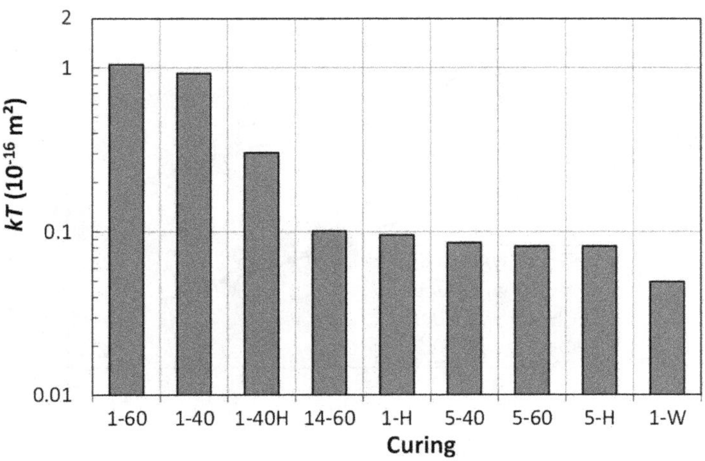

Figure 6.47 Effect of curing regime on *kT*; data from Kurashige and Hironaga (2010).

One of the earliest investigations on the effect of curing on *kT* was reported by Kubens et al. (2003). They subjected specimens, made with two OPC concrete mixes (200 kg/m³; $w/c=0.87$ and 320 kg/m³; $w/c=0.55$) to five different curing conditions, from worst to best:

- *A*: no curing, storage at 30°C/40% RH until the testing age
- *B*: 5 minutes in water, three times a day, for 3 days and thereafter at 30°C/40% RH until testing age
- *C*: 5 minutes in water, three times a day, for 6 days and thereafter at 30°C/40% RH until testing age
- *D*: 6 days in water and thereafter at 30°C/40% RH until testing age
- *E*: 28 days in water and thereafter at 30°C/40% RH until the testing age

Tests were performed at 91 days and included, besides *kT*, also compressive strength, chloride migration (ASTM C1202) and accelerated carbonation (7 days at 5% CO_2; 30°C/50% RH).

Figure 6.48 shows the impact of the curing schedule on *kT* and accelerated carbonation depth. The plots confirm the significant benefits of a continued moist curing on both *kT* and carbonation depth. It is worth noting that the richer mix, totally deprived of moist curing (A), shows almost the same permeability *kT* and carbonation depth as the poorer mix cured 28 days in water (E). This means that not curing the concrete was equivalent to reducing the cement content of the mix by 38% (120 kg/m³), with respect to standard curing. As becomes evident from Figure 6.48, the logarithm of *kT* showed a very good correlation with the accelerated carbonation depth.

In another relevant investigation (Nakarai et al., 2019), different curing conditions were applied to six large concrete mock-up blocks (1.5×1.5 m),

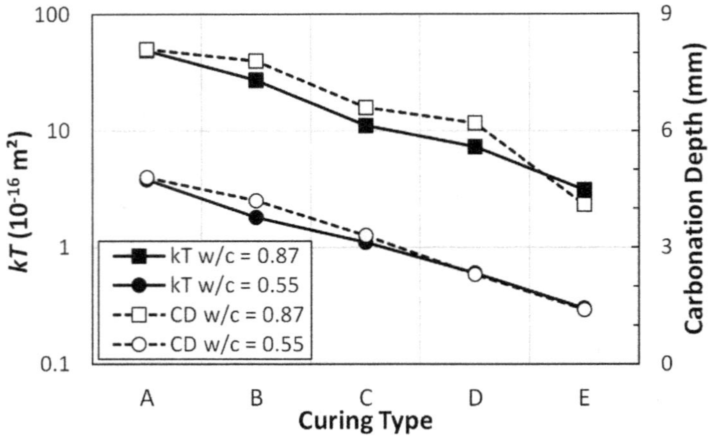

Figure 6.48 Effect of curing schedule on *kT* and accelerated carbonation depth; data from Kubens et al. (2003).

with thickness of 0.6 m for N mix and 0.4 m for B mix. The blocks were representative of companion box culvert structures. The N mix contained 295 kg/m³ of OPC with w/c=0.55, whilst the B mix contained 303 kg/m³ of BFSC and w/c=0.525. Each mock up block received a different curing: 1 day in the mould (N-1d and B-1d); 5 and 7 days in the mould (N-5d and B-7d) and 3 months protected with a plastic sheet (N-3m and B-3m); the latter curing was also applied to the real box culverts. After finishing the corresponding curing period, the mock-up elements were exposed to the field conditions (in average 16°C/62% RH, with minimum RH around 35%) protected from rain, with the lateral faces sealed, with the box culverts exposed to the same climatic conditions. The 1.5×1.5 m faces of the mock-up blocks were used for testing; regarding the N and B real box culverts, the tests were conducted on the inner walls, at a height of 1 m from the floor.

The change of air-permeability and surface moisture (electrical impedance-based method) was periodically monitored for each mock up element and box culvert from 1.3 to 39 months. The results of *kT* are plotted as function of exposure time in Figure 6.49a and b for the N elements and the B elements, respectively (data at 2.6 months, not shown, removed as outliers, as shown in Bueno et al. (2021)).

The beneficial effect of initial curing, at least of 5 days for N mix and of 7 days for B mix, is clear from Figure 6.49; the higher sensitivity to moist curing of the concrete B, made with BFSC, is also evident.

The advantages of wet curing, with respect to membrane and dry curing, on the air-permeability *kT* of concrete, were also confirmed by Mu et al. (2015), with special relevance for concretes containing PFA and/or GGBFS as OPC replacements.

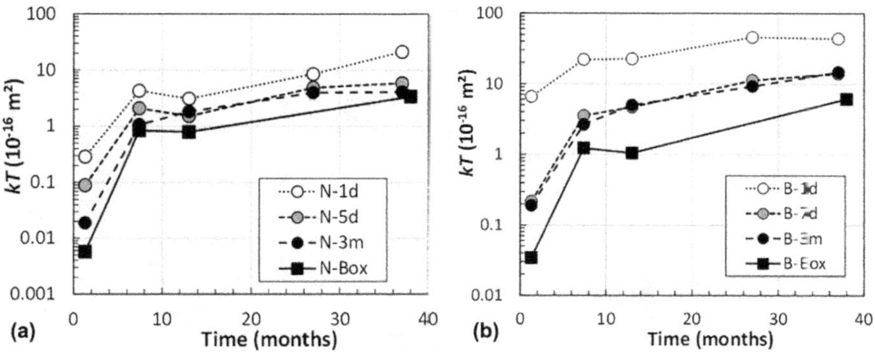

Figure 6.49 Effect of curing on evolution of air-permeability *kT* for (a) Mix N and (b) Mix B; data from Nakarai et al. (2019).

Another test method applied to evaluate the effect of curing duration on the water-permeability of the elements (Nakarai et al., 2019) was the WIST, Section 4.2.2.6, which yielded similar results, regarding the positive contribution of curing in reducing concrete surface permeability, especially for a mix with BFSC.

Results of air-permeability *kT* of concretes made with three different cements and cured in the lab (20°C/95% RH) and outdoors were reported by de Schutter (2016). The cements, described by its designation after EN 197-1 Standard, were CEM I 52,5 N (OPC); CEM III/B 42,5 (BFSC, with 66%–80% slag) and CEM III/C 32,5 (BFSC, with 81%–95% slag). Figure 6.50 presents the data, showing that outdoors curing doubles the permeability of the OPC concrete, but increases ten-fold that of the concretes made with both BFSC. This confirms the higher sensitivity of cements containing active mineral additions to the lack of moist curing, already shown and discussed in connection with Figures 6.41–6.44, showing results obtained with other test methods.

A curing method using thermoplastic, water-repellent sealing sheets that are attached to the formwork and left in place after demoulding (for periods of weeks and even months), was studied by Nukushina et al. (2015). The method proved to be effective, especially for concretes containing PFA and/or GGBFS.

The investigations discussed above consisted in measuring the effect of the duration of moist curing, be it by external contact with water or humid air or by sealing, on *kT* of conventional concretes.

6.7.3.2 Self-Curing

An investigation on the effect of internal curing on the performance of high-strength concretes was reported by Zhutowsky and Kovler (2012). For that,

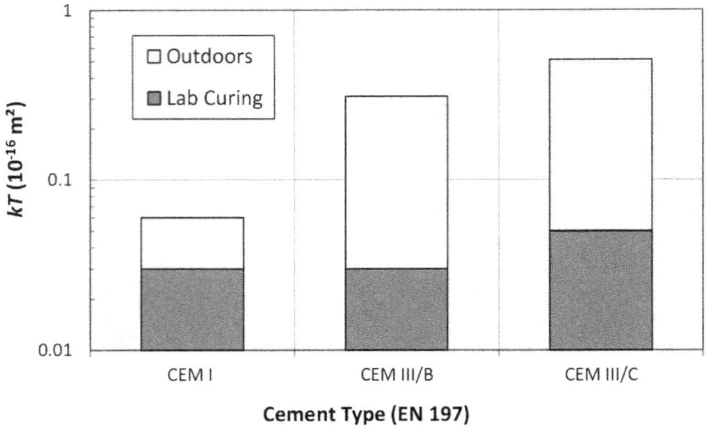

Figure 6.50 Effect of curing on concretes made with different cement types; data from de Schutter (2016).

OPC concrete mixes with *w/c*=0.21, 0.25 and 0.33 were prepared with normal weight aggregates (mixes 0.21, 0.25 and 0.33, respectively) and also replacing the fraction 2.4–4.8 mm of the sand with water-saturated pumice particles (mixes 0.21 L, 0.25 L and 0.33 L, respectively). Pumice sand had a 1 hour vacuum absorption of 73% and the water carried by them was meant to compensate for the chemical shrinkage of the cement paste at 7 days.

The specimens were seal-cured at 30°C and, at ages of 1, 7 and 28 days, oven-dried at 60°C till constant weight and cooled to ambient temperature. Then, *kT* was measured, as well as water sorptivity and chloride migration (ASTM C1202). Figure 6.51 shows the effect of the internal curing on *kT*. It can be seen that, except for the specimens, oven-dried at 1 day, the *kT* of the specimens with internal curing (ending with L in Figure 6.51) was lower than the reference samples, with a more significant effect at 28 days. This effect was confirmed by the Coulombs measured (ASTM C1202) for mixes 0.25 and 0.33 L, but not for 0.21 L; on the contrary, the sorptivity tests showed higher values for the internally cured samples.

6.7.3.3 Accelerated Curing

Another investigation compared the effect of two accelerated curing methods on the properties of a concrete (*w/c*=0.43; OPC=387 kg/m³), namely: steam curing (SC) and microwave heating (MH) of steel moulds (Choi et al., 2019). Different cycles of both SC and MH (variable pre-curing, heating and cooling times) with maximum temperatures within the range of 50°C–65°C were investigated on specimens and mock-up elements. The response of the

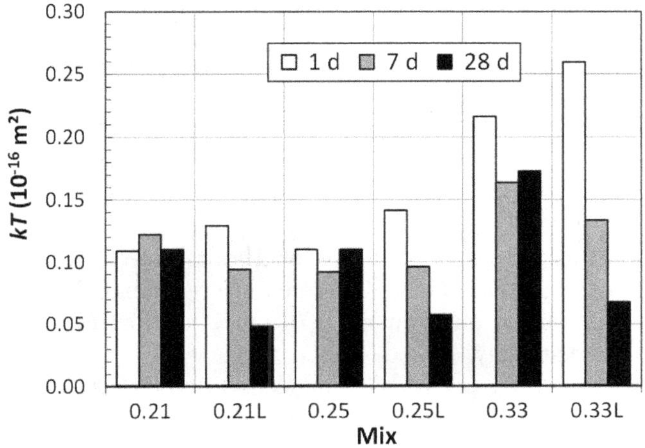

Figure 6.51 Effect of internal curing (*L*) on concretes of *w/c* ratios 0.21, 0.25 and 0.33; data from Zhutowsky and Kovler (2012).

material in terms of compressive strength, MIP, scanning electron microscopy (SEM), free and restrained shrinkage and air-permeability *kT* were measured. They found that both SC and MW curing led to similar, quite low values of *kT* ($<0.1\times10^{-16}$ m²). The good performance of the MW-cured concrete and the potential reduction in curing costs and CO_2 emissions open interesting perspectives for industrial application.

The effect of steam curing on concretes made with OPC and with 50% replacement by two types of GGBFS (one containing 4.5 % gypsum as activator) was investigated by Nguyen et al. (2020). All three mixes contained 50 kg/m³ of a calcium sulphoaluminate-based expansive agent, intended to compensate an alleged susceptibility of slag-containing concrete to cracking. For all three mixes the steam curing improved the compressive strength at 1 day, compared to standard curing at 20°C, but impaired it at later ages. Regarding permeability, both the air-permeability *kT* and the water sorptivity of the three mixes were severely increased by the application of steam curing, compared with standard curing.

The effect of secondary curing, after steam curing, on *kT* and other transport properties of mortars was investigated by Azuma et al. (2016). The five mortars tested had *w/c* ratios within 0.39 and 0.49, with binders made with 100% OPC and with the following replacements: 50% GGBFS+9% LF, 20% PFA and 3% and 5% of SF. Five different curing conditions were applied to the specimens (150×150×530 mm prisms for *kT*), namely:

St: permanently in the dry room (20°C/60% RH) after the steam curing
StW3: 3 days wet mat curing after the steam curing, thereafter in the dry room

StW7: 7 days wet mat curing after the steam curing, thereafter in the dry room
StS7: 7 days sealed curing after the steam curing, thereafter in the dry room
W: reference curing, no steam-cured but cured under water after demoulding (1 day) until the age of 21 days, afterwards stored in the dry room

The steam curing consisted in a cycle of 3 hours pre-curing at 20°C, followed by a heating phase at a 15°C/h rate until reaching 50°C (kept during 5 hours) and a cooling phase at a 5°C/h rate. In all cases, the specimens were demoulded at 1 day.

For all five mortars, curing St produced the higher *kT* values; almost invariably the best *kT* results were obtained with curing StW7, followed by the reference curing W. Only in the case of BFS binder W curing was slightly better than StW7. In terms of compressive strength at 28 days, the best results were obtained invariably with curing W. Regarding the binder composition, the *kT* values after curing StW7 were very similar for all binders, except BSF that showed higher values. On the contrary, chloride penetration and diffusion tests showed a better performance of all blended binders compared to the 100% OPC. It is worth mentioning that curing StS7 reduced the air-permeability *kT* with respect to steam curing St, but significantly less than curing StW7.

The test results indicate the advantage of secondary curing to enhance the durability performance of steam-cured concrete.

6.7.3.4 "3M-Sheets" Curing

This section deals with an innovative solution for curing concrete developed by the company 3M. It is described in a company's brochure in Japanese (3M, 2019), from which the following aspects were extracted. It consists of a polyolefin-based sheet, 0.4 mm thick, containing fine grooves that are applied on the flatwork surface. For vertical and bottom faces, a version exists equipped with adhesive tape that can be fixed on and removed from the surface. Water supplied at a certain location is spread horizontally and also vertically by capillarity with a maximum recommended usage period of 3 months. According to the manufacturer's brochure, the air-permeability *kT* achieved by the use of this sheet is two orders of magnitude (two classes) lower than the same concrete subjected to the standard curing of 7 days with wet mats.

In a field investigation during the construction of Mukaisadanai bridge, located on National Route 283 in Kamaishi City, having a length of 44.5 m, the above-described sheet was used (Sakakibara et al., 2018; Iwaki, 2020). The concrete used had a $w/b = 0.42$, with a binder composed of

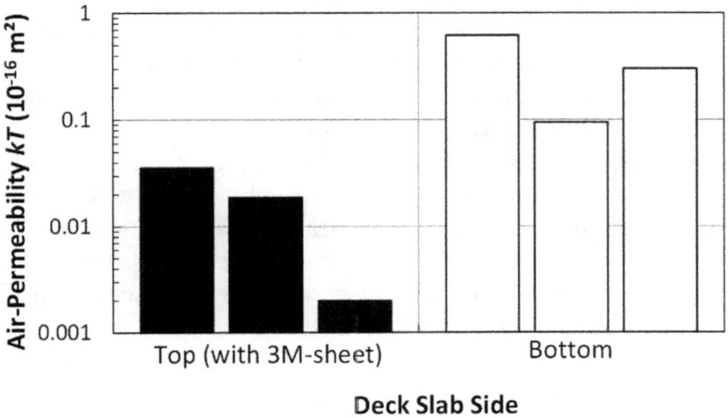

Figure 6.52 Effect of 3M curing sheet (a) compared with curing compound (b) on the air-permeability of bridge deck; data from Sakakibara et al. (2018).

283 kg/m³ OPC, 62.5 kg/m³ of Noshiro PFA and 20 kg/m³ of an expansive agent, displaying a slump of 120 mm and 6.0% entrained air. In order to fully exploit the beneficial effect of PFA on durability (resistance to de-icing salts penetration), curing of the concrete was crucial, both in terms of moisture and temperature. Here the focus will be placed on the bridge deck, the bottom and lateral sides of which were cured with a curing compound of the water-based paraffin wax type. On the contrary, the upper side of the deck was wet-cured by applying the so-called FTF (Fluid Transfer Film), which was actually the 3M sheet described above (Iwaki, 2020). The curing time was extended for 3 months. On top of the curing systems, insulating mats were applied, so as to offer better conditions for the binder hydration and reduce the risk of thermal cracks (the ambient temperature during the first 5 days after concreting was 5°C, whilst the concrete temperature was kept above 10°C). At the age of 14 months, air-permeability *kT* tests were performed on the top and bottom sides of the deck with the results shown in Figure 6.52. It can be seen that the air-permeability of the bottom side of the deck is about 20 times higher than the top side, which seems to confirm the positive effect of the 3M curing sheet.

6.8 EFFECT OF TEMPERATURE ON PERMEABILITY

It is generally assumed that exposing concrete to temperatures between 0° and 50°C does not affect its microstructure and pore structure. Hence, within that range of temperatures, that covers the normal exposure conditions of most structures, the only factor that affects the flow of water and gas through concrete is the viscosity of the fluid. As shown in Table 3.2, the viscosity of water drops by a factor of about 3 when the temperature rises from 0°C to 50°C; on the contrary, the viscosity of gases is almost unaffected by the temperature. Indeed, the results presented in Figure 5.14 showed that the coefficient of air-permeability *kT* is virtually the same when measuring the same concretes at 5.7°C, 8.6°C and 23.4°C, even when using the same default value of the viscosity of air in the calculations (2.0×10^{-5} Pa.s). In the case of water-permeability, Darcy Eq. (3.21) will yield different values of the coefficient of water-permeability at different temperatures, whilst Hagen-Poiseuille-Darcy Eq. (3.19) will yield the same coefficient of water-permeability, provided the correct value of the viscosity of water, at the test temperature, is introduced.

Regarding below-zero temperatures, it is not possible to measure representative values of water-permeability; regarding gas-permeability, it might be possible on dry concrete, but data are lacking.

Regarding high temperatures, say above 60°C, the problem is more complex because high temperatures produce alterations and damage to the concrete microstructure that change significantly the permeability of the material. Knowledge of the gas-permeability of concrete at high temperatures is very important in avoiding explosive spalling of the concrete cover when exposed to fire, as discussed in detail in Chapter 10.

A good review of the effect of high temperatures on the gas-permeability of concrete can be found in Van der Merwe (2019), from which some aspects are discussed here. At temperatures exceeding 100°C, an increase in gas-permeability with concrete temperature has been observed; Fischer (1967) reported a steep increase in permeability between 450°C and 600°C. This temperature range is typically associated with substantial differential thermal expansion between coarse aggregates and the surrounding cement matrix, which could explain his observation.

A 100-fold steady increase in gas-permeability from ambient to 600°C temperature was reported by Schneider and Herbst (1989), although a more complex phenomenon was observed as they reduced the conditioning time at high temperatures from 50 to 10 hours. When shortening the conditioning time to 10 hours, they observed a region of reduced permeability between 150°C and 200°C, as shown in Figure 6.53. An explanation for this phenomenon was provided by Klingsch (2014) who attributed it to the dehydration of bound water from the cement matrix. This suggests that the increased free moisture content following such dehydration reduces concrete gas-permeability. Once this free moisture evaporates, permeability continues to increase as the temperature rises.

Fares et al. (2009) investigated the effect of high temperatures on the gas-permeability of a conventional vibrated concrete (VC) of 41 MPa compressive strength at 90 days and of two self-consolidating concretes: SCC1 of 37 MPa and SCC2 of 54 MPa at same age. For that, Ø150×300 mm cylinders were heated in a furnace at different temperatures, kept 1 hour at the maximum temperature of the cycle and then cooled down slowly to

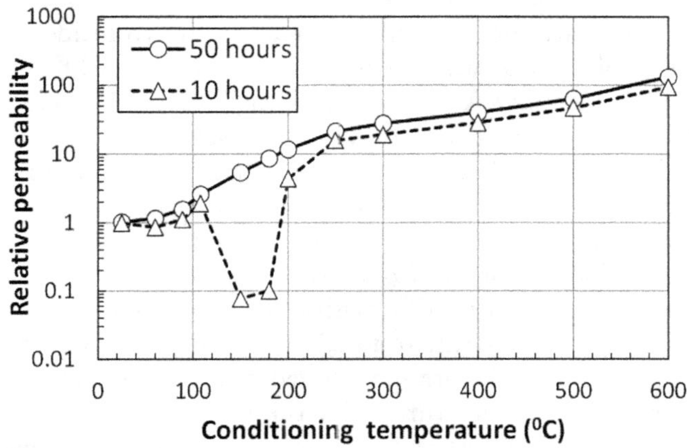

Figure 6.53 Gas-permeability measured at high temperatures, after different conditioning times; data from Schneider and Herbst (1989).

ambient temperature. Four Ø150×50 mm discs were cut from the central part of the cylinder, on which the residual coefficient of gas-permeability was measured applying the Cembureau method. The results obtained for the three mixes are shown in Figure 6.54, indicating that the behaviour of SCCs is almost identical to that of conventional vibrated concrete. Between 75°C and 600°C, the gas-permeability increases some four orders of magnitude, a more marked effect to that shown in Figure 6.53.

A much smaller effect of high temperatures on the air-permeability *kT* of 70 MPa strength concretes was measured by Li et al. (2008). They heated

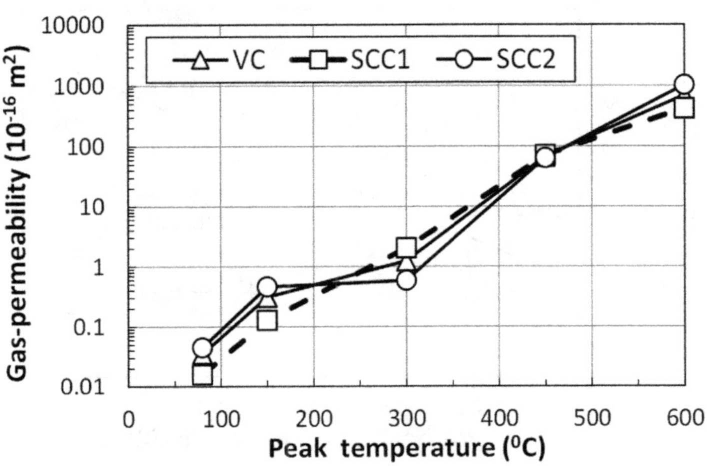

Figure 6.54 Residual gas-permeability of conventional and SCC, previously subjected to indicated peak temperatures; data from Fares et al. (2009).

concretes of different strengths up to 1,000°C, keeping the specimens 1 hour at the peak temperature before letting them cool down inside the furnace. Then air-permeability *kT*, water sorptivity (undescribed CPK method) and chloride ion permeability (Chinese Standard JTJ270–98) were measured. Between room temperature and 1,000°C, a seven-fold increase in *kT* was recorded. The increase of chloride-permeability and water sorptivity was 9-fold and 30-fold, respectively, between room temperature and 800°C.

Zeiml et al. (2006) tested a conventional concrete mix (w/b=0.49), without and with 1.5 kg/m^3 of short polypropylene fibres (PPF), heated at different temperatures up to 600°C, with a 90-hour retention time at each temperature. After cooling, the residual air-permeability was measured by a decay pressure device; Figure 6.55 shows their results. The air-permeability increases with temperature, with the mix containing PPF showing higher permeability at all temperatures and a discontinuity around 150°C, discussed later in this section.

Lu (2015) investigated the air-permeability of two high-strength concrete mixes (w/c=0.20), containing 1 (M1P1) and 2 (M1P2) kg/m^3 of PPF, at high temperatures ("Hot" permeability) using an adaptation of the Cembureau method (KA) suitable for testing Ø150×40mm discs inside a furnace at temperatures up to 300°C. The residual permeability was measured by the Torrent method (*kT*) after cooling the specimens to room temperature. Both "Hot" and "Residual" permeabilities are plotted in Figure 6.56, for both mixes, as function of the temperature.

What becomes evident in Figure 6.56 is that both test methods, applied under different conditions, provide a similar picture of the changes in permeability with temperature. Both permeabilities are very similar and grow

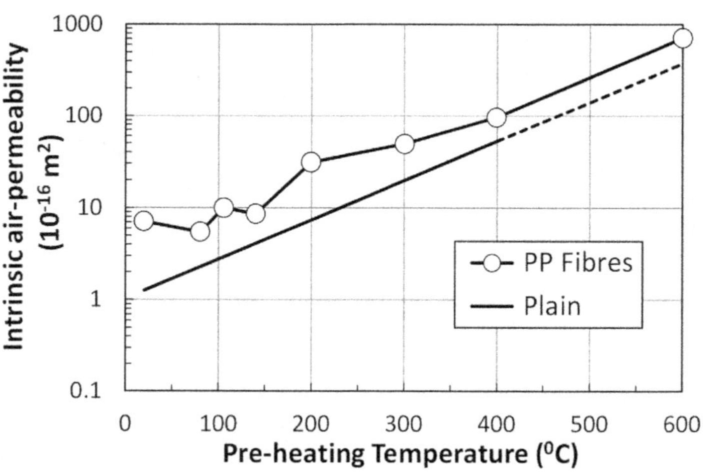

Figure 6.55 Effect of temperature and PP fibres on "residual" air-permeability; data from Zeiml et al. (2006).

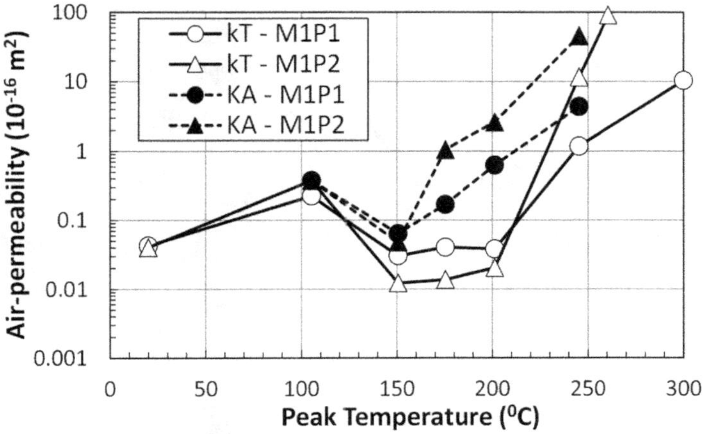

Figure 6.56 Effect of temperature and content of PP fibres on "hot" (KA) and "residual" air-permeability *kT*; data from Lu (2015).

until a temperature of 105°C. Beyond that temperature and up to 200°C for "Residual permeability" (a bit lower for "Hot permeability"), both permeabilities decrease, with the "Residual permeability" becoming lower than the "Hot permeability". This decrease in both permeabilities is attributed to an increase in moisture content due to dehydration. At temperatures above 150°C, the PPF start to soften and melt, which explains the steep increase in permeability, an increase that is more marked for the concrete mix with a higher content of PP fibres (M1P2). As discussed in Chapter 10, this increase in gas-permeability at high temperatures of concretes containing PPF is made good use of to prevent explosive spalling of high-strength concretes under fire.

A comprehensive investigation was performed by Niknezhad et al. (2019) on the effect of temperature (between 20°C and 500°C) on the behaviour of a plain self-compacting concrete (SCC) and of another SCC reinforced with PPF with 170°C melting point. Cylinder and slab specimens were cast and moist cured for 28 days, on which the following properties of both SCCs were measured after reaching 20°C, 200°C, 300°C, 400°C and 500°C: compressive strength, dynamic *E*-Modulus, nitrogen-permeability *kN* (Cembureau), Torrent air-permeability *kT*, water and MIP porosity and chloride migration coefficient, complemented by SEM observations.

The temperature was raised gradually to the maximum level, with a 1-hour retention time and later cooled down inside the furnace to a temperature of 20°C or 80°C, at which the tests were performed. Summarizing the test results:

- exposition to high temperatures led to higher water and MIP porosity and size of the pores; this effect was more marked for the SCC containing PPF

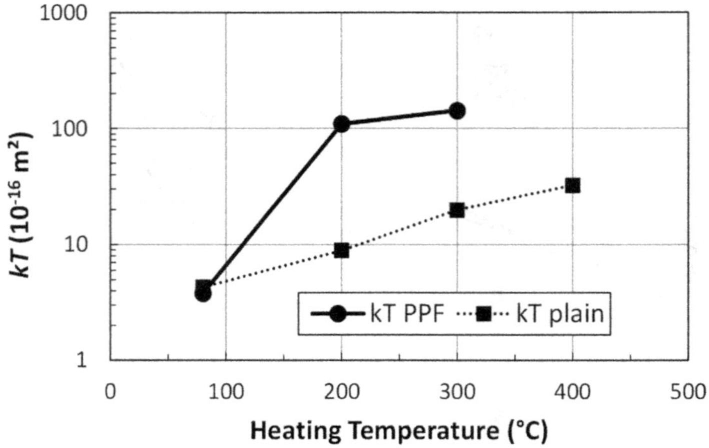

Figure 6.57 Effect of temperature and PP fibres on "residual" air-permeability *kT* (Niknezhad et al., 2019).

- the compressive strength of plain SCC decreases 40% between 20°C and 500°C, whilst that of the PPF reinforced SCC drops 50% (from about same initial level as the plain SCC)
- the dynamic modulus E_{dyn} decreased ≈70% between 20°C and 500°C
- the intrinsic N_2 permeability was similar at 80°C (0.5 and 0.3×10^{-16} m^2 for plain and fibre-reinforced SCC, respectively. After heating to 400°C, the corresponding values were 2.4 and 24×10^{-16} m^2, respectively. This behaviour is explained by the melting of the PPF at 170°C, which increases the interconnectivity between the pores. The same effect was observed for the air-permeability *kT*, as shown in Figure 6.57, as well as by the migration test
- the changes in transport properties are well correlated with the damage factor (relative decrease in E_{dyn})
- regarding microstructural changes observed by SEM, two relevant aspects were found: the disappearance (melting) of the PPF at 200°C and the appearance of matrix cracks at 400°C

The results obtained by Niknezhad et al. (2019) are in close agreement with those reported by Zhang et al. (2018), discussed in detail in Chapter 10.

6.9 EFFECT OF MOISTURE ON PERMEABILITY

All transport mechanisms taking place in concrete are affected, to a greater or lesser extent, by the degree of water saturation of the pores and permeability is not an exception. For a more general treatment of this topic, the reader can consult Section 4.3.2 of RILEM TC 116-PDC (1995) and

Chapter 6 of RILEM TC 189-NEC (2007). As a result, most permeability tests performed in the laboratory require that the specimens are preconditioned with respect to moisture content, as mentioned in Section 4.3.1.1.

As discussed in Chapter 6 of RILEM TC 189-NEC (2007), the moisture content of the cover concrete under field conditions can vary across a wide range, depending on the season and on the daily conditions (e.g. during or immediately after a rainfall). Therefore, for tests performed in the field, some information on the moisture content of the concrete, at the time of testing, is essential for a proper interpretation of the test results.

When dealing with gas-permeability it is obvious that the drier the concrete, more pores will be open to the passage of the gas, resulting in higher permeability. In the case of water-permeability, a steady-state flow of water can only be achieved after all the pores in concrete are filled with water. Therefore, in non-saturated concrete, the water penetrating under pressure will be first accommodated in the free pore space before a true flow through the element can take place.

Figure 6.58a shows the variation in relative humidity (RH) inside four Ø20 mm cavities drilled in a 300×300×200 mm concrete block that was subjected, after 28 days moist curing, deliberately to wetting and drying cycles along 1 year (Parrott & Hong, 1991). In parallel, measurements of air-permeability by the Hong-Parrott method (see Section 4.3.2.2) were conducted in the same cavities, with the results shown in Figure 6.58b.

The air-permeability results mirror very well the RH values, confirming the strong effect the latter has on the former.

Although not truly representative of the conditions of structural concrete, an investigation conducted by Jacobs et al. (1994) is worth describing. In it, the influence of the degree of water saturation on a pervious mortar, consisting of 2–3 mm sand particles, glued together by cement paste, was investigated. The amount of cement paste of the mortar was adjusted so as not to fill the voids between the sand grains. Two mortar samples were manufactured in transparent plastic tubes (25 mm in diameter and 1,000 mm in length). They differed slightly in the composition and pore size distribution. The several months old specimens (tubes filled with mortar) were stored for 28 days in water, before the tests were conducted. During the experiment different degrees of saturation were achieved by varying the pressure of inflowing gas between 1 and 30 mbar. The higher the gas inlet pressure, the more water was expelled from the sample and the lower the resulting degree of water saturation. The gas and water flow through the samples was monitored in parallel. The higher the degree of water saturation, the higher the water-permeability and the lower the gas-permeability, as shown in Figure 6.59.

The moisture content has a strong impact on water capillary suction, because both the driving force (capillary pressure) and the coefficient of permeability are influenced by the saturation degree of the concrete pores. Figure 6.60 shows data obtained by Lunk (1997), who produced concrete

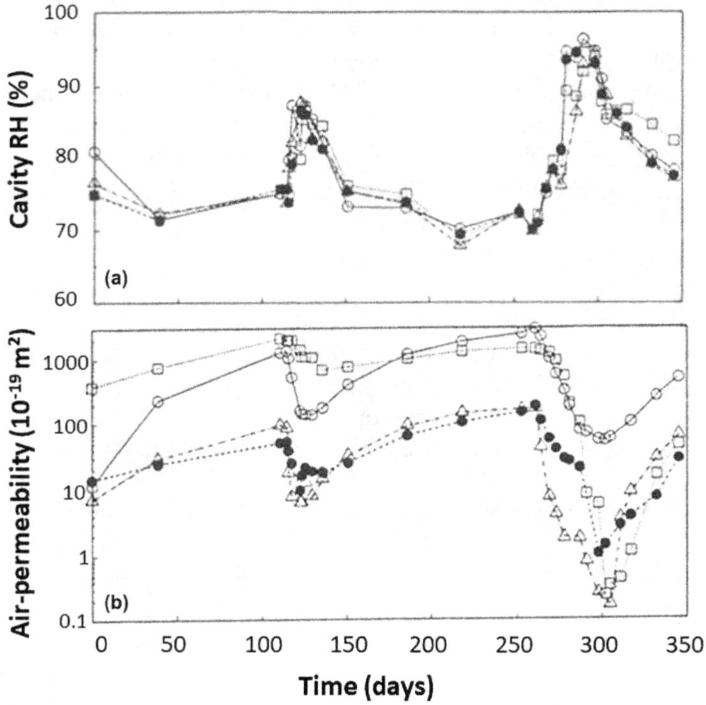

Figure 6.58 Variation of (a) relative humidity and (b) air-permeability, measured inside four cavities, due to drying and wetting (Parrott & Hong, 1991).

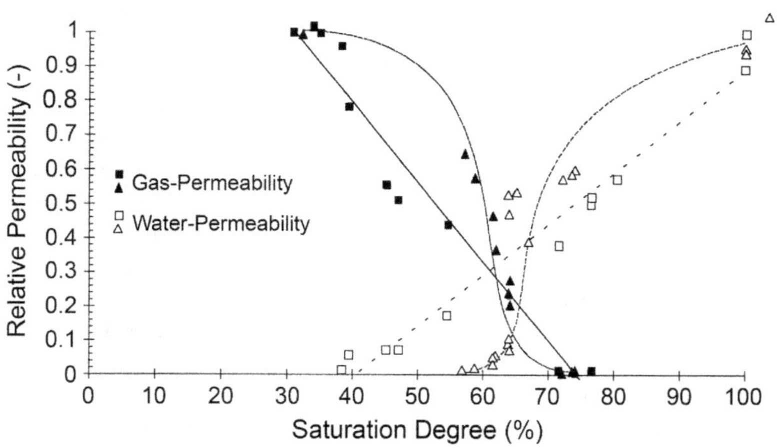

Figure 6.59 Effect of saturation degree on the gas- and water-permeability of porous mortars (Jacobs et al., 1994). Reproduced with permission of Nagra.

samples (w/c=0.30–0.70) which, after demoulding, were wrapped with wet burlap for 7 days and afterwards stored for 1 year at 20°C and different relative humidity (33%, 60% and, 90%) to achieve different moisture contents. The specimens were coated with epoxy resin on all surfaces, except the bottom and top surface, put in contact with 3 mm of water and weighed at defined intervals. Figure 6.60a shows the uptake of water with time of the mix with w/c = 0.40 for the three storage conditions. The results show that the lower the moisture content in the samples due to drying, the higher and faster the capillary uptake of water.

Figure 6.60b shows the effect of w/c ratio on the water uptake of concrete under RH storage of 90%. Comparing the two adjacent figures, it can be seen that the effect of moisture content (between 1.6 and 3.9 wt.%) on the capillary suction was similar to that of the w/c ratio (between 0.30 and 0.70). This highlights the importance of a strict moisture content control to obtain meaningful and comparable capillary suction test results.

Another important investigation on the effect of moisture on gas-permeability was conducted by Abbas et al. (1999). In it, Ø150×50 mm discs were cast from a mix with 270 kg/m³ OPC, w/c=0.67 and cylinder strength $f'c_{28d}$=33 MPa. After 1 day curing in the mould, the discs were demoulded and kept wrapped in plastic sheeting and aluminium foil for 27 days at 20°C. Periodically, the discs were oven-dried at 50°C until losing a predefined mass of water, sealed again and kept at 50°C for 72 hours to homogenize their moisture content and allowed to cool to 20°C during 24 hours. Then, the coefficient of O_2-permeability kO (Cembureau method) was measured under different applied pressures to quantify the Klinkenberg effect (Section 3.6). As a result, a plot of the intrinsic permeability K_{int} as function of the degree of saturation of the discs could be built, as shown in Figure 6.61.

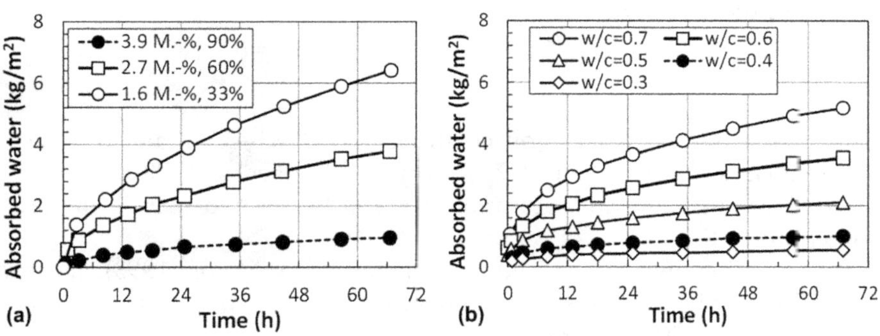

Figure 6.60 Effect of moisture (a) and *w/c* ratio (b) on water capillary suction of concrete; data from Lunk (1997).

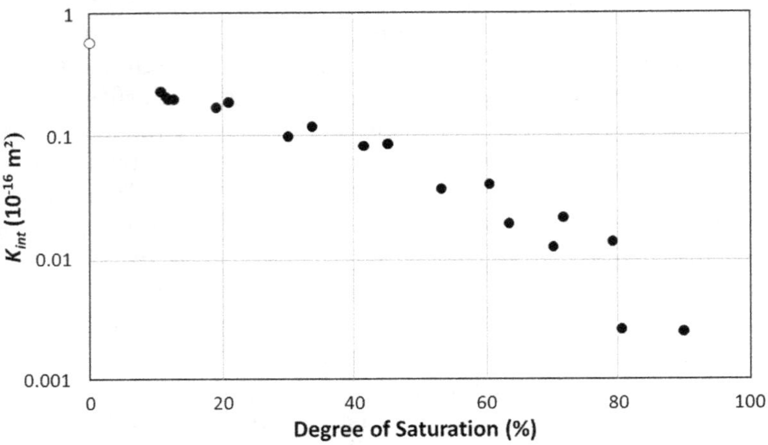

Figure 6.61 Effect of the degree of saturation on the gas-permeability of concrete; data from Abbas et al. (1999).

It can be seen from Figure 6.61 that K_{int} is reduced by two orders of magnitude when measured near saturation compared to when measured in a dry state.

One of the most comprehensive investigations on the effect of the degree of water saturation on gas-permeability was conducted and reported by Jacobs (1998). In this investigation, three concrete mixes with *w/c* of 0.45, 0.60 and 0.80 (57.2, 33.3 and 27.6 MPa cube $f'c_{28d}$, respectively) were carefully preconditioned during 2 years, when they were tested for H_2-permeability (kH). The specimens were discs saw-cut from Ø150 mm cores drilled from larger blocks. Different degrees of saturation were achieved by letting the discs absorb different quantities of water, keeping them in watertight plastic bags for 8 weeks, so as to homogenize the moisture. A linear regression between the logarithm of kH and the degree of saturation S was fit to the results, of the form:

$$\log kH = b - m \,.\, S \tag{6.7}$$

with relatively high correlation coefficients R (0.83–0.89) and not very different slopes m. Equation (6.7) is proposed by Jacobs (1998) to correct gas-permeability data for saturation degrees different to that of the test conditions.

Based on his own results and those reported by Major (1993), Jacobs (1998) built the chart shown in Figure 6.62, valid for concretes within the range of 0.40–0.80 in *w/c* ratio.

The inclined lines in Figure 6.62 indicate the RH of storage conditions that led to the corresponding degrees of saturation.

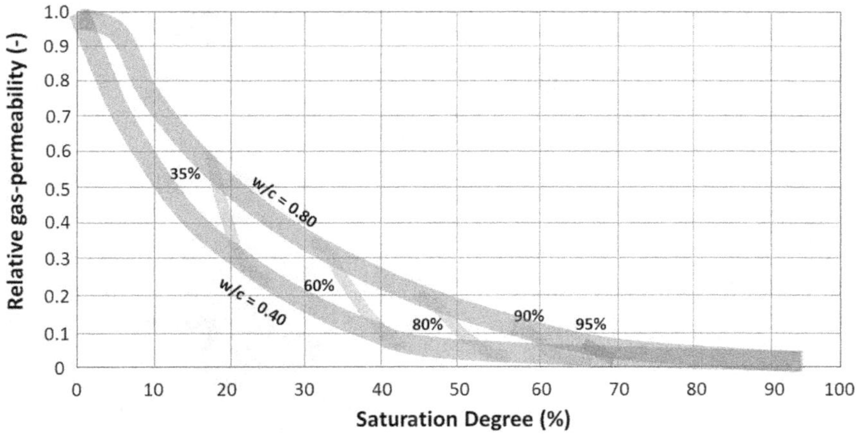

Figure 6.62 Relation between relative gas-permeability and saturation degree (obtained at various RH storages); data from Jacobs (1998).

Figure 6.63 presents data of air-permeability *kT* measured on 150 mm cubes as function of the degree of saturation *S* of the cubes, after different periods of permanence in a dry room (18°C–23°C; 50%–65% RH) and of oven-drying at 50°C and finally at 105°C (framed symbols). The data correspond to eight widely different concretes more details on the experiment and mixes in Section 5.7.2.1 and in Torrent et al. (2014). The results depart from a strict linear relationship between logarithm of *kT* and saturation degree *S*, as postulated by Jacobs (1998), which can be due to the fact that *kT* is measured on the surface of the cubes, whilst the degree of saturation *S* corresponds to the cube in bulk. In this case, the change of *kT* with the degree of saturation spans three to four orders of magnitude.

In a comprehensive research, Gui et al. (2016) investigated the relation between the intrinsic N_2-permeability K_g of concretes of *w/c* ratios 0.30, 0.40, 0.50 and 0.60, made with an OPC alone, with 30% PFA and 50% GGBFS, at different degrees of saturation. After 90 days (OPC) and 180 days (FA, GGBFS) moist curing 50 mm slices were saw-cut from Ø150 mm cylinders and oven-dried at 60°C until reaching predefined saturation degree levels. Then, the slices were sealed and kept in the oven during 14 days to homogenise their moisture. Figure 6.64 shows the data of K_g values (relative to the value for 0 saturation degree), obtained on the OPC concrete samples, to which the Van Genuchten-Mualen (VGM) model was fitted (lines in Figure 6.64). A similar analysis was conducted for the electrical conductivity of the same concrete mixes.

Similar relation between the air-permeability, measured under pressure and under vacuum, and the saturation degree was reported in Sogbossi et al. (2019).

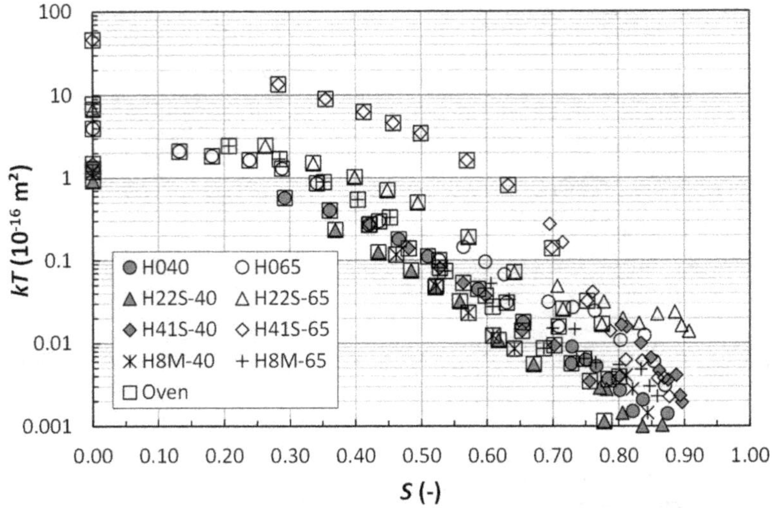

Figure 6.63 Relation between air-permeability *kT* and saturation degree *S* of 150mm concrete cubes (Torrent et al., 2014).

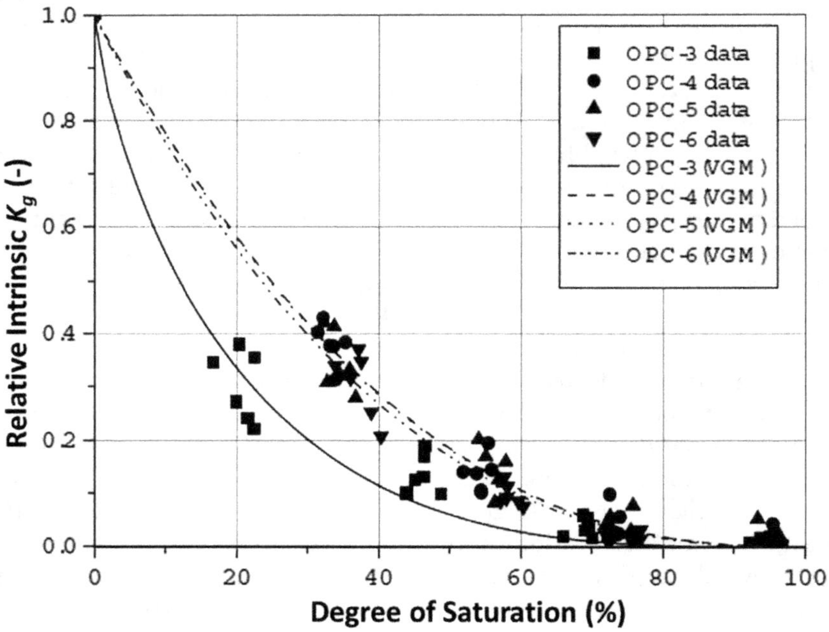

Figure 6.64 Relation between intrinsic N_2 permeability K_g and saturation degree (Gui et al., 2016).

The results shown in this section confirm quantitatively the intimate relation existing between both gas- and water-permeability of concrete and its degree of saturation. With respect to the effect of moisture on air-permeability *kT* and its compensation, see Section 5.7.2.

6.10 EFFECT OF APPLIED STRESSES ON PERMEABILITY

When testing concrete for permeability in the laboratory, almost invariably this is done on specimens which are free of external loads (stresses). This is not the case when permeability of concrete is measured on site on real structures, especially if they are fully loaded. For instance, columns will be subjected predominantly to compressive stresses whilst bent elements may be subjected to tensile stresses close to or exceeding the tensile strength of concrete. In the latter case, cracks will develop; the effect of cracks on permeability will be discussed in Section 6.11.

Therefore, for a better interpretation of permeability measurements performed on site, an understanding of the effect of stresses on the permeability of concrete is of importance.

6.10.1 Effect of Compressive Stresses

A good review of the effect of stresses (predominantly compressive) on concrete permeability can be found in Hoseini et al. (2009). The authors reviewed different testing techniques for measuring concrete permeability, both under the effect of mechanical loads and also after the loads were removed from the sample (residual permeability).

There is a certain agreement (Li, 2016) in that, at relatively low level of compressive stresses (up to about 50% of the compressive strength), the gas-permeability of concrete slightly decreases. This is attributed to the effect of consolidation and closure of micro-cracks (mostly originated by shrinkage) that already exist in unloaded concrete (Shah & Slate, 1968). Beyond 50% of the maximum load, gas-permeability starts to slightly increase, until the critical stress level is reached (≈75%–80% of the compressive strength), after which a very sharp increase in permeability is observed. This critical stress level corresponds to the onset of unstable coalescence of matrix cracks, such that if that load is sustained, fracture will happen after a certain period (due to tertiary creep). Figure 6.65 presents results reported by Li (2016), originated by Meziani and Skoczylas (1999) and Banthia and Barghava (2007), that confirm that behaviour.

The coupled effect of compressive stress and high temperature on the gas-permeability of concrete was investigated by Choinska et al. (2007). For that, they cast hollow cylinders (Ø110×220 mm) of a 26 MPa concrete,

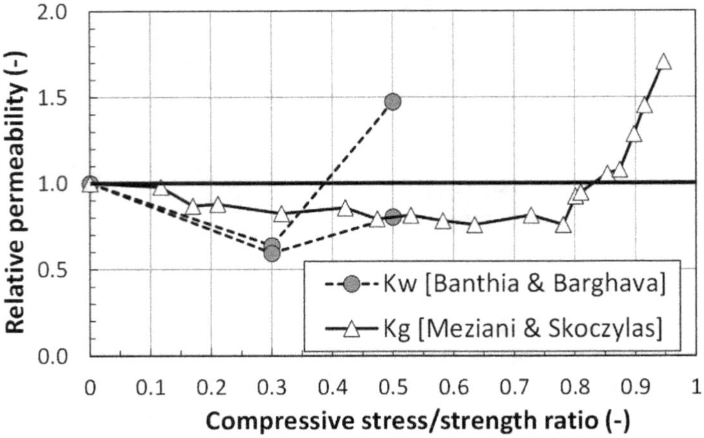

Figure 6.65 Effect of the applied compressive stress on the permeability of concrete to gas K_g and water K_w; data from Meziani and Skoczylas (1999), Banthia and Barghava (2007) apud (Li, 2016).

that were subjected to different levels of axial stresses until 90% of the peak stress (compressive strength); each stress level was maintained during 30 minutes to allow the measurement of the gas-permeability under load. Gas-permeability was measured injecting N_2 at five different relative pressures through the central hole (Ø14 mm) and measuring the gas flow rate at the inlet. At each pressure, the coefficient of gas-permeability was computed with the Hagen-Poiseuille-Darcy equation adapted to radial flow which, through the application of Klinkenberg formula (Section 3.6), yielded a value of the intrinsic coefficient of gas-permeability. The tests were performed after stabilizing the temperature at 20°C, 105°C and 150°C. Figure 6.66 (Choinska et al., 2007) shows the relation between the relative intrinsic gas-permeability (ratio between that measured under stress and unloaded) and the relative applied compressive stress level, for the three temperature levels investigated. At all temperatures the behaviour is similar, showing a decrease in gas-permeability up to about 70% of the peak stress, followed by a sharp increase at higher load levels. The effect of temperature does not seem relevant, except that at 150°C the increase in permeability happens at a slightly lower level of stress.

Van der Merwe (2019) investigated the effect of compressive stress (up to 50% of the maximum load) on air-permeability *kT*, on three cubes each cast with six different concrete mixes (M1-M6), with 28-days cube strengths (MPa) of M1: 42; M2: 45; M3: 59; M4: 69; M5: 87 and M6: 97. After 28 days of moist curing, the cubes were moved to the laboratory environment; prior to testing, the cubes were oven-dried at 105°C to constant mass.

The cubes were placed in a testing machine, where increasing compressive loads were applied in steps; at each step the air-permeability *kT* of

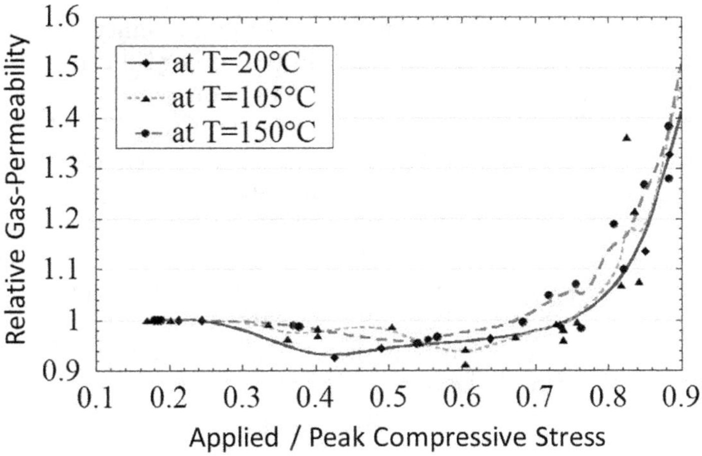

Figure 6.66 Effect of applied compressive stress on concrete permeability at different temperatures (Choinska et al., 2007). Courtesy of G. Pijaudier-Cabot.

each cube was measured on two opposite sides. In all but one case, a small reduction in permeability was observed between 0% and 50% of the maximum load; in the case of mix M5 a very small increase in permeability was observed. Figure 6.67a shows the average results obtained on the six mixes; top of the black bar is unloaded *kT* and top of the white bar is *kT* at 50% of the maximum load.

Performing an ANOVA, Van der Merwe (2019) proved that there is no statistically significant effect of compressive load on the dried intrinsic permeability of concrete within load levels expected during typical service conditions, and can therefore be ignored in practice.

It is interesting to see that, when analysing the residual permeability, i.e. measured after releasing the applied compressive load, the decrease in permeability at low stress levels is not detected (Picandet et al., 2001).

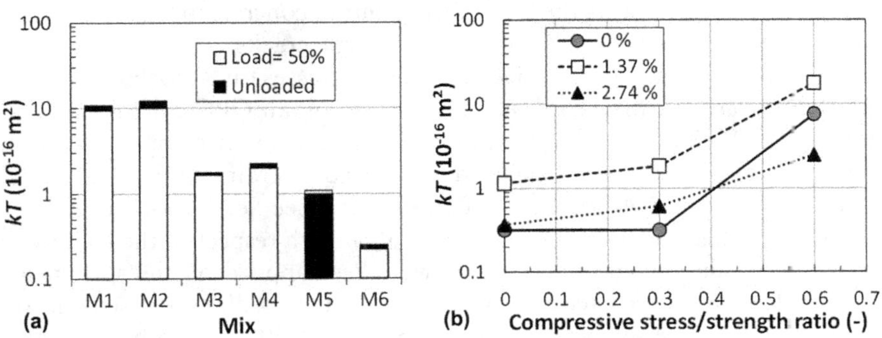

Figure 6.67 Effect of compressive stress on air-permeability *kT*. (a) data from Van der Merwe (2019); (b) data from Paulík and Hudoba (2009).)

With a similar scheme as that used by Van der Merwe (2019), Paulík and Hudoba (2009) investigated the change in air-permeability *kT* with compressive stresses of a plain and two steel-fibre reinforced concretes, with StF contents of 0%, 1.37% and 2.74% by mass of concrete, respectively. The corresponding cube strengths at 28 days were 51, 61 and 52 MPa, respectively. Two cubes of each mix were placed in the testing machine and the air-permeability *kT* was measured at applied stresses of 0%, 30% and 60% of the compressive strength of each mix. Figure 6.67b shows the effect of the applied stress on *kT* for the three mixes; in the case of the plain concrete, *kT* remains unaltered until a stress of 30% of the strength, increasing significantly when reaching 60% of the strength. The behaviour of the fibre reinforced concretes is similar only that, at 30% of the strength, already a small increase in *kT* could be observed, with a larger increase when reaching 60% of the compressive strength.

6.10.2 Effect of Tensile Stresses

Gérard et al. (1996) report data of water-permeability, measured on 40 mm thick slices cut from concrete cylinders, cured 45 days in water at 20°C. Steel plates were glued to the specimens so as to apply a controlled tensile load even after the peak load; the deformations were measured with glued strain-gauges. Holes in the centre of the opposite steel plates allowed the application, on one side, of water at 0.2 MPa pressure, that was collected and accurately measured on the opposite side by means of a suitable device. Figure 6.68 shows the gradual increase of water-permeability K_w with the strain to which the specimen was stretched, for three mixes of different compressive strengths (indicated in the figure). Just as reference, the peak stress corresponds to strains between 200 and 300 μstrains.

What Figure 6.68 shows is that, until the maximum load has been reached (corresponding to ≈250 μstrains), a moderate increase in water-permeability is observed, the latter growing several orders of magnitude in the post-peak period. This reflects the brittle behaviour of concrete in tension, which fails by the propagation of just one or few single cracks.

The effect of both compressive and tensile stresses on Autoclam air-permeability was measured by Tang et al. (2018) on an OPC concrete of *w/c* ratio = 0.60. They tested 100×100×400 mm prisms in compression and on dumbbell-shaped specimens in direct tension. Strains, ultrasonic pulse velocity (UPV) and Autoclam air-permeability (see Section 4.3.2.7) were measured at load levels of 0%, 30% and 60% with respect to the compressive and tensile strengths, loads that were kept applied on the specimens during 2, 4, 8 and 16 weeks. Confirming what was already discussed in Section 6.10.1, the air-permeability at 30% and 60% compressive loads was smaller and higher, respectively, than that measured on the unloaded specimen. In the case of tensile loads, there was an increase in permeability

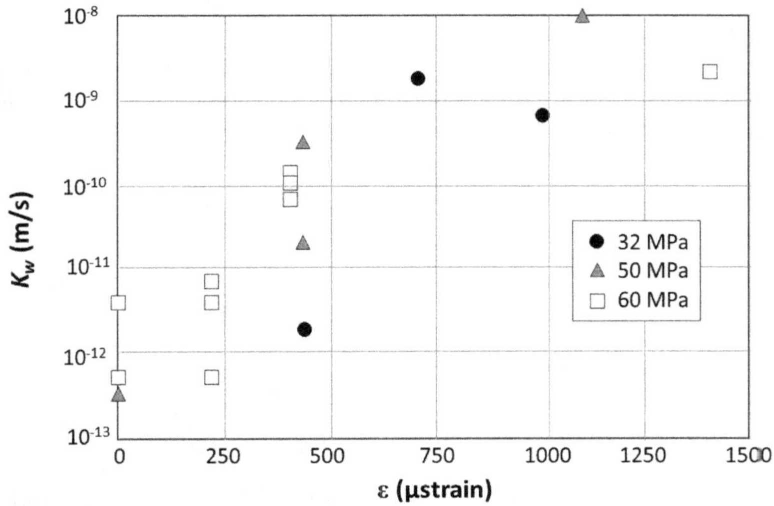

Figure 6.68 Effect of tensile strain on the water-permeability K_w of three concretes of different strengths; data from Gérard et al. (1996).

at the 30% level (compared to the unloaded specimen) and a further significant jump when the loaded was risen to the 60% level. A linear relation was established between a damage factor (computed from the UPV measurements) and air-permeability.

6.11 PERMEABILITY OF CRACKED CONCRETE

6.11.1 Permeability through Cracks: Theory

Most investigations on the permeability of concrete are conducted on laboratory specimens which are essentially free of cracks. As discussed in the previous section, site concrete in service often presents cracks due to loading, to restrained deformations originated by settlement or moisture or thermal changes. In aged structures, cracks may arise due to restrained delayed deformations, deleterious internal reactions, e.g. alkali-silica reaction (ASR) or delayed-ettringite formation (DEF), external attack (e.g. sulphate attack), steel corrosion, applied loads, etc.

The presence of cracks drastically changes the resistance of the material to the transport of fluids, a fact often neglected by durability design models. For instance, Blagojevic (2016) demonstrated that chloride ingress is accelerated by cracks at an early stage of formation.

Prof. Li Kefei has been concerned with the effect of cracking on permeability and other transport properties and, more generally, on the durability of reinforced concrete structures, discussing the topic in detail in Li (2016).

Interesting discussion on the role of cracks in concrete permeability can also be found in Hoseini et al. (2009).

From the theoretical point of view, the flow rate of a liquid through a smooth parallel plate crack of opening w and side length b can be estimated by Eq. (3.25), derived in Section 3.5.3 and recalled here as Eq. (6.8)

$$Q = \frac{a}{12} \cdot \frac{w^3}{\mu} \cdot \frac{\Delta P}{L} \tag{6.8}$$

where

Q=flow rate (m³/s)
a=crack side length (m)
w=crack width/opening (m)
μ=viscosity of the liquid (Pa.s)
$\Delta P/L$=gradient of pressure (Pa/m)

Equation (6.8) indicates that a single wide crack will conduct more fluid that several cracks of smaller widths, totalling the same added width. Dealing theoretically with multiple cracks is a much more complex challenge that has been addressed by Li and Li (2019).

If we compute the flow area as $A_f = w.\ a$, we can express Eq. (6.8) as:

$$Q = \frac{A_f}{12} \cdot \frac{w^2}{\mu} \cdot \frac{\Delta P}{L} \tag{6.9}$$

Comparing Eq. (6.9) with the Hagen-Poiseuille-Darcy law (Eq. 3.19), we conclude that the coefficient of permeability k_c of a smooth parallel plate crack is

$$k_c = \frac{w^2}{12} \tag{6.10}$$

In reality, cracks in concrete never look like smooth parallel plates, but present irregular shapes and rough surfaces that can be taken into consideration by introducing a tortuosity and roughness factor α in Eq. (6.10), such that

$$k_c = \alpha \frac{w^2}{12} \tag{6.11}$$

Factor α lies, according to Tsukamoto and Wörmer (1991), typically between 0.01 and 0.15.

There seems to be a threshold crack opening around 50–100 μm, after which the permeability starts to grow sharply; in addition, the permeability tends to decrease with time, due to self-healing (Hoseini et al., 2009).

From the form of Eqs. (6.10) and (6.11), it is obvious that the permeability of concrete should be strongly affected by the presence and opening of cracks. Indeed, water and air-permeability tests are listed in Sections 2.3.1.2 and 2.3.1.3 of RILEM TC 221-SHC (2013) as suitable indirect methods to investigate the formation and healing of cracks. Other methods, like optical microscopy, are also discussed in the above-mentioned reference. In the following, some experiences in applying permeability tests to assessing crack formation and healing will be discussed.

6.11.2 Effect of Cracks on Permeability

A quantification of the effect of crack opening on water- and gas-permeability is of importance when assessing permeability data measured on site since, very often and particularly for aged structures, the investigated concrete may contain cracks (see Section 11.4.2).

A compilation of data on the relation between water-permeability and crack width was made by Bentur and Mitchell (2008), based on data provided by Schiessl and Raupach (1997), Aldea et al. (1999, 2000), Edvardsen (1999), Reinhardt and Jooss (2003) and Burlion et al. (2003). The compiled data are presented in Figure 6.69, showing the strong effect of the crack width on water-permeability, relative to the uncracked concrete (in log-scale).

A very elaborated experiment was reported by Rastiello et al. (2014), aimed at verifying the validity of Eq. (6.10), by measuring the water flow through cylinders (of different sizes) of a single concrete mix subjected to a splitting tension test. They could relate, in real time, the water-permeability and the crack opening displacement (COD); moreover, they could relate COD with the crack surface percolated by the fluid, measured on separate specimens by a digital image technique.

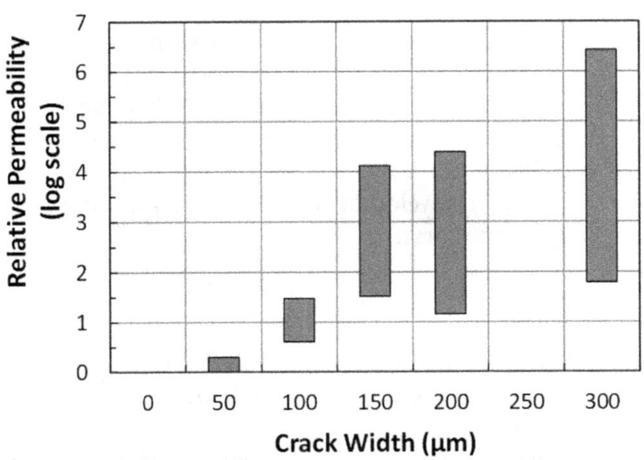

Figure 6.69 Compilation of data on the effect of crack width on concrete water-permeability; data from Bentur and Mitchell (2008).

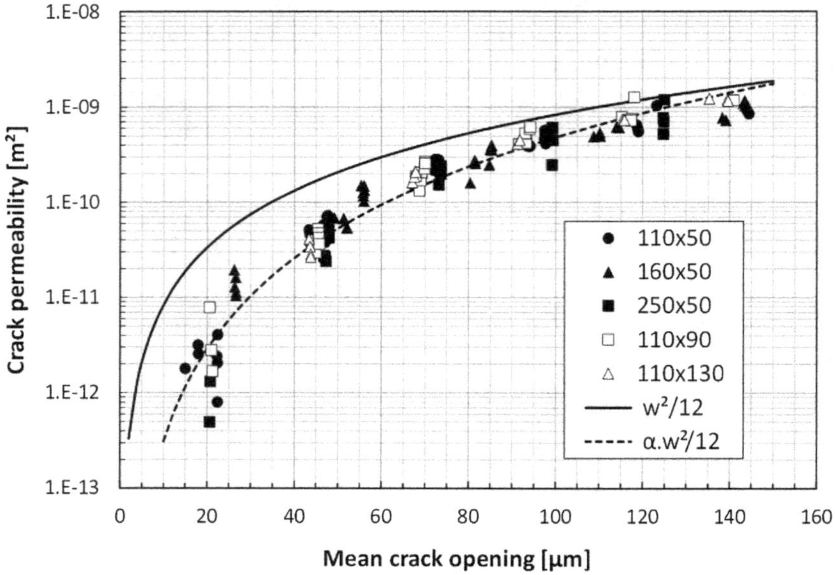

Figure 6.70 Experimental vs. theoretical relation between crack water-permeability and crack opening (Rastiello et al., 2014, 2018). Data courtesy of G. Rastiello.

Figure 6.70 presents the results obtained by Rastiello (2013) and Rastiello et al. (2014), where the dots correspond to the permeability measured on different cylinder sizes (diameter×thickness) as function of the mean crack opening, whilst the full line corresponds to the theoretical value expressed by Eq. (6.10).

The fact that the experimental values fall below the theoretical quadratic curve can be attributed to the roughness and tortuosity of the crack (factor α in Eq. (6.11)). It can be seen from Figure 6.70 that the departure from theory is more marked for small crack openings. Indeed, Rastiello et al. (2014) calculated that factor α grows from 0.1 for w=20 µm to almost 1.0 for w=160 µm. In a later development, Rastiello et al. (2018) proposed the following function for factor α:

$$\alpha = 3.33 \times 10^4 \ . \ w^{1.19} \tag{6.12}$$

where w=crack opening (m).

The dotted line in Figure 6.70 corresponds to Eq. (6.11), with factor α computed by Eq. (6.12).

The results of Rastiello et al. (2014) were confirmed by Zeng et al. (2020) on pre-cracked concretes with and without fibres, subjected to different degrees of frost damage.

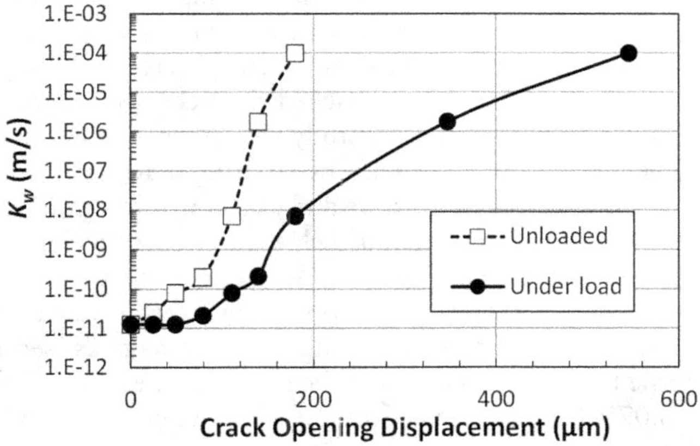

Figure 6.71 Effect of COD (under load or unloaded) on the water-permeability coefficient K_w; data from Wang et al. (1997).

A similar study was conducted by Wang et al. (1997), who studied the effect of the COD, under a controlled splitting tensile test, on water-permeability. The Ø100×25 mm specimens were loaded to a predetermined crack width and unloaded, recording the COD under load and after unloading. The water-permeability was measured using a variable water column permeameter. Just for reference, an unloaded COD of 50 μm corresponds to a crack length of about 20 mm and one of 150 μm to a crack length of 75–80 mm. Figure 6.71 shows the water-permeability coefficient K_w as function of the COD under loading and after unloading.

Using the same technique, Aldea et al. (2000) compared the experimental water flow rate with the theoretical one given by Eq. (6.8), assuming the crack width w to be equal to the mean COD. They obtained results very similar to those in Figure 6.70, i.e. showing experimental flow rates lower than that predicted by theory, a fact also attributed to the roughness factor α.

Similar experiments were conducted to study the effect of cracks on gas-permeability. For instance, Picandet and Khelidj (2003) and Picandet et al. (2009) applied splitting loads to Ø110×50 mm disks, made of concretes with 28-day cylinder compressive strengths of 65, 110 and 130 MPa, measuring the COD, associated with the diametrical displacement. Between 4 and 11 load cycles were applied to the cylinders. After unloading, the intrinsic N_2-permeability was measured on the discs by the Cembureau test method. Different pressures were applied, so as to calculate the intrinsic permeability applying the Klinkenberg correction (also taking turbulent flow into consideration). The same disc was reloaded to the next step, taking care of applying the load always at the same points. They found a monotonically increase in intrinsic N_2-permeability with the COD for all three concretes, which grew three to four orders of magnitude between 0 and about 250 μm of COD. After the final load step, the water-permeability was also measured, giving values of the same order of magnitude as gas-permeability, quite differently than for uncracked concrete (see Section 3.5.5).

Stehlik et al. (2015) attempted to study the changes in gas-permeability *kT* with the opening of tensile cracks, by splitting diagonally 150 mm concrete cubes of a conventional concrete and of a concrete replacing 2/3 of its coarser fraction (4–8 mm) by expanded clay. Some cubes were reinforced to ensure a more controlled crack opening. Simultaneously they were measuring the crack opening with a microscope. Stehlik et al. (2015) concluded that the test method is suitable to evaluate the damage caused by microcracks with the limitation that, when the crack reaches a maximum opening around 0.075–0.1 mm, the amount of air flowing into the vacuum cell is such that tends to detach it from the specimen.

Rzezniczak (2013) subjected concrete slabs (plain OPC concrete and with 8% SF) to four-point bending in a servocontrolled testing machine, so as to create tensile cracks, measuring the air-permeability *kT* of uncracked and cracked concrete. She observed that the uncracked SF concrete had a smaller *kT* than the OPC, attributed to the more compact microstructure of the former. After cracking, the average *kT* values increased by a factor of 8 and 52 for the OPC and SF concretes, respectively. This reflects the wider cracks observed in the SF concrete (0.25 mm) than in the OPC concrete (0.17 mm). Later, the cracks were injected with epoxy resin and a flexible polyurethane grout, as well as with a polymer-modified, cementitious waterproof and protective slurry, used as a repair overlay system. The injections, particularly with epoxy resin, were the most successful techniques to restore the tightness of the cracks, as measured by the *kT* air-permeability test method. The performance of the repaired samples subjected to thermal cycles and freezing and thawing (unloaded and loaded) is also reported, showing no relevant damage in terms of their measured *kT* values.

An elaborated experiment was reported by Malek et al. (2018), with the aim of establishing the degree of damage of confined reinforced cylindrical columns, loaded to failure, on the basis of changes in oxygen permeability. Three columns (Ø500×1500 mm) were cast from the same load of ready-mixed concrete, reinforced with the same longitudinal steel bars, but having low, medium and high confinement levels, obtained by reducing the separation of helical stirrups. After failure, each column was carefully saw-cut into pieces, from which two Ø93×430 mm cores were drilled, one for compression test and the other for oxygen-permeability test. The permeability core was further saw cut into 25 mm thick discs which, after drying at 50°C to constant mass, were subjected to the South African OPI (oxygen permeability index) test, described in Section 4.3.1.3. The permeability values k_{OPI} measured away from the failure zone were of 22 and 35 ($\times 10^{-11}$ m/s) for the low and medium confinement columns, which correspond to a good quality concrete (see Table 4.2). Tests conducted closer and closer (from below and from above) to the failure plane showed a monotonic increase of the k_{OPI} values, reaching values around 700 $\times 10^{-11}$ m/s at the failure plane that, in this way, could be clearly identified. Unfortunately, for the high confinement column, the amount of damage near the failure plane was too high to allow the determination of k_{OPI}. A damage factor is proposed based on k_{OPI}.

Giaccio et al. (2015) investigated the damage generated by ASR in concrete mixes made with a highly reactive rock, used as 40% replacement of the coarse aggregate. Mix R contained no fibres, *S* contained 40 kg/m³ of hooked StF, M and m contained 3 and 1 kg/m³ of synthetic micro-fibres, respectively. In Group 1, the concrete contained the natural NaO_{eq}=2.8 kg/m³ supplied by the constituents, whilst for Group 2, Na(OH) was added to the mixing water to raise the NaO_{eq} to 4.0 kg/m³. Prisms were cast with the eight concretes that were kept in a moist environment, on which linear expansion, air-permeability *kT* and crack density and opening were measured at approximately 6 months of age. Figure 6.72a and b shows the measured air-permeability *kT* as function of the linear expansion and the crack density, respectively.

A good relationship was obtained, Figure 6.72a, between the expansion due to ASR and the air-permeability *kT*, which is explained by the excellent relationship found between *kT* and the crack density, Figure 6.72b, observed in the specimens. The maximum crack width for which *kT* could be measured was 0.5 mm.

In another experiment by the same research group (Giaccio et al., 2019), a significant increase in *kT* with time was found in concrete blocks (0.7×0.4×0.4 m) exposed outdoors, made with reactive coarse aggregate (with and without fibres), containing 4.0 kg/m³ of NaO_{eq}, which was not observed in the control mix with just 2.8 kg/m³ of NaO_{eq}. An interesting conclusion of the paper is: "The deformations and the air-permeability measured on the blocks are consistent with the cracking level, but the changes in permeability started earlier, probably detecting incipient micro-cracking".

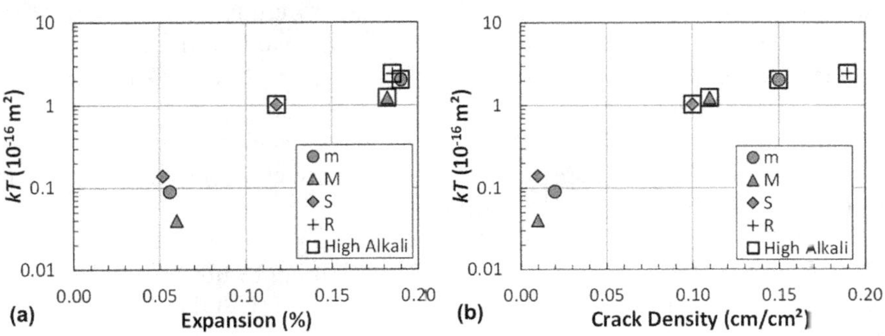

Figure 6.72 Effect of (a) ASR expansion and (b) crack density on *kT* (Giaccio et al., 2015).

6.11.3 Self-Healing of Cracks and Permeability

Self-healing is a phenomenon by which concrete cracks are sealed naturally by hydration of unreacted cement or hydraulic additions, with hydration products growing from the crack walls that gradually seal the opening. The effect can be enhanced by adding suitable inorganic or organic compounds in the concrete mix. There are several comprehensive reviews and state-of-the-art reports dealing with this subject (Li & Herbert, 2012; Mihashi & Nishiwaki, 2012; RILEM TC 221-SHC, 2013; Van Tittelboom & De Belie, 2013).

One of the first attempts to study the self-healing phenomenon through permeability tests was reported by Edvardsen (1999), who investigated the causes for the decrease in permeability of cracked concrete, which she attributed to the precipitation of $CaCO_3$ crystals in the cracks, proposing also the underlying reaction mechanisms.

A water-permeability test was used by Ahn and Kishi (2010) to study a self-healing concrete containing a Type K expansive agent coupled with a montmorillonite-containing geomaterial, aimed at producing swelling compounds that can help in closing cracks. The resulting self-healing concrete showed a water-permeability three times less than the control concrete.

Fibres are known for their positive effect in keeping the crack opening narrower, through their mechanical action. Nishiwaki et al. (2012) advanced the theory that their effect is not just mechanical, but physical as well, due to their action as nuclei for the precipitation of $CaCO_3$ crystals. In this respect, the chemical composition of the fibre plays a role. To prove this theory, they conducted two series of experiments, in which they measured the volume of $CaCO_3$ precipitated around the fibres, complemented by microscopic observation of the crack surfaces. They also measured the water-permeability K_w on fibre-reinforced mortar ($w/b=0.45$) plates, containing 2% vol. of six different synthetic fibre types, to evaluate the self-healing capability of each system. Table 6.5 describes the six fibre types investigated by Nishiwaki et al. (2012).

The fibre-reinforced mortar plates were subjected to direct tension, reaching crack openings (microscopic observation) ranging typically between 100 and 400 μm, that were maintained by fixing plates and nuts. A water-permeability test was applied immediately and again after 3, 14 and 28 days of immersion under water.

Figure 6.73a shows the relative change of water-permeability (relative to that measured immediately after generating the cracks) with time under water, as a measure of the self-healing effect.

It can be seen from Figure 6.73a that mortars containing PVA fibres showed a relatively fast healing, followed by that containing deformed PPF (C-PP). The mortar containing ethylene vinyl alcohol (EVOH) fibres showed a slower and less effective healing, whilst the mortars containing circular PPF and polyacetal fibres (POM) showed virtually no self-healing effect.

Table 6.5 Characteristics of the six fibre types investigated by Nishiwaki et al. (2012)

Code	*Fibre type*	*Density (g/cm³)*	*Tensile strength (MPa)*	*Length (mm)*	*Diameter (μm)*	*Cross-section*
PP	Polypropylene	0.91	760	6	11	Circular
C-PP					18	Deformed
POM	Polyacetal	1.41	135	10	48	Circular
EVOH	Ethylene vinyl alcohol copolymer	1.04	231	5	15	Circular
PVA-I	Polyvinyl alcohol	1.30	1,600	6	14	Circular
PVA-II					37	

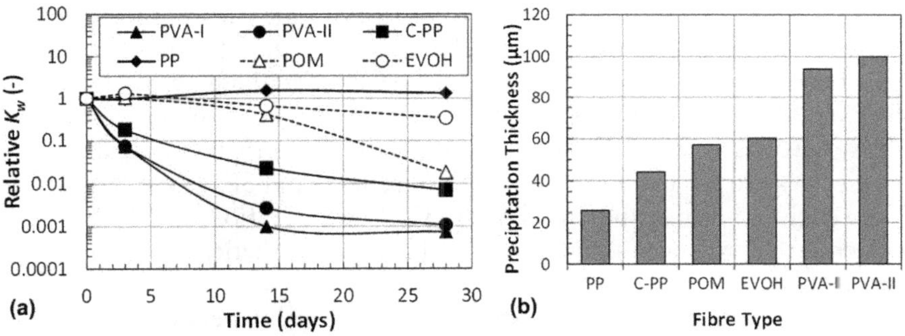

Figure 6.73 (a) Change in water-permeability K_w of healing pre-cracked mortars reinforced with different fibre types and (b) thickness of precipitated healing products on different fibre types; data from Nishiwaki et al. (2012).

An explanation for these different behaviours is derived from the observation of Figure 6.73b that shows the average thickness of precipitation products around the different types of fibres. It is shown (Nishiwaki et al., 2012) that the amount (thickness) of precipitated products increases with the polarity strength of the synthetic composite (e.g. OH^- ions), which tends to attract Ca^{2+} ions and precipitate $CaCO_3$. It is also concluded that longer fibres, of more complex shapes (like C-PP fibres), also generate larger quantities of self-healing products. This provides an explanation for the different healing effects of mortars reinforced with the various fibre types represented in Figure 6.73a.

The effect of self-healing on an Ultra High-Performance Strain-Hardening Cementitious Composite (UHP-SHCC) was investigated by Kunieda et al. (2012). The material under investigation contained over 1,500 kg/m³ of binder (low heat cement and 15% of SF), 155 kg/m³ of fine quartz sand and 1.5% vol. of high-strength polyethylene fibres, with *w/b* ratio = 0.22 and tensile strength of 6.8 MPa. Plates (200×150×30 mm) were cast, and at an age of 36–37 days they were tensioned through embedded steel bars until reaching 0.1% and 0.2% tensile strains so as to crack them. The coefficient of air-permeability *kT* was measured immediately before tensioning the plates and immediately after unloading them. A water-flow rate test, similar to Karsten tube (see Section 4.2.2.2), was performed after the *kT* test. Then the plates were kept in either water or air during periods of 20, 90 and 360 days. At the end of those periods, *kT* and the water flow rate were measured again, to investigate the self-healing process. In all cases, after removing the samples from water, the plates were kept for 2 days in a dry environment (at 20°C) before starting the permeability tests.

Figure 6.74a shows the change in air-permeability *kT* before the load (uncracked), after applying a tensile strain of 0.1% (cracked) and after 20, 90 and 360 days of water- and air-curing. Figure 6.74b is similar but plotting the results of water flow rate. Each bar represents the average of the results of four (for water curing) or three (for dry air curing) specimens.

Figure 6.74a and b shows a similar picture: a significant reduction in permeability of the plates stored under water after being cracked, whilst the plates kept in dry air did not show significant reduction in permeability. This indicates that self-healing can only take place if the material is kept moist, so as to facilitate the hydration of cement and SF. The system tested has a high self-healing potential due to the low *w/b* ratio (much unhydrated cement), the presence of highly active SF and of synthetic fibres, all helping the process.

Another experiment to investigate the self-healing effect of an Ultra High-Performance Hybrid-Fibre Reinforced Cementitious Composite (UHP-HFRCC) was conducted by Kwon et al. (2013a). The material tested was a self-compacting composite with $w/c=0.143$, made with commercial cement composed of 82% low heat cement and 18% SF. It contained a mix of fine silica sand and wollastonite as fine aggregate (average size 0.212 mm) and two types of StF. It contained 1.0%vol. of short (6 mm) and fine (Ø0.16 mm) straight fibres together with 1.5%vol. of longer (30 mm) and thicker (Ø0.38 mm) hooked-end fibres. Two special plate specimens (200×150×30 mm), arranged for applying direct tensile loads, were cast

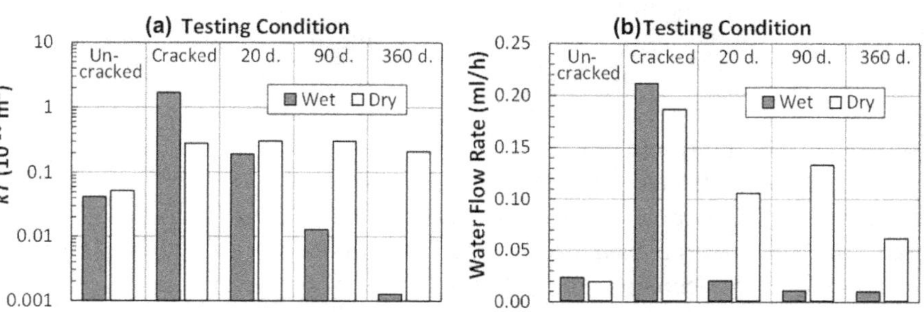

Figure 6.74 Change in (a) air-permeability *kT* and (b) water flow rate, for healing under water and in dry air (Kunieda et al., 2012). Data courtesy of M. Kunieda.

and steam cured (T_{max}=90°C) for 24 hours and later kept in a moist room (20°C/95% RH) until the moment of test. The compressive strength of the composite was 182 MPa; more details on the mechanical performance of the composite can be found in Kwon et al. (2013b). At the age of 7 days (time 0) the coefficient of air-permeability *kT* was measured, yielding a very low value ($kT<0.001\times10^{-16}$ m²), as expected given the characteristics of the material and its moisture condition. Immediately after, the plate was loaded in tension until reaching 0.2% strain and subsequently unloaded. Then, the specimen was inspected with an optical microscope to detect the maximum crack width (76 μm) and the air-permeability *kT* was measured again in the centre of the specimen, where multiple cracks were observed. Afterwards, the specimen was kept under water and the evolution of microscopy crack width and permeability *kT* was measured, removing them from the water after 3, 7, 14, 21 and 28 days. Figure 6.75 shows the self-healing effect, as measured by the coefficient of air-permeability *kT*.

Figure 6.75 shows the very high values of *kT* measured on the freshly strained specimens (time=0) and their sharp reduction after just 3 days of self-healing, process that seems completed already at 7 days. Of course, the self-healing conditions of the UHP-HFRCC are extremely favourable, due to its very low *w/c* ratio that leaves considerable amount of unhydrated cement and SF which, in contact with water and finding room to grow hydrated products in the cracked space, rapidly fills the crack. The StF help in developing multiple closely spaced cracks of small openings, that facilitated their rapid self-healing.

Beglarigale et al. (2021) studied the self-healing capability of UHPFRC of *w/b*=0.22, made with different binder types and contents. Specimens

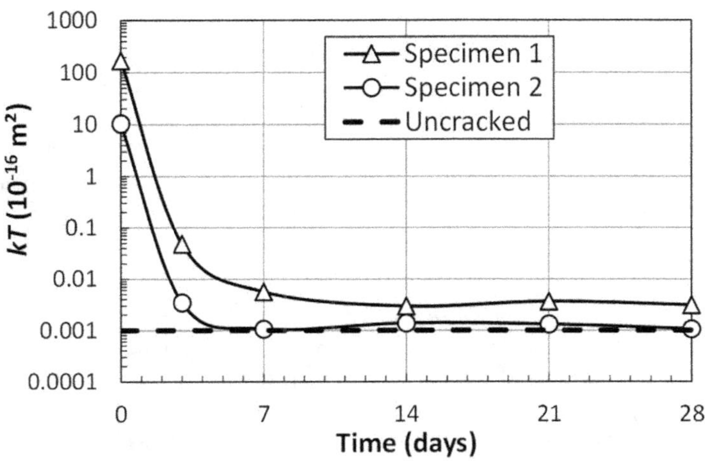

Figure 6.75 Effect of self-healing on air-permeability *kT* of a UHP-HFRCC; data from Kwon et al. (2013a).

were loaded up to near the compressive strength, unloaded and stored under water for 30 days. Despite the fact that a special loading device was purposely designed, the air-permeability *kT* could not be measured on the cracked specimens and yielded erratic values after self-healing, thus precluding the assessment of the self-healing effect. A water-permeability test, similar to the Karsten tube (see Section 4.2.2.2), yielded better results.

A relatively new field of research is the inclusion of microbial adjuvants into the concrete mix, with the aim of improving its natural tendency to self-healing of cracks. One such investigation was reported by Nguyen et al. (2019b) in which the healing capacity of three concrete mixes (351 kg/m³ of binder composed of cement and LF, $w/b=0.52$ and about 200 mm slump) was compared. Control mix (C) contained no adjuvants, a second mix (N) contained as adjuvants just the "nutrients" and the third (B) the nutrients (1.82 kg/m³ peptone+0.91 kg/m³ yeast) and 1.82×10^{-10} cells/m³ of *Bacillus subtilis.* This family of bacteria can remain dormant in the highly alkaline environment of concrete for over 50 years and is capable of producing enough urease enzyme, resulting in calcium carbonate precipitation through urea hydrolysis. With the three mixes, Ø150×50 mm discs were cast, demoulded at 24 hours and kept under water until the age of 28 days. Then, the discs were subjected to a controlled splitting force, so as to generate crack opening widths between 80 and 400 μm. The position of the cracks was marked, their opening measured with an optical microscope and, then, the specimens were immersed in water at 30°C. At intervals, the crack opening was monitored with the microscope; a good colour picture in Nguyen et al. (2019b) shows the more effective healing of Mix B after 23 and 44 days of immersion, compared with mixes C and N that, still, show some but more limited natural self-healing. The bacterial-healing effect is attributed (based on EDS analysis) to the enhanced formation of $CaCO_3$. Water and gas-permeability of the discs were measured to follow the healing process of the three mixes. Figure 6.76a shows the amount of water

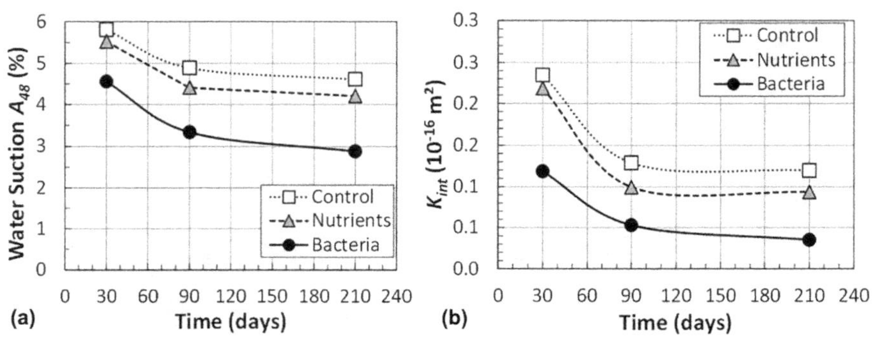

Figure 6.76 Effect of bacterial-healing on (a) water suction A_{48} and (b) intrinsic air-permeability K_{int}; data from Nguyen et al. (2019b).

absorbed by capillary suction after 48 hours A_{48}, and Figure 6.76b shows the intrinsic coefficient of gas-permeability K_{int} (Cembureau method), measured at four different pressures and applying the Klinkenberg correction (Section 3.6).

The positive contribution of the bacteria to the self-healing effect is evident in Figure 6.76, already after 30 days of immersion in water at 30°C.

REFERENCES

3M (2019). "3M™ concrete curing control sheet water transport & supply". https://multimedia.3m.com/mws/media/1201556O/itd-266.pdf.

Abbas, A., Carcasses, M. and Ollivier, J-P. (1999). "Gas-permeability of concrete in relation to its degree of saturation". *Mater. & Struct.*, v32, 3–8.

Adámek, J. and Juránková, V. (2009). "Durability of concrete as function of the properties of the concrete layer". *Trans. Transp. Sci.*, v2, n4, Prague, Czech Rep., 188–195.

Ahn, T-H. and Kishi, T. (2010). "Crack self-healing behavior of cementitious composites incorporating various mineral admixtures". *J. Adv. Concr. Technol.*, v8, n2, 171–186.

Alamri, A.M. (1988). "Influence of curing on the properties of concretes and mortars in hot climates". PhD Thesis, University of Leeds, October, 287 p.

Aldea, C.-M., Ghandehari, M., Shah, S.P. and Karr, A. (2000). "Combined effect of cracking and water-permeability of concrete". EMC 2000, 6 p.

Aldea, C.-M., Shah, S.P. and Karr, A. (1999). "Permeability of cracked concrete". *Mater. & Struct.*, v32, n218, 370–376.

Azuma, Y., Mori, H. and Tada, K. (2016). "Influences of secondary wet curing on strength, durability and microstructure of steam cured mortar". *Cement Sci. & Concr. Technol.*, v70, n1, 405–412 (in Japanese).

Bahurudeen, A. and Santhanam, M. (2014). "Performance evaluation of sugarcane bagasse ash-based cement for durable concrete". *4th International Conference Durability of Concrete Structures*, Purdue Univ., West Lafayette, IN, July 24–26, 275–281.

Bai, J., Wild, S. and Sabir, B.B. (2002). "Sorptivity and strength of air-cured and water-cured PC-PFA-MK concrete and the influence of binder composition on carbonation depth". *Cem. & Concr. Res.*, v32, n11, 1813–1821.

Balayssac, J.P., Détriché, C.H. and Grandet, J. (1995). "Effects of curing upon carbonation of concrete". *Constr.& Build. Mater.*, v9, n2, 91–95.

Banthia, N. and Barghava, A. (2007). "Permeability of stressed concrete and role of fiber reinforcement". *ACI Mater. J.*, v104, 70–77.

Beglarigale, A., Eyice, D., Tutkun, B. and Yazici, H. (2021). "Evaluation of enhanced autogenous self-healing ability of UHPC mixtures". *Constr. & Bldg. Mater.*, v280, 122524, 16 p.

Beglarigale, A., Ghajeri, F., Yigiter, H. and Yazici, H. (2014). "Permeability characterization of concrete incorporating fly ash". *ACE 2014*, Istanbul, Turkey, October 21–25, 7 p.

Bentur, A. and Mitchell, D. (2008). "Materials performance lessons". *Cem. & Concr. Res.*, v38, 259–272.

Bhutta, M.A.R., Imamoto, K. and Ohama, Y. (2009). "Air-permeability of hardener-free epoxy-modified mortars as repair materials". Concrete Repair, Rehabilitation and Retrofitting II, Alexander et al. (Eds).

Blagojevic, A. (2016). "The influence of cracks on the durability and service life of reinforced concrete structures in relation to chloride-induced corrosion". PhD Thesis, TU Delft, NL.

Bueno, V., Nakarai, K., Nguyen, M.H., Torrent, R.J. and Ujike, I. (2021). "Effect of surface moisture on air-permeability kT and its correction". *Mater. & Struct.*, v54, 89, 12 p.

Burlion, N., Skoczylas, F. and Dubois, T. (2003). "Induced anisotropic permeability due to drying of concrete". *Cem. & Concr. Res.*, v33, 679–687.

Bustos, F., Videla, C., López, M. and Martinez, P. (2015). "Reducing concrete permeability by using natural pozzolans and reduced aggregate-to-paste ratio". *J. Civil Eng. Manag.*, v21, n2, 165–176.

Carcasses, M. and Ollivier, J.-P. (1999). "Transport by permeation". Engineering and Transport Properties of the Interfacial Transition Zone in Cementitious Composites – State-of-the-Art Report of RILEM TC 159-ETC and 163-TPZ.

Cather, R. (1994). "How to get better curing". *Concrete*, v26, n5, 22–25.

CEB-FIP (1991). "CEB-FIP model code 1990 Final Draft". *CEB Bulletin d'Information* N° 203, 204 and 205, Lausanne, Switzerland, July 1991.

Chmielewska, B. (2013). "Effect of polymer additive on gas-permeability coefficient of concrete". *Adv. Mater. Res.*, v687, 191–197.

Choi, H., Koh, T., Choi, H. and Hama, Y. (2019). "Performance evaluation of precast concrete using microwave heating form". *Materials*, v12, 1113, 18 p.

Choinska, M., Khelidj, A., Chatzigeorgiou, G. and Pijaudier-Cabot, G. (2007). "Effects and interactions of temperature and stress-level related damage on permeability of concrete". *Cem. & Concr. Res.*, v37, n1, 79–88.

Conciatori, D. (2005). "Effet du microclimat sur l'initiation de la corrosion des aciers d'armature dans les ouvrages en béton armé". Ph.D. Thesis N° 3408, EPFL, Lausanne, 264 p.

de Schutter, G. (2016). "The quest for absolute concrete durability performance criteria". *Key Eng. Mater.*, v711, 599–606.

Ebensperger, L. and Olivares, M. (2017). "Efecto de la razón agua/cemento, condiciones de curado y edad en el desempeño de concretos durante su vida útil". *CONPAT 2017*, Asunción, Paraguay, 9 p.

Edvardsen, C. (1999). "Water permeability and autogenous healing of cracks in concrete". *ACI Mater. J.*, v96, n4, 448–454.

Ewertson, C. and Petersson, P.E. (1993). "The influence of curing conditions on the permeability and durability of concrete. Results from a field exposure test". *Cem. & Concr. Res.*, v23, n3, 683–692.

Fares, H., Noumowe, A. and Remond, S. (2009). "Self-consolidating concrete subjected to high temperature. Mechanical and physicochemical properties". *Cem. & Concr. Res*, v39, 1230–1238.

Fernández Luco, L. (2004). "Hormigón de Contracción Compensada". *Hormigones Especiales*, Capítulo 6, AATH, Buenos Aires, August, 185–213.

Fernández Luco, L. (2010). "Propuesta de indicadores de la eficacia del curado en obra". *Concreto y Cemento. Invest. y Desarrollo*, México, v1, n2, 17–29.

Fernández Luco, L., Pombo, R. and Torrent, R. (2003). "Shrinkage-compensating concrete in Argentina". *Concr. Intern.*, May, 49–53.

fib (2012). "Model code 2010 – Final draft". *fib Bulletins* 65 & 66, April 2012.

Fischer, R. (1967). "Über das Verhalten von Zementmörtel und Beton bei Höheren Temperaturen". *DAfStb*, Heft 212.

Gérard, B., Breysse, D., Ammouche, A., Houdusse, O. and Didry, O. (1996). "Cracking and permeability of concrete under tension". *Mater. & Struct.*, v29, April, 141–151.

Giaccio, G., Bossio, M.E., Torrijos, M.C. and Zerbino, R. (2015). "Contribution of fiber reinforcement in concrete affected by alkali–silica reaction". *Cem. & Concr. Res.*, v67, 310–317.

Giaccio, G., Torrijos, M.C., Milanesi, C. and Zerbino, R. (2019). "Alkali-silica reaction in plain and fibre concretes in field conditions". *Mater. & Struct.*, v52, n31, March, 15 p.

Glinicki, M.A. and Nowowiejski, G. (2013). "Strength and permeability of concrete with CEM II and CEM V cements containing high calcium fly ash". *SCTM3 Conference*, Kyoto, Japan, August 18–21.

Gonçalves, A. (1999). "Durabilidede real e potencial do betão". LNEC, Lisbon, Portugal.

Gonçalves, A., Salta, M. and Neves, R. (2000). "Caracterização da durabilidade do betão". Encontro Nacional Betão Estrutural, 85–94.

Gonen, T. and Yazicioglu, S. (2007). "The influence of compaction pores on sorptivity and carbonation of concrete". *Constr. & Build. Mater.*, v21, n5, 1040–1045.

Gopinath, R. (2020). "Concrete carbonation prediction for varying environmental exposure conditions". PhD Thesis, Univ. Cape Town, South Africa, 256 p.

Gowripalan, N., Cabrera, J.G., Cusens, A.R. and Wainwright, P. (1990). "Effect of curing on durability". *Concr. Internat.*, v12, n2, 47–54.

Gräf, H. and Grube, H. (1984). "The influence of curing on the gas-permeability of concrete with different compositions. *Proceedings on RILEM Seminar on the Durability of Concrete Structures Under Normal Outdoor Exposure*, Univ. Hanover, March, 68–79.

Gui, Q., Qui, M. and Li, K. (2016). "Gas permeability and electrical resistivity of structural concretes: Impact of pore saturation". *PRO 112: International RILEM Conference on Materials, Systems and Structures in Civil Engineering Conference – Moisture in Materials and Structures*, Hansen, K.K., Rode, C. and Nilsson, L-O. (Eds.), ISBN: 978-2-35158-178-0, RILEM Publications, 44–53.

Güneyisi, E., Özturan, T. and Gesoğlu, M. (2007). "Effect of initial curing on chloride ingress and corrosion resistance characteristics of concretes made with plain and blended cements". *Build. Environ.*, v42, n7, 2676–2685.

Gurdián Curran, F.H. (2016). "Caracterización de hormigones ecológicos con adiciones puzolánicas y árido reciclado y su protección frente a la corrosión de armaduras". PhD Tesis, Univ. Alicante, Spain, 250 p.

Hansen, T. C. (1986). "Physical structure of hardened cement paste. A classical approach". *Mater. & Struct.*, v19, n6, 423–436.

Harashina, T., Imamoto, K. and Kiyohara, C. (2018). "Fundamental study on carbonation characteristics of the concrete using non-iron slag fin aggregate". Annual Meeting of Architectural Inst. of Japan (in Japanese).

Hayakawa, K. and Kato, Y. (2010). "An experimental study for the quality variation of cover concrete by construction works". *Cem. Sci. & Concr. Technol.*, v64, 421–427 (in Japanese).

Hayakawa, K. and Kato, Y. (2012). "Effects of construction works and mix proportions on quality of cover-concrete". *J. Japan Soc. Civil Eng.*, v68, n4, 399–409 (in Japanese).

Hayakawa, K., Mizukami, S. and Kato, Y. (2012). "A fundamental study for quality evaluation of structural cover-concrete by surface air-permeability". *J. Japan Soc. Civil Eng.*, v68, n4, 385–398 (in Japanese).

Hooton, D. (2011). "Canadian use of ground granulated blast-furnace slag as a supplementary cementing material for enhanced performance of concrete". *Canadian J. Civil Eng.*, v27, n4, 754–760.

Hoseini, M., Bindiganavile, V. and Banthia, N. (2009). "The effect of mechanical stress on permeability of concrete: A review". *Cem. & Concr. Composites*, v31, 213–220.

Ichiro, P., Yoshida, A. and Kishi, T. (2009). "Influence of curing, water content and water-cement ratio on the permeability of the concrete surface". *Proc. Japan Concrete Institute*, v31, n1, 757–762 (in Japanese).

Imamoto, K., Shimozawa, K., Nagayama, M., Yamasaki, J. and Nimura, S. (2009). "Air-permeability of concrete cover and its relationship with carbonation progress under long-term exposure test in Japan". *RILEM Proceedings Pro 63*, 508–514.

Iwaki, I. (2020). Personal communication to Prof. K. Imamoto.

Jacobs, F. (1998). "Permeability to gas of partially saturated concrete". *Mag. Concr. Res.*, v50, n2, 115–121.

Jacobs, F. (2006). "Luftpermeabilität als Kenngrösse für die Qualität des Überdeckungsbetons von Betonbauwerken". Office Fédéral des Routes, VSS Report 604, Bern, Switzerland, 85 p.

Jacobs, F., Denarié, E., Leemann, A. and Teruzzi, T. (2009). "Empfehlungen zur Qualitätskontrolle von Beton mit Luftpermeabilitätsmessungen". Office Fédéral des Routes, VSS Report 641, December, Bern, Switzerland, 53 p.

Jacobs, F., Mayer, G. and Wittmann, F.H. (1994). "Hochpermeable, zementgebundene Verfüllmörtel für SMA Endlager". Nagra Technical Report NTB 92-11, Wettingen, Switzerland, 123 p.

Jiménez, L.F. and Moreno, E.I. (2015). "Durability indicators in high absorption recycled aggregate concrete". *Adv. Mater. Sci. & Eng.*, v2015, Article ID 505423, 8 p.

Kato, Y. (2013). "Characteristics of the surface air-permeability test and the evaluation of quality variation in cover concrete due to segregation of concrete". *J. Adv. Concr. Technol.*, v11, 322–332.

Kato, Y. and Hayakawa, K. (2013). "Durability design of structural cover concrete based on bleeding rate". *Life-Cycle of Structural Systems*. Furuta, Hitoshi, Frangopol, Mitsuyoshi, and Akiyama, Dan M. (Eds.). Taylor & Francis, London, 1335–1340.

Kattar, J.E., Abreu, J.V. and Cruz, L.O. (1995). "Concreto de alto desempenho modificado con polímero para pisos industriais". *37ª Reunião Anual do IBRACON*, Goiânia, Brazil, Julho 3–7, 15 p.

Kawaai, K., Ujike, I., Kunikata, S. and Okasaki, S. (2015). "Effect of curing condition on air-permeability coefficient measured by in-situ test method". *IALCCE 2014*, Tokyo, Japan, 1207–1214.

Klingsch, E.W.H. (2014). "Explosive spalling of concrete in fire". Ph.D. Thesis, ETHZ, Zürich, Switzerland.

Koh, K-T., Kang, S-T., Park, J-J. and Ryu, G-S. (2006). "Evaluation on durability of high performance concrete with expansive additive and shrinkage reducing admixture". *J. Korean Concr. Inst.*, v18, n2, 205–211 (in Korean).

Krobba, B., Bouhicha, M., Kenai, S. and Courard, L. (2018). "Formulation of low cost eco-repair mortar based on dune sand and Stipa tenacissima microfibers plant". *Constr. & Build. Mater.*, v171, 950–959.

Kubens, S., Wassermann, R. and Bentur, A. (2003). "Non destructive air-permeability tests to assess the performance of the concrete cover". *15th ibausil Intern. Baustofftagung*, Bauhaus, Univ. Weimar, September 24–27.

Kubissa, W. (2020). "Air-permeability of air-entrained hybrid concrete containing CSA cement". *Buildings*, v10, 119, 13 p.

Kubissa, W. and Jaskulski, R. (2019). "Improving of concrete tightness by using surface blast-cleaning waste as a partial replacement of fine aggregate". *Periodica Polytechnica Civil Engineering*, v63, n4, December, 1193–1203.

Kumar, R. and Bhattacharjee, B. (2004). "Assessment of permeation quality of concrete through mercury intrusion porosimetry". *Cem. & Concr. Res* , v34, n2, 321–328.

Kunieda, M., Choonghyun, K., Ueda, N. and Nakamura, H. (2012). "Recovery of protective performance of cracked ultra high performance-strain hardening cementitious composites (UHP-SHCC) due to autogenous healing". *J. Adv. Concr. Technol.*, v10, June, 313–322.

Kurashige, I. and Hironaga, M. (2010). "Nondestructive quality evaluation of surface concrete with various curing conditions". *CONSEC'10*, Mérida, México, June 7–9.

Kwon, S., Nishiwaki, T., Kikuta, T. and Mihashi, H. (2013a). "Experimental study on self-healing capability of cracked ultra-high-performance hybrid-fiber-reinforced cementitious composites". *SCTM3 Conference*, Kyoto, Japan, August 18–21, Paper e271, 9 p.

Kwon, S., Nishiwaki, T., Kikuta, T. and Mihashi, H. (2013b). "Tensile behavior of ultra high performance hybrid fiber reinforced cement-based composites". *VIII International Conference on Fracture Mechanics of Concrete and Concrete Structures, FraMCoS-8*, Toledo, Spain.

Larbi, J.A. (1993). "Microstructure of the interfacial zone around aggregate particles in concrete". *Heron*, v38, 69 p.

Leemann, A. and Moro, F. (2017). "Carbonation of concrete: the role of CO_2 concentration, relative humidity and CO_2 buffer capacity". *Mater. & Struct.*, v50, 30, 14 p.

Li, K. (2016). *Durability Design of Concrete Structures. Phenomena, Modeling and Practice.* John Wiley & Sons, Singapore, 280 p.

Li, M., Kao, H. and Qian, C. (2008). "Degradation of permeability resistance of high strength concrete after combustion". *Front. Archit. Civ. Eng. China*, v2, n3, 281–287.

Li, K. and Li, L. (2019). "Crack-altered durability properties and performance of structural concretes". *Cem. & Concr. Res.*, v124, July, 105811.

Li, V.C. and Herbert, E. (2012). "Robust Self-Healing Concrete for Sustainable Infrastructure". *J. Adv. Concr. Technol.*, v10, June, 207–218.

Liang, J., Maruya, T., Sakamoto, J., Matsumoto, J., Shimomura, T. and Takazawa, M. (2013). "Method for compactability evaluation of concrete based on compaction completion energy". *J. Japan Soc. Civil Eng.*, v69, n4, 433–449 (in Japanese).

Lu, F. (2015). "On the prediction of concrete spalling under fire". PhD Diss. ETH No. 23092, ETHZ, Zürich, 139 p.

Lunk, P. (1997). "Capillary suction of water and salt solution in concrete" (in German). *Building Materials Report No. 8*, Aedificatio Publishers, IRB Verlag, 198 p.

Major, J. (1993). "Permeabilität des Betons als Funktion der Porosität und des Wassergehaltes". Institute for Building Materials, ETH Zurich, Internal Report.

Malek, A., Andisheh, K., Scott, A., Pampanin, S., MacRae, G. and Palermo, A. (2018). "Residual capacity and permeability-based damage assessment of concrete in damaged RC columns". *J. Mater. Civ. Eng.*, v30, n6: 04018104.

Maslehuddin, M., Ibrahim, M., Shameem, M., Ali, M.R. and Al-Mehthel, M.H. (2013). "Effect of curing methods on shrinkage and corrosion resistance of concrete". *Constr.& Build. Mater.*, v41, 634–641.

Mathur, V.K., Verma, C.L., Gupta, B.S., Agarwal, S.K. and Kumar, A. (2005). "Use of high-volume fly ash in concrete for building sector". Report No. T(S)006, Central Build. Res. Inst., Roorkee, India, January, 35 p.

Meeks, K.W. and Carino, N.J. (1999). "Curing of high-performance concrete: Report of the state-of-the-art". NIST Report 6295, Gaithersburg, March 1.

Metha, P.K. (1986). *Concrete: Structure, Properties and Materials.* Prentice Hall, England Cliffs, NJ.

Mehta, P.K. and Monteiro, P.J.M. (2006). *Concrete. Microstructure, Properties and Materials.* 3rd Ed., McGraw-Hill, New York, Chicago, San Francisco, Athens, London, Madrid, Mexico City, Milan, New Delhi, Singapore, Sydney, Toronto, 660 p.

Meziani, H. and Skoczylas, F. (1999). "An experimental study of the mechanical behaviour of a mortar and of its permeability under deviatoric loading". *Mater. & Struct.*, v32, 403–409.

Mihashi, H. and Nishiwaki, t. (2012). "Development of engineered self-healing and self-repairing concrete - state-of-the-art report". *J. Adv. Concr. Technol.*, v10, May, 170–184.

Mittal, T., Borsaikia, A. and Talukdar, S. (2006). "Effect of silica fume on some properties of concrete". IIT, Guwahati, India, 7 p.

Mohr, P., Hansen, W., Jensen, E. and Pane, I. (2000). "Transport properties of concrete pavements with excellent long-term in-service performance". *Cem. & Concr. Res.*, v30, 1903–1910.

Moreno, E.I., Varela-Rivera, J., Solís-Carcaño, R. and Sánchez-Pech, O. (2013). "Efecto de las fibras poliméricas en la permeabilidad y características mecánicas del concreto con agregado calizo de alta absorción". *Ingeniería*, v17, n3, September–December, 205–214.

Moro, F. (2013). Personal communication.

Moro, F. and Torrent, R. (2016). "Testing *fib* prediction of durability-related properties". *fib Symposium 2016*, Cape Town, South Africa, November 21–23.

Mu, S., Wu, Y., Jiang, Q. and Shi, L. (2015). "Study on transient air permeability of concrete under different curing conditions". *Key Engng. Mater.*, v629–630, 223–228.

Nakarai, K., Shitama, K., Nishio, S., Sakai, Y., Ueda, H. and Kishi, T. (2019). "Long-term permeability measurements on site-cast concrete box culverts". *Constr. & Build. Mater.*, v198, 777–785.

Neves, R., Santos, J. V. and Monteiro, A. (2011). "Estudo sobre a influência da compactação na resistência do betão à carbonatação". 6° Congresso Luso – Moçambicano de Engenharia.

Neves, R. and Torrent, R. (2016). "Using an air-permeability test to assess curing influence on concrete durability". COST TU 1406.

Neves, R., Vicente, C., Castela, A. and Montemor, M.F. (2015). "Durability performance of concrete incorporating spent fluid cracking catalyst". *Cem & Concr. Composites*, v55, 308–314.

Neville, A. (2003). "Neville on concrete – An examination of issues in concrete practice". ACI, Farmington Hill, MI.

Nguyen, H.V., Nakarai, K., Pham, K.H., Kajita, S. and Sagwa, T. (2020). "Effects of slag type and curing method on the performance of expansive concrete". *Constr. & Build. Mater.*, v262, 120422.

Nguyen, T.H., Ghorbel, E., Fares, H. and Cousture, A. (2019b). "Bacterial self-healing of concrete and durability assessment". *Cem. & Concr. Composites*, v104, 103340, 15 p.

Nguyen, M.H., Nakarai, K., Kubori, Y. and Nishio, S. (2019a). "Validation of simple nondestructive method for evaluation of cover concrete quality". *Constr. & Build. Mater.*, v201, 430–438.

Niknezhad, D., Bonnet, S., Leklou, N. and Amiri, O. (2019). "Effect of thermal damage on mechanical behavior and transport properties of self-compacting concrete incorporating polypropylene fibers". *J. Adhesion Sci. & Technol.*, v33, n23, 2535–2566.

Nishimura, K., Kato, Y. and Mita, K. (2015). "Influence of construction work conditions on the relationship between concrete carbonation rate and the air-permeability of surface concrete". *International Conference on Regeneration and Conservation of Concrete Structure*, Nagasaki, Japan, 1–3. Paper R1–4.

Nishiwaki, T., Koda, M., Yamada, M., Mihashi, H. and Kikuta, T. (2012). "Experimental study on self-healing capability of FRCC using different types of synthetic fibers". *J. Adv. Concr. Technol.*, v10, June, 195–206.

Nsama, W., Kawaai, K. and Ujike, I. (2018). "Influence of bleeding on modification of pore structure and carbonation-induced corrosion formation". *4th International Conference on Service Life Design for Infrastructures (SLD4)*, Delft, Netherlands, August 27–30, 674–685.

Nukushina, T., Watanabe, K., Fujioka, S., Murata, K., Sakai, G. and Sakata, N. (2015). "New sheet curing method for enhancing durability of concrete". *Annual Report of Kajima Techn. Res. Inst.*, 103–108 (in Japanese).

Okasaki, S., Ujike, I. and Kasuga, S. (2015). "Effect of admixtures and curing condition on the corrosion resistance properties of reinforced concrete". *IALCCE 2014*, Tokyo, Japan, 1453–1458.

Ollivier, J.P., Maso, J.C. and Bourdette, B. (1995). "Interfacial transition zone in concrete". *Adv. Cement Based Mater.*, v2, 30–38.

Page, C.L., Short, N.R. and El Tarras, A. (1981). "Diffusion of chloride ions in hardened cement pastes". *Cem. & Concr. Res.*, v11, n3, 395–406.

Panesar, D.K. and Churchill, C.J. (2010). "The influence of design variables and environmental factors on life-cycle cost assessment of concrete culverts". *Struct. Infrastruct. Eng.*, v9, n3, 201–213.

Parrott, L. and Hong, C.Z. (1991). "Some factors influencing air permeation measurements in cover concrete". *Mater. Struct.*, v24, 403–408. https://doi.org/10.1007/BF02472013.

Pasupathya, K., Berndta, M., Sanjayana, J. and Pathmanathana, R. (2016). "Durability performance of concrete structures built with low carbon construction materials". *Energy Procedia*, v88, 794–799.

Paulík, P. and Hudoba, I. (2009). "The influence of the amount of fibre reinforcement on the air-permeability of high performance concrete". *Slovak J. Civil Eng.*, n2, 1–7.

Paz Montes de Oca, J.F. (2016). "Evaluación de la permeabilidad al aire en especímenes de hormigón elaborados con cemento de bajo carbono (LC^3)". Trabajo de Diploma, Univ. Central "Marta Abreu" de Las Villas, Santa Clara, Cuba, 99 p.

Picandet, V. and Khelidj, A. (2003). "Gas and water permeability of cracked concrete". *ICPCM – A New Era of* Building, Cairo, Egypt, February 18–20, 10 p.

Picandet, V., Khelidj, A. and Bastian, G. (2001). "Effect of axial compressive damage on gas-permeability of ordinary and high-performance concrete". *Cem. & Concr. Res.*, v31, 1525–1532.

Picandet, V., Khelidj, A. and Bellegou, H. (2009). "Crack effects on gas and water permeability of concretes". *Cem. & Concr. Res.*, v39, 537–547.

Positieri, M., Oshiro, A., Helene, P. and Baronetto, C. (2011). "Influencia de la adición de pigmentos en la durabilidad del hormigón". *CONPAT 2011*, Antigua Guatemala, Octubre 4–6, 30 slides.

Powers, T.C., Copeland, L.E., Hayes, J.C. and Mann, H.M. (1954). "Permeability of Portland cement paste". *ACI J. Proc.*, v51, 285.

Quoc, P.H.D. and Kishi, T. (2008). "Effect of curing condition on air-permeability of cover concrete". *3rd ACF International Conference-ACF/VCA*.

Quoc, P.H.D. and Kishi, T. (2009). "Strong effect of curing conditions on air-permeability of concrete". *Concr. Plant Intern.*, n4, August, 72–76.

Ramezanianpour, A.A. and Malhotra, V.M. (1995). "Effect of curing on the compressive strength, resistance to chloride-ion penetration and porosity of concretes incorporating slag, fly ash or silica fume". *Cem. & Concr. Composites*, v17, n2, 125–133.

Rasheeduzzafar, A.S. and Al-Saadoun, S.S. (1989). "Influence of construction practices on concrete durability". *ACI Mater. J.*, v86, n6, 566–575.

Rastiello, G. (2013). "Influence de la fissuration sur le transfert de fluides dans les structures en béton: stratégies de modélisation probabiliste et étude expérimentale". PhD Thesis, Univ. Paris-Est, May 6, 192 p.

Rastiello, G., Boulay, C., Dal Pont, S., Tailhan, J-L. and Rossi, P. (2014). "Real-time water-permeability evolution of a localized crack in concrete under loading". *Cem. & Concr. Res.*, v56, February, 20–28.

Rastiello, G., Dal Pont, S., Tailhan, J-L. and Rossi, P. (2018). "On the threshold crack opening effect on the intrinsic permeability of localized macro-cracks in concrete samples under Brazilian test conditions". *Mechanics Res. Commun.*, v90, 52–58.

Reinhardt, H.W. and Jooss, M. (2003). "Permeability and self-healing of cracked concrete as a function of temperature and crack width". *Cem. & Concr. Res.*, v33, 981–985.

RILEM TC 116-PCD (1995). *Performance Criteria for Concrete Durability*. RILEM Report 12, E & F Spon, London, 327 p.
RILEM TC 189-NEC (2007). "Non-destructive evaluation of the penetrability and thickness of the concrete cover". Torrent, R. and Fernández Luco, L. (Eds.), RILEM Report 40, May, 223 p.
RILEM TC 221-SHC (2013). "Self-healing phenomena in cment-based materials". RILEM Report v11, de Rooij, M., Van Tittelboom, K., De Belie, N. and Schlangen, E. (Eds.), 241 p.
RILEM TC 224-AAM (2014). "Alkali activated materials". Provis, J.L. and van Deventer, J.S.J. (Eds.), RILEM Report Vol. 13, 506 p.
Rodríguez de Sensale, G. (2006). "Strength development of concrete with rice-husk ash". *Cem. & Concr. Composites*, v28, n2, February, 158–160.
Rodríguez de Sensale, G., Rodríguez Viavaca, I., Rolfi, R., Segura-Castillo, L. and Fernández, M.E. (2018). "Nuevos hormigones para premoldeados en Uruguay". *HAC 2018*, Valencia, Spain, Marzo 5–6, 219–228.
Romer, M. (2005). Personal communication and supply of test data.
Romer, M. and Leemann, A. (2005). "Sensitivity of a non-destructive vacuum test method to characterize concrete permeability". *ICCRRR*, Cape Town, November 21–23.
Roy, S., Miura, T., Nakamura, H. and Yamamoto, Y. (2018). "Investigation on applicability of spherical shaped EAF slag fine aggregate in pavement concrete – Fundamental and durability properties". *Constr. & Build. Mater.*, v192, 555–568.
Rzezniczak, A-R. (2013). "Durability of repair techniques of fine cracks in concrete". MSc Thesis, McMaster Univ., Canada, 127 p.
Sakakibara, N., Tanka, Y., Sato, K. and Iwaki, I. (2018). "Empirical research on performance evaluation of highly durable rc road bridge deck at construction process". *Cem. Sci. & Concr. Technol.*, v27, n1, 277–284.
Sánchez Stasiw, C. (2008). "Estudio experimental del empleo de diatomita en la producción de concreto de alto desempeño". Tesis de Ing. Civil, Univ. Peruana de Ciencias Aplicadas, Lima, Perú, 117 p.
Schiessl, P. and Raupach, M. (1997). "Laboratory studies and calculations on the influence of crack width on chloride-induced corrosion of steel in concrete". *ACI Mater. J.*, v94, n1, 56–62.
Schneider, U. and Herbst, H.J. (1989). "Permeabilität und Porosität von Beton bei hohen Temperaturen". *DAfStb*, Heft 403.
Scrivener, K.L., Crumbie, A.K. and Laugesen, P. (2004). "The interfacial transition zone (ITZ) between cement paste and aggregate in concrete". *Interface Sci.*, v12, 411–421.
Sezer, G.I. and Gülderen, M. (2015). "Usage of steel slag in concrete as fine and/or coarse aggregate". *Ind. J. Eng. & Mater. Sci.*, v22, June, 339–344.
Shah, S.P. and Slate, F.O. (1968). "Internal microcracking, mortar-aggregate bond and the stress-strain curve of concrete". *The Structure of Concrete and its Behavior under Load, Proceedings International Conference*, Brooks, A.E. and Newman, K. (Eds.), Cem. & Concr. Assoc., London, September, 82–92.
Shekarchi, M., Bonakdar, A., Bakhshi, M., Mirdamadi, A. and Mobasher, B. (2010). "Transport properties in metakaolin blended concrete". *Constr. & Build. Mater.*, v24, 2217–2223.

Sogbossi, H., Verdier, J. and Multon, S. (2019). "New approach for the measurement of gas permeability and porosity accessible to gas in vacuum and under pressure". *Cem. & Concr.* Composites, v103, 59–70.

Solís-Carcaño, R. and Alcocer-Fraga, M.A. (2019). "Durabilidad del concreto con agregados de alta absorción". *Ingeniería Investigación y Tecnología*, vXX, n4, octubre–diciembre, 1–13.

Solís-Carcaño, R.G. and Moreno, E.I. (2009). "Durabilidad del concreto en clima cálido subhúmedo: Efecto del curado". *CONPAT 2009*, Valparaíso, Chile, 29 de Setiembre al 2 de Octubre.

Song, M., Ye, W., Qian, J. and Liang, S. (2014). "Study on transient air-permeability of concrete under different curing conditions". *Key Eng. Mater.*, v629–630, October, 223–228.

Starck, S. (2013). "The integration of non-destructive test methods into the South African durability index approach". MSc Dissertation, Univ. Cape Town, March, 188 p.

Starck, S., Beushausen, H., Alexander, M. and Torrent, R. (2017). "Complementarity of in situ and laboratory-based concrete permeability measurements". *Mater. & Struct.*, v50, 177–191.

Stehlik, M., Helmánková, V. and Vitek, L. (2015). "Opening of microcracks and air-permeability of concrete". *J. Civil Eng. Manag.*, v21, n2, January, 177–184.

Surana, S., Pillai, R. and Santhanam, M. (2017). "Performance evaluation of field curing methods using durability index tests". *Ind. Concr. J.*, v91, July, 1–14.

Tang, G., Yao, Y., Wang, L., Cui, S. and Cao, Y. (2018). "Relation of damage variable and gas permeability coefficient of concrete under stress". *J. Wuhan Univ. of Technol.-Mater. Sci. Ed.,* Dec., 14S1–14S5.

Torrent, R. and Ebensperger, L. (1993). "Methoden zur Messung und Beurteilung der Kennwerte des Überdeckungsbetons auf der Baustelle". Office Fédéral des Routes, Rapport No. 506, Bern, Switzerland, Januar, 119 p.

Torrent, R. and Frenzer, G. (1995). "Methoden zur Messung und Beurteilung der Kennwerte des Ueberdeckungsbetons auf der Baustelle -Teil II". Office Fédéral des Routes, Rapport No. 516, Bern, Suisse, October, 106 p.

Torrent, R. and Gebauer, J. (1992a). "Influence of cement characteristics on the protective value of the concrete cover". 9th *International Congress Chemistry of Cement*, New Delhi, Vol. V, 67–73.

Torrent, R. and Gebauer, J. (1992b). "Durability aspects of high-strength concretes". *Joint Intern. Ready Mixed Concr. Congress*, Madrid, June 23–26, 18 p.

Torrent, R. and Jornet, A. (1990). "Covercrete study: Part I". Report MA90/3815/E, "Holderbank" Managmenet and Consulting Ltd.". Holderbank, Switzerland, 48 p.

Torrent, R. and Jornet, A. (1991). "The quality of the covercrete of low- medium- and high-strength concretes". *ACI SP 126*, 1147–1162.

Torrent, R., Moro, F. and Jornet, A. (2014). "Coping with the effect of moisture on air-permeability measurements". *International Workshop on Performance-based Specification and Control of Concrete Durability*, Zagreb, Croatia, June 11–13, 489–498.

Tsukamoto, M. and Wörmer, J.D. (1991). "Permeability of cracked fiber-reinforced concrete". Ann. *Concr. Concr. Struct.*, Darmstadt, v6, 123–135

USBR (1975). *Concrete Manual*, 8th Ed., U.S. Bureau of Reclamation, p.37.

Van, L.H., Nakarai, K., May, N-H., Kubori, Y., Matsuyama, T., Kawakane, H. and Tani, S. (2019). "Air permeability coefficients of expansive concrete confined by rebars". *Lecture Notes in Civil Engng.*, v54, 6 p.

Van der Merwe, J. (2019). "Constitutive models towards the assessment of concrete spalling in fire". PhD Diss. ETH No. 26205, ETHZ, Zürich, 235 p.

Van Eijk, R.J. (2009). "Evaluation of concrete quality with Permea-TORR, Wenner Probe and Water Penetration Test". KEMA Report, Arnhem, July 8, 46 p. (in Dutch).

Van Tittelboom, K. and De Belie, N. (2013). "Self-healing in cementitious materials—A review". *Materials*, v6, 2182–2217.

Vizcaino, L., Antoni M., Alujas, A., Martirena, J.F. and Scrivener, K. (2015). "Industrial manufacture of a low-clinker blended cement using low-grade calcined clays and limestone as SCM: The Cuban experience". *RILEM Bookseries 10: Calcined Clays for Sustainable Concrete.* Scrivener, K. and Favier, A. (Eds.), Springer, Lausanne, 347–358.

Wang, A., Deng, M., Sun, D., Mo, L., Wang, J. and Tang, M. (2011). "Effect of combination of steel fiber and MgO-type expansive agent on properties of concrete". *J. Wuhan Univ. of Techn., Mater. Sci. Ed.*, 786–790.

Wang, K., Jansen, D.C., Shah, S. and Karr, A.F. (1997). "Permeability study of cracked concrete". *Cem. & Concr. Res.*, v27, n3, March, 381–393.

Watson, A.J. and Oyeka, C.C. (1982). "Oil permeability of hardened cement paste and concrete". *Mag. Concr. Res.*, v41, n147, 87–97.

WBCSD (2009). "Recycling concrete". *The Cement Sustainability Initiative*, World Business Council for Sustainable Development, July, 42 p.

Yamamoto, M. (2016). "Valoración del perfil sostenible para hormigones elaborados con áridos reciclados a partir del diseño de hormigones con iso-prestaciones en durabilidad". Disertación Ing. Civil, Univ. de Buenos Aires, Argentina, Junio, 88 p.

Yokoyama, Y., Sakai, Y., Nakarai, K. and Kishi, T. (2017). "Change in surface air-permeability of concrete with different mix designs and curing". *Cem. Sci. & Concr. Technol.*, v71, n1, 410–417 (in Japanese).

Zhang, D., Dasari, A. and Tan, K.H. (2018). "On the mechanism of prevention of explosive spalling in ultra-high performance concrete with polymer fibers". *Cem. & Concr. Res.*, v113, 169–177.

Zeiml, M., Leithner, D., Lackner, R. and Mang, H.A. (2006). "How do polypropylene fibers improve the spalling behavior of in-situ concrete?" *Cem. & Concr. Res.*, v23, 929–942.

Zeng, W., Ding, Y., Zhang, Y. and Dehn, F. (2020). "Effect of steel fiber on the crack permeability evolution and crack surface topography of concrete subjected to freeze-thaw damage". *Cem. & Concr. Res.*, v138, 106230, 13 p.

Zhutowsky, S. and Kovler, K. (2012). "Effect of internal curing on durability-related properties of high-performance concrete". *Cem. & Concr. Res*, v42, 20–26.

Zong, L., Fei, Z. and Zhang, S. (2014). "Permeability of recycled aggregate concrete containing fly ash and clay brick waste". *J. Clean. Prod.*, v70, 175–182.

Zürich City (2019). "A low carbon, circular economy approach to concrete procurement". *GPP in* Practice, European Commission, n88, May, 5 p.

Chapter 7

Why durability needs to be assessed on site?

7.1 THEORECRETE, LABCRETE, REALCRETE AND COVERCRETE

When discussing durability performance, four different concepts of concrete are worth considering:

- *Theorecrete*: it is the concrete as specified by the designer, traditionally including minimum compressive strength, prescribed limits to the composition (e.g. maximum *w/c* ratio, cement type and content), some eventual durability performance tests and a nominal cover thickness
- *Labcrete*: it is the concrete tested in the laboratory, on specimens cast at the concrete plant or at the jobsite on delivery, typically fully compacted and moist cured for 28 days. It is expected to comply with the measurable requirements of the specifications (Theorecrete)
- *Realcrete*: it is the same concrete mix, tested as *Labcrete*, but transported and conveyed to the placement point, poured, compacted, finished and cured under the prevailing jobsite conditions
- *Covercrete*: it is the external layer of the *Realcrete*, say 50±25 mm thick; for reinforced structures, it involves the concrete cover protecting the reinforcing steel

The four categories encompass entirely different qualities of concrete, in terms of composition and performance, as discussed in detail in the following sections.

The terminology used is an extension of that previously coined by Newman (1987).

7.1.1 Theorecrete

At the design stage, the structural and concrete engineers are faced with the need to specify some key parameters to ensure that the structure will reach its intended service life (obviously, also its required strength). Typically,

DOI: 10.1201/9780429505652-7

the engineers rely on the codes' and standards' prescriptions to achieve that goal. Most standards consider, explicitly (EN, Australian, Spanish, Argentine, Swiss) a service life of 50 years, whilst ACI 318 does not give an explicit value for the service life. For exposures involving risk of reinforcement corrosion, EN Standards (EN 1992-1-1, 2004) allow longer service lives, merely by raising the cover thickness (e.g. +10 mm for 100 years), approach challenged by Torrent (2018).

Most codes and standards follow a prescriptive or 'deemed to satisfy' (Andrade, 2006) approach, by which minimum specified cover thicknesses are accompanied by limits set to the concrete mix composition (maximum *w/c* ratios, sometimes coupled with minimum binder contents), together with minimum strength grades.

Regarding durability, almost invariably (Australian Standards is one exception) a maximum value of the water/cement or water/binder ratio (*w/c* or *w/b*) is specified, a limit that decreases with increasing aggressiveness of the exposure. These limits are specified in most codes, although the values are not the same in all of them. Even within European Standard EN 206 (EN 206, 2013), large discrepancies are found among the different countries (national annexes), see Chapter 3 of RILEM TC 230-PSC (2016) or the original source (CEN, 2007).

The limits are, in most cases, "expert opinion" of the codes' writers and the discrepancies can be attributed to this and also to country-to-country differences in climatic conditions, design and construction traditions, etc.

Considerations on the unsuitability of either *w/c* or *w/b* as durability indicator were given in Section 1.6.2, in terms of its relation with the parameters that govern the degradation mechanisms and, in more practical terms, on the virtual impossibility of controlling compliance with the specified limits.

One of the authors has conducted numerous audits in ready-mixed concrete plants all over the world and has not found a single case in which the *w/c* ratio was measured, either at the plant or at the jobsite. In a few cases (Germany, Indonesia, Switzerland), just the determination of the water content of the mix by drying a sample was observed, using the same principle of Annex H of Swiss Standard (SIA 262/1, 2019) which prescribes how to measure the water content by drying ≈10 kg of fresh concrete in a "*paella*-looking" pan.

The fact remains that possibly the most critical weakness of the use of *w/c* ratio as durability indicator is the impossibility of checking compliance by the user. Specifying a characteristic that cannot be measured is clearly meaningless. Regrettably, a not so uncommon fraudulent practice of dishonest ready-mixed concrete producers is to deliberately trespass the maximum *w/c* limit, complying just with the specified compressive strength (usually less demanding) under the certainty that the unconformity will not be detected by the customer.

Therefore, prescriptive specifications remain a *Theorecrete*, a wishful expression of specifiers in their attempt to achieve a durable concrete structure.

7.1.2 Labcrete

Some countries have started, rather timidly in most cases, to incorporate performance durability requirements in their standards. Table 1.2 summarizes the tests involved and their corresponding limiting values (function of the severity of the exposure) included in some national standards.

To check conformity, the concrete producer and/or user takes samples regularly from the delivered fresh concrete and cast specimens, typically cylinders (Ø100×200 or Ø150×300 mm) or 100/150 mm cubes. The fresh concrete samples can be collected at the concrete plant or during discharge at the jobsite, the latter being more representative of the final quality of the product.

These specimens are cast, compacted, finished and kept under optimal conditions, demoulded at 24 hours and moist cured for 28 days at constant temperature (20°C in Europe, 23°C in the USA, 27°C in India), moment at which the tests are initiated.

The concrete resulting from this process is called *Labcrete*, indicating that the concrete was processed under laboratory conditions which are favourable for the development of the material's performance. Very often, especially for strength, the result of these tests is considered as a "potential" property (i.e. achievable under quasi-ideal conditions).

7.1.3 Realcrete

The same concrete characterized by the *Labcrete* tests ends up being placed in a jobsite under far from ideal *Realcrete* conditions. Figure 7.1 compares some *Labcrete* (top row) and *Realcrete* (bottom row) concreting conditions.

The difference between *Labcrete* and *Realcrete* is further illustrated in Figure 7.2.

Figure 7.1 (a) *Labcrete* and (b) *Realcrete* operations (l. to r.): placement, compaction, finishing and curing; courtesy of José Marques from Betão Liz.

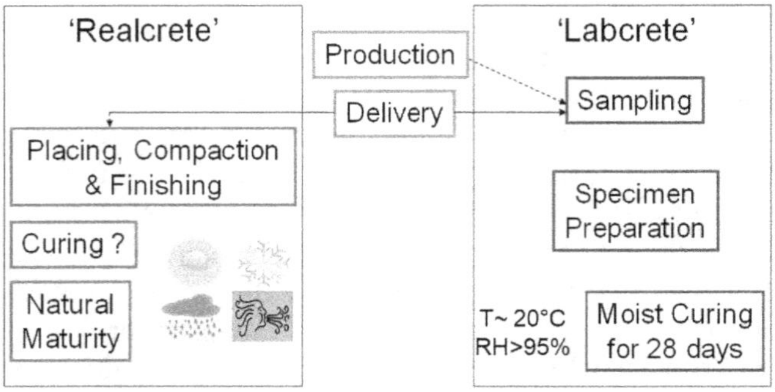

Figure 7.2 Parallel processes undergone by concrete under *Labcrete* (*r.*) and *Realcrete* (*l.*) conditions.

In general, the quality of the *Realcrete* is worse than that of the *Labcrete*, due to the practical impossibility of reproducing, at the jobsite, the ideal *Labcrete* conditions. In terms of strength, this is recognized by European and North-American codes and standards by accepting a 15% reduction in compressive strength when testing drilled cores relative to cast specimens; the tolerated reduction is 13% in Australia and 20% in South Africa. As discussed in the next section, the reduction in quality tends to be more severe in terms of durability performance than of compressive strength.

7.1.4 Covercrete

When dealing with durability, it is very important to realize that the quality of the *Realcrete* and, to a lesser extent of the *Labcrete*, is not homogeneous.

The well-known "wall effect" by which the surfaces of the specimens and elements cast against a form or finished are always covered by a thin layer of paste, is visually evident for everybody. Possibly the first research on the composition and properties of this layer, called the "skin" of concrete, was conducted by Kreijger (1984) on *Labcrete* specimens, later complemented by Mayer (1987) and Newman (1987).

When dealing with *Realcrete*, this heterogeneity is aggravated by the size and shape (especially depth and width) of the forms, the presence of steel reinforcement that constitutes an obstacle to the free flow of the fresh concrete, complicating its consolidation (especially in the gap between the rebars and the form) and the endemic absence or insufficiency of moist curing. Not uncommon incorrect finishing practices (e.g., spraying water and or cement on the surface of slabs), and/or excessive bleeding, may further weaken the upper layer of slabs.

Another weakening factor of *Realcrete* surface layers is the appearance of cracks, some load-induced and others induced by restrained volume changes of concrete, in the plastic or hardened state.

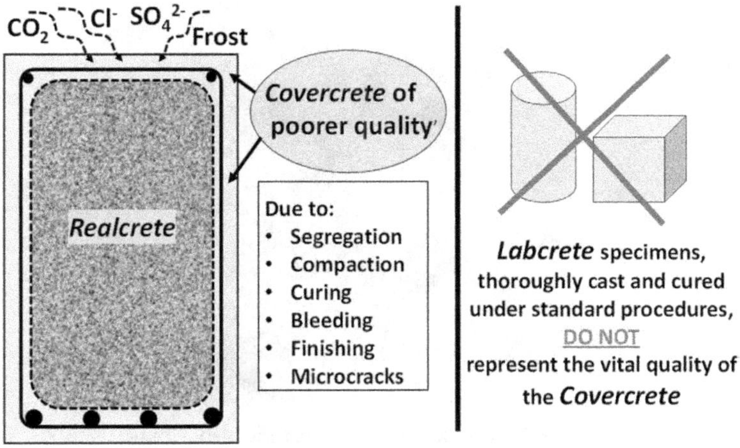

Figure 7.3 Concept of *Covercrete*.

The heterogeneity described in the preceding paragraphs calls for a special designation of the surface layer of concrete elements, the role of which is, for reinforced concrete, to provide a protective cover to the steel bars, sufficiently thick and of low "penetrability"[1]: the so-called *Covercrete*.

The *Covercrete* concept is illustrated in Figure 7.3, indicating that the quality of the outer layers (*Covercrete*) is usually poorer than that of the core of the structural element and, yet, that weaker layer is the defence barrier of the system against the penetration of aggressive agents, including frost.

So, we have the undesirable combination that the defensive barrier is the weakest link of the system, which explains to a large extent – together with the lack of cover thickness – the corrosion durability problems faced nowadays. Moreover, our quality assurance system is exclusively based on testing *Labcrete* specimens, the quality of which does not represent at all the quality of the vital *Covercrete*. This was clearly stated by Angst (2018): "It is impossible to mimic the jobsite conditions in the laboratory". So, even cast specimens kept close to the real structure are still *Labcrete*, not *Realcrete* (see real case in 7.1.5.2).

In Section 7.1.3, we referred to the loss of strength of *Realcrete* in comparison to that of *Labcrete*. Let us look at this issue from the viewpoint of durability, using the *Covercrete* concept, as illustrated in Figure 7.3.

1 "Penetrability" is a general term that expresses the easiness with which aggressive agents penetrate concrete by different mechanisms (permeability, diffusivity, etc.)

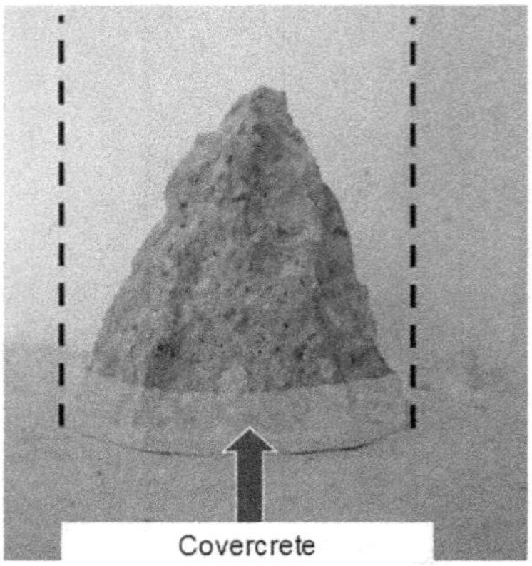

Figure 7.4 Undamaged *Covercrete* in compression test (Fernández Luco, 2009).

Figure 7.4 shows a drilled core tested for strength. When tested, the weaker *Covercrete* remains in the area of the specimen under triaxial compression stress state, due to the friction with the stiffer test machine steel plates, area unaffected by the test. Therefore, the *Covercrete* weakness is not revealed by the compression test; this is shown vividly in Figure 7.4, highlighting the unscathed *Covercrete* after a compression test (Fernández Luco, 2009).

> "This confirms the statement of CEB-FIP Model Code 1990 (MC90, 1991), quoted at the beginning of Section 1.6.1."

7.1.5 Quality Loss between Covercrete and Labcrete

It is of interest to know the relation that exists between the durability performance of *Labcrete* specimens and the same concrete placed in a real structure (*Realcrete*), e.g. for mix design purposes.

This was investigated within the frame of a research project, sponsored by ASTRA (Swiss Federal Highway Administration), and published in the report (Torrent & Frenzer, 1995). The investigation covered two jobsites: a tunnel (Bözberg Tunnel) and a bridge (Schaffhausen Rheinbrücke N4), which will be described in some detail below; further information can be found in Torrent (1999).

7.1.5.1 Bözberg Tunnel

It is a 3.7 km long road tunnel on Swiss Motorway A3, located in Canton Aargau, incidentally not far from Holcim Technology laboratories at Holderbank.

The object of the investigation was the deck slab of the tunnel. The declared mix design had a *w/c* ratio of 0.45, 325 kg/m³ of OPC, was air-entrained and showed a mean cube compressive strength (28 days) of 50.5 MPa. The investigation covered four concrete stages (blocks), 12.5 m long each. For each block, fresh concrete samples were taken from three trucks bringing concrete to the tunnel, totalizing ten samples.

In Bözberg, the following methodology was adopted:

a. The delivered concrete was sampled at the jobsite, carefully identifying the location where it was poured
b. *Labcrete Preparation*: 120×250×360 mm specimens were cast, kept protected 24 hours at the jobsite, sent to Holcim technology lab, stripped, moist cured (20°C/>95% RH) during 6 days and thereafter kept in a dry room (20°C/50% RH) for 21 days, moment at which they were tested
c. *Labcrete Testing*: at 28 days, the following ND tests were applied on the *Labcrete* specimens (all tests described in Chapters 4 and 5 and Annex A): Torrent air-permeability kT, TUD N_2-permeability time t and Wenner electrical resistivity ρ. In addition, Ø150×50 mm cores were drilled and saw-cut from the specimens to measure Cembureau O_2-permeability kO (after 6 days 50°C oven-drying followed by 1 day cooling in desiccator) and Ø50×50 mm cores to measure water sorptivity at 24 hours a_{24} (after 2 days 50°C oven-drying followed by 1 day cooling in desiccator)
d. *Realcrete Testing*: at an age between 28 and 35 days, the same ND tests were applied directly on the six Blocks tunnel floor; afterwards, cores were drilled from four of them to get samples for kO and a_{24}, tested under exactly the same conditions as for the *Labcrete* slabs

The results obtained on the sampled Blocks are plotted in Figure 7.5, with the grey bars corresponding to the *Labcrete* and the black bars to the *Realcrete/Covercrete*.

Figure 7.5 shows values of air-permeability kT measured directly on the cast specimens (*Labcrete*) and directly on the deck (*Realcrete*) and of O_2-permeability kO, measured on cores drilled from the specimens (*Labcrete*) and from the deck (*Realcrete*). Except for Block 10, the *Realcrete*'s kT is higher than the *Labcrete*'s; in the case of kO, in all cases the values for the *Realcrete* were higher than for the *Labcrete*. On average, the kT values of the *Realcrete* were 3.0 times that of the *Labcrete*, whilst for kO the ratio was 1.76.

7.1.5.2 Schaffhausen Bridge

It is a 152 m long cable-stayed bridge on the N4 Motorway, linking Swiss cantons Schaffhausen in the N and Zürich in the S, over the Rhine River (see Figure 7.6). The pylon is inclined 20° and is 51 m high.

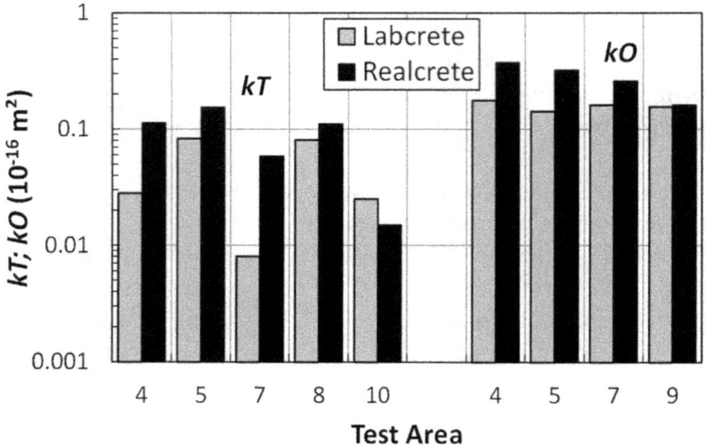

Figure 7.5 Quality of *Labcrete* and *Realcrete* assessed by air- (*kT*) and O_2- (*kO*) permeability.

Figure 7.6 Picture of Schaffhausen Road Bridge.

The objects of the investigation were the deck slab and the pylon. The deck slab was built with an OPC concrete with measured 28-day cube compressive strength of 50.0 MPa (designated according to old Swiss Standard SIA 162 as B45/35 mix, meaning 45 MPa mean and 35 MPa characteristic cube strength); for the pylon, a Silica Fume Cement was used and the measured cube strength was 75.5 MPa (designated as B55/45 mix).

In Schaffhausen, same durability testing procedure as for Bözberg Tunnel was adopted, with some modifications, as the owner did not want to apply intrusive tests nor drill cores from the structural elements. Therefore, only NDT kT and ρ were applied directly on the structure at ages between 28 and 150 days. Cubes (1 m) were cast and kept for 25 days at the jobsite under nominally the same conditions as the *Realcrete*; then, they were transported to Holcim Technology laboratories for testing.

Due to the low ambient temperature, thermocouples were installed in the centre of 1 m cubes, of the pylon (~3×4 m) and of 360×250×120 mm slabs, to monitor the thermal evolution of the concrete, with the results shown in Figure 7.7.

The plots shown in Figure 7.7 correspond to the B 55/45 mix, measured on the Pylon (1), field-cured 1 m cube (2) and slab specimen (3) and to the B45/35 mix, measured on a 1 m cube (4). The mean air temperature during the period of recording was 4°C. We can see that important temperature differences (ΔT>50°C) were built up between the centre and the surface of the pylon and moderate, but not negligible, for the cubes (ΔT>25°C), with the consequent risk of thermal cracking. The different thermal evolution of cast specimens, cubes and the pylon, confirms the impossibility of mimicking the *Realcrete* by keeping elements (even as large as 1 m cubes) close to the real structure. Also, the thermal volume changes are differently restrained in the *Realcrete* than in the cubes.

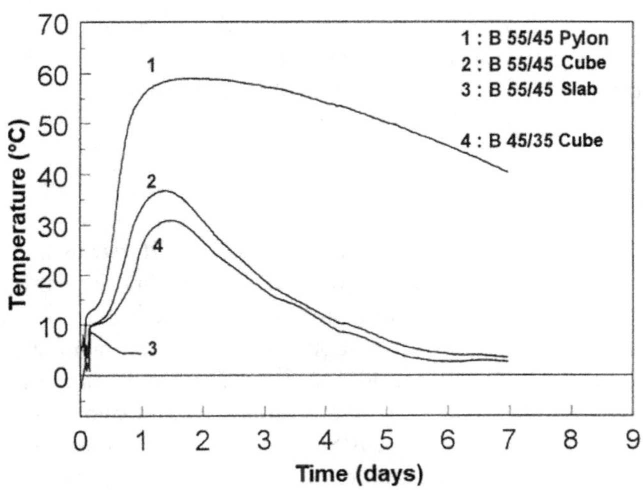

Figure 7.7 Thermal evolution of 250×360×120 mm laboratory slabs, 1 m field cubes and pylon (Schaffhausen bridge).

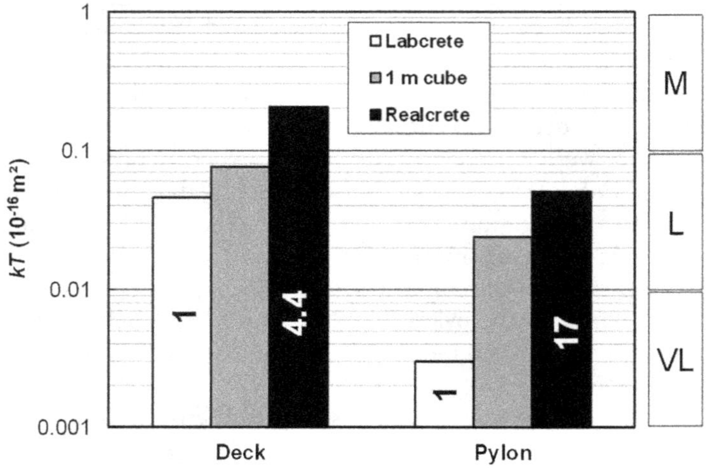

Figure 7.8 Air-permeability *kT* of *Labcrete*, 1 m cube and *Realcrete* of Schaffhausen Bridge Deck and Pylon.

The *kT* test results are summarized in Figure 7.8 (Torrent, 1999) for the *Labcrete* slab, the 1 m cube and the *Realcrete* of both Bridge Deck and Pylon. To the right of the chart, the Permeability Classes (see Table 5.2) "Moderate" (M), "Low" (L) and "Very Low" (VL) are indicated.

As indicated along the bars, the *kT* of the *Realcrete* was 4.4 times higher than *Labcrete*'s for the Deck, and 17 times higher for the Pylon. In the latter, thermal cracks that were not visible at the time of the tests but became evident later, required a coating surface treatment. The large difference between *Realcrete* and *Labcrete* is attributable to the cracks. It is interesting to note that, even with that large loss in performance, the *Realcrete kT* of the Pylon is significantly lower than that of the Deck, due to the much higher quality of its mix design.

7.1.5.3 Lisbon Viaduct

The difference between the *kT* of *Labcrete* and *Realcrete* was also investigated by Neves and Santos (2008) during the construction of a motorway viaduct in Lisbon. It is a 770 m long and 32.4 m wide box girder, with spans ranging from 50 to 105 m, simply supported by double-headed pylons. The investigation comprised concreting surveillance of three segments of the box girder, one pylon, two sets of prefabricated struts and two phases of one of the abutments. For each, a 120×250×360 mm specimen for air-permeability testing and three 150 mm cubes (for compressive strength testing) were cast. The location where the concrete from the corresponding truck was poured was pinpointed, to carry out *in situ* testing in the same spots afterwards.

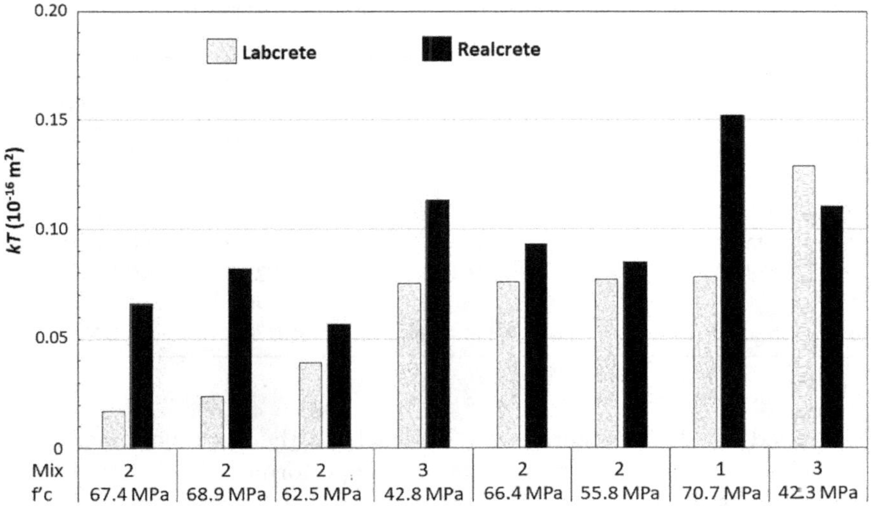

Figure 7.9 Air-permeability *kT* of *Labcrete* and *Realcrete* of Lisbon's Viaduct (Neves & Santos, 2008).

After casting, the cubes were moved to the jobsite laboratory, stripped at 24 hours and cured in water until the testing age (28 days), while the slab specimens were cured in water for 6 days and then kept at 22°C and 60% RH until the testing age. Three different concrete mixes were applied in the several monitored concreting operations: (1): C40/50.S4.D25, (2): C40/50.S3.D25 and (3) C30/37.S3.D25, following EN 206-1 (EN 206, 2013) notation. The slab specimens and the corresponding structural elements were both tested when the concrete was 28 days old. Figure 7.9 summarizes the test results, revealing a worse performance of *Realcrete* compared with *Labcrete*, as on average the *kT* of *Realcrete* was nearly twice that of *Labcrete*. Interestingly, trends were noticed for higher ratios (*Realcrete kT*/*Labcrete kT*) in troubled concreting segments of the box girders, carried out in the evening, than for quieter concreting such as for the prefabricated struts.

7.1.5.4 Swiss Bridges' Elements

A recent investigation conducted in Switzerland for the Federal Highway Administration (Jacobs et al., 2018) adds more information on the difference between *Realcrete* and *Labcrete* in terms of durability. It consisted in testing, in parallel, cast specimens (*Labcrete*) and the *Realcrete* of different elements belonging to six structures, as described in Table 7.1.

Labcrete 150 mm cube specimens were cast with the fresh concrete destined to the elements and moist cured for 28 days. Cores were drilled from the *Labcrete* cubes, on which mechanical and durability tests were

Table 7.1 Characteristics of the concretes of the six structures investigated (Jacobs et al., 2018)

Site	*Elements*	*Binder*	*Strength class*	*EN exposure class*
Ha	Parapets; abutment wall	CEM II/B-M (T-LL)	C30/37	XC4, XD3, XF4
Gn	Parapets; abutment wall			XC4, XD3, XF2
Ep	Walls; decks			XC4, XD3, XF2
Ma	Retaining wall	CEM II/B-M (S-T)	C25/30	XC4, XF2
Gr	Walls; decks	CEM II/A-LL+PFA	C25/30 C30/37	XC4, XD1, XF4 XC4, XD3, XF4
Po	Shoulder curb	CEM I+PFA	C30/37	XC4, XD3, XF4

performed. At 28 days, cores were drilled from the corresponding structural elements, on which the same tests were performed (*Realcrete*).

The results can be summarized as follows:

- the *Realcrete* density was generally 1%–2% lower than *Labcrete*'s
- the *Realcrete* 28-day compressive strength was ~20% lower than *Labcrete*'s
- the *Realcrete* chloride migration coefficient M_{Cl} was 45% higher than *Labcrete*'s (see Figure 7.10)
- the *Realcrete* freeze-thaw-salts resistance was, in average, similar to *Labcrete*'s although with very large scatter
- the *Realcrete* carbonation rate was ≈ 40% higher than *Labcrete*'s (determined only on Ep jobsite)

These results confirm that the impact of the lower quality of *Realcrete/Covercrete* is much more acute for durability tests than for strength tests.

Figure 7.10 shows that the coefficient of chloride migration M_{Cl} of the *Realcrete* is almost invariably higher than *Labcrete*'s, at an average 45% higher. Swiss Standard (SIA 262/1, 2019) specifies that the value of M_{Cl} (Tang-Nilsson test method described in A.2.1.2) of the *Labcrete* shall not exceed 10×10^{-12} m²/s (for moderate and severe chloride exposures). Figure 7.10 shows that 24 out of 26 results (92%) conform to that limit. The same standard specifies a maximum limit of 12×10^{-12} m²/s for tests made on cores drilled from the structures (*Realcrete*). Figure 7.10 shows that now 19 out of 26 results comply with that limit; the percentage of conformity drops from 92% to 73%.

The results discussed in this section, obtained from Bözberg Tunnel, Schaffhausen Bridge, Lisbon Viaduct and from the six other structures investigated in Switzerland, confirm that the durability performance of the *Realcrete/Covercrete* is significantly worse than *Labcrete*'s, a fact also acknowledged in the literature (Bouwer, 1998; Beushausen & Alexander,

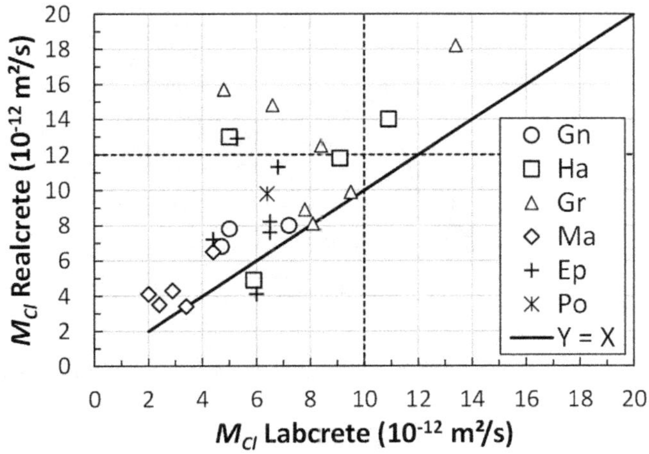

Figure 7.10 Chloride migration M_{Cl} of *Realcrete* vs *Labcrete* for different structures; data from Jacobs et al. (2018).

2009). This fact stresses the need to include not just the *Labcrete*, but especially the *Realcrete*, in performance specification and quality assurance schemes.

7.2 ACHIEVING HIGH COVERCRETE'S QUALITY

Section 7.1 was devoted to demonstrating the typically lower quality of the *Covercrete*, compared to the *Labcrete*. This section, in turn, deals with means to achieve a *Covercrete* of high quality (low "penetrability").

7.2.1 Mix Design and Curing

The simplest way of improving the "penetrability" of the *Covercrete* is to build with higher quality concretes, which involves the correct selection of the cement/binder, the design of mixes with sufficiently low water/binder ratios and the application of sound concreting practices: good consolidation without segregation and, especially, adequate and sufficiently long moist curing conditions (see Sections 6.2–6.7).

7.2.2 UHPFRC

Ultra High-Performance Fibre-Reinforced Composites (UHPFRC) are cementitious composite materials made with extremely low *w/b* ratios (0.15 and lower), silica fume, well-graded fine aggregate (typically below 0.5 mm) and large volume of steel fibres (StF) and superplasticizers. Different

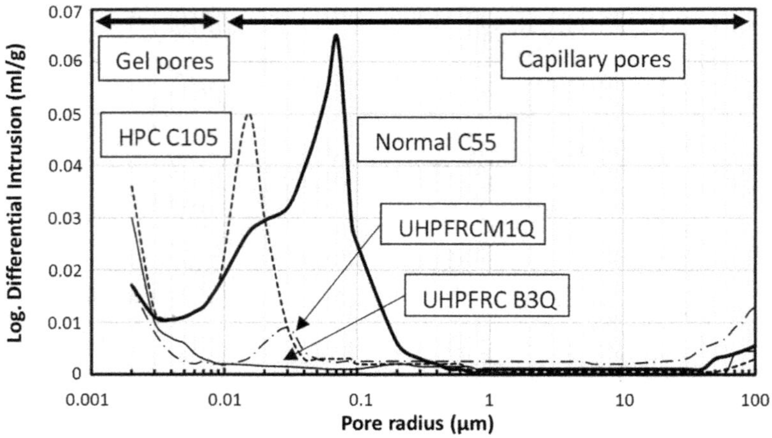

Figure 7.11 Pore size distribution of conventional, HPC and two UHPFRC (Fehling et al., 2005).

concepts have been developed worldwide, some known by their abbreviation or commercial brands (SIFCON, SIMCON, CEMTEC, Ductal, etc.). These materials show an extremely low and fine porosity, see Figure 7.11 in which the pore size distributions of two UHPFRC mixes (B3Q and M1Q) with compressive strengths above 200 MPa, are compared with those of a Normal Concrete (cube strength 55 MPa) and a High-Performance Concrete (cube strength 105 MPa) (Fehling et al., 2005). It is expected that such tight material as UHPFRC should show a low "penetrability" to fluids and ions.

Within their very comprehensive investigation (Fehling et al., 2005), performance tests were conducted on conventional and UHPFRC mixes, namely, migration, gas-permeability and water sorptivity. In all cases, the UHPFRC mixes showed much better performance than the reference conventional mix (cube strength class C45). Figure 7.12 presents the test results of Nitrogen-permeability kN_2 (Cembureau test, see 4.3.1.2) for the conventional concrete and for two UHPFRC formulations. For the latter, the indication WL denotes storage under water while 90°C stands for the temperature of a thermal treatment applied to the specimens.

In Figure 7.12, it can be seen that UHPFRC mixes have significantly lower permeability to gas than a normal conventional concrete, something expected given their tighter pore structure, shown in Figure 7.11. Results of water absorption coefficient (DIN 52617), reported by Fehling et al. (2005), confirm the results of gas-permeability.

This type of UHPFRC has been used in Switzerland (Brühwiler, 2007) to repair concrete bridges, applying it selectively in the most vulnerable areas, as described in detail in Section 11.3.7. Information on the spatial variability of tensile strength and air-permeability kT of such material, within

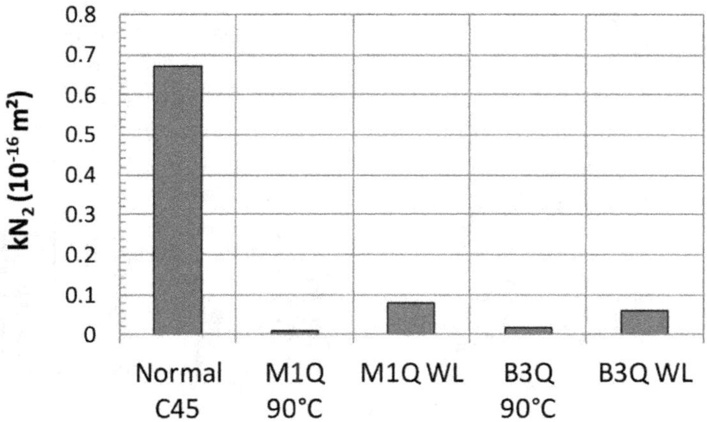

Figure 7.12 Permeability to N_2 of UHPFRC compared to conventional concrete; data from Fehling et al. (2005).

a 1.5×3.0 m panel (42 mm thick), can be found in Oesterlee et al. (2009), showing values within the range 0.001–0.01 (× 10^{-16} m²), for the extremely low geometric mean value of 0.0046×10^{-16} m².

A different concept of UHPFRC was developed in Japan, embedding high-strength polyethylene fibres (Ø12 μm×6 mm in size) in a cementitious composite of w/b=0.18–0.22, containing approximately 1300, 230 and 155 kg/m³ of cement, silica fume and fine sand, respectively, plus a high dosage of superplasticizer and an air-remover agent. The resulting material was designated as UHP-SHCC (Ultra High-Performance Strain-Hardening Cement Composite) and was intended as a repair material (Kunieda et al., 2011; 2012). The durability of the composite was measured in terms of its air-permeability kT and its water sorptivity, using a test method similar to the Karsten tube (Section see 8.3.3.3). The measured values of air-permeability kT of the composites were rated within the "Very Low" Permeability Class (see Table 5.2); the water sorptivity was also very low.

7.2.3 Controlled Permeable Formwork (CPF) Liners

7.2.3.1 Action Mechanism of CPF Liners

Controlled permeable formwork (CPF) liners are sheets of synthetic fibres fabric, capable of retaining cement-sized and larger particles, but of allowing water and air to flow through it under hydrostatic pressure, intensified during vibration. They are attached onto the formwork inner surface. As schematized in Figure 7.13a, under the vibration action the surface concrete layers lose water and air and are enriched by cement dragged from the inner layers (Leow, 2004), resulting in a local lowering

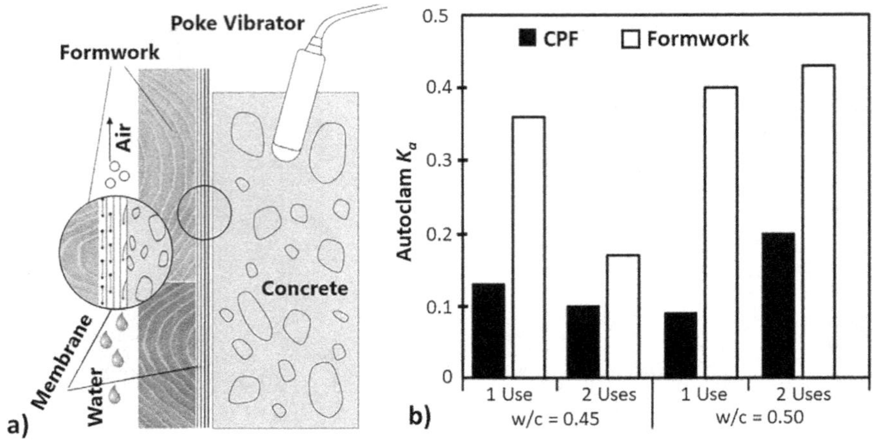

Figure 7.13 (a) CPF action with vibrated concrete; (b) effect of CPF on *Autoclam* air-permeability index (Basheer et al., 2005).

of the *w/c* ratio. The effect of the CPF liners is achieving a *Covercrete* that is less penetrable than the core, thus offering a better protection to the embedded steel. The effect of CPF liners in reducing the amount of bug-holes from the concrete surface was discussed in Section 5.7.5.4 and shown visually in Figure 5.28.

The efficiency of CPF to reduce the "penetrability" of the treated surface has been proved by several investigations (Cullen, 1998; Malone, 1999; COWI-Almoayed Gulf WLL, 2002; Leow, 2004; Basheer et al., 2005; Law et al., 2012; Adam et al., 2009; Tanaka et al., 2012; Torrent et al., 2012; Ohta et al., 2019).

7.2.3.2 *Impact of CPF on the "Penetrability" of the Covercrete*

Several investigations were carried out to assess the potential benefit of using CPF on the quality of the *Covercrete*. One of them was carried out at Queen's University Belfast (UK) (Basheer et al., 2005) to assess the performance of "Formtex", a CPF liner manufactured by a Danish company called Fibertex. Prisms measuring 250×750×150 mm were cast with two concrete mixes of *w/c*=0.45 and 0.50 (50 and 40 MPa cube strength at 28 days, respectively). One of the surfaces (250×750 mm) was cast against the natural plywood form, whilst the opposite against a Formtex CPF liner stapled onto the plywood form. After 24 hours, the formwork was stripped and the prisms were stored 28 days at 20°C and 75% RH, moment in which they were tested for: Pull-off strength, Air-Permeability and Water Absorption (both with Autoclam system, see Section 4.1.2.2), Accelerated Carbonation and Chloride ingress (after 100 days immersion in salt solution).

The test results showed an increase in pull-off strength of 30%–33% on the CPF face, compared with the natural surface. Figure 7.13b shows the effect of the CPF liner on Autoclam's Air-permeability Index (K_a), comparing the results with those obtained on the surface cast against the plywood form, for each *w/c* ratio. The test was repeated reusing the liner once. The results show clearly the beneficial effect of the CPF liner in reducing air-permeability, even when reusing the liner once. This lower air-permeability was accompanied by significantly lower water sorptivity and carbonation depths and moderate lower chloride penetration depths for the surfaces cast against the CPF liner.

The suitability of "Formtex" CPF liner in improving the tightness of the *Covercrete* was confirmed by Holčapek (2011), who found an approximately ten-fold reduction in air-permeability *kT* on the surface treated with the CPF liner.

Another study was conducted at Holcim Technology Ltd, Switzerland, but on a concrete aged 16 years (to test the lasting effect of the CPF) (Torrent et al., 2012). A panel measuring 600×500×200 mm was cast with a concrete of *w/c* ratio = 0.55 and 28-day cube compressive strength = 42.0 MPa. The concrete was poured inside a wooden formwork, one of its 600×500 mm faces covered with the "Zemdrain Classic" CPF Liner, manufactured by Dupont, and carefully compacted with internal poke vibration. During and immediately after casting, it was possible to observe water being drained to the floor through the CPF. The panel was stripped at 24 hours and stored permanently in a dry room (20°C/50% RH), till the age of test (16 years).

At that age, the following NDTs were applied on both faces of the panel: Rebound hammer *R* and Air-Permeability coefficient *kT*. In addition, Ø50×200 mm cores were drilled through the entire panel thickness to measure the carbonation depth *CD* and chloride migration coefficient M_{Cl} through the depth of chloride penetration X_d of both surfaces (Tang-Nilsson method, see A.2.1.2). Table 7.2 summarizes the recorded results.

The results of Table 7.2 indicate that, after 16 years of conservation in a dry room, the performance of the CPF face is significantly superior to that of the Formwork face, both in terms of hardness *R* and air-permeability

Table 7.2 Test results obtained on the two opposite faces of the panel (Torrent et al., 2012)

Property	*Rebound R*		*kT (10^{-16} m²)*		*CD (mm)*		*M_{Cl} (10^{-12}m²/s)*		*X_d (mm)*	
Face	*Mean*	*s*	*kT_{gm}*	*s_{LOG}*	*Mean*	*Max*	*Mean*	*s*	*Mean*	*s*
Formwork	46	4.8	6.6	0.16	23	27	28.8	3.3	43	4.4
CPF	54	2.1	0.79	0.14	15	18	17.5	1.5	26	2.2

kT. An analysis of the differences in R and kT values measured on both faces indicates that the CPF has reduced the w/c ratio of the CPF surface by around 0.15.

The application of the CPF liner reduced the Carbonation Depth CD by 35% and the Chloride Migration M_{Cl} and Penetration X_d by 40%, in both cases compared to the wooden formwork face.

One important question regarding the performance of CPF liners is to which depth the dewatering effect is noticeable; this information is also required if service life models are to be applied. To assess the influence depth of "Zemdrain", 25 mm thick slices were saw-cut from seven drilled cores, at different depths from the "Z" face (in contact with the CPF liner). A slice was also obtained from the other extreme of the core to test the "F" face (in contact with the formwork). These slices were oven-dried for 2 days at 50°C and then put in contact with 3 mm of water for a water suction test (Swiss Standard SIA 262/1:2019, Annex A). The coefficient of water absorption at 24 hours a_{24}, at different depths, is plotted in Figure 7.14 (black dots with full line).

Figure 7.14 illustrates the average water sorptivity of the four cores tested at each depth; it can be seen that the effect of the CPF is stronger at the surface, but that it is still noticeable at a depth of 25 mm from the CPF liner face. For layers beyond 50 mm, the sorptivity does not differ significantly from that of the "F" formwork surface. These results are in agreement with the 20–50 mm range of action of the CPF, indicated by Malone (1999). Based on the a_{24} profile, tentative profiles for M_{Cl} and kT have been built (the values reported in Table 7.2 are shown with symbols at both faces of the panel) that could be used for modelling purposes.

Another investigation with a similar aim was carried out in Japan (Ohta et al., 2019) in which four reinforced concrete columns (0.4×0.4×1.5 m)

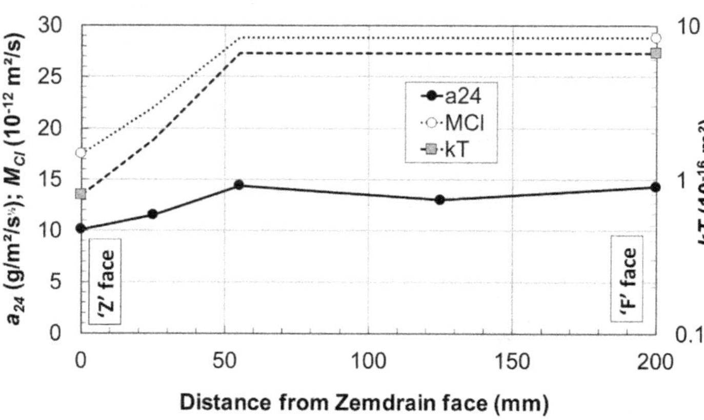

Figure 7.14 Water sorptivity of concrete slices at different depths from "Zemdrain" face (Torrent et al., 2012).

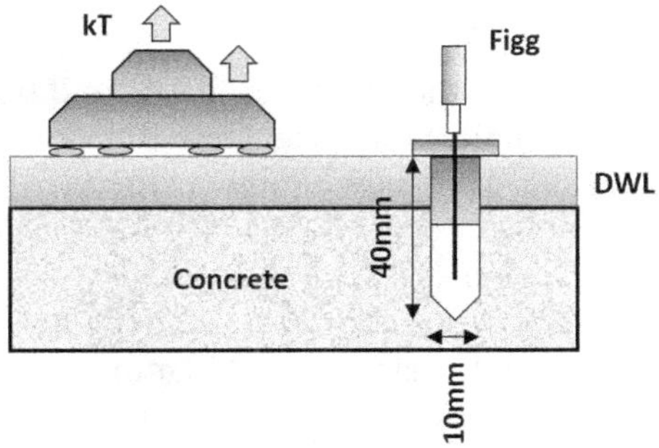

Figure 7.15 *kT* and Figg tests applied on Dewatered Layer of concrete (Ohta et al., 2019).

were cast with concrete mixes of design strengths 21 and 30 MPa, with opposite faces cast against the formwork and against CPF liners. The columns were exposed to outdoor conditions in Japan for 17 years.

The following tests were conducted on the Formwork and CPF faces: air-permeability *kT* and Figg method (Section 4.3.2.1). Both methods are sketched in Figure 7.15, applied on the Dewatered Layer (DWL), being worth mentioning that *kT* is measured right on the surface whilst Figg's Air-Permeability Index is measured inside a Ø10×40 mm hole drilled from the surface.

In addition, cores were drilled from the columns to measure the carbonation depth at 17 years, expressed as carbonation rate K_c (carbonation depth/square root of exposure time). Figure 7.16 shows the K_c values measured on both faces (CPF and formwork) for the two mixes, as function of (a) the air-permeability *kT* and (b) of Figg's Air-Permeability Index.

Figure 7.16 clearly shows that the carbonation rate on the surfaces cast against the CPF liner is negligible, compared with the low value of 1.1 mm/y$^{1/2}$

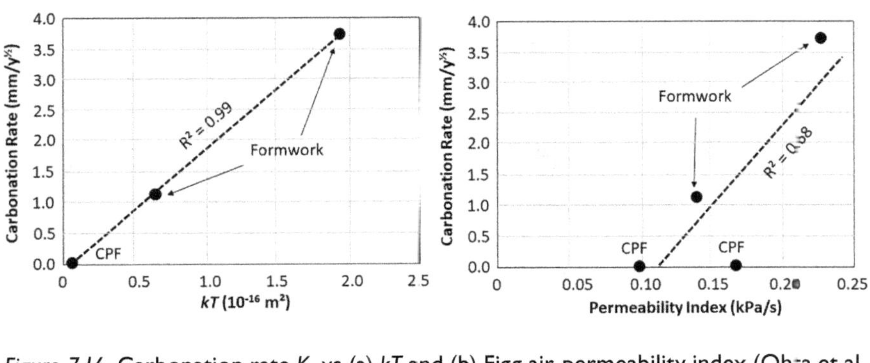

Figure 7.16 Carbonation rate K_c vs (a) *kT* and (b) Figg air-permeability index (Ohta et al., 2019).

for the 30 MPa mix and moderate value of 3.7 mm/y^{½} for the 21 MPa mix, values measured on the surfaces cast against the formwork. There is an excellent correlation between the carbonation rate and the air-permeability coefficient kT (Figure 7.16a), whilst Figg test did not yield such a good correlation (Figure 7.16b). This may be due to the fact that Figg intrusive method explores concrete at a depth of 20–40 mm, where the effect of the liner is not so strong (see Figure 7.15).

The above described investigation by Ohta et al. (2019) was completed by sealing the curved surfaces of Ø50 mm cores drilled from surfaces cast against the conventional formwork and against CPF liner with epoxy resin, leaving only the external surface exposed. The sealed cores were immersed in a 10% NaCl solution for 28 days. After that, the cores were saw-cut into 10 mm slices, the concentration of chlorides of which was measured according to JIS 1154 (Japanese industrial standard) test method. Figure 7.17 shows the chloride profiles obtained. The chloride profile for the concrete in contact with the conventional formwork presents much higher chloride concentration and a higher slope of the profile compared with the concrete in contact with the CPF, indicating a higher coefficient of chloride-diffusion.

In another investigation, Tanaka et al. (2012) studied the effect of the type of form (wooden or metal) and of two types of CPF liners applied on them, on several concrete properties: air-permeability kT, rebound number, water-permeability plus carbonation depth and chloride penetration, both in accelerated tests. The investigated surface conditions are described in Table 7.3 (the tested mix had 291 kg/m^3 of OPC and w/c=0.57).

In general, no significant difference was found between the concrete surfaces cast directly in contact with the wooden and the metal forms (some tests indicated a slightly better performance for the metal form). On the

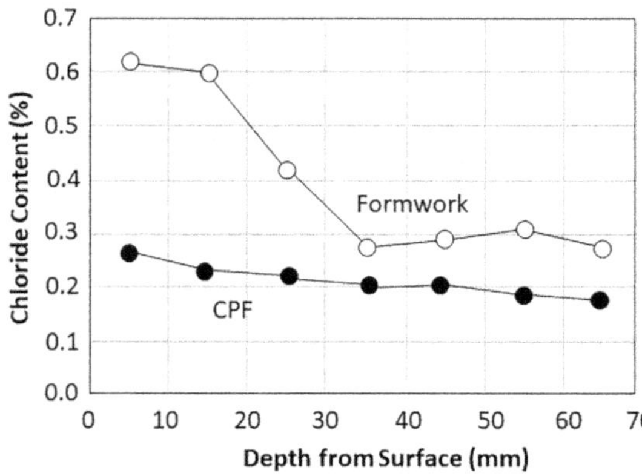

Figure 7.17 Chloride profiles of cores exposed to NaCl solution, with and without CPF liner (Ohta et al., 2019).

Table 7.3 Surface conditions investigated by Tanaka et al. (2012)

Code	*NW*	*NM*	*AW*	*AM*	*BW*	*BM*
Form	Wooden	Metal	Wooden	Metal	Wooden	Metal
CPF liner	None	None	Type A	Type A	Type B	Type B

contrary, all test methods found a significant reduction in the "penetrability" of the concrete surfaces cast against both CPF liners (without a significant effect of the type of liner or of the substrate on which they were applied). Tanaka et al. (2012) studied the profile of the properties with height on the tested blocks measuring 0.48 × 0.48 m by 1.2 m high. All tests, except water-permeability showed higher "penetrability" of the concretes cast on the CPF liners near the top of the blocks compared with that measured at lower positions, which is due to the lack of sufficient hydraulic head to push the water through the fabric. This is shown in Figure 7.18a for air-permeability *kT* and in Figure 7.18b for chloride ion penetration after immersion of cores with the external faces exposed to 10% NaCl solution. This is important, as some measures are required to compensate for that effect (perhaps a richer mix just for the top 250 mm of the element) so as to ensure the tightness of the whole structural element.

It can be concluded that CPF liners are a good solution to improve the quality of the *Covercrete*. In particular, for massive structures that only require a low *w/c* near the surfaces, the solution offers technical (less heat in the concrete mass and less susceptibility to thermal cracks) and economic advantages (the cost of the CPF liner is offset by the reduction in cement content of the mass). Some economic considerations on the use of the CPF are presented in Torrent et al. (2012), related to the surface/volume ratio of the structure.

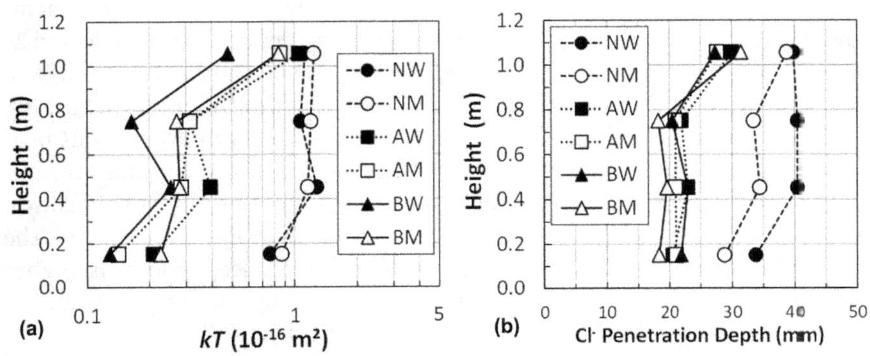

Figure 7.18 Profiles of (a) *kT* and (b) chlorides penetration with columns height (Tanaka et al., 2012).

It is worth mentioning that the effect of CPF liners for self-compacting concrete is not so significant as with vibrated concrete, due to the lack of hydrostatic pressure to force the water through the fresh concrete of high viscosity, typical of that type of concrete, as experienced by one of the book's authors and as reported by Barbhuyia et al. (2011).

7.2.4 Shrinkage-Compensating Concrete

Shrinkage-Compensating Concrete (ShCC) can be defined as a concrete that, when properly restrained by reinforcement or other means, expands an amount equal to, or slightly greater than, the anticipated drying shrinkage. Subsequent drying shrinkage will reduce these expansive strains but, ideally, a residual expansion will remain in the concrete, thereby eliminating shrinkage cracking.

To achieve that, ShCC must contain a suitable expansive component, that may be a special expansive cement or a combination of a conventional cement with a suitable expansive agent (EA) which is added at the concrete mixer. ShCC finds applications in jointless industrial floors (Fernández Luco et al., 2003), in hydraulic and containment structures, nuclear power plants, bridge decks, etc. (ACI 223R, 2010).

The positive effect of adding an expansive agent on the quality of the *Covercrete* was discussed in Section 6.5.4.

7.2.5 Self-Consolidating Concrete

Conventional concrete is usually consolidated in the forms by means of vibration, be it internal, through poker vibrators, or external, through vibrators attached to the forms. Standards and Recommendations (e.g. EN 13970, 2009; ACI 309R, 2005), dealing with the execution of concrete structures describe the correct techniques for consolidating concrete applying vibration. In real practice, vibration is seldom applied according to the recommended good practices, with gross errors becoming visually evident. These malpractices lead to under-consolidation in certain areas (reflected as honeycombing in extreme cases), over-consolidation in others and usage of the vibrator to move concrete horizontally (both resulting in segregation), etc. Even when properly applied, vibration is "*per se*" a discontinuous process, with the concrete directly under the influence of the vibrator consolidating differently than concrete in between successive points of application of the vibrator. In addition, as a result of settlement of fresh concrete, the consolidation of the bottom layers, particularly in deep pours, is higher than in the upper layers, the latter showing usually higher permeability. These aspects have been discussed in detail in Section 6.6.

Self-consolidating (or self-compacting) concrete, or SCC, when properly designed and produced, does not need consolidation by external means, except natural gravity.

In a laboratory experiment (Assié et al., 2007), the permeability of SCCs was compared to that of conventional vibrated concretes VC of similar strength classes (20, 40 and 60 MPa). The permeability tests applied were O_2-permeability *kO* (Cembureau) and water sorptivity at 24 hours a_{24}. These tests were complemented by porosity, chloride migration, accelerated carbonation, MIP and leaching by NH_4NO_3. The relative merits of both sets of concretes vary according to the test applied but not to an extent to claim superiority of one over the other.

A comprehensive investigation was reported by Fornasier et al. (2003), in which the permeability of several SCC mixes was compared with that of conventional concretes CC in the laboratory. Here we will refer only to Group A mixes; Table 7.4 summarizes their characteristics.

Three permeability tests were included in the laboratory investigation: water sorptivity at 24 hours a_{24}, water penetration under pressure W_p, and air-permeability *kT*. Figure 7.19a shows the results of a_{24} and W_p (maximum penetration), whilst Figure 7.19b presents the *kT* results obtained at 28 and 56 days of age.

Figure 7.19a shows that the water sorptivity a_{24} of all SCC mixes (especially SCC3) is lower than that of the conventional concrete CC. Regarding W_p only SCC1 shows a better performance than CC, with SCC2 showing a poorer performance. Table 7.4 shows that the 28-day compressive strength of CC is also higher than that of all SCCs.

Table 7.4 Main characteristics of Group A mixes investigated by Fornasier et al. (2003)

Mix	*CC*	*SCC1*	*SCC2*	*SCC3*
Composite cement (kg/m³)	420	430	370	250
GBFS (kg/m³)	-	-	-	200
Limestone filler (kg/m³)	-	-	240	-
Water/powder ratio	0.40	0.44	0.28	0.40
Cylinder $f'c_{28d}$ (MPa)	55.1	45.7	47.6	50.6

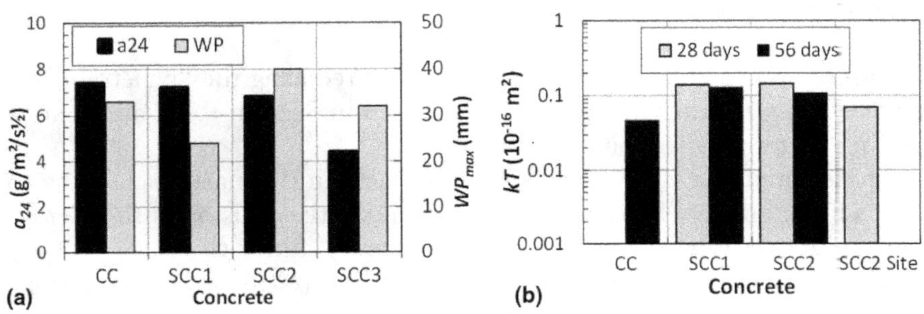

Figure 7.19 Performance comparison of SCC vs. CC in terms of (a) a_{24} and W_p and (b) *kT*; data from Fornasier et al. (2003).

In the case of *kT*, Figure 7.19b presents results of SCC obtained on laboratory specimens, as well as results obtained on site with mix SCC2. Indeed, SCC2 was used for the construction of "E-shaped" walls (*H*: 4.2 m, *T*: 0.25 m and *L*: 4 and 10 m) in an industrial complex, being poured from the top. The data shown in Figure 7.19b indicate that the *kT* of the three SCCs is higher than that of the CC. It is interesting to observe that the *kT* value obtained on site is about half that obtained in the laboratory on the same mix. This may be due to good quality of the applied concreting practices, but also to higher moisture (not reported) of the concrete on site, compared to the 50°C oven-dried (5 days) laboratory specimens.

As a conclusion, although experimental confirmation is still lacking, especially based on large scale site tests, the level and homogeneity of consolidation expected for SCC should be higher than for vibrated concrete. In addition, thanks to its high viscosity, the settlement of SCC is lower compared to conventional concrete, adding to the homogeneity of the finished structure. These advantages of SCC over conventional concrete are highlighted in Section 5.11 of ERMCO (2005).

7.2.6 Permeability-Reducing Agents

According to ACI 212.3R (2016), permeability-reducing admixtures (PRAs) are a class of materials developed to improve concrete durability through controlling water and moisture movement, as well as by reducing chloride ion ingress and permeability. PRAs encompass a range of materials of various performances.

Permeability-reducing admixtures (PRAs) typically include, but are not limited to, the following categories:

a. Hydrophobic water repellents
b. Polymer products
c. Finely divided solids
d. Hydrophobic pore blockers
e. Crystalline products
f. Pore-sealing bacteria

Regarding polymers, their beneficial effect in reducing the "penetrability" of the *Covercrete* has already been discussed in Section 6.5.3 and that of sealing bacteria in Section 6.11.

The amount and types of materials available in the market, claiming to act as PRA, are very large and growing, making the decision of the user on whether to use a PRA, and which one, extremely difficult.

The following recommendation of ACI 212.3R (2016) is worth quoting:

> "Users of a PRA should evaluate performance of the product in concrete based on the application requirements. A commercial PRA can

> include components from several material categories, making classification based strictly on terminology or chemistry inaccurate. The final selection should be based on the project requirements and the performance of the PRA based on appropriate testing, as described in 15.3. The PRA manufacturer is responsible for conducting tests to demonstrate the PRA is suitable for its recommended application. The PRA's performance should be evaluated over a sufficient amount of time to demonstrate the long-term performance of the product, as some PRAs have an extended history of successful use."

In short, the use of these products should be decided strictly on the basis of testing with the materials and proportions intended for the project. For instance, a report by a well-reputed laboratory in Italy indicates that the performance of a pore blocker was very good in reducing the water penetration under pressure of concretes with *w/c* ratio 0.50–0.60, but its effect on a concrete with *w/c*=0.45 was not significant.

Another important issue in the decision is the duration of the PRA's effect and the possibility of reapplication, if the concerned structure is to be protected during a long service lifetime.

The test methods presented in Chapter 4 and in Annex A offer a good choice to be used with the purpose of assessing the efficiency of PRAs, with the following remarks:

- tests applied on the surface are better suited, since intrusive tests may be testing the quality of the concrete beyond the depth of influence of the PRA (Figure 7.15)
- PRA intended to reduce the penetration of water or liquid solutions are to be tested by water-permeability or sorptivity tests, because they may be permeable to gases (e.g. water vapour)
- PRA intended to reduce the penetration of gases (e.g. CO_2) are to be tested by gas-permeability or gas-diffusion tests

A comprehensive assessment of coating agents and impregnation materials, applying different test methods, can be found in Misono et al. (2014).

Not just the quality of the product matters, but also the application technique and skills are very important to ensure a reliable protection of the structure. Figure 7.20 (l) shows the variability of the coefficient of air-permeability *kT* on one side of a 6-year old concrete wall (Quoc & Kishi, 2006), whilst Figure 7.20 (r.) shows the *kT* values measured on the opposite side that was treated with a coating (undisclosed type).

Figure 7.20 (l.) shows a large variability of *kT* values in the uncoated wall, with some values very high, illustrating the heterogeneity often found in the field. On the contrary, Figure 7.20 (r.) shows a much more homogeneous set of significantly lower *kT* values, demonstrating the positive effect of the coating even after 6 years of application. Yet, Figure 7.20 (r.) still

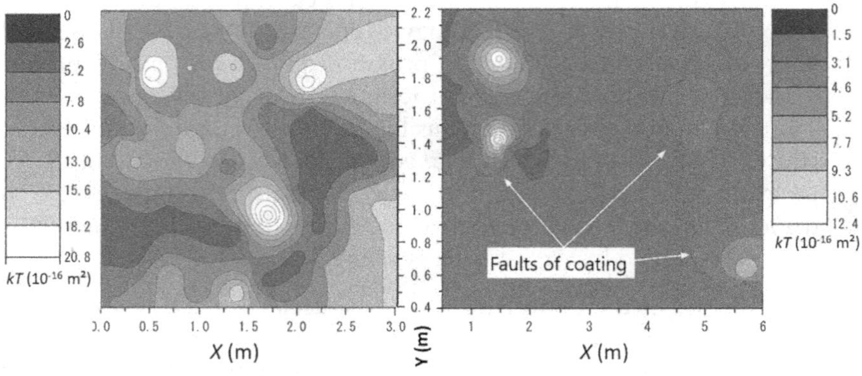

Figure 7.20 Contour *kT* map of (l) uncoated wall and (r) coated wall. Original in colour in Quoc and Kishi (2006).

reveals some isolated spots of high permeability, due to faulty application of the coating, spots that constitute vulnerable areas of the wall. This confirms the need to check that the application of the PRAs is executed correctly, without leaving vulnerable areas.

7.3 COVER THICKNESS

Although beyond the scope of this book, the relevance of this topic is so great that cannot be overlooked. The thickness of the concrete cover is a very important durability indicator for the deterioration of structures due to steel corrosion. Let us recall the solution of the second Fick's law of diffusion presented in Section 3.4.1:

$$C(x,t) = C_0 + (C_s - C_0) . \left[1 - \text{erf}\left(\frac{x}{\sqrt{4.D.t}}\right)\right] \tag{7.1}$$

Let us make $C(x, t) = C_{cr}$ (critical concentration of the aggressive agent, CO_2 or Cl^-, that triggers corrosion). If we take the radical of the error function, the rate of penetration of aggressive agents (front of CO_2, front of critical Cl^- concentration) into concrete follows a "square root" law:

$$x = R.\sqrt{t} \tag{7.2}$$

where

x = penetration of critical front of the agent (mm)
R = rate of penetration (mm/year^{½})
t = age (years)

If we calculate the time t_i taken for the critical front to reach the steel, we have from (7.2):

$$t_i = \left(\frac{d}{R}\right)^2 \tag{7.3}$$

where
t_i=corrosion initiation time (years)
d=cover depth (mm)

In theory, both the second Fick's diffusion law (through the argument of the error function solution) and capillary suction theory (see Sections 3.4.1 and 3.7.1) predict a progress of the penetration front of carbonation, chlorides and water with the square root of time. This means that the time of initiation of corrosion can be considered proportional to the cover depth squared, Eq. (7.3). As a result, a reduction of 10% in the cover thickness would mean a reduction of about 20% in corrosion initiation time t_i.

Figure 7.21 shows the combined effect of changes in the cover thickness d and of the air-permeability kT on the corrosion initiation time, taking as reference the Swiss specification for corrosion induced by de-icing salts (exposure class XD3). The central curve shows a conformity situation for a concrete of $kT = 0.1 \times 10^{-16}$ m^2; it can be assumed that the corrosion initiation time will reach the expected 50 years if the cover thickness equals the nominal value specified (55 mm), situation represented by the black dot on the central curve.

If, instead of 55 mm, the cover thickness happens to be just 50 mm, the service life is reduced from 50 to 41 years, situation represented by the white square on the central curve. The upper and lower curves in Figure 7.21 represent the evolution of the critical chloride front with time, for kT values

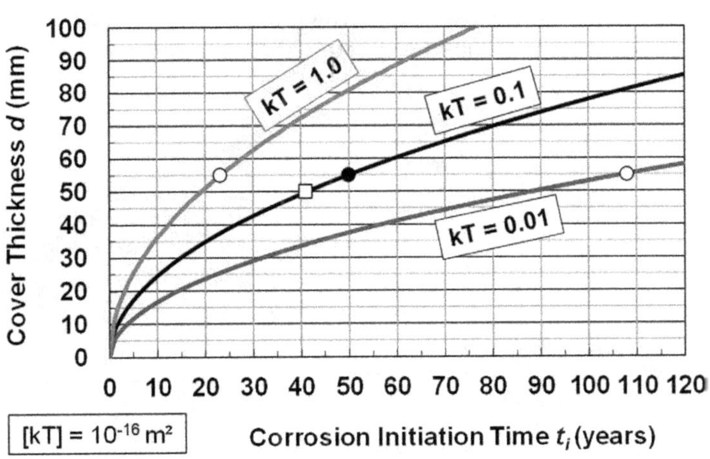

Figure 7.21 Effect of cover thickness d and kT on corrosion initiation time t_i.

one order of magnitude higher and lower, respectively, from the central value (assuming that R in Eq. (7.3) is proportional to $kT^{1/3}$, see Section 9.4.3.1). It can be seen that the service life (white dots on each curve) is reduced and increased, respectively, by a factor of ~2 with respect to the expected 50 years.

Despite the progress made on instruments capable of assessing, non-destructively, the cover depth quite accurately, their use is not forcibly specified in the standards. It should be reminded that the cover thicknesses, measured prior to placing the concrete, may be modified during the concreting operations, often resulting in insufficient covers of the steel in the end-product (Neville, 1998). Neville also coined the term "negative cover" when the steel happens to protrude from the concrete surface.

There are many examples reporting lack of conformity with specified cover thickness. For instance, Figure 7.22a shows the histogram of measured cover thickness (*Realcrete*) in Norwegian 1981 Gymsøystraumen Bridge, which reflects a significant lack of compliance with the specified value (*Theorecrete*) (Gjørv, 2014). Figure 7.22b shows several cases, reported by Lim (2013) for Malaysian jetties. The horizontal black lines indicate the specified nominal cover (60 or 75 mm, depending on the type of element, with −5 mm tolerance), whereas the vertical bars indicate the minimum and maximum covers measured on site with covermeters, calibrated with direct physical measurements. It can be seen that in most cases the *Realcrete* measured cover is in defect with respect to the *Theorecrete* specified value; in particular, in structures 1, 6 and 9, none of the measurements reached the specified value. This was considered the primary cause for the unsatisfactory performance of the jetties (Lim, 2013).

A survey in the UK showed that 77 out of 200 bridges had too low cover thickness (Wallbank, 1989) that resulted in spalling of the cover and rust.

It must be mentioned that there are in the market electromagnetic covermeters capable of assessing the cover thickness to ±10% accuracy (RILEM TC 189-NEC, 2007; Fernández Luco, 2005). A relatively old, possibly

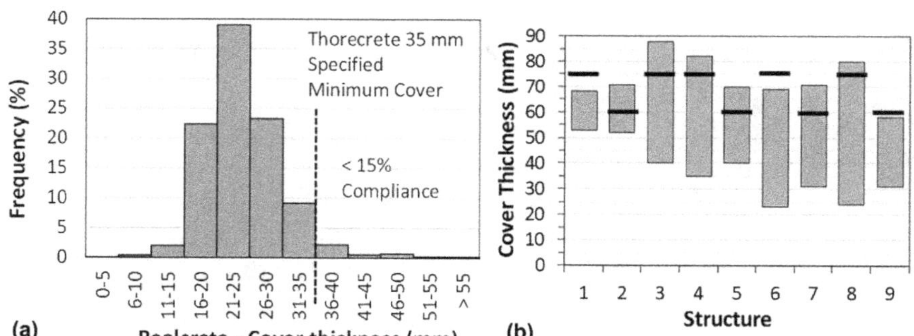

Figure 7.22 Measured (*Realcrete*) vs. specified (*Theorecrete*) covers: (a) Norway and (b) Malaysia; data taken from Gjørv (2014) and Lim (2013), respectively.

outdated standard indicates how to do it (BS 1881-204, 1988). For very deep bars or complex reinforcement patterns, the modern ground penetrating radar (GPR) instruments complement well the capabilities of the electromagnetic instruments.

More up-to-date documents on the topic are Chapter 9 of RILEM TC 230-PSC (2016) and DBV (2015).

For durability and/or service life assessment of new or old reinforced concrete structures under risk of steel corrosion deterioration, measuring the thickness of the cover is as important as measuring its permeability ("penetrability").

7.4 SPACERS AND PERMEABILITY

Spacers are small pieces used to position and fix the steel bars at the required distance from the exposed surfaces, so as to ensure that the proper cover thickness is achieved. The more popular spacers are made of mortar/concrete and plastics, although stainless steel is also used.

Inevitably, the spacers intrude the *Covercrete* and establish a link between the reinforcement and the environment to which the structural element is exposed. As spacers are placed every meter or so and remain permanently in place, some concern exists on their influence on the potential corrosion risk of the supported steel (Alzyoud et al., 2016), concern supported by cited reports and field investigations.

An investigation was conducted by Alzyoud et al. (2016), in which the effect of spacers on transport properties (O_2-diffusivity, O_2-permeability and water sorptivity) was investigated. The variables considered were the type of spacer (none, plastic, concrete and steel), the aggregate size (D_{max}=10 and 20 mm), curing (3 and 28 days sealed curing) and the pre-conditioning of the specimens (20°C/75% RH; 20°C/55% RH; oven-dried at 50°C) on the above-mentioned variables. Complementary techniques were applied on the samples, namely: chloride-diffusion, microscopy of fluorescent epoxy impregnated surfaces and backscattered electron imaging.

Cylindrical specimens (Ø100×25 or 50 mm) were cast (D_{max}=10 mm) or cored from 1,500×600×50 mm slabs (D_{max}=20 mm). The specimens contained in the centre a spacer, kept firmly in place during casting; a control specimen without spacer was also prepared. After curing, the specimens were subjected to the pre-conditioning regimes described above until constant mass prior to conducting the transport tests.

Figure 7.23 presents the results of (a) O_2-permeability and (b) water sorptivity a7 (at 7 hours) of the samples, relative to the values measured on the control sample (no spacer). The spacer codes are the following:

- *SS*: stainless steel spacer
- *CS*: concrete spacer
- *PS*: plastic spacer

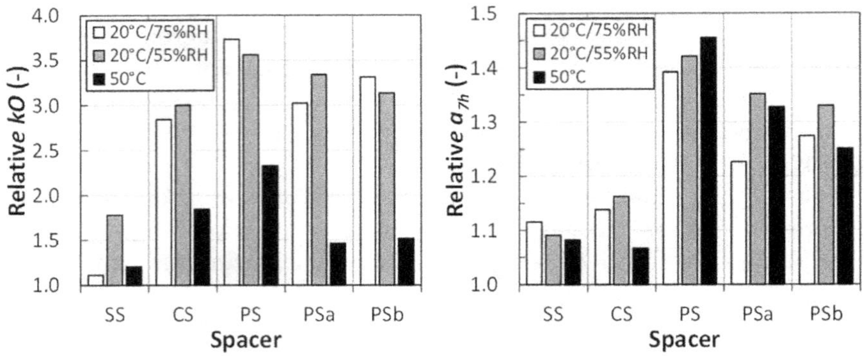

Figure 7.23 Effect of spacer type and pre-conditioning regime on relative (a) O_2-permeability and (b) water sorptivity at 7 hours of 10 mm D_{max} concretes, cured 28 days (Alzyoud et al., 2016).

- *PSa*: PS plastic spacer ground with 120-grit SiC paper
- *PSb*: PS plastic spacer with scoring notches on the main flange

Figure 7.23 shows clearly that all the samples containing spacers, irrespective of the pre-conditioning treatments, present higher gas- and water-permeability than the control sample. However, the performance of the different spacers is not the same, with the negative effect of spacers on the permeability increasing in the order: steel → concrete → plastic spacers. Regarding the latter, the performance improves when the smooth surfaces are roughened (PSa and PSb) but not to the extent of equalling the performance of either concrete or steel spacers.

It is interesting to remark that the effect of spacers on permeability can only be realistically revealed and quantified through site testing of the finished structure.

7.5 CONCLUDING REMARKS

The objective of this chapter was to highlight that a large difference in quality frequently exists between the concrete in the structure (*Realcrete*), with respect to that tested in the laboratory on cast specimens (*Labcrete*) and even more to that conceived and specified by the designer (*Theorecrete*). In the case of durability that relies to a large extent on the performance (and thickness) of the few cm/in constituting the near-surface layer (*Covercrete*) the impact is significantly higher than for bulk properties, such as strength. According to Polder and Rooij (2005), this difference between the intended and effective quality and thickness of *Covercrete* is the major cause for premature deterioration of reinforced concrete structures.

It has been shown that it is impossible to reproduce the *Realcrete* conditions by casting specimens, even as large as 1 m thick, to be stored near the real structure, not to mention small specimens kept under laboratory conditions.

The conclusion is that the quality (and thickness) of the vital *Covercrete* can only be assessed realistically by means of testing the end-product, i.e. site testing of the actual structural elements. This can (and should) be done preferably by non-destructive methods or, alternatively/complementarily, by laboratory testing of specimens prepared from cores drilled from the structure. This holds true also when surface treatments, intended to reduce the "penetrability" of the *Covercrete*, are applied on the exposed concrete.

It follows that service life assessment of concrete structures can only be realistically done by involving direct measurements of the *Covercrete*, which reflect the true spatial variability of both "penetrability" and thickness of the cover concrete. Without them, current deterministic or probabilistic models, based on *Theorecrete* or *Labcrete* definition of the key materials' parameters, will always have a large degree of inaccuracy. Chapter 9 deals more extensively with this matter.

REFERENCES

ACI 212.3R (2016). "Report on chemical admixtures for concrete", 76 p.

ACI 223R (2010). "Guide for the use of shrinkage-compensating concrete", 20 p.

ACI 309R (2005). "Guide for consolidation of concrete".

Adam, A.A., Law, D.W., Molyneaux, T., Patnaikuni, I. and Aly, T. (2009). "The effect of using controlled permeability formwork on the durability of concrete containing OPC and PFA". Techn. Letter, Institut. of Engs., Australia, 1–12 p.

Alzyoud, S., Wong, H.S. and Buenfeld, N.R. (2016). "Influence of reinforcement spacers on mass transport properties and durability of concrete structures". *Cem. & Concr. Res.*, v87, 31–44.

Andrade, C. (2006). "Multilevel (four) methodology for durability design". RILEM Proceedings *PRO 47*, 101–108.

Angst, U. (2018). "Battling infrastructure corrosion". Keynote Lecture, 4th *International Conference on Service Life Design for Infrastructures (SLD4)*, August 27–30, Delft, The Netherlands.

Assié, S., Escadeillas, G. and Waller, V. (2007). "Estimates of self-compacting concrete 'potential' durability". *Constr. & Build. Mater.*, v21, 1909–1917.

Barbhuiya, S.A., Jaya, A. and Basheer, P.A.M. (2011). "Influence of SCC on the effectiveness of controlled permeability formwork in improving properties of cover concrete". *The Indian Concr. J.*, v85, n2, February, 43–50.

Basheer, P.A., Basheer, L., Bailie, R. & Nanukuttan, S.V. (2005). "An investigation into the performance of Formtex controlled permeability formwork and effects of its re-use". Report, Queen's University Belfast, UK, September 2005, 49 p.

Beushausen, H. and Alexander, M. (2009) "Application of durability indicators for quality control of concrete members – A practical example". *RILEM Conference* 'Concrete in *Aggressive Aqueous Environments* – Performance, Testing, and Modeling', June 3–5, Toulouse, France.

Bouwer, S. (1998). "Practical implementation of index tests for assessment and control of potential concrete durability". MSc Thesis, University of Stellenbosh.

Brühwiler, E. (2007). "Lifetime oriented composite concrete structures combining reinforced concrete with Ultra-High Performance Fibre Reinforced Concrete". 3rd *International Conference on Lifetime-Oriented Design Concepts*, November 12–14, Ruhr-Universität Bochum, Germany, 10 p.

BS 1881-204 (1988). British Standard "Testing concrete. Recommendations on the use of electromagnetic covermeters". August, 14 p.

CEN (2007). "Survey of national provisions for EN 206-1". CEN/TC 104/SC 1, N 485, January 30, 148 p.

COWI-Almoayed Gulf WLL (2002). "Application of Formtex controlled permeability formwork in the Arabian Gulf". Report 201689-1, December 9, 48 p.

Cullen, D.A. (1998). "Evaluation of the effects of Formtex CPF on the surface properties of concrete". Taywood Engng. Ltd., Report 1304/98/10115, June 1998, 50 p.

DBV (2015). "DBV-Merkblatt, Betondeckung und Bewehrung nach Eurocode 2". DBV 2015-12 (in German).

EN 13970 (2009). "Execution of concrete structures". European Standard.

EN 1992-1-1 (2004). "Eurocode 2: Design of concrete structures – Part 1-1: General rules and rules for buildings". December.

EN 206 (2013). "Concrete – Specification, performance, production and conformity". European Standards, December, 93 p.

ERMCO (2005). "The European guidelines for self-compacting concrete: Specification, production and use". BIBM, Cembureau, EFCA, EFNARC and ERMCO, May, 68 p.

Fehling, E., Schmidt, M., Teichmann, T., Bunje, K., Bornemann, R. and Middendorf, B. (2005). "Entwicklung, Dauerhaftigkeit und Berechnung Ultrahochfester Betone (UHPC)". Forschungsbericht DFG FE 497/1-1, Kassel University, Kassel, 132 p.

Fernández Luco, L. (2009). "Importancia del curado en la calidad del hormigón de recubrimiento. Parte II: Métodos experimentales para identificar o prevenir el curado deficient". Cemento Hormigón, n926, 30–41.

Fernández Luco, L., Pombo, R. and Torrent, R. (2003). "Shrinkage compensating concrete in Argentina". *Concr. Intern.*, May, 49–53.

fib (2010). "Model code 2010". 1st Complete Draft, v1, March.

Fornasier, G., Fava, C., Fernández Luco, L. and Zitzer, L. (2003). "Design of self compacting concrete for durability of prescriptive vs. performance-based specifications". ACI SP 212, 197–210

Gjørv, O.E. (2014). *Durability Design of Concrete Structures in Severe Environments*. 2nd Ed., Taylor & Francis, UK, 254 p.

Holčapek, O. (2011). "Influence of surface layer on the permeability of concrete". JUNIORSTAV *2011*, Brno, Czech Rep., February 4, 4 p.

Jacobs, F., Hunkeler, F. and Mühlan (2018). "Prüfung und Bewertung der Betonqualität am Bauwerk". Office Fédéral des Routes, Rapport No. 691, Bern, Switzerland, Juli, 106 p.

Kreijger, P.C. (1984). "The skin of concrete: Composition and properties". Mater. & Struct., v17, n100, 275–283.

Kunieda, M., Shimizu, K., Eguchi, T., Ueda, N. and Nakamura, H. (2011). "Fundamental properties of ultra high performance-strain hardening cementitious composites and usage for repair". J. Japan Soc. Civil Engs., Ser. E2 (Materials and Concrete Structures), v67, n4, 508–521 (in Japanese).

Kunieda, M., Choonghyun, K., Ueda, N. and Nakamura, H. (2012). "Recovery of Protective Performance of Crack Ultra High Performance-Strain Hardening Cementitious Composites (UHP-SHCC) Due to Autogenous Healing". J. Adv. Concr. Technol., v10. Sept., 313–322.

Law, D.W., Molyneaux, T., Patnaikuni, I. and Adam, A.A. (2012). "The site exposure of concrete cast using controlled permeability formwork". *Australian J. Civil Eng.*, v10, n2, 163–176.

Leow, C-H. (2004). "Surface enhancements of concrete through use of controlled permeability formwork (CPF) liners". *International Conference on Bridge Engineering & Hydraulic Structure*, Selangor, Malaysia, July 26–28.

Lim, C.C. (2013). "An investigation into the deterioration of reinforced concrete jetties in Malaysia". CONSEC13, September 23–25, Nanjing, China.

Malone, Ph.G. (1999). "Use of permeable formwork in placing and curing concrete". US Army Corps of Eng., Techn. Re-port SL-99-12, October 1999, 53 p.

Mayer, A. (1987). "The importance of the surface layer for the durability of concrete structures". ACI SP-100, v1, 49–61.

MC90 (1991). "CEB-FIP model code 1990 – Final draft". Bulletin d'Information CEB *203*, Lausanne, July.

Misono, M., Imamoto, K., Nagai, K. and Kiyohara, C. (2014). "Evaluation of moisture and gas permeability of surface treated concrete under accelerated weathering conditions towards conservation of reinforced concrete buildings". *International Workshop on Performance-based Specification and Control of Concrete Durability*, Zagreb, Croatia, June 11–13, 361–368.

Neves, R. and Santos, J.V. (2008). "Air permeability assessment in a reinforced concrete viaduct". SACOMATIS 2008, September 1–2, Varenna, Italy.

Neville, A. (1998). "Concrete cover to reinforcement — Or cover up?" *Concr. Intern.*, v20, n11, November, 25–29.

Newman, K. (1987). "Labcrete, realcrete, and hypocrete. Where we can expect the next major durability problems". *ACI SP-100*, v2, 1259–1283.

Oesterlee, C., Denarié, E. and Brühwiler, E. (2009). "Strength and deformability distribution in UHPFRC panels", *ConMat'09*, Nagoya, Japan, 24–26 Aug. 2009, 390–397.

Ohta, Y., Kiyohara, C., Shimozawa, K. and Imamoto, K. (2019). "Quality of cover concrete with controlled permeable formwork exposed outdoor condition for 17 years". Annual convention of Architectural Institute of Japan: Kanto branch, (in Japanese).

Polder, R. and Rooij, M. (2005). "Durability of marine concrete structures – Field investigations and modeling". *HERON*, v50, 133–153.

Quoc, P.H.D. and Kishi, T. (2006). "Measurement of air permeation property of cover concrete". *Proceedings on JSCE Annual Meeting*, v61, Disk 2, 2 p.

RILEM TC 189-NEC (2007). "Non-destructive evaluation of the penetrability and thickness of the concrete cover". R. Torrent and L. Fernández Luco (Eds.), RILEM Report 40, 223 p.

RILEM TC 230-PSC (2016). "Performance-based specifications and control of concrete durability". H. Beushausen and L. Fernández Luco (Eds.), *RILEM Report 18*, 373 p.

SIA 262/1 (2019). "Concrete construction – Complementary specifications". Swiss Society of Engineers and Architects.

Tanaka, R., Habuchi, T., Amino, T. and Fukute, T. (2012). "A study on improvement and its evaluation for the surface layer of concrete placed with permeable form". *Intern. J. Modern Physics: Conference Series*, v6, 664–669.

Torrent, R. (1999). "The gas-permeability of high-performance concretes: Site and laboratory tests". *ACI SP-186*, Paper 17, 291–308.

Torrent, R. (2018). "Bridge durability design after EN standards: Present and future". *Struct. & Infrastruct. Eng.*, v15, n5, 1-13.

Torrent, R. and Frenzer, G. (1995). "Methoden zur Messung und Beurteilung der Kennwerte des Ueberdeckungsbetons auf der Baustelle -Teil II". Office Fédéral des Routes, Rapport No. 516, Bern, Switzerland, October, 106 p.

Torrent, R., Griesser, A., Moro, F. and Jacobs, F. (2012). "Technical-economical consequences of the use of Controlled Permeable Formwork". ICRRR, Cape Town, South Africa, September 2–5.

Wallbank, E.J. (1989). "The performance of concrete in bridges". HMSO, London, April 1989, 96 p.

Chapter 8

Why air-permeability *kT* as durability indicator?

8.1 INTRODUCTION

The durability of concrete structures is closely dependent on the resistance of concrete to the penetration of aggressive agents (CO_2, chlorides, sulphates, etc.) as well as of substances that, although not being aggressive *per se*, play a role in the degradation process. Typical examples of the latter are water, the presence of which in sufficient quantities is required for most deterioration processes to progress (expansion of ASR gel, frost damage, steel corrosion, etc.) and O_2 (also required for steel corrosion).

As discussed in Chapter 3, the penetration of these substances happens by different mechanisms, through the interconnected network of pores present in the hardened cement paste and in the ITZ. In Chapter 3, the close connection between the coefficients ruling the different transport properties (permeability, sorptivity, diffusion, migration) was discussed. In addition, as treated in Chapter 7, what matters more in terms of durability is the resistance of the *Covercrete* to the penetration of external agents.

All these concepts are superbly summarized in p. d-14 of CEB-FIP Model Code 1990 (CEB-FIP, 1991):

> "The durability of concrete is understood to be its resistance to physical and chemical attack such as frost or elevated temperatures, carbonation, sulphate attack etc. The resistance of concrete to such actions is governed primarily by its resistance to the ingress of aggressive media and thus by the capillary porosity of the hydrated cement paste as well as by entrapped air. A dense paste with a low capillary porosity is in most instances more durable than a paste with a high capillary porosity and a coarser pore system.
>
> There is no generally accepted method to characterize the pore structure of concrete and to relate it to its durability. However, several experimental investigations have indicated that concrete permeability both with respect to air and to water is an excellent measure for the resistance of concrete against the ingress of aggressive media in the gaseous or in the liquid state and thus is a measure of the potential durability of a particular concrete.

DOI: 10.1201/9780429505652-8

> There are at present no generally accepted methods for a rapid determination of concrete permeability and of limiting values for the permeability of concrete exposed to different environmental conditions. However, it is likely that such methods will become available in the future allowing the classification of concrete on the basis of its permeability. Requirements for concrete permeability may then be postulated; they would depend on exposure classes i.e. environmental conditions to which the structure is exposed.
>
> Though concrete of a high strength class is in most instances more durable than concrete of a lower strength class, compressive strength per se is not a complete measure of concrete durability, because durability primarily depends on the properties of the surface layers of a concrete member which have only a limited effect on concrete compressive strength."

Regarding the third paragraph of the above quote, written around 1990, the situation at that time has changed radically and, nowadays, as discussed in Chapter 4, several rapid test methods to measure concrete permeability have been developed, some of them having been standardized and for which limiting values have even been established. The Torrent method to measure the coefficient of air-permeability *kT* is one of such standardized test methods; this method is especially suitable for non-destructive site testing of the end-product (the finished structure), its results reflecting the better or worse contribution of all the players along the concrete construction chain (owners, specifiers, materials producers, contractors, inspection). This is in line with the last paragraph of the text cited above.

The rest of this chapter is devoted to presenting and discussing experimental data, obtained both in the laboratory and on site, with the aim of evaluating the credentials of *kT* as durability indicator. When dealing with the experimental data, it has to be borne in mind that when measuring *kT*, it is the *Covercrete* (see Chapter 7) which is being tested, affected not just by the mix design but also by factors such as placing, consolidation, segregation (including natural settlement and bleeding), curing, age and testing conditions (factors discussed in Chapter 6).

8.2 RESPONSE OF *kT* TO CHANGES IN KEY TECHNOLOGICAL PARAMETERS OF CONCRETE

The effect of several key technological parameters of concrete (*w/c* ratio, compressive strength, curing, compaction/segregation, etc.) on the permeability of concrete was discussed in detail in Chapter 6. Chapter 6 presents abundant experimental evidence on the impact of such technological parameters on the permeability of concrete to water and gases, measured by several test methods, including *kT*, the one here discussed.

Table 8.1 Sections dealing with the response of *kT* to key technological parameters

Parameter	*w/c & f′c*	*Binder type*	*Aggr. type*	*Special raw materials*	*Compaction, segregation and bleeding*	*Curing*	*Temperature and moisture*	*Stresses*	*Cracks*
Section	6.2.2.3	6.3.2	6.4.1	6.5	6.6	6.7	6.8 & 6.9	6.10	6.11

Table 8.1 indicates the different Sections of Chapter 6 where the response of the coefficient of air-permeability *kT* to changes in different key technological parameters of concrete is discussed.

The experimental evidence provided in the sections indicated in Table 8.1 confirms that the coefficient of air-permeability *kT* is very sensitive to changes in *w/c* ratio and compressive strength of concrete as well as to the lack of moist curing at early ages. It also detects insufficient compaction and is negatively affected by excessive bleeding and by cracking, detecting even incipient ASR cracks. Compressive stresses up to 50% of the maximum load have virtually no effect on *kT*. The not so relevant effect of temperature (if not extremely low or high) and the strongly relevant effect of moisture on *kT* are discussed in detail in Sections 5.7.1 and 5.7.2, respectively. An approach to compensate *kT* for surface moisture content is presented in Section 5.7.2.2.

It can be concluded that *kT* is a good indicator for variations in quality or condition of the *Covercrete* due to the factors listed in Table 8.1.

8.3 CORRELATION WITH OTHER DURABILITY TESTS

Being a good indicator for variations in *Covercrete* quality is not enough for *kT* to qualify as a suitable durability indicator, because it must be proved that *kT* is also related to the resistance of concrete to the penetration of species by the different transport mechanisms discussed in Chapter 3.

In Chapter 3, the close relation between the pore structure of concrete and its permeability was demonstrated in theoretical terms, confirmed experimentally for the case of air-permeability *kT* in Section 5.5.2. Since other transport mechanisms (sorptivity, diffusion) are also related, albeit differently, to the pore structure (Section 3.8), some relations between the transport parameters are expected, as theoretically formulated in Section 3.9.

In the rest of this section, experimental evidence on the relations existing between the coefficient of air-permeability *kT* and other transport parameters is presented and discussed, to assess its potential as sound durability indicator. This experimental evidence has been obtained from worldwide series of data sets reported in the literature or accessible to the authors, as detailed in Table 8.2.

The relations between variables, presented in this chapter, have been established on the basis of the data reported in the documents listed in Table 8.2, with reference to the corresponding data sets. Please bear in mind that the 47 sets of data were generated by many laboratories across the globe (17 countries involved), applying different instruments, personnel and testing protocols.

The main test methods (mostly standardized) mentioned in this chapter (described in Chapter 4 and Annex A) are identified by the following symbols:

- a_{24} = coefficient of water capillary absorption at 24 hours
- a_o = coefficient of water capillary absorption at unknown ages
- a_K = Karsten tube water sorptivity test
- CCI = chloride-conductivity Index, South African method
- D_P = coefficient of chloride-diffusion by ponding (AASHTO T259)
- D_I = coefficient of chloride-diffusion by immersion (ASTM C1556 or similar)
- D_S = coefficient of chloride-diffusion obtained on site
- δm = mass loss after 30 freeze/thaw/salts cycles (SIA 262/1)
- F_a = Figg air-permeability test
- F_w = Figg water sorptivity test
- K_c = natural carbonation in laboratory or on site
- K'_c = natural carbonation from accelerated test
- kO = oxygen-permeability test, Cembureau method
- k_{OPI} = coefficient of oxygen-permeability, OPI (oxygen-permeability index) index, South African method
- kT = air-permeability test (Torrent) using second or later generation instruments
- kT_3 = air-permeability using first generation prototype, $kT = 1.846 * kT_3$
- M = coefficient of chloride-migration (Tang-Nilsson method)
- N_{50} = number of freeze/thaw cycles for a reduction of 50% in E-modulus
- OD = coefficient of oxygen-diffusion (D_o)
- Q = charge passed in Rapid Chloride Permeability Test (ASTM C1202)
- ρ_s = surface electrical resistivity (Wenner)
- S_{Cl} = site chlorides
- T = TUD oxygen-permeability test
- WP = maximum penetration of water under pressure (EN 12390-8 or DIN 1048)

8.3.1 Gas Permeability

8.3.1.1 Cembureau Test

Figure 8.1 shows the results of parallel measurements of the coefficients of air-permeability kT and of oxygen-permeability kO from 11 data sets (described in Table 8.2), involving 141 pairs of data.

A very good correlation ($R = 0.89$) exists between both coefficients of gas-permeability, through the following regression, shown in Figure 8.1 as a full line:

$$kO = 1.23\ kT^{0.57} \quad kT,\ kO\ \left(10^{-16}\ \mathrm{m}^2\right) \tag{8.1}$$

The dotted line in Figure 8.1 shows the equivalence line ($Y = X$), for the case in which the results had been the same for both methods. It is clear that both test methods yield well-correlated but significantly different results. The kO results span four orders of magnitude, whilst kT results span five

Table 8.2 Details on data sets used in this chapter and their sources

Data set	*Source*	*Applied tests*	*Country code*	*Brief description*
1	Torrent and Ebensperger (1993); see Section 6.2.1.1	kT_3, kO, a_{24}, T, F_a, F_w	CH	Table 3.1-III reports five different concretes mixes that, after 0, 7, or 28 days moist curing were stored in a dry room (20°C/50% RH) till the moment of test (age ≈ 1.5 years).
2		kT_3, kO, a_{24}, D_P, δm, F_a, F_w, T, K_c		Table 3.2-IV reports ten mixes, moist cured 0 or 7 days, followed by storage in the same dry room until the age of test (28 days). Prisms made with four concrete mixes, subjected to 28 days moist curing, were exposed to an ambient of 20°C/57% RH for 2 years.
2a		kT_3, ρ_s, kO, a_{24}, F_a, F_w, T,		Table 4-III reports three mixes, moist cured 90 days, thereafter kept in a dry room (20°C/50% RH), measuring kT and ρ_s, and kO and a_{24} on cores drilled after 0, 19, 68 and 111 days drying.
3	Torrent and Frenzer (1994); Section 6.2.1.1	kT, kO, a_{24}		Table III reports durability performance data of three Mexican cements, same procedure as for data set 2.
4	Torrent and Frenzer (1995); Sections 6.2.1.1 and 6.2.1.2	kT, kO, a_{24}		Table 1.2.1.1 reports data of lab specimens cast from nine batches of ready mixed concrete to be placed in Bözberg Tunnel (Section 7.1.5.1). The specimens were moist cured 7 days, followed by storage in dry room (20°C/50% RH) until the age of test (28 days).
5				Table 1.2.1.2 reports measurements made on the deck of the Tunnel in correspondence with some of the mixes tested in data set 3.
6		kT, kO, a_{24}, D_P, δm		Tables 2.2.2.1, 2.2.4.2, 2.2.4.3 report data of lab specimens cured and stored like for data set 3 and of 1 m cubes that were field-cured and tested at ages within 28–35 days. Two concrete qualities from Schaffhausen Bridge (Section 7.1.5.2) were investigated.
7		kT, kO, a_{24}, N_{50}, K_c, K'_c		Table 3.2.1.1 and p. 95 report data of lab cubes (0.5 m), moist cured for 0 and 7 days and stored in the laboratory. The experiment is described in Section 5.6.2. From certain locations where kT was measured, specimens were drilled and saw-cut for measuring other properties. Prisms made with several concrete mixes, subjected to different curing conditions, exposed to an ambient of 20°C/50% RH for 2 years.

(*Continued*)

Table 8.2 (Continued) Details on data sets used in this chapter and their sources

Data set	*Source*	*Applied tests*	*Country code*	*Brief description*
8	Kattar et al. (1995)	kT_3, Q, WP_D	BR	Data of a conventional concrete made with a triple-blend cement (containing slag and silica fume) and to four mixes where different vinyl-based polymers were added (*w/c* ratio between 0.33 and 0.48).
9	Kattar et al. (1999)	kT_3, Q		Data of three mixes of 26, 30 and 36 MPa cylinder strength at 28 days. The specimens were moist cured up to 25 days of age, to be tested at 28 days of age.
10	Roelfstra et al. (1999)	kT, D	CH	Modelling chloride-induced corrosion in reinforced concrete. For the modelling, three *Covercrete* classes are defined as function of the kT (0.1, 1.0 and 10. 10^{-16} m^2). Values of water and chloride-diffusion coefficients are associated with those classes.
11	Andrade et al. (2000)	kT, kO, a_{24}, Q, K_c	SP, CH	Round robin tests within RILEM TC 116-PCD, consisting in testing Ø150 × 50 mm discs of five concrete mixes: four binders, *w/b* = 0.4 and 0.7, sealed cured 3 and 7 days and pre-conditioned at 50°C.
12	Mohr et al. (2000)	kT, Q, a_o	US	Tests on saw-cut slices of cores drilled from old concrete pavements with compressive strengths in the range of 40–80 MPa.
13 14	Denarié et al. (2003)	kT, WP	CH	kT measured on panels made with OPC mixes with *w/c* ratios of 0.41 and 0.59. Different finishing techniques of the top surface and a "Zemdrain" permeable formwork liner were applied. *WP* measured on cores drilled from the top and lateral surfaces. WP of set 13 by DIN test and of set 14 by EN test method.
15	Fornasier et al. (2003)	kT, WP, a_{24}, Q, D_l	AR	Results obtained at Loma Negra laboratory in Argentina on three concretes, two of them self-compacting with *w/c* between 0.40 and 0.44.
16	Kubens et al. (2003)	kT, Q, K'_c	IL	Cubes and prisms made with two concrete mixes, subjected to different curing conditions, stored in an ambient of 30°C/40% RH for 90 days. Companion specimens were also exposed to accelerated carbonation: 5% CO_2, 30°C/50% RH for 7 days.
17	Romer and Leemann (2005)	kT, kO, k_{OPI}, CCI	CH	200 mm cubes from six concretes of *w/c* ratios between 0.35 and 0.62 (OPC+limestone filler cement) stored at 35%, 70% and 90% RH (20°C). Direct measurements of kT followed by measurements of k_{OPI} and CCI on drilled cores were made at 365 days. kT values at 90% RH not considered due to high moisture content.

(*Continued*)

Table 8.2 (Continued) Details on data sets used in this chapter and their sources

Data set	Source	Applied tests	Country code	Brief description
18	Fernández and Revuelta (2005)	kT, WP	SP	Tests on pozzolanic cement concretes with $w/c = 0.40$ and 0.45, intended for a LNG tank in México.
19	Rodríguez de Sensale et al. (2005)	kT, WP	UY	Tests on self-compacting concretes (28 days cylinder compressive strength within 35 and 66 MPa), with and without electrofilter powder as filler.
20	Mathur et al. (2005)	kT, Q	IN	Reference and high-volume fly-ash concretes (30%–50% of PFA) of three w/b ratios between 0.33 and 0.65 were tested for durability by the Rapid Chloride Permeability Test (ASTM C1202) and kT.
21	Jacobs (2006)	kT, a_{24}, K_c	CH	Comprehensive survey of data from old concrete structures.
22	RILEM TC 189-NEC (2007)	kT, kO, k_{OPI}, a_{24}, Q, M, CCI	CH, PT, ZA	Tables B.1–B.5 report results of site kT tests made on ten panels made with OPC and GBFS mixes of different w/c, curing and testing conditions and lab tests on drilled and saw-cut specimens. Several labs involved.
23	Di Pace and Calo (2008)	kT, WP	AR	Report site (kT) and laboratory (kT, WP_D) test results of concrete of Buenos Aires Metro. Improved mix design and construction practices led to an enhanced water tightness of the underground construction.
24	Imamoto et al. (2008)	kT, K_c	JP	Exposed panels made with three concrete mixes, subjected to two different curing conditions, to an ambient of 20°C/60% RH for 3.5 years.
25	Jacobs (2008)	kT, a_0, K_c, S_{Cl}	CH	Site kT on 12 bridges 30 years old, ranging 0.001–10 (10^{-16} m²). No correlation kT vs 28 days cube strength. Good correlation kT vs cores' sorptivity. Carbonation and chlorides generally higher for higher kT.
26	Van Eijk (2009)	kT, WP, ρ_s	NL	Experiment described in Section 6.2.1.5
27	Teruzzi (2009)	kT, SC	CH	55 parallel measurements of kT on site and carbonation depth on a building in Canton Ticino, Switzerland. Model to predict carbonation rate from kT.
28	Jornet et al. (2011)	kT, a_0, M	CH	Performance comparison of concretes made with limestone filler cement and OPC; $w/c = 0.40$–0.60. Different curing conditions applied. Durability tests and petrographic analysis.

(Continued)

Table 8.2 (Continued) Details on data sets used in this chapter and their sources

Data set	*Source*	*Applied tests*	*Country code*	*Brief description*
29	Zhutowsky and Kovler (2012)	kT, Q, a_o	IL	Comparative effect of internal curing by inclusion of pre-saturated lightweight aggregates on mechanical and durability performance of HPC, w/c between 0.21 and 0.33.
30	Imamoto et al. (2012)	kT, K_c	JP	One-hundred and eleven parallel ND measurements of kT and cover thickness d on Tokyo's Museum of Western Art. Few small cores to establish regression kT vs SC. Model to predict service life from $kT \rightarrow K_c$ and d
31	Neves (2012)	kT, K_c	PT	Comprehensive site investigation of relation between kT and K_c.
32	Starck (2013), Starck et al. (2017); Section 6.2.1.4	kT, k_{OPI}	ZA	Cubes made with six different concretes, exposed to three different conditions, were tested for kT and, later for k_{OPI} on cores drilled and cut from the cubes and oven-dried at 50°C for 7 days.
33	Imamoto et al. (2014)	kT, K_c, K'_c	JP	Reports parallel measurements of carbonation and kT on old structures and on new concretes measured in the laboratory (accelerated carbonation).
34	PC (2014)	kT, Q, ρ_s	PA	Values measured at 28, 56 and 90 days during quality control of concretes for the Panama Canal, Pacific side.
35				Ibid, Atlantic Side
36	Imamoto et al. (2016)	kT, K_c	JP, PT, CH	Compilation of parallel site tests of kT and carbonation measured on 14 concrete structures in Japan, Portugal and Switzerland.
37	Maître (2012), Jacobs (2006), Bisschop et al. (2016)	kT, S_{Cl}	CH	Investigation fully described in Section 11.2.2.
38	Moro and Torrent (2016)	kT, kO, a_{24}, Q, M, OD, WP	CH	Data from investigation described in Section 6.3.2. Both kT and other properties measured on drilled + cut specimens. Values obtained after 28 days (some also at 91 days) of moist curing are reported here.
39	Beglarigale et al. (2014)	kT, a_{24}, Q, ρ_s	TR	Study on the effect of replacing 15% and 30% of OPC by high Ca fly ash on various transport properties of concrete; the reported W_p data are the mean values.
40	Bahurudeen and Santhanam (2014)	kT, k_{OPI}, Q, CCI, WP	IN	Study on the effect of replacing 10%, 15% and 20% of OPC by sugar cane bagasse ash on various transport properties of concrete.

(Continued)

Table 8.2 (Continued) Details on data sets used in this chapter and their sources

Data set	*Source*	*Applied tests*	*Country code*	*Brief description*
41	Neves et al. (2015)	kT, K'_c	PT	Parallel measurements of kT and accelerated carbonation (20°C/65% RH, 5% CO_2) of concretes (w/c = 0.39–0.70) made with OPC, LF cement and PFA cement.
42	Nishimura et al. (2015)	kT, K'_c	JP	Relation between kT and accelerated carbonation rate (20°C/60% RH, 5% CO_2) of concretes with w/b = 0.40, 0.50 and 0.60, made with OPC and with binders containing 30%, 45% and 60% of blastfurnace slag, subjected to three different curing conditions. Other factors were also studied.
43	Wang et al. (2014)	kT, ρ_s	CN	Site measurements of cover thickness, kT and Wenner electrical resistivity in precast segments for submerged tunnel of Hong Kong–Zhuhai–Macao link.
44	Kurashige and Hironaga (2010)	kT, K'_c	JP	Effect of several curing conditions on kT and accelerated carbonation (20°C/60% RH, 5% CO_2) of OPC concrete with w/c = 0.50
45	Ebensperger and Olivares (2019)	kT, Q, M, WP, ρ_s	CL	Effect of w/c ratio (0.40; 0.50; 0.60; 0.70), curing and age of concretes made with High Strength Pozzolanic Cement on transport properties. The data included in the charts correspond to concretes moist cured 28 days, later stored in a dry room (23°C/50% RH) and tested at 28 and 91 days of age (taken from table in p. 14). In the case of kT, the specimens were oven-dried 3 days at 50°C prior to testing.
46	(Park et al., 2004)	kT, D_b	KR	Effect of binder type (plain OPC and with addition of PFA, GGBFS and SF) on w/b=0.55 concretes tested for kT and chloride-diffusion D_l (immersion in 3.6% NaCl solution). Specimens cured 28 days in water (kT cubes dried 48 h at 60% RH).
47	(Akiyama et al., 2010)	kT, K'_c	JP	kT and accelerated carbonation (20°C/60% RH, 5% CO_2) of 12 concretes mixed with various binders and w/c ratios, subjected to 3 curing conditions.

orders of magnitude. Up to a kT value of ≈2×10^{-16} m² the kT values are smaller than the kO values, situation that is reversed above that value. The reasons for this discrepancy are varied, among them:

- the different preconditioning of some samples: the moisture content of the kT slabs, especially for low-permeability concretes, was higher than for the kO drilled and oven-dried discs
- the different volumes explored by the tests: Ø150 × 50 mm for kO and Ø50 × L mm for kT (L is the penetration depth of the test which, for low-permeability concretes may reach only few mm)

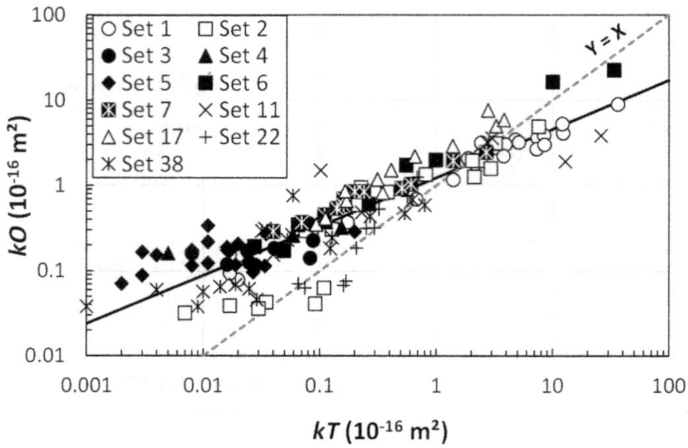

Figure 8.1 Relation between Cembureau oxygen-permeability *kO* and *kT*.

- the "wall-effect"; if *kO* is measured on cast discs or discs cut of longer cylinders, the surface layers close to the curved walls are richer in paste and may conduct more gas flow than the core. This effect is absent in the *kT* tests
- steady-state conditions for *kO* against non steady-state for *kT* leading to simplifications in the model under which *kT* is calculated
- the Klinkenberg effect (Section 3.6). *kO* is measured under positive relative pressures, whilst *kT* under negative relative pressures
- the default value of 0.15 for the porosity ε of the concrete. Concretes of low permeability have also lower porosities than 0.15 (see Figure 5.8) which, according to the correction in Eq. (5.43), would yield higher *kT* values than those reported. Similarly, concretes of high permeability would yield lower *kT* values, making the correlation line rotate towards the "Equality" line

8.3.1.2 South-African OPI

A very comprehensive research purposely looking for a possible correlation between k_{OPI} and *kT* was conducted at the University of Cape Town (Beushausen et al., 2012; Starck, 2013; Starck et al., 2017), data set 32 in Table 8.2, investigation described in detail in Section 6.2.1.4.

Figure 8.2 presents Starck's data as black dots, which yield an excellent correlation ($R = 0.97$) for the regression line shown in Figure 8.2:

$$k_{OPI} = 3.1\ kT^{0.31} \quad k_{OPI}\left(10^{-10}\ \text{m/s}\right);\ kT\left(10^{-16}\ \text{m}^2\right) \tag{8.2}$$

Data sets 17, 22 and 40 (Table 8.2) show a different trend (white symbols in Figure 8.2). The reason for this discrepancy is attributed (Starck, 2013)

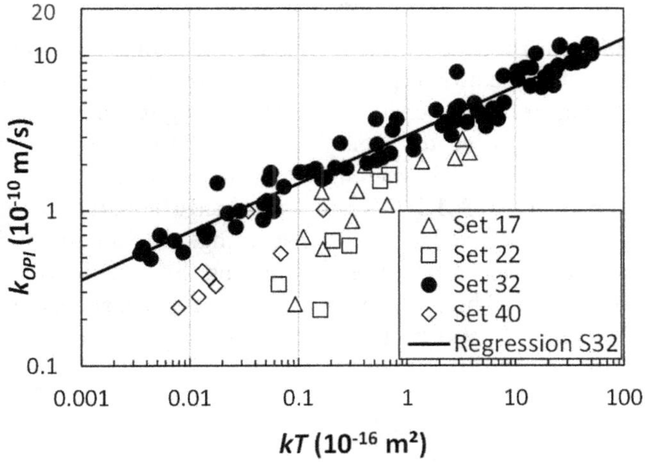

Figure 8.2 Correlation between k_{OPI} (South African oxygen-permeability index) and *kT*.

to the different experimental conditions under which data sets 17 and 22 were obtained.

8.3.1.3 Figg Air and TUD Permeability

Figg air-permeability test method was described in Section 4.3.2.1; although not standardized, the method has achieved certain acceptance. Figure 8.3a shows an excellent (negative) correlation found between Figg's air time (required to raise the relative pressure from −0.45 to −0.35 bar in an evacuated hole) and *kT*, based on test results from data sets 1 and 2 (upper face and bottom face of slabs as cast). TUD method, described in Section 4.3.2.4, works on similar principles as Figg. Figure 8.3b shows the excellent correlation between TUD time (required for a decay of the pressure in a pressurized hole from 11.0 to 10.5 bar) and *kT* for sets 1 and 2.

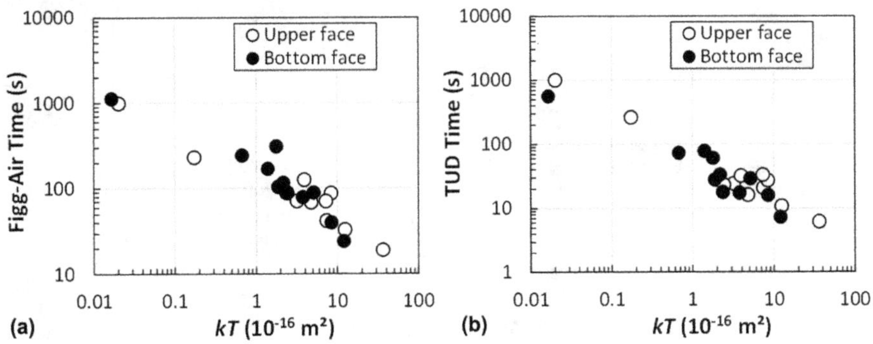

Figure 8.3 Correlation between Figg air (a) and TUD (b) test time and *kT*.

8.3.2 Oxygen-Diffusivity

Parallel results of the coefficient of oxygen-diffusion D_O and kT were obtained in the investigation reported by Moro and Torrent (2015), data set 38, plotted in Figure 8.4. The values corresponding to nine different binder types are coded according to Table 5.4. The kT values were measured at Holcim Technology laboratory whilst D_O was measured at EMPA, by the test method described in A.1.1.1. Results obtained after 28 days and 1 year of moist curing are plotted indistinctly, constituting $N = 36$ pairs of data (measurements were conducted after preconditioning the specimens at 50°C).

Despite the variety of binders and ages, a very good correlation ($R = 0.89$) was found between D_O and kT for the regression:

$$D_O = 3.46\ kT^{0.55} \quad D_O\ (10^{-8}\ \text{m}^2/\text{s});\ kT\ \left(10^{-16}\ \text{m}^2\right) \tag{8.3}$$

It is interesting to note the closeness of the exponent in Eq. (8.3) to the theoretical one (0.50), as per Eq. (3.62).

8.3.3 Capillary Suction

8.3.3.1 Coefficient of Water Absorption at 24 Hours

Figure 8.5 presents parallel test results of the coefficient of water absorption at 24 hours a_{24} and kT, for ten data sets, as described in Table 8.2.

Considering the disparity of the sources and test procedures, a reasonably good correlation ($R = 0.73$) was obtained for the $N = 133$ pairs of data, according to the regression:

$$a_{24} = 12 + 1.6.\ln kT \quad a_{24}\left(\text{g/m}^2/\text{s}^{½}\right);\ kT\ \left(10^{-16}\ \text{m}^2\right) \tag{8.4}$$

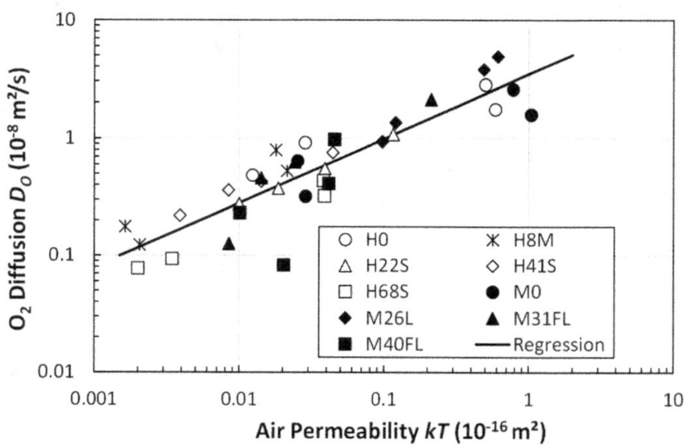

Figure 8.4 Correlation between oxygen-diffusivity D_O and kT.

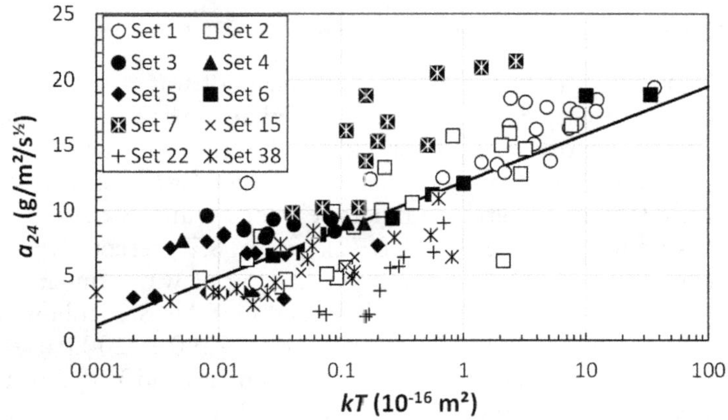

Figure 8.5 Correlation between water sorptivity a_{24} and *kT*.

8.3.3.2 Figg Water

Similar to Section 8.3.1.3, parallel results of *kT* and Figg test, this time Figg-Water test (Section 4.2.2.3), are plotted in Figure 8.6a. The result of the Figg-Water test is the time required for the concrete to absorb 0.01 mL of water. A good (negative) correlation is observed between the results of both tests.

8.3.3.3 Karsten Tube

The Karsten tube test (Section 4.2.2.2) was also applied to some of the specimens reported in Sections 8.3.1.3 and 8.3.3.2, as shown in Figure 8.6b. A good correlation between the rate of water absorption by Karsten and *kT* exists which, incidentally, fits well to the regression of Eq. (8.4), also plotted in Figure 8.6b.

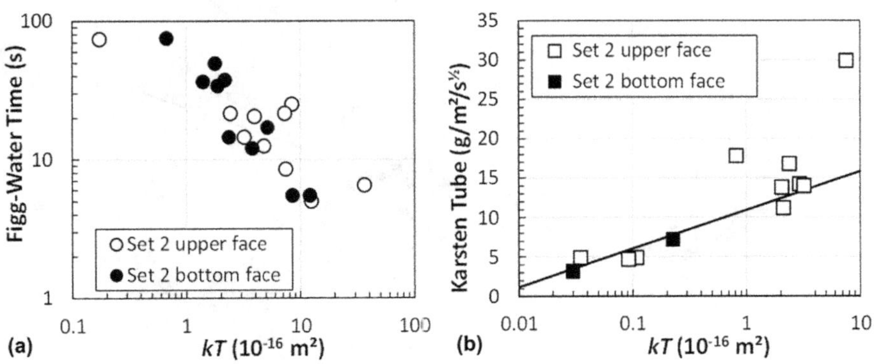

Figure 8.6 Correlation of *kT* with (a) Figg time to absorb 0.01 mL of water and (b) Karsten tube absorption rate.

8.3.4 Water-Permeability and Penetration under Pressure

A very good correlation between the coefficients of water-permeability K_w and of air-permeability kT was reported by Sakai et al. (2013), as shown in Figure 3.12. The regression equation indicated in the figure or Eq. (2.1) can be used to convert kT experimental data into equivalent K_w estimated values.

More abundant are experimental data relating kT and Water Penetration under Pressure tests (see Section 4.1.1.2). Figure 8.7 presents $N = 96$ parallel measurements of the maximum penetration of water under pressure WP_{max} and kT, from eight data sets described in Table 8.2, following two similar test methods: DIN 1048 (1978) and EN 12390-8 (2009).

Despite the huge diversity of laboratories, countries and WP_{max} test methods involved, a quite good correlation ($R = 0.73$) is obtained according to the regression line also shown in Figure 8.7:

$$WP_{max} = 65 + 8.82.\ln kT \quad WP_{max}(\text{mm}); kT\left(10^{-16}\,\text{m}^2\right) \tag{8.5}$$

The dotted-lined boxes shown in Figure 8.7 correspond to equivalent permeability classes, associated with both test methods (Tables 4.1 and 5.2). The fact that the regression line crosses the boxes near their intersection points indicates that both test methods tend to judge the quality of the *Covercrete* quite coherently.

8.3.5 Migration

In this section, the relation between the output of several migration tests and kT is presented. It has to be borne in mind that kT depends exclusively on the pore structure, whilst the results of migration tests also depend

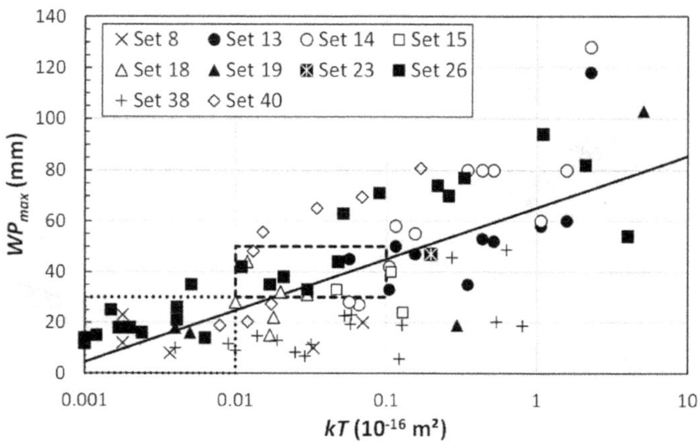

Figure 8.7 Correlation between maximum water penetration under pressure WP_{max} and kT.

strongly on the ionic composition of the pore solution, with or without the incorporation of foreign ions from outside (typically Cl^- ions)., Section 3.4.2

8.3.5.1 Rapid Chloride Permeability Test ("RCPT" ASTM C1202)

Figure 8.8 shows $N = 114$ parallel results of the current passed Q (coulomb) in the "RCPT" ASTM C1202 and kT, reported in 14 data sets of Table 8.2.

A good correlation ($R = 0.77$) exists between both variables according to the regression (full line in Figure 8.8):

$$Q = 5{,}117\,kT^{0.38} \quad Q(\text{Coulomb});\; kT\left(10^{-16}\ \text{m}^2\right) \tag{8.6}$$

The dotted line boxes in Figure 8.8 represent areas where both test methods judge the permeability of the concrete to chlorides (Q) and to air (kT) as very low (VL), low (L), moderate (M) and high (H). The fact that the regression line intersects the boxes almost exactly at the crossing points indicates that both test methods tend to judge the quality of the *Covercrete* quite coherently. The large scatter observed is to be expected, since kT measures just the "openness" of the pore structure whilst Q (same as other migration tests) is also strongly affected by the ionic composition (particularly by the content of OH^- ions) of the pore solution (Andrade, 1993; Shi, 2003).

8.3.5.2 Coefficient of Chloride Migration (NT Build 492)

This migration test, that can be considered as an improvement over ASTM C1202 "RCPT", is the Chloride Migration Test, developed by Tang and Nilsson (1992), standardized in Scandinavia (NT Build 492), in Switzerland

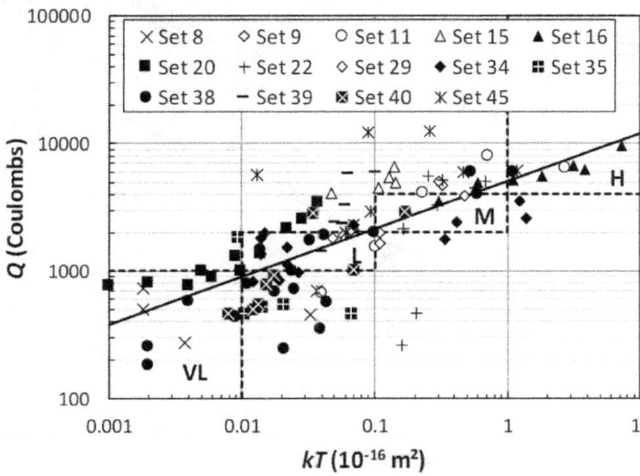

Figure 8.8 Correlation between electric charge Q passed in ASTM C1202 "RCPT" test and kT.

(SIA 262-1, Annex B) and lately in Europe (EN 12390-18). The test method is described in Annex A.2.1.2 giving as test result a coefficient of chloride migration M which, in theory (Section 3.4.2), should be equal to the coefficient of chloride-diffusion D.

Figure 8.9 presents $N = 60$ parallel results of M and kT reported in data sets 22, 28, 45 and 38, all, but in particular the latter, covering a wide range of binder types (Section 6.3.2).

A very good correlation ($R = 0.83$) exists between both variables, according to the regression (plotted in Figure 8.9 as full line):

$$M = 40.5\ kT^{0.58} \quad M\left(10^{-12}\,\mathrm{m^2/s}\right); kT\ \left(10^{-16}\,\mathrm{m^2}\right) \tag{8.7}$$

Just as a reference, a dotted line is added in Figure 8.9, corresponding to the relation between the coefficient of diffusion D and kT, established completely independently in Section 8.3.6.1, Eq. (8.9). It can be seen that the values of M, predicted from Eq. (8.7), are 2–3 times higher than the D values predicted by Eq. (8.9), for the same kT, except for concretes of very low permeability ($kT < 0.01 \times 10^{-16}$ m^2). This is in agreement with the findings of Li et al. (2015), who recommend design values of M that are twice those corresponding to D, confirmed experimentally by Ren et al. (2021). Again, the exponent in Eq. (8.7) is close to the theoretical 0.50.

8.3.5.3 Electrical Resistivity (Wenner Method)

There are many parallel results of electrical resistivity (Wenner method) ρ and kT, particularly from early investigations (data sets 1–7), when ρ was used as estimator of the moisture content of the concrete in order to compensate kT results for its effect (see Section 5.7.2). However, these results

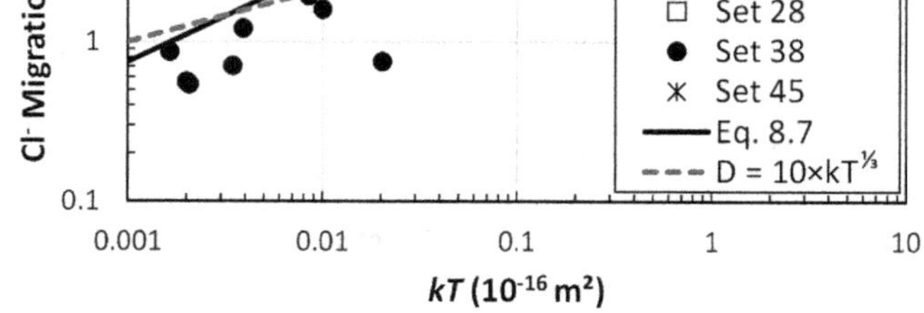

Figure 8.9 Correlation between chloride migration M (Tang-Nilsson test) and kT.

are misleading in establishing a relation between both variables because, in order for ρ to act as durability indicator, it should be applied on saturated specimens (opposite to *kT*, applicable on rather dry concrete).

Another factor worth considering when comparing data of ρ and *kT* is that the former is strongly influenced by the composition of the electrolyte (pore solution), as discussed in Section 3.4.2, whilst *kT*, being governed by the pore structure of the concrete, is not.

Figure 8.10 shows *kT* results from data set 26 reported by Van Eijk (2009), measured on panels as described in Section 6.2.1.5 and of ρ, measured on cubes cured under water for 14 days and then stored at 65% RH until the age of test (56 days). If ρ could be measured also in the panels, the results are not reported. The panels and cubes were cast with mixes of $w/c = 0.40$ and 0.57 made with two different cements: an OPC (CEM I) and a GBFS cement (CEM III).

It can be seen that *kT* judges the mixes in decreasing order of quality (from left to right in Figure 8.10: 1, 3, 4, 2) whilst ρ in decreasing order of quality 3, 4, 1, 2; i.e. *kT* gives precedence to the *w/c* ratio and ρ to the cement type.

In Van Eijk (2009), it is stated (translation from Dutch)

> "It is known that measurement results of the Wenner Probe are sensitive to the type of cement: Portland cement or GGBFS cement. The absolute results must therefore always be compared with values measured with the same type of cement. Measurement values for GGBFS cement are not directly comparable with those for Portland cement."

8.3.5.4 South African Chloride Conductivity Index

The reciprocal of ρ is the electrical conductivity, typically expressed in Siemens ($1/\Omega$) per m. Some parallel data exist between the South African Chloride Conductivity Index *CCI* (described in Section A.2.2.3) and *kT*,

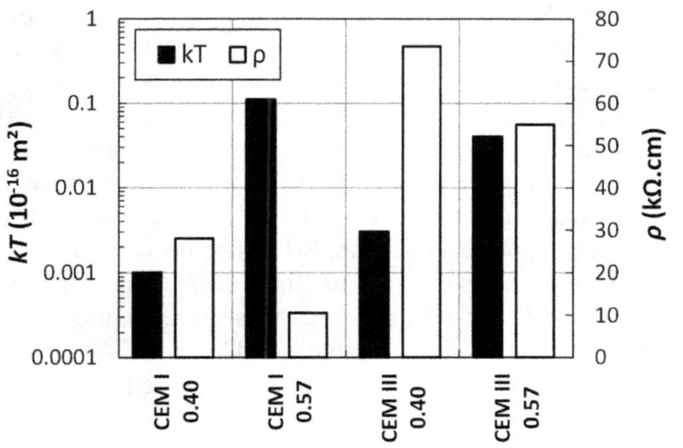

Figure 8.10 Values of *kT* and ρ for the four mixes tested; data from Van Eijk (2009).

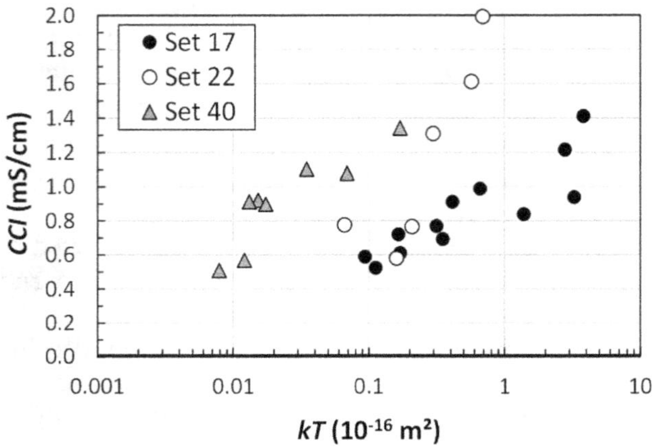

Figure 8.11 Correlation between South African Chloride Conductivity Index *CCI* and *kT*.

data sets 17, 22 and 40. They are plotted in Figure 8.11, showing that, for each data set, *CCI* increases with *kT*, as expected, but following different trends for the three data sets. This may be due to the different concretes (binders) tested and/or to different testing conditions.

8.3.6 Chloride-Diffusion

8.3.6.1 Laboratory Diffusion Tests

Few Relatively few results have been found in the literature, reporting parallel measurements of the coefficient of chloride-diffusion *D* and of air-permeability *kT*. This may be due, in part, to the duration (months) of the existing standardized tests to measure *D*, be it by immersion (ASTM C1556) or by ponding (AASHTO T259), coupled to the high costs associated with performing a chloride profile analysis of the cores. Both test methods are described in Section A.1.2. Nowadays, the *D* tests are being gradually replaced by migration tests, ASTM C1202 or NT Build 492, although the former requires a calibration with long-term immersion tests for validation.

The parallel experimental results of *D* and *kT* found in the literature correspond to data sets 2 and 6 ($N = 14$), set 15 ($N = 3$) and set 46 ($N = 34$) and are plotted as black symbols in Figure 8.12. The broken line in Figure 8.12 plots the values reported as data set 10 which correspond, strictly speaking, to relations between *D* and *kT* proposed in a service life model developed at EPFL (Roelfstra et al., 1999), see Table 9.4. The model derives values of *D* (and of the water diffusion coefficient as well) on the basis of *kT* measurements and is described in Section 9.4.1.

In a very comprehensive research, conducted in the USA, (Olek et al. (2002) established a correlation between the coefficient of chloride-diffusion

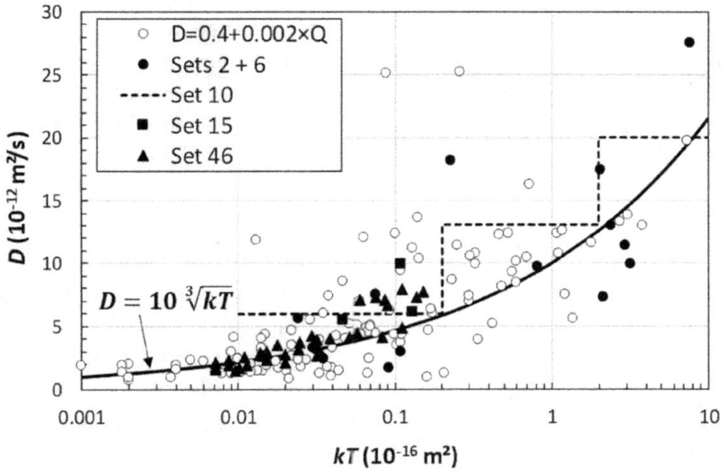

Figure 8.12 Correlation between coefficient of chloride-diffusion *D* and *kT*.

D (AASHTO T259) and the charge *Q* passed in ASTM C1202 "RCPT" of the form:

$$D = 0.4 + 0.002Q \quad D\left(10^{-12}\,\text{m}^2/\text{s}\right); Q\ (\text{Coulombs}) \tag{8.8}$$

A validation of Eq. (8.8) was made using data of *D* (ASTM C1556) and *Q* (ASTM C1202) independently obtained by Alexander and Thomas (2015). The agreement is very good, especially within the usual range of *Q* values (100–10,000 Coulomb).

This allows, by applying Eq. (8.8), converting all the *Q* values in Figure 8.8 into *D* values; the resulting values are plotted as white circles in Figure 8.12.

It can be seen that the converted values (white circles) merge quite well with the experimental results (black symbols) constituting a total sample of $N = 171$ pairs of data. A general relation (shown as full line in Figure 8.12) can be established between the coefficient of chloride-diffusion *D* and of air-permeability *kT* of the form:

$$D = 10\sqrt[3]{kT} \quad D\left(10^{-12}\,\text{m}^2/\text{s}\right); kT\left(10^{-16}\,\text{m}^2\right) \tag{8.9}$$

The standard error for the estimation of *D* is $\approx 3.5 \times 10^{-12}$ m²/s, with a determination coefficient $R = 0.80$. Figure 8.12 shows that the scatter in the area of interest for durability design ($kT < 0.1 \times 10^{-16}$ m²) is smaller than for the whole set of data, especially when considering the results involving direct measurements of *D* (black symbols). An independent regression obtained by Park and Kim (2000) on 45 parallel measurements of *D* (immersion) and kT, for OPC concretes, yielded almost the same exponent (0.328) as in Eq. (8.9) but with a factor of 19.3 instead of 10.

Figure 8.9 also confirms that the relation between D and kT, expressed by Eq. (8.9), is supported by the migration M test results (known to be higher than D). The relationship expressed by Eq. (8.9) is used in the "Exp-Ref" Model for service life assessment for chloride-induced steel corrosion, presented in Section 9.4.3.1.

8.3.6.2 Site Chloride Ingress in Old Structures

Parallel measurements of chloride ingress and air-permeability kT, obtained on site, are scarce. Figure 8.13 shows results from data set 25, plotting the chloride content at the level of the reinforcement and kT, both measured on bridges about 30-years old located along a Swiss Motorway. It can be seen that for kT values below 0.1×10^{-16} m², the chloride content has not yet reached the critical level of 0.6% of the cement weight (sometimes a more conservative value of 0.4% is adopted) expected to initiate corrosion. But a bridge should last at least 75 years, so one of the six bridges with $kT < 0.1 \times 10^{-16}$ m² is not on the safe side.

Other results showing the relation between chloride ingress rate and kT can be found in Section 11.2.2 (especially Figure 11.8b).

8.3.7 Carbonation

8.3.7.1 Laboratory Tests (Natural Carbonation)

Data sets 2, 7, 11, 16 and 24 report parallel data of kT and the carbonation rate K_c measured after at least 1 year of exposure to a dry room, where

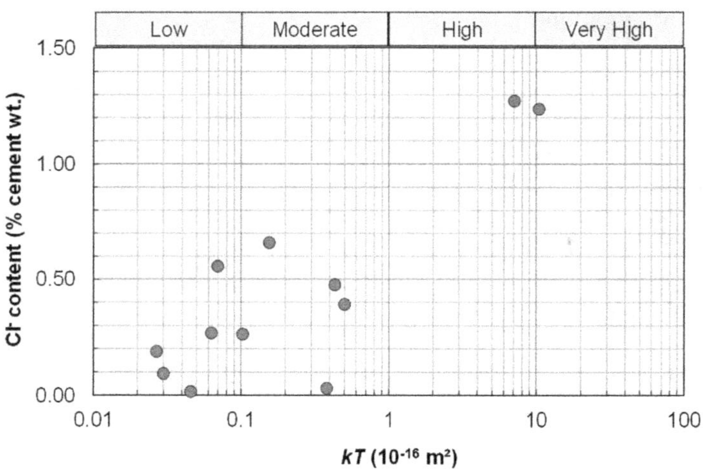

Figure 8.13 Relation between chloride content at steel bar level and kT for 30 years old Swiss bridges; data from Jacobs (2008).

carbonation proceeds at maximum rates in natural air. The concretes tested were made with OPC, GBFSC, FAC, SFC and the storage conditions and duration varied. For sets 2 and 7 it was 20°C/50% RH during 210 days and 2 years storage, respectively. For set 16, it was 90 days at 30°C/40% RH and for set 24 specimens were stored 2 years at 20°C/60% RH and thereafter 2 years indoors in the laboratory. For Set 11, it was 1 year under undisclosed conditions. In all cases, *kT* was measured before the specimens were stored in the dry rooms and K_c was calculated as the carbonation depth (measured by phenolphthalein method) divided by the square root of the exposure time; the results are shown in Figure 8.14.

Despite the different storage conditions and binder types, there is a good agreement among all sets (perhaps less for Set 11) and a good overall correlation ($R = 0.82$) between the variables, according to the regression (shown in Figure 8.14):

$$K_c = 1.67 \ln\left(\frac{kT}{0.006}\right) \quad K_c\left(\text{mm/y}^{1/2}\right);\, kT\left(10^{-16}\,\text{m}^2\right);$$

$$\text{valid for } kT \geq 0.006 \times 10^{-16}\,\text{m}^2 \tag{8.10}$$

The results in Figure 8.14 show that for *kT* below 0.006×10^{-16} m^2, the carbonation rate becomes negligible. The relationship expressed by Eq. (8.10) is used in the "Exp-Ref" Model for service life assessment for carbonation-induced steel corrosion, presented in Section 9.4.3.2.

8.3.7.2 Laboratory Tests (Accelerated Carbonation)

Data sets 2, 7, 16, 41, 42, 44 and 47 report parallel data of *kT* and accelerated carbonation tests. In the latter, concrete specimens are exposed to environments of controlled temperature and (typically low) relative humidity,

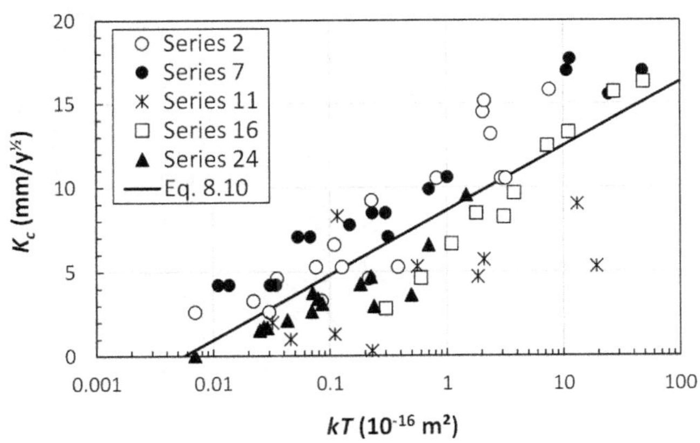

Figure 8.14 Correlation carbonation rate K_c vs *kT* for natural carbonation under laboratory conditions.

enriched in CO_2 concentration. The natural concentration of CO_2 in air is ≈ 0.04%; in some tests (Sets 16, 41, 42, 44 and 47) the specimens were stored, during different periods, in a chamber the CO_2 concentration of which was kept at 5%, whilst in others (Sets 2 and 7) the specimens were kept, for different periods, at a CO_2 concentration of 90% and 100%, respectively. It is worth mentioning that some standardized test methods for accelerated carbonation specify a 3% CO_2 environment (ISO 1920-12, 2015) or a 4% CO_2 environment (Annex I of SIA 262/1 (2019)).

To analyze the accelerated carbonation test results obtained under such disparate conditions (time and CO_2 concentration), Eq. (8.11) will be used, taken from Annex I of SIA 262/1 (2019).

$$K_c^{'} = 1.36 \cdot \sqrt{\frac{0.04}{CO_2}} \cdot K_a \tag{8.11}$$

where

$K_c^{'}$ = equivalent natural carbonation rate

CO_2 = CO_2 concentration (%) in the test chamber

K_a = accelerated carbonation rate, equal to the measured carbonation depth divided by the square root of the time of permanence in the accelerated test chamber

Figure 8.15a shows the results of K'_c, calculated with Eq. (8.11), as function of the kT values measured before introducing the specimens in the accelerated carbonation test chamber; the regression line corresponding to Eq. (8.12) is also plotted (full line), with a coefficient of correlation (R = 0.75), lower than for the natural carbonation tests, Eq. (8.10). This can be due to the widely different test methods, especially of the CO_2 concentrations in the test chambers, but also because different cement types were tested.

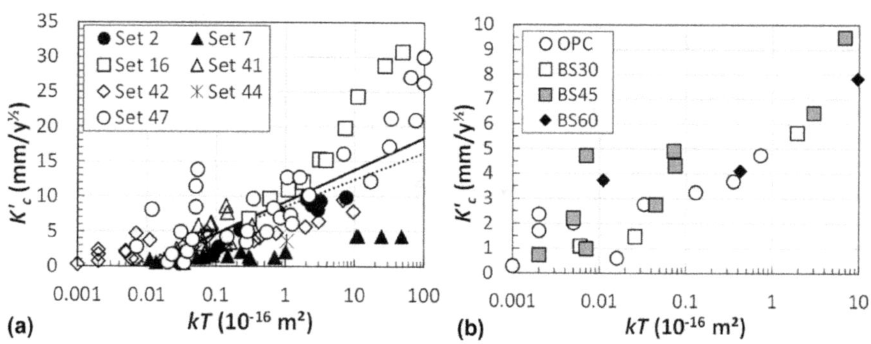

Figure 8.15 Correlation natural carbonation rate $K_c^{'}$ vs kT from accelerated carbonation tests; (a) data from several sources; (b) data from Nishimura et al. (2015).

$$K_c' = 1.99 \cdot \ln\left(\frac{kT}{0.0096}\right) \qquad K_c'\left(mm/y^{1/2}\right); kT\left(10^{-16}\,\mathrm{m}^2\right);$$

$$\text{valid for } kT \geq 0.0096 \times 10^{-16}\,\mathrm{m}^2 \tag{8.12}$$

In Figure 8.15, the dotted line corresponds to Eq. (8.10); the similarity between Eqs. (8.12) and (8.10) is remarkable.

Figure 8.15b presents the results of data set 42, covering four orders of magnitude of kT values, obtained through changes in *w/b* ratio, in cement type (binders containing 0%, 30%, 45% and 60% of blast-furnace slag) and in curing conditions. A tenuous trend, not clearly defined, of higher values of K_c' for the same kT can be detected for slag-containing binders. The research corresponding to Data Set 47 (Akiyama et al., 2010) showed a strong influence of the curing conditions (moist, sealed, wind) on the relation between K'_c and kT.

8.3.7.3 Site Carbonation in Old Structures

This topic will be dealt with in more detail in Section 9.5. Here suffice to say that a large number of parallel tests of kT and carbonation depth CD were measured (data set 36) on several old structures located in Japan, Portugal and Switzerland. After measuring kT, cores were drilled at the same spots to measure CD by the phenolphthalein method. The carbonation rate is computed as $CR = CD/\text{age}^{1/2}$. The results obtained in Switzerland, Japan and Portugal are shown in Figure 8.16.

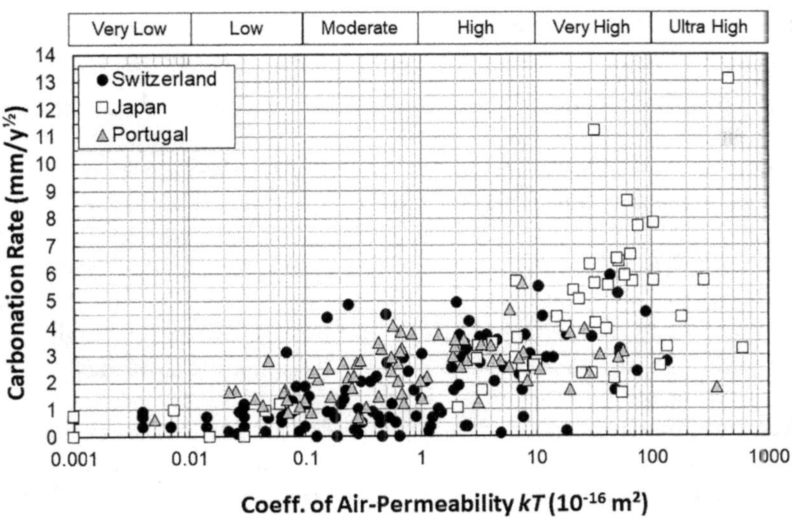

Figure 8.16 Relation carbonation rate vs. *kT* in old Swiss, Japanese and Portuguese structures.

In some of the structures, the values of kT measured on a single structure span 5 and 4 Permeability Classes (Imamoto et al., 2016). This might be attributed to the test method; however, the carbonation rates also cover a wide range, from nearly 0 up to values exceeding 5 mm/y^{½}. That means that the high variability in kT and CR are predominantly a consequence of the heterogeneity of the material and micro-exposure conditions, resulting from the service loads and weathering impact to which the structures have been subjected, see also Section 11.4.2.

Figure 8.16 contains a compilation of the results obtained in the three countries, showing a large degree of consistency, despite the different climates, materials and construction practices prevailing in them. Good use is made of this consistency of data to develop a method to estimate carbonation-induced corrosion initiation time in old structures, based on kT (Section 9.5.2). Please notice that same as for laboratory tests (Figure 8.14), the carbonation rate becomes negligible for kT values below $\approx 0.006 \times 10^{-16}$ m^2.

A research studied the protective effect of coating materials for textured finishes on the carbonation of concrete (Karasawa & Matsuda, 2011; Karasawa et al., 2011). Cores were drilled from a structure several decades old, on the surface of which coating materials for textured finishes had been applied. Both air-permeability kT and accelerated carbonation tests were applied on the cores, and carbonation depth was estimated from the kT values applying a model. A comparison showed that the results of the predictive model, with due consideration given to the ageing of the coating, agreed well with the measured values.

8.3.8 Frost Resistance

Results of scaling frost-thaw-salts tests of concretes, the air-permeability kT of which had been previously measured, are presented and discussed. The frost test (old Swiss Standard SIA 162/1:2003, Test No. 9) consists in ponding the investigated surface of the concrete sample with a 3% NaCl solution and subjecting it to 30 frost (–12°C) and thaw (+20°C) cycles, collecting the loose material after every ten cycles (which take 7 days). The result of the test is the total loose mass δm_{30} collected after 30 cycles, referred to the ponded surface area of the specimen. The standard gives an indication that concretes with $\delta m_{30} \leq 600$ g/m^2 have a "high" frost-thawing salt resistance, whilst those with $\delta m_{30} \geq 3{,}600$ g/m^2 have a "low" frost-thawing salt resistance.

Another frost resistance criterion included in old Swiss Standard SIA 162/1:2003, Test No. 6, was the determination of the spacing factor (AF), which is the maximum distance from any point in the cement paste to the nearest air bubble, obtained by optical microscopy. The standard gives an indication that concretes with $AF \leq 200$ µm have a "high" frost resistance, whilst those with $AF \geq 250$ µm have a "low" frost resistance.

Since the results of these investigations (data sets 2 and 6) were not published before, the details of the mixes used as well as the results obtained on the hardened concrete samples are presented in Table 8.3, including *kT* and standard cube strength. The investigation of data set 2 consisted in casting 360 × 250 × 120 mm slabs with ten concrete mixes, testing the upper 360 × 250 mm surface, as cast, of samples kept permanently in a dry room at 20°C/50% RH (samples Ao) and the opposite bottom surface of slabs moist cured during 7 days prior to storage in same dry room (Samples Bu). In both cases, the samples were demoulded at 24 hours and the test was initiated at 28 days of age.

The investigation of data set 6 consisted in casting, on site, 1 m cubes, with the same concretes used to build Schaffhausen Bridge (Switzerland), see Section 7.1.5.2. Cube 2 was cast with the mix used for the Pylon (made with a Silica Fume cement type CEM II/A-D 52.5) and Cube 4 with the mix used for the Deck (made with an OPC type CEM I 42.5), see Table 8.3. The cubes were demoulded and kept near the bridge until 25 days of age when they were moved to Holcim Laboratory for testing at ages between 28 and 35 days. Measurements of frost-thaw salts resistance were made on Ø150 mm cores drilled from two sides of each cube (2–1 and 2–4 for Cube 2 and 4–1 and 4–4 for Cube 4). The recorded data of *kT*, *AF* and δm_{30} are reported in Table 8.3.

Table 8.3 Characteristics of the mixes and properties of hardened concrete (data sets 2 and 6)

Mix no.	Cement (kg/m³)	Silica fume (kg/m³)	w/b (kg/kg)	Air[a] (%)	Aggr. type	$f'c_{cube}$ 28 days (MPa)	**kT** (10^{-16} m²) Ao	Bu	δm_{30} (g/m²) Ao	Bu	**AF** (µm) Ao	Bu
Data set 2, p. 70, Tables 3.2-IV (a)–(c) (Torrent & Ebensperger, 1993)												
1	250	-	0.60	4.7	Z1	25.4	7.61	0.227	1,120	85	80	70
2	325	-	0.46	4.8		43.0	2.39	0.127	819	-	100	-
3	400	-	0.42	4.5		42.9	2.94	0.085	672	-	99	-
4	325	26	0.39	*2.0*		74.2	0.109	0.007	34	-	300	-
5	400	32	0.32	*1.8*		79.7	0.035	0.030	14	35	245	222
6	325	-	0.47	6.1	Z2	38.0	2.05	0.214	872	-	106	-
7	325	-	0.47	5.7	Z3	37.4	3.21	0.382	1,665	-	204	-
8	Commercial bagged repair mortar						0.092	0.017	6	-	324	-
9	320	-	0.41	9.5	Z4	35.2	2.12	0.076	138	-	87	-
10	325	-	0.50	*1.2*	Z1	48.2	0.818	0.022	4,086	-	150	-
Data set 6, p. 82 and 89 (Torrent & Frenzer, 1995)												
2–1	Ready-mixed concrete			2.2		79.6	0.003		27		275	
2–4	B55/45						0.019		15			
4–1	Ready-mixed concrete			3.8		51.1	0.004		29		144	
4–4	B45/35						0.200		7			

[a] Mixes 4, 5 and 10 (Air values in italics) did not contain an air-entraining agent.

Figure 8.17a presents the scaling mass loss δm_{30} as a function of *kT*, differentiating the samples containing air ≥ 4.5% (white symbols) and air ≤ 3.0% (black symbols). A clear trend of higher mass loss δm_{30} for higher *kT* can be observed, indicating that it is not enough to entrain air to achieve high frost resistance, but that the concretes shall have a low permeability as well. This is evident in the points linked by the arrow, corresponding to the same Mix No. 1, with similar *AF* (see Table 8.3), but one of high *kT* (upper face of dry-cured sample) and the other of low *kT* (bottom face of sample moist-cured 7 days).

Another piece of evidence of the effect of *kT* on frost resistance is produced by data set 7, corresponding to an experiment conducted at the Swiss Federal Polytechnic University in Zürich (ETHZ), within the frame of a project financed by the Swiss Federal Highway Administration (Torrent & Frenzer, 1995). In this case, another frost test (old Swiss Standard SIA 162/1:2003, Test No. 8) was applied, in which the number of frost-thaw salt cycles N_{50} leading to a 50% reduction in the modulus of elasticity of the sample is measured. If $N_{50} \geq 100$, the concrete is judged as having "high" frost resistance and if $N_{50} \leq 20$ as having "low" frost resistance. The test details are described below.

Concrete cubes (0.5 m) were cast with four different concrete mixes with a wide range of characteristics (w/c = 0.3–0.75; OPC = 200–450 kg/m³; $f'c$ = 14–66 MPa), made with the same constituents, see Table 5.3. Two cubes were cast with each mix, one of which was moist cured for 7 days (B), whilst the other was totally deprived of moist curing (A).

At the age of 28 days, the *kT* of the cubes was measured by Holcim Technology personnel, without knowing the identity of the eight cubes (blind test). The tests were conducted using a *TPT*, on two opposite faces of each cube, five tests on each face following a pattern like number 5 of a dice; the reported results in Table 5.3 correspond to the geometric mean of the ten tests conducted on each cube. After finishing the *kT* tests,

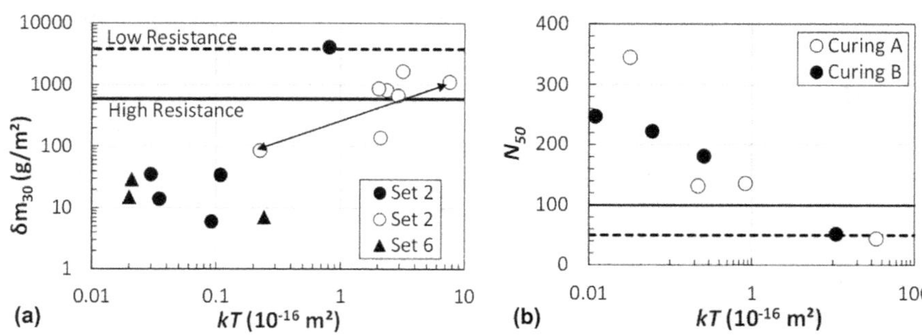

Figure 8.17 Relation between *kT* and (a) mass loss δm_{30} after 30 frost-thaw cycles scaling test and (b) number of frost-thaw-salts cycles for 50% reduction of *E*-modulus N_{50}.

3 Ø50 × 120 mm cores were drilled from each face tested for *kT* for the determination of N_{50}. The samples were kept under water at 20°C for 5 days, moment at which the initial static modulus of elasticity E_0 was measured. Then the samples were immersed in a NaCl (1,290 kg/m³ concentration) and subjected to repeated cycles of freezing (44 minutes at −20°C) and thawing (22 minutes at +20°C). Periodically, the samples were tested for static *E*-modulus, until its reduction (with respect to E_0) exceeded 50%. Then, by interpolation, the number of cycles N_{50} causing a 50% reduction in *E*-modulus was computed.

Figure 8.17b shows the very good correlation between N_{50} and *kT* (measured before freezing), confirming the beneficial effect of having a low-permeability concrete for achieving a high frost-thaw salts resistance.

It is worth mentioning a research (Choi et al., 2017) aimed at studying the damaging effect of early freezing on Ø100 × 200 mm mortar and concrete cylinders (OPC; *w/c* = 0.50). The cylinders were exposed to freezing temperature (−20°C) for 15 and 24 hours, at variable times after casting (Choi et al., 2017). In the mortar program (Series I), 15 hours freezing started 2 hours after casting and parts of the specimens were protected by an insulating material, leaving different exposed lengths to freezing (from 0 to 200 mm). In the concrete program, 24 hours freezing started 2, 6, 12, 24 and 48 hours after casting and the entire specimen length (200 mm) was exposed to freezing. Immediately after finishing the freezing process, the specimens were stored in a dry room (20°C; 65% RH). The moisture content and air-permeability *kT* of the frozen surface of the samples were monitored between 1 and 28 days of storage in the dry room.

Figure 8.18a shows the *kT* values measured on mortar specimens after 1 day in the dry room as function of the freezing exposed length of the specimens, indicating with shades the degree of damage observed visually. The broken line represents the value measured on a companion specimen not subjected to early freezing.

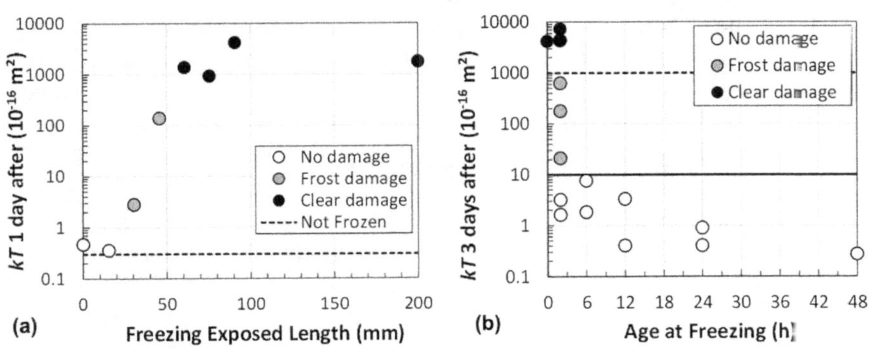

Figure 8.18 (a) Effect of exposed length and degree of damage on *kT* of mortars tested 1 day after freezing, and (b) effect of freezing age and degree of damage on *kT* tested 3 days after freezing; data from Choi et al. (2017).

The early frost damage induces a huge increase in *kT*; similar results were obtained for *kT* measured 28 days after freezing. This behaviour is explained by the fact, confirmed by MIP analysis, that freezing coarsens the pore structure, which reflects in an increased air-permeability *kT*.

Figure 8.18b shows the effect of the age at which both mortar and concrete specimens were subjected to freezing and of the subsequent observed damage on *kT*. The researchers recommend measuring *kT* after 3 days of freezing as a good indicator of the degree of damage caused by early freezing. They suggest two thresholds ($kT = 10$ and $1{,}000 \times 10^{-16}$ m^2) delimiting zones of no damage, damage and obvious damage by early freezing, indicated in Figure 8.18b.

Zhang et al. (2019) tested the frost-thaw-salts resistance of concrete mixes subjected to different curing conditions, with and without the inclusion of a controlled permeable formwork (CPF) liner (see Section 7.2.3); prior to initiating the frost tests, *kT* was measured. They found that extended curing and the presence of the CPF liner reduced the surface air-permeability, thus resulting in lower scaling mass loss.

It can be concluded that air-permeability *kT* is, as transport property, a useful indicator of the frost resistance of concrete, but by no means the only one. Liu et al. (2014) and Liu and Hansen (2015) present data with good correlations between the scaling mass loss in frost-thaw-salts tests and water sorptivity.

8.4 SOME NEGATIVE EXPERIENCES

So far, with the possible exception of the lack of correlation with electrical resistivity, positive cases of correlation between *kT* and many other transport and durability-related properties of concrete have been reported. As with all test methods, sometimes, abnormal measurement results happen, be them due to shortcomings of the test method itself, to wrong operation (e.g. lack of conditioning or inaccurate calibration of the instrument) or to extreme conditions of the concrete (surface temperature and moisture, coatings, microcracks, poorly bonded surface layers, etc.). In this section, reported cases where abnormal or out of expectations results were obtained are described.

8.4.1 Tunnel in Aargau, Switzerland

This case refers to 16 parallel measurements of *kT*, conducted on the walls of a Tunnel by F. Jacobs (using a *TPT* instrument) and by R. Torrent (using a *PermeaTORR* instrument). The measurements were performed exactly on the same spots with both instruments (after a delay of at least 30 minutes); the results were reported in Jacobs et al. (2009). The black squares in Figure 5.13 show the excellent correlation obtained between both sets of measurements, which covered three different Permeability Classes (orders of magnitude). On the initiative of Materials Advanced Services SRL, Ø50 mm cores were

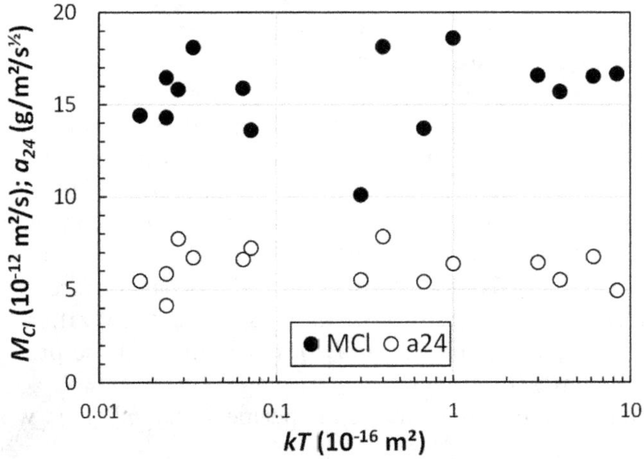

Figure 8.19 Relation between kT measured on site in a Tunnel and M_{Cl} and a_{24}, measured in the lab on cores drilled from same locations.

drilled from 15 of the locations where kT had been measured and saw-cut to a thickness of 50 mm. The specimens were tested for coefficient of water absorption at 24 hours (a_{24}) and for coefficient of chloride migration (M_{Cl}), according to Annexes A and B, respectively, of Swiss Standard SIA 262/1. It was expected that the significant differences in kT would be reflected in similar differences in both a_{24} and M_{Cl}; the results presented in Figure 8.19 indicate that that was not the case. Indeed, the results of both laboratory tests do not show any correlation with the kT values measured on site (the results obtained with the *PermeaTORR* are plotted along the x-axis).

It is worth mentioning that the migration test results are rather high, with 12×10^{-12} m²/s being the upper limit of the Swiss Standards for tests made on cores drilled from structural elements exposed to chlorides (just one out of the 15 tests fell below that limit).

Two possible explanations for this behaviour are: (1) kT is being affected by some very superficial effect (coating, delamination, etc.) that is not affecting the laboratory results (the penetration of chlorides in the migration test reached depths between 25 and 32 mm), and (2) there was some problems with the samples and/or the laboratory tests (the kT site tests are assumed to be correct, as they were confirmed by two sets of measurements performed by different operators and instrument brands).

8.4.2 Wotruba Church, Vienna, Austria

A condition assessment was made by Matea Ban (Institute of Conservation and Restoration, University of Applied Arts Vienna, Austria) on the iconic example of modern Austrian heritage known as Wotruba Church (Ban, 2013, 2014).

A visual inspection showed the following reported results: weathered cement skin, eroded edges, efflorescence, microbiological growth and consequent surface corrosion. Due to the architectural value of the building, only low invasive examinations were allowed (samples for carbonation depth and for thin sections preparation). The following NDTs were performed: water sorptivity (Karsten Tube, Section 4.2.2.2), rebound hammer, cover depth and air-permeability *kT*. All tests were conducted at an age of 38 years (Ban, 2013, 2014).

The Karsten tube tests yielded moderate to high rates of water absorption on the E, S and W faces (values between 8 and 17 $g/m^2/s^{½}$) and extremely high on the N face (28–62 $g/m^2/s^{½}$). The carbonation rate (carbonation depth divided by the square root of 38 years) yielded low values on the surfaces exposed to rain (1.6–3.2 $mm/y^{½}$), but high on those protected from rain (7.3–8.1 $mm/y^{½}$).

Regarding the measurements of air-permeability *kT*, it is worth citing (Ban, 2014)

> "The air permeability values were unfortunately unusable; the data gained showed extremely high kT [$\times 10^{-16}$ m^2] values, not comparable with standard values. However, the values received indicated that the yellowish concrete was more permeable than the grey concrete. Even though it was not possible to take examples from the façade, where the yellow concrete was applied, it is known from the architect in charge Fritz G. Mayr that unintentionally different aggregates with high clay content were used."

Mrs. Ban was thoroughly trained in the use of the *PermeaTORR* (instrument she used for the survey), jointly by experts Dr. F. Jacobs and Dr. R. Torrent, so neither the instrument nor the operator can be blamed for the high results (not disclosed). Yet, the high water sorptivity and carbonation rate measured, at least in parts of the structure, may correspond to concrete of very high air-permeability.

8.4.3 Ministry of Transport, Ontario, Canada

An investigation on the "penetrability" of precast concrete barrier walls was conducted by the Ministry of Transport (MoT), Ontario, Canada (Berszakiewicz & Konecny, 2008). The experiment is not well described, but it seems that the walls were cast with mixes of *w/c* ratios 0.30, 0.45 and 0.60 (possibly made with OPC). Two test methods were applied directly on the walls, namely, air-permeability *kT* and a version of water sorptivity ISAT test, modified by the University of Toronto.

Unfortunately, the test results are not presented but the following comments were made by the researchers on the *kT* test

> "A series of tests results were carried out under laboratory conditions on slab specimens with different water/cement ratios (0.60; 0.45 and 0.30), at different concrete ages. Analysis of test results showed that

> there was no clear relationship between air permeabilities and concrete of different *w/c* ratios. When tested in the field, test could not clearly differentiate between the permeabilities of normal and the high performance concrete. The air permeability tests had repeatability even lower than the sorptivity measurements, in the range of the coefficient of variation of 50 to 70 percent. Thus, it is possible that additional factors, other than those already listed with respect to sorptivity, may have had an effect on air permeability results."

On the more positive side

> "The Torrent test method provided interesting results when used to comparatively evaluate the air permeability of formed concrete surface with different finishes ... While the test did not measure different air permeability levels of the two types of surface finishes for the 50 MPa concrete, it showed a consistent difference in air permeability for the 30 MPa concrete. The surface of 30 MPa concrete formed with the liner had an air permeability 3 times lower than the surface of the same concrete formed without the liner. At the same time the air permeability of the surface of 30 MPa concrete formed with the liner was close to the permeability of the 50 MPa concrete surfaces. This leads to the conclusion that the surface of the 50 MPa concrete formed with or without the liner, had a pore system consistent with low air permeability, while the use of the form liner for the 30 MPa concrete lowered the permeability to air, thus improving the quality of the concrete surface."

The water sorptivity test performed better in differentiating between mixes of different *w/c* ratios (0.60, 0.45 and 0.30). The investigation also included electrical resistivity tests (Wenner Probe) that, when applied on site, showed the typical obstacle of the effect of moisture on the readings. More information on the experience of Ontario's MoT with *kT* test can be found in Ip et al. (1998).

8.4.4 Mansei Bridge, Aomori, Japan

This bridge, built in 1955, was demolished in 2010 because of operational limitations, without significant deterioration of the main girder and slab. Before demolition, a condition survey was conducted including the site measurement of *kT* at different points of the piers, girders and slab (Watanabe et al., 2012). Later, cores were drilled near the *kT* testing points for measuring: *E*-modulus and compressive strength, carbonation depth, chloride ion penetration, MIP and scaling resistance, showing interesting results. Regarding air-permeability *kT*, the measurement points were selected to be free of surface damage (scaling, cracks, etc.). Yet, *kT* could not be measured (too high permeability), a fact attributed by the researchers to some subsurface damage (possibly freeze-thaw delamination), not visible on the surface.

8.4.5 Tests at FDOT Laboratory

A series of 7-year-old concrete cylinders (Ø100 × 200 mm), stored under water at the Florida Department of Transportation (FDOT) lab in Gainesville, FL, USA, were selected for a pilot test with a *PermeaTORR* instrument (Torrent & Armaghani, 2011). The concrete mixes involved had *w/c* ratios between 0.28 and 0.49, most of them of *w/c* = 0.35 and were prepared with a large variety of binders (including plain OPC and double and triple blends with PFA, SF, GGBFS and Metakaolin). The specimens were saw-cut by halves, leaving four plane faces for testing *kT*. Due to the small dimensions of the specimens, a special cell reduced to 70% the size of the standard cell had to be manufactured and installed in the instrument. Huge differences were found between the *kT* measurements obtained on the four faces of each specimen, which did not follow any systematic pattern (for instance, the top surface as cast was the best and sometimes the worst of the four surfaces, in terms of *kT*). This variability precluded yielding any conclusion from the tests performed. One possible explanation for the high variability observed could be the influence of coarse aggregate particles on the reduced size of the internal chamber of the small cell (Ø = 35 mm).

8.5 AIR-PERMEABILITY *kT* IN STANDARDS AND SPECIFICATIONS

8.5.1 Swiss Standards

Swiss Standards present the most complete and comprehensive approach regarding the measurement of *kT* on site, which started in 2003 with the following statements in standard SIA 262 "Concrete Construction", Swiss version of Eurocode 2, still included in version 2013 of the same standard (SIA 262, 2013): "The impermeability of the cover concrete shall be checked by means of permeability tests (e.g. air-permeability measurements) on the structure or on core samples taken from the structure". In parallel, a complementary standard, describing tests not covered by EN standards, was issued in 2003, included the air-permeability test *kT* as Annex E: "Air-Permeability on site". The current version is largely improved (SIA 262/1, 2019).

The standard provides instructions on how to calibrate the instrument and run the tests and sets several conditions for the measurement:

- age of the structure between 28 and 120 days
- temperature of the concrete ≥ 5°C
- surface moisture of the concrete (electrical impedance method) ≤ 5.5%
- cover thickness when measuring in coincidence with rebar: ≥ 20 mm
- it also recommends limiting values of *kT* to be specified (kT_s) for different exposure conditions, see Table 8.4, taken from Version 2008 of SIA 262/1

Table 8.4 Limiting values kT_s as function of the exposure conditions, taken from SIA 262/1:2013

	Concrete type						
Description	*A*	*B*	*C*	*D*	*E*	*F*	*G*
Strength classes[a]	C20/25	C25/30	C30/37	C25/30	C25/30	C30/37	C30/37
Exposure classes[b]	XC1 XC2	XC3	XC4 XF1	XC4 XD1 XF2	XC4 XD1 XF4	XC4 XD3 XF2	XC4 XD3 XF4
Minimum cement content (kg/m³)	280	280	300	300	300	320	320
Maximum *w/c* ratio	0.65	0.60	0.50	0.50	0.50	0.45	0.45
Air-permeability kT_s (10^{-16}m²)	-	-	2.0	2.0	2.0	0.5	0.5

[a] The indicated values correspond to the required characteristics strength (MPa) at 28 days, measured on cylinders/cubes.

[b] Correspond to the exposure classes defined in European Standard EN 206-1. The combinations of exposures are those typically found in Switzerland. The limits for XD classes can be applied to equivalent XS classes for marine environments, absent in Switzerland.

In SIA 262/1 (2019), the way of grouping structural elements cast with same mixes and subjected to similar concrete practices is explained, defining a lot (within each group) as the minimum exposed surface area of the following two alternatives:

- 500 m² of exposed surface
- three days of concreting

From each lot, six *kT* tests are performed at locations randomly selected with the following conformity conditions:

1. Not more than one out of the six individual *kT* tests performed shall exceed kT_s
2. If just two out of the six individual *kT* tests performed exceed kT_s, it is possible to perform another set of six tests on different randomly located places. Not more than one out of the new six individual *kT* tests performed shall exceed kT_s

If none of the conditions (1) and (2) is met, the Lot is regarded as non-compliant with the corresponding specified air-permeability limit kT_s.

The Operating Characteristic (O-C) curve of the above-mentioned compliance criterion is presented in Figure 8.20, thus giving a clearer

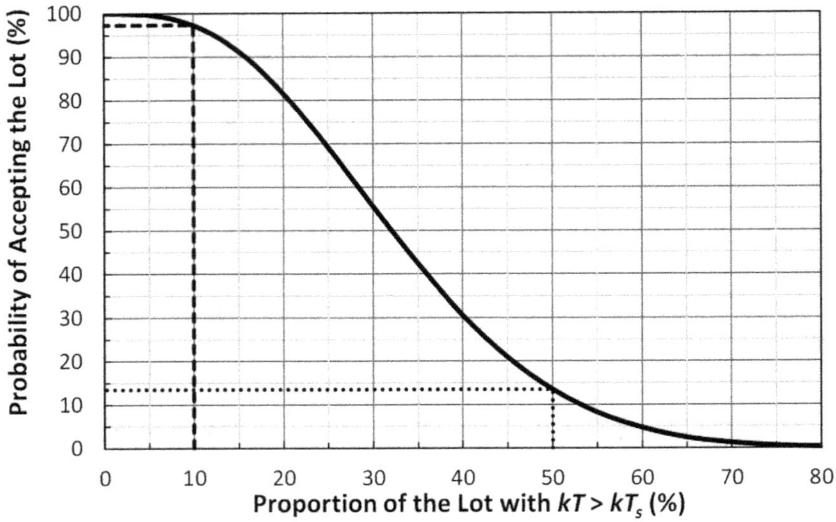

Figure 8.20 O-C curve of the conformity criterion of Swiss Standard (SIA 262/1, 2019) for *kT* tests on site.

probabilistic meaning to the kT_s value; derivation (with wrong Figure D-5) in Jacobs et al. (2009). The O-C curve carries in abscissas the proportion of the concrete surface in a Lot or Element with *kT* higher than the specified value kT_s, i.e. the proportion of "defective" concrete. In ordinates, the chart presents the probability of accepting a lot (applying conformity rules (1) and (2) above) containing a given proportion of "defectives".

From Figure 8.20, we can see that a lot containing 10% of "defectives" will have 97% probability of being accepted (broken line), whilst for one with 50% "defectives", the probability drops to just 13% (dotted line).

A translation into English of the parts of SIA 262/1 (2019) relevant to *kT* test can be found in www.m-a-s.com.ar.

Version 2019 of Swiss Standard SIA 262/1 does not cover the application of the test method in the laboratory, which is very much needed in order to better compare the results obtained in different laboratories. The Argentine Standard IRAM 1892 (see the next section) covers that gap.

8.5.2 Argentina

IRAM Standard 1892:2021 covers the application of air-permeability *kT* both in the laboratory and on site (the latter based on SIA 262/1) . The standard, in the preparation of which Dr. Torrent was actively involved, updates Swiss Standard SIA 261/1:2019 but, more important, covers the use of the test method in the laboratory. The Model Standard in Annex B is based on IRAM 1892.

8.5.3 Chile

A Highways Manual (MdC, 2017), elaborated by the Chilean Highways Administration, Ministry of Public Works, includes in the specifications for precast reinforced concrete box sections for culverts, storm drains, and sewers, upper limits for the value of *kT* of 0.01×10^{-16} m² for severe exposures and of 0.1×10^{-16} m² for moderate exposures (exposures defined in the same document).

In addition, the new Chilean Concrete Standard (NCh170, 2016) includes, in its Annex B3, the *kT* test method, referring directly to the Swiss Standard SIA 262/1 for its application.

8.5.4 China

The test method is included in Jiangsu Prov. Chinese Standard (DGJ32/TJ 206, 2016).

8.5.5 India

The *kT* test method is included in Section 2.9 "Permeability Test" of ES 103 (2009), dealing with non-destructive testing of bridges.

8.5.6 Japan

In Appendix 1 of JCI (2014), the air-permeability *kT* test method is included within "Detailed survey methods and setting of characteristic evaluation values" for existing structures.

NETIS (New Technology Information System) is a database owned by the Japanese Ministry of Land, Infrastructure and Transport. NETIS shares cutting-edge technologies applicable to construction with the private sector. Technologies registered to NETIS will be disclosed in the database so the information and evaluation on the technology will be available to all sectors. On 16 March 2020, the instrument *PermeaTORR AC* was registered in NETIS under No. QS-150029-VE.

Furthermore, in 2020, the test method was standardized by The Japan Society for Non-destructive Inspection (JSNDI) as NDIS 3436-2 "Non-destructive testing of concrete - Air permeability testing method Part 2: Double chamber method".

8.6 CREDENTIALS OF AIR-PERMEABILITY *kT* AS DURABILITY INDICATOR

In Chapter 6, it was shown that the coefficient of air-permeability *kT* responds sensitively, and according to the expectations, to changes in key technological parameters affecting the quality of concrete, such as *w/b* ratio, compressive strength, binder and aggregate types, compaction, segregation and bleeding, curing, applied stresses and cracks (see Table 8.1).

In this chapter, from good to excellent correlations with other transport and durability-related tests were presented, including: other gas-permeability and gas-diffusivity tests, water-permeability tests (including penetration under pressure and capillary suction), chloride-diffusion and migration tests, natural and accelerated carbonation and freezing and thawing tests. The only property with which a good correlation could not be established was electrical resistivity, which should therefore be considered a complementary rather than an alternative test.

Over 430 publications worldwide (list available at www.m-a-s.com.ar) present positive applications of the test method to investigate the potential durability of concrete structures, with just a handful of negative experiences as those described in Section 8.4. There may be more such cases, as often negative experiences tend to be omitted or not published; yet the number of positive cases is overwhelmingly high.

Timidly, the test method starts to be included in concrete standards and specifications, a trend likely to intensify in the future.

All these evidences lead to the conclusion that the coefficient of air-permeability *kT* constitutes a suitable durability indicator for concrete material and structures.

REFERENCES

Akiyama, H., Inoue, S. and Kishi, T. (2010). "Effect of curing conditions on the relation between air permeability and carbonation resistance of surface concrete". Seisan Kenkyu, v62, n6, 603–604 (in Japanese).

Alexander, M. and Thomas, M. (2015). "Service life prediction and performance testing — Current developments and practical applications". *Cem. & Concr. Res.*, v78, 155–164.

Andrade, C. (1993). "Calculation of chloride diffusion coefficients in concrete from ionic migration measurements". *Cem. & Concr. Res.*, v23, 724–742.

Andrade, C., González Gasca, C. and Torrent, R. (2000). "The suitability of the TPT to measure the air-permeability of the covercrete". ACI SP-192, 301–318.

Bahurudeen, A. and Santhanam, M. (2014). "Performance evaluation of sugarcane bagasse ash-based cement for durable concrete". 4th *International Conference on Durability of Concrete Structure*, Purdue Univ., West Lafayette, IN, USA, July 24–26, 275–281.

Ban, M. (2013). "Aspects of conserving exposed concrete architecture with Wotruba Church as an example". RILEM Proc. PRO 89, 549–556.

Ban, M. (2014). "Wotruba Church and Cologne Opera: Aspects of concrete aging". Concrete Solutions. Grantham et al. (Eds.), 619–625.

Beglarigale, A., Ghajeri, F., Yigiter, H. and Yazici, H. (2014). "Permeability characterization of concrete incorporating fly ash". *ACE* 2014, Istanbul, Turkey, October 21–25, 7 p.

Berszakiewicz, B. and Konecny, J. (2008). "In search of reliable in situ test methods for development of performance-based specifications for concrete in highway structures". Session "Bridges – Links to a Sustainable Future (B)" of the 2008 Annual Conference and Exhibition of the Transportation Association of Canada – Transportation: A Key to a Sustainable Future, Ottawa, Ontario.

Beushausen, H., Starck, S. and Alexander, M. (2012). "The integration of non-destructive test methods into the South African durability index approach". *Microdurability 2012*, Amsterdam, 11-13 April, Paper 113.

Bisschop, J., Schiegg, Y. and Hunkeler, F. (2016). "Modelling the corrosion initiation of reinforced concrete exposed to deicing salts". Bundesamt für Strassen, Bericht Nr. 676, Switzerland, February, 91 p.

BS 103 (2009). "Guidelines on non-destructive testing of bridges". Government of India, Ministry of Railways, August, 133 p.

CEB-FIP (1991). "CEB-FIP model code 1990". Final Draft, CEB Bulletin d'Information N° 203, 204 and 205, Lausanne, Switzerland, July 1991.

Choi, H., Zhang, W. and Hama, Y. (2017). "Method for determining early-age frost damage of concrete by using air-permeability index and influence of early-age frost damage on concrete durability". *Constr. & Build. Mater.*, v153, 630–639.

Denarié, E., Conciatori, D. and Simonin, P. (2003). "Essais comparatifs de caractérisation de bétons d'enrobage - phase I: bétons de laboratoire". Rapport d'essais MCS 02.12.07-1, EPFL, Lausanne, Switzerland, 21 p.

DGJ32/TJ 206 (2016). "Technical specification for quality control of high performance concrete in urban rail transit construction". Jiangsu Province Standard, R.P. China.

DIN 1048 (1978). "Prüfverfahren für Beton - Bestimmung der Wassereindringtiefe".

Di Pace, G. and Calo, D. (2008). "Assessment of concrete permeability in tunnels". *SACoMaTIS* 2008, Varenna, Italy, September 1–2, v1, 327–336.

Ebensperger, L. and Olivares, M. (2019). "Envejecimiento a mediano plazo de probetas de concreto: factor incidente en las estimaciones de vida útil". *CONPAT* 2019, Chiapas, México, v1, 13 p.

EN 12390-8 (2009). "Testing hardened concrete – Part 8: Depth of penetration of water under pressure".

Fernández Luco, L. and Revuelta Crespo, D. (2005). "Ensayo de penetración de agua bajo presión y Ensayo de permeabilidad al aire, método de Torrent, sobre probetas de hormigón de 150 ×300 mm" (in Spanish). Informe N° 18'728, Instituto Eduardo Torroja, Madrid, Spain, July, 8 p.

Fornasier, G., Fava, C., Fernández Luco, L. and Zitzer, L. (2003). "Design of self compacting concrete for durability of prescriptive vs. performance-based specifications". ACI SP 212, 197–210.

Imamoto, K., Neves, R. and Torrent, R. (2016). "Carbonation rate in old structures assessed with air-permeability site NDT". *IABMAS* 2016, Paper 426, Foz do Iguaçú, Brazil, June 26–30.

Imamoto, K., Shimozawa, K., Nagayama, M., Yamasaki, J. and Nimura, S. (2008). "Threshold values of air permeability of concrete cover – A case study in Japan". SACoMaTIS 2008, v1, 169–177.

Imamoto, K., Shimozawa, K., Nagayama, M., Yamasaki, J. and Tanaka, A. (2014). "Relationship between air-permeability and carbonation progress of concrete in Japan". *International* Workshop on Performance-based Specification and Control of *Concrete* Durability, Zagreb, Croatia, June 11–13, 325–333.

Imamoto, K., Tanaka, A. and Kanematsu, M. (2012). "Non-destructive assessment of concrete durability of the National Museum of Western Art in Japan". Paper 180, *Microdurability* 2012, Amsterdam, April 11–13.

Ip, A., Berszakiewicz, B. and Pianca, F. (1998). "Nondestructive test methods for evaluating durability of concrete highway structures: experience of Ontario Ministry of Transportation". Proc. Struct. Mater. Technol. III: An NDT Conference, Eds: R.D. Medlock; D.C. Laffrey, v3400, 270–280.

ISO 1920-12 (2015). "Testing of concrete – Part 12: Determination of the carbonation resistance of concrete accelerated carbonation method". ISO International Standard, 20 p.

Jacobs, F. (2006). "Luftpermeabilität als Kenngrösse für die Qualität des Überdeckungsbetons von Betonbauwerken". Bundesamt für Strassen, Bericht Nr. 604, September, 100 p.+Anhänge.

Jacobs, F. (2008). "Beton zerstörungsfrei untersuchen". *der Bauingenieur*, n3, 24–27.

Jacobs, F., Denarié, E., Leemann, A. and Teruzzi, T. (2009). "Empfehlungen zur Qualitätskontrolle von Beton mit Luftpermeabilitätsmessungen". Office Fédéral des Routes, VSS Report 641, December, Bern, Switzerland, 53 p.

JCI (2014). "Performance evaluation guidelines of existing concrete structures". *Japan Concr. Inst.*, 459 p. (in Japanese), I–17–18.

Jornet, A., Corredig, G. and Mühlethaler, U. (2011). "Concretes made with CEM II/A-LL: Relationship between microstructure and properties". 13th *Euroseminar on Microscopy Applied to Building Materials*, Ljubljana, Slovenia, June 14–18.

Karasawa, T. and Masuda, Y. (2011). "Carbonation supressing effects of coating materials for textured finishes based on the research of air permeability coefficient and carbonation depth of existing structures". J. Struct. Eng., AIJ, v76, n661, 449–454 (in Japanese).

Karasawa, T., Masuda, Y. and Lee, YR. (2011). "Research on carbonation supressing effect of coating materials for textured finishes and air permeability coefficient based on the result of the survey of an existing structure". J. Struct. Eng., AIJ, v76, n669, 1885–1890 (in Japanese).

Kattar, J.E., Abreu, J.V. de and Cruz, L.O. (1995). "Concreto de alto desempenho modificado con polímero para pisos industriais". *37ª Reunião Anual do IBRACON*, Goiânia, Brazil, Julho 3–7, 15 p.

Kattar, J., Abreu, J.V. and Regattieri, C.E.X. (1999). "Inovações na metodologia para avaliação da permeabilidade por difusão ao ar". 41° Congresso do IBRACON, Salvador, Bahia, Brazil.

Kubens, S., Wassermann, R. and Bentur, A. (2003). "Non destructive air permeability tests to assess the performance of the concrete cover". 15th *ibausil Intern. Baustofftagung*, Bauhaus, Univ. Weimar, September 24–27.

Kurashige, I. and Hironaga, M. (2010). "Nondestructive quality evaluation of surface concrete with various curing conditions". *CONSEC'10*, Mérida, México, June 7–9.

Li, K., Li, Q., Zhou, X. and Fan, Z. (2015). "Durability design of the Hong Kong–Zhuhai–Macau Sea-Link project: Principle and procedure". J. Bridge Eng., ASCE, 04015001, 11 p.

Liu, Z. and Hansen, W. (2015). "Sorptivity as a measure of salt frost scaling resistance of air-entrained concrete". Key Engng. Mater., v629–630, 195–200.

Liu, Z., Hansen, W. and Wei, Y. (2014). "Concrete sorptivity as a performance-based criterion for salt frost scaling resistance". *RILEM International Symposium on Concrete Modelling*, Beijing, China, October 12–14, 487–496.

Maître, M. (2012). "Tunnel de Naxberg - Perméabilité à l'air du béton d'enrobage (méthode Torrent)". EPFL, Rapport d'essais n° MCS 02.09-01, Lausanne, November, 9 p.

Mathur, V.K., Verma, C.L., Gupta, B.S., Agarwal, S.K. and Kumar, A. (2005). "Use of high-volume fly ash in concrete for building sector". Report No. T(S)006, Central Build. Res. Inst., Roorkee, India, January, 35 p.

MdC (2017). "Manual de Carreteras – Especificaciones Técnicas Generales de Construcción". MOP, Dirección de Vialidad, Sección 5.612.201 'Cajones Prefabricados, de Hormigón Armado'. Chile.

Mohr, P., Hansen, W., Jensen, E. and Pane, I. (2000). "Transport properties of concrete pavements with excellent long-term in-service performance". *Cement & Concr. Res.*, v30, 1903–1910.

Moro, F. and Torrent, R. (2016). "Testing fib prediction of durability-related properties". *fib Symposium* 2016, Cape Town, South Africa, November 21–23.

NCh170 (2016). "Hormigón – Requisitos generales". Norma Chilena, 4ª Ed., 25 May, 44 p.

Neves, R.D. (2012). "A Permeabilidade ao Ar e a Carbonatação do Betão nas Estruturas". PhD Thesis, Universidade Técnica de Lisboa, Instituto Superior Técnico, Portugal, 502 p.

Neves, R., Sena da Fonseca, B., Branco, F., de Brito, J., Castela, A. and Montemor, M.F. (2015). "Assessing concrete carbonation resistance through air permeability measurements". *Constr. & Build. Mater.*, v82, 304–309.

Nishimura, K., Kato, Y. and Mita, K. (2015). "Influence of construction work conditions on the relationship between concrete carbonation rate and the air permeability of surface concrete". *International Conference* Regeneration and Conservation of *Concrete Structures*, Nagasaki, Japan, June 1–3, Paper R1–4, 8 p.

Olek, J., Lu, A., Feng, X. and Magee, B. (2002). "Performance-related specifications for concrete bridge superstructures, volume 2: High-performance concrete". Purdue Univ. – Joint Transportation Research Program Technical Report Series, 215 p.

Park, J-J., Koh, K-T., Kim, D-G. and Kim, S-W. (2004). "The Chloride Diffusion Properties of Concrete with Mineral Admixtures". Korean Soc. *Concr.* Diagnosis, v8, n4, 239–246 (in Korean).

Park, S-B. and Kim, D-G. (2000). "A experimental study on the chloride diffusion properties in concrete". J. Korea Concr. Inst., v12, n1, 33–44 (in Korean).

PC (2014). Personal communication from ACP (Panama Canal Authority).

RILEM TC 189-NEC (2007). "Non-destructive evaluation of the penetrability and thickness of the concrete cover". RILEM Report 40, May, 223 p.

Rodríguez de Sensale, G., Sabalsagaray, B.S., Cabrera, J., Marziotte, L. and Romay, C. (2005). "Effect of the constituents on the properties of SCC in fresh and hardened state". fib Symp*osium on* "Structural Concrete and Time", La Plata, Argentina, September.

Roelfstra, G., Adey, B., Hajdin, R. and Brühwiler, E. (1999). "The condition evolution of concrete bridges based on a segmental approach, non-destructive test methods and deterioration models". 78th *Annual Meeting Transportation Research Board*, Denver, April, 13 p.

Romer, M. and Leemann, A. (2005). "Sensitivity of a non-destructive vacuum test method to characterize concrete permeability". ICCRRR, Cape Town, November 21–23.

Shi, C. (2003). "Another look at the rapid chloride permeability test (ASTM C1202 or AASHTO T277)". FHWA Resource Center, Baltimore, MD, 15 p.

SIA 262 (2013). "Betonbau". Swiss Soc. of Engineers and Architects (in French, German and Italian).

SIA 262/1 (2019). "Concrete construction – Complementary specifications". Swiss Soc. of Engineers and Architects (in French and German).

Starck, S. (2013). "The integration of non-destructive test methods into the South African durability index approach". MSc Dissertation, Univ. Cape Town, March, 188 p.

Starck, S., Beushausen, H., Alexander, M. and Torrent, R. (2017). "Complementarity of in situ and laboratory-based concrete permeability measurements". *Mater. & Struct.*, v50, 177–191.

Tang, L. and Nilsson, L.-O. (1992). "Rapid determination of chloride diffusivity of concrete by applying an electric field". *ACI Mater. J.*, v49, n1, 49–53.

Teruzzi, T. (2009). "Estimating the service-life of concrete structures subjected to carbonation on the basis of the air permeability of the concrete cover". *EUROINFRA* 2009, Helsinki, October 14–15.

Torrent, R. and Ebensperger, L. (1993). "Methoden zur Messung und Beurteilung der Kennwerte des Überdeckungsbetons auf der Baustelle". Office Fédéral des Routes, Rapport No. 506, Bern, Switzerland, Januar, 119 p.

Torrent, R. and Frenzer, G. (1994). "Durabilidad de concretos elaborados con cementos Tipo I, Puzolánico y de Escoria. Estudio Comparativo". "Holderbank" Report MA-94-3246-S, June 15, 20 p.

Torrent, R. and Frenzer, G. (1995). "Methoden zur Messung und Beurteilung der Kennwerte des Ueberdeckungsbetons auf der Baustelle -Teil II". Office Fédéral des Routes, Rapport No. 516, Bern, Suisse, October, 106 p.

Van Eijk, R.J. (2009). "Evaluation of concrete quality with Permea-TORR, Wenner Probe and Water Penetration Test". KEMA Report, Arnhem, July 8, 46p. (in Dutch).

Wang, Y.F., Dong, G.H., Deng, F. and Fan, Z.H. (2014). "Application research of the efficient detection for permeability of the large marine concrete structures". *Appl. Mechanics Mater.*, v525, 512–517.

Watanabe, K., Sakoi, Y., Aba, M., Kamiharako, A. and Tsukinaga, Y. (2012). "Durability investigation of RC bridge after 56 years". 37th *Conference on Our World in Concrete & Structure*, August 29–31, Singapore.

Zhang, M., Sakoi, Y., Aba, M. and Tsukinaga, Y. (2019). "Effect of initial curing conditions on air permeability and de-icing salt scaling resistance of surface concrete". *J. Asian Concr. Federation*, v5, n1, June, 56–64.

Zhutowsky, S. and Kovler, K. (2012). "Effect of internal curing on durability-related properties of high-performance concrete". *Cem. & Concr. Res.*, v42, 20–26.

Chapter 9

Service life assessment based on site permeability tests

9.1 INTRODUCTION

Traditionally, concrete codes and standards have applied and still apply the "deemed-to-satisfy" approach (Andrade, 2006) to specify durability requirements. Based on the accumulated experience in many countries, a set of primarily prescriptive rules have been established which, when rigorously observed, are expected to result in a service life typically of 50 years (e.g. Eurocode 2 (EN 1992-1-1, 2004)).

As they are used later, a classification of exposure environments, concerning just reinforcement corrosion, is provided in Table 9.1 (including carbonation- and chloride-induced corrosion). It corresponds to European Standards (EN 206, 2013; EN 1992-1-1, 2004); the maximum w/c ratios and minimum cover thickness d stipulated in those standards are also indicated in Table 9.1. It is assumed ("deemed-to-satisfy") that, if the maximum water-cement ratio w/c_{max} has been observed by the concrete producer and the contractor has executed the concreting operations correctly, e.g. by following (EN 13670, 2009), complying with the minimum cover thickness d_{min}, the structure will reach its expected service life t_{SL} of 50 years.

The limitations of this approach have been highlighted in Section 1.6.

Recently, codes and standards have been moving towards the "Performance Indicators" approach, by which the prescriptive rules have been replaced by performance requirements (see Table 1.2). For instance, Canadian Standard (CSA A23.1, 2006) establishes limiting values for the charge passed in "RCPT" test (ASTM C1202, 2010) for normal and extended service lives in a chloride-rich environment. Swiss Standard (SN EN 206, 2016) establishes limiting values for water sorptivity, chloride migration, accelerated carbonation and frost-thaw-salt scaling resistance tests, applicable to the corresponding exposure conditions. Swiss concrete producers shall prove that their concretes pass the relevant tests, conducted on specimens obtained from samples taken from their regular production. The limitations of this *Labcrete* approach have been highlighted in Chapter 7. Swiss Standard (SIA 262/1, 2019) complements the previously mentioned standard by

DOI: 10.1201/9780429505652-9

Table 9.1 EN exposure classes and main requirements for steel corrosion induced by carbonation and chlorides

Class designation	*Description of environment*	w/c_{max}	d_{min} *(mm)*
Corrosion induced by carbonation			
XC1	Dry Permanently wet	0.65	15
XC2	Wet, rarely dry	0.60	25
XC3	Moderate humidity	0.55	25
XC4	Cyclic wet and dry	0.50	30
Corrosion induced by chlorides other than from sea water			
XD1	Moderate humidity	0.55	35
XD2	Wet, rarely dry	0.55	40
XD3	Cyclic wet and dry	0.45	45
Corrosion induced by chlorides from sea water			
XS1	Exposed to airborne salt	0.50	35
XS2	Permanently submerged	0.45	40
XS3	Tidal, splash and spray zones	0.45	45

establishing limiting values for the coefficient of air-permeability *kT*, measured on site, thus evaluating the quality of the end-product, the *Realcrete*. It also offers the alternative of performing the above-mentioned laboratory tests on cores drilled from the structures, with more lenient requirements for the *Realcrete* compared with those for the *Labcrete*. These Swiss limiting values for *Labcrete* and *Realcrete* apply (in conjunction with the minimum cover thickness requirements) for an expected service life of 50 years.

The situation is that, nowadays, many important structures are designed for service lives of 100, 150 or even more years, which clearly exceed the reach of existing codes and, therefore, requires some extrapolation or prediction via modelling.

Moreover, in the past, the burden of maintenance and repair costs of structures fell predominantly on the shoulders of the owner, with other players (designers, contractors, materials suppliers) assuming the responsibility for durability for a relatively short period (typically 5–10 years). The advent of Design, Build and Operate contracts, whereby a private organization designs, builds and operates the facility for a period of several decades has changed the picture. Now, contractors have a direct interest in the durability of the construction, since maintenance and repair costs plus eventual penalties for reduced operability of the facility will be borne by them. Moreover, often, the transfer price of the facility to the final owner is associated with its residual service life that needs to be fairly established.

These examples show the increasing economic relevance of having tools capable of reliably predicting the service life of concrete structures that are:

- *accurate*: the prediction is close to the service life actually reached
- *meaningful*: based on sound principles
- *realistic*: take into consideration relevant parameters of the end-product
- *objective*: contain few (if any) hardly measurable parameters that could be freely and subjectively chosen

Various service life prediction models/methods have been developed recently. From them, there are two that have gained wide acceptance: DuraCrete (DuraCrete, 2000) in Europe, on which the *fib* has based its own model (*fib*, 2006), and Life-365 (Life-365, 2018) in North America. DuraCrete deals with steel corrosion induced either by carbonation or by chlorides, whilst Life-365 deals just with chloride-induced corrosion. Regarding chloride-induced steel corrosion, both methods assume (with some differences) a purely diffusive process, the key concrete property used as input being the coefficient of chloride-diffusion D_0 at 28 days.

When modelling service life, it is important to define which is the Limit State considered by the model, for which Tuutti's model (Tuutti, 1982) is of help (Figure 1.4). Tuutti's model consists, basically, in dividing the deterioration process of a concrete structure into two well-differentiated phases: Incubation and Propagation, as already discussed in Section 1.3.

In the Incubation phase (often called Initiation phase), no visible damage can be detected in the structure but relevant processes are taking place, e.g. the penetration of the carbonation or critical chlorides front towards the reinforcement in the case of steel corrosion. At a certain time, called Initiation Time, the reinforcing steel becomes depassivated and the true damaging action starts to take place, with the amount of steel corrosion products being of such magnitude that stains or micro-cracks appear on the concrete surface. If these deleterious reactions are allowed to continue (Propagation phase), the damage increases until it reaches a certain critical level that puts the serviceability or safety of the structure at risk.

Therefore, the service life t_{SL} of a concrete structure can be defined as

$$t_{SL} = t_i + t_p \tag{9.1}$$

where t_i is the initiation and t_p is the propagation time of the damage.

In modelling chloride-induced corrosion, the propagation period is often disregarded, or assumed to be rather short (6 years in the case of Life-365), as the moist environments rich in chloride ions are usually favourable to the propagation of corrosion. In the case of carbonation-induced corrosion, the situation is somewhat different because in dry environments the Incubation phase may be rather short but the Propagation phase can be very long.

In the next section the general principles of modelling or assessing the corrosion initiation time t_i will be discussed.

9.2 GENERAL PRINCIPLES OF CORROSION INITIATION TIME ASSESSMENT

Of the different deterioration mechanisms affecting reinforced concrete structures, steel corrosion is the most insidious, be it induced by carbonation or by chlorides. As both mechanisms involve the penetration of CO_2 and chloride ions through the pore system of the *Covercrete*, they are particularly suited to be modelled on the basis of its permeability. Strictly speaking, the penetration of CO_2 into concrete happens predominantly by gas diffusion, whilst the penetration of chlorides happens by mix modes (permeability, sorptivity and diffusion), see Figure 1.1. We have seen in Chapter 3 that all transport mechanisms depend on the pore structure of the concrete, with theoretical relations between the parameters governing them, some of which have been confirmed experimentally (Section 8.3).

In order to fully understand the models to be presented, a brief description of the general principles involved in the modelling of corrosion Initiation Time of steel due to carbonation and chlorides is provided in the following sections.

9.2.1 Carbonation-Induced Steel Corrosion

Carbonation is a physical-chemical process generated by the penetration of CO_2 into concrete by gas diffusion. In the presence of sufficient moisture, CO_2 reacts with Ca-bearing hydrated cement phases, predominantly $Ca(OH)_2$, to produce $CaCO_3$ in a process known as "carbonation" (Eq. 9.2). The $CaCO_3$ crystals partially fill the capillary pores, densifying the concrete cover and increasing its hardness.

$$\mathrm{Ca(OH)_2 + CO_2 \rightarrow CaCO_3 + H_2O} \tag{9.2}$$

Figure 9.1 presents a scheme of the carbonation process along an element of surface area S assuming that at time t the carbonation front has penetrated a depth x. We assume that the concentration of CO_2 at the concrete surface is C_s, decreasing linearly to a value of 0 at depth x. If we assume that CO_2 diffuses into the carbonated zone following Fick's first law (see Section 3.4.1), the differential amount of CO_2 that will penetrate into the concrete element in a time differential dt, under a concentration gradient C_s/x, will be (see Eq. 3.2):

$$d\mathrm{CO_2} = D \cdot \frac{C_s}{x} \cdot S \cdot dt \tag{9.3}$$

where D is the coefficient of diffusion of CO_2 through concrete.

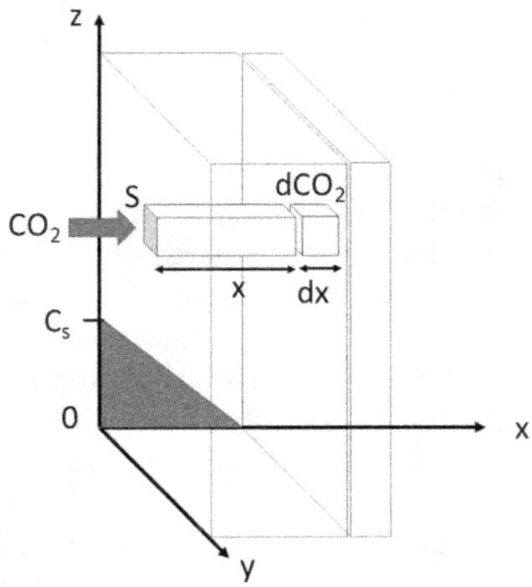

Figure 9.1 Schematic description of carbonation progress (AIJ, 2016).

The amount of CO_2 that penetrates the concrete element will neutralize an amount H of $Ca(OH)_2$ existing in the extra volume of concrete to be carbonated $S.\ dx$, or:

$$dCO_2 = H \cdot S \cdot dx \tag{9.4}$$

where H is the amount of $Ca(OH)_2$ per m^3 of concrete, assuming that 1 g of CO_2 neutralizes 1 g of $Ca(OH)_2$ which, stoichiometrically, is a reasonable approximation.

Equating Eqs. (9.3) and (9.4), we get the following differential equation:

$$x\ .\ dx = \frac{D \cdot C_s}{H}\,dt \tag{9.5}$$

which, after integrating both members, yields

$$CD = \sqrt{\frac{2 \cdot D \cdot C_s}{H}}\,\sqrt{t} \tag{9.6}$$

where CD is the carbonation depth at time t.

In general, Eq. (9.6) is expressed as

$$CD = K_c \cdot \sqrt{t} \tag{9.7}$$

where

CD = carbonation depth (mm)

t = time of exposure to the CO_2-bearing atmosphere (years)

K_c = carbonation rate ($mm/y^{1/2}$)

Therefore, the carbonation rate K_c depends on the actual diffusion coefficient D of CO_2 through the concrete, on its content of carbonatable material H, and on the CO_2 concentration of the environment C_s. In addition, the actual diffusion coefficient D depends on the moisture content of the concrete. Concretes of high *w/c* ratios and poorly cured will carbonate faster (due to a more porous microstructure). Concretes containing pozzolanic or latent hydraulic mineral additions (e.g. PFA or GBFS) tend to carbonate faster (due to having less carbonatable material H). The CO_2 concentration in the air is increasing markedly due to growing emissions. Carbonation in tunnels, car parkings, traffic-intensive urban areas or industrial areas is faster due to the higher CO_2 concentration in the surrounding air. The carbonation rate is extremely low for dry concrete (no moisture for chemical reaction) and also for concrete near saturation (diffusion of CO_2 blocked by water in the pores) and maximum for RH around 50%–60% (see broken line of Figure 9.2).

More details on this topic can be found in Chapter 5 of Bertolini et al. (2004), in Chapter 1.6 of Böhni (2005), in Chapter 1 of Li (2016) and in Section 7.6 of Alexander et al. (2017).

Equation (9.6) was derived by Hamada (1968) and now this square root theory is widely used. Although determining D, H and C_s is difficult, through the measurement of carbonation depth CD at a certain age t, the carbonation rate K_c can be obtained and future progress of carbonation can be predicted applying Eq. (9.7). The carbonation depth can be easily measured spraying a phenolphthalein solution on a freshly broken surface (CPC-18, 1988).

For durability, the main consequence of carbonation is the decrease in alkalinity of the pore solution, that drops the pH from around 13 for non-carbonated concrete, to less than 9 for carbonated concrete. If the carbonation front reaches the embedded steel, this change in pH alters the latter's thermodynamic equilibrium, producing its "depassivation" and making it vulnerable to corrosion. Once the carbonation front has reached the location of the steel bar, the metal corrosion process may start, the rate of which is strongly dependent on the moisture conditions to which the structure is exposed.

Figure 9.2 presents the relative carbonation rate (broken line, left axis) and the carbonation-induced corrosion rate (full line, right hand axis) as function of the RH of the environment (Parrott, 1994; Hunkeler et al., 2013).

Figure 9.2 shows that in environments with relative humidity around 50%–60%, carbonation progresses very fast but the corrosion rate is negligible; this may be the case of a concrete element located indoors (dry). The corrosion rate reaches its peak at around 95% RH, where the carbonation rate is rather low, but not negligible. Carbonation-induced corrosion happens more frequently in elements that are exposed to wetting and drying cycles, typically outdoors in temperate and tropical climates.

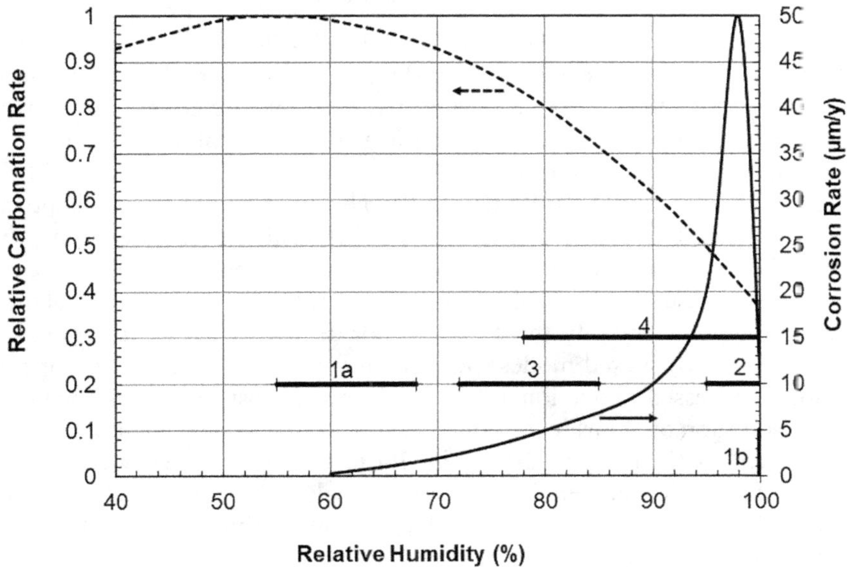

Figure 9.2 Relative rate of carbonation and rate of carbonation-induced corrosion as function of the relative humidity of the air (Parrott, 1994; Hunkeler et al. 2013).

Figure 9.2 indicates that both the initiation and the propagation periods in Tuutti's model (Figure 1.4) deserve consideration when modelling carbonation-induced corrosion. The thick segments indicated in Figure 9.2 correspond to the RH associated with the carbonation exposure classes previously defined in Table 9.1.

The initiation time of carbonation-induced corrosion happens when the carbonation depth reaches the position of the steel, i.e. when $CD = d$, where d = cover thickness or, from Eq. (9.7):

$$t_i = \left(\frac{d}{K_c}\right)^2 \tag{9.8}$$

The simplest manner to account for the propagation time is to assume it as the time required for the corrosion process to have penetrated 100 μm from the steel surface, which corresponds approximately to the appearance of visible cracks on the surface (Parrott, 1994). Therefore

$$t_p = \frac{100}{CR} \tag{9.9}$$

where CR is the corrosion rate (μm/y) that can be obtained from the full line in Figure 9.2 as function of the RH of the environment.

9.2.2 Chloride-Induced Steel Corrosion

This phenomenon may occur whenever the chloride concentration at the location of the steel reaches an elusive threshold value, very difficult to guess (Angst et al., 2009). This may happen if the concrete constituents carry enough chlorides to reach that value (e.g. by using chloride-rich admixtures, unwashed sea sand or mixing sea water), which is nowadays unlikely due to strict regulations in that respect. The most common case happens when a concrete structure is exposed to a chloride-rich environment, the most classical example being marine environment, but also when in contact with chloride-bearing solutions (de-icing salts, water treatment plants, swimming pools, etc.). In these cases, chloride ions penetrate into the concrete element by mixed modes (permeability, capillary suction and/or diffusion), addressed in Section 1.2.2. To simplify, most models assume that the penetration of chloride ions into concrete happens through a purely diffusive process (DuraCrete, 2000; *fib*, 2006; Life-365, 2018). More details can be found in Chapter 6 of Bertolini et al. (2004), in Chapter 1.7 of Böhni (2005), in Chapter 2 of Li (2016) and in Sections 5.5.1.2 and 7.5 of Alexander et al. (2017).

As discussed in Section 3.4.1, the diffusion process is governed by Fick's second law, expressed by

$$\frac{\partial C}{\partial t} = D\frac{\partial^2 C}{\partial x^2} \tag{3.4}$$

which, assuming that the coefficient of chloride-diffusion D and the surface chloride concentration C_s are constant, has an explicit solution

$$C(x,t) = C_0 + (C_s - C_0)\cdot\left[1 - erf\left(\frac{x}{\sqrt{4\cdot D\cdot t}}\right)\right] \tag{3.5}$$

$C\ (x,\ t)$ = chloride concentration at distance x from the surface and time t
C_s = chloride concentration at the surface ($x = 0$)
C_0 = initial chloride concentration in the concrete, before being exposed
D = chloride-diffusion coefficient [m²/s] or [mm²/y]
erf = error function

The experimental evidence indicates that neither D nor C_s is constant, but that they change with time. This is properly taken into account by Life-365 (2018) that operates solving Eq. (3.4) numerically with values of D and C_s that are changed for each time step of the calculation.

In the case of DuraCrete (2000), *fib* (2006), C_s is assumed as constant but D is assumed to decrease with time (in Life-365 too) according to

$$D(t) = D_0 \left(\frac{t_0}{t}\right)^n \quad \text{for } t \le t_d \tag{9.10}$$

where

$D(t)$ = coefficient of chloride-diffusion at time t

D_0 = coefficient of chloride-diffusion measured at time t_0 (typically 28 days)

n = exponent indicating the decay rate of D, with $0 < n < 1$

t_d = duration of the decay period: elusive variable, 25 years for Life-365 (2018), t_i for DuraCrete (2000) and 10 years for Andrade (2014)

Although not strictly correct, Eq. (9.10) is inserted within Eq. (3.5), despite the fact that the latter was derived assuming a constant D.

$$C(x,t) = C_0 + (C_s - C_0) \cdot \left\{1 - erf\left[\frac{x}{2\sqrt{D_o \cdot \left(\frac{t_0}{t}\right)^n \cdot t}}\right]\right\} \tag{9.11}$$

The chloride-induced corrosion initiation time t_i is reached when the chloride content at the location of the steel (d) is

$$C(d,t_i) = C_{cr} \tag{9.12}$$

where d is the cover thickness and C_{cr} the elusive critical chloride concentration capable of depassivating the steel and initiating its corrosion process. Introducing Eq. (9.12) into (9.11), assuming for simplicity that $C_0 = 0$ and operating, we get

$$t_i = \frac{d^2}{4 \cdot D_0 \cdot (t/t_0)} A^2 \quad \text{with } t = \text{MIN}(t_i; t_d) \tag{9.13}$$

and

$$A = \frac{1}{erf^{-1}\left(1 - \frac{C_{cr}}{C_s}\right)} \quad erf^{-1} = \text{inverse error function} \tag{9.14}$$

Although Eqs. (9.13) and (9.14) look of simple resolution, the values of the parameters involved, namely C_s, C_{cr}, t_d and n, are extremely difficult to predict, some (especially exponent n and t_d) having a crucial influence on the predicted initiation time t_i. This topic is beyond the scope of this book, but the reader can find more information in Gulikers (2006) and Torrent (2017).

Since concrete exposed to chlorides is often wet or subject to wetting-drying cycles, t_p is often neglected ($t_p = 0$) or assumed to be rather short, e.g. 6 years for Life-365 (2018). This may be too conservative for structures built with high-performance concretes, likely to be used when long design service lives are specified (Di Pace & Torrent, 2020).

The matter gets more complicated because, in the case of chloride-induced corrosion, the corrosion of the steel is not generalized and uniform (as is the case for carbonation-induced corrosion) but is concentrated in certain areas (pit corrosion). Data on corrosion rates (for different exposure conditions) and pitting factors can be found in CONTECVET (2001), even including central values and scatter allowing a probabilistic prediction of t_p.

9.3 SERVICE LIFE ASSESSMENT OF NEW STRUCTURES WITH SITE PERMEABILITY TESTS

Here, different service life models, based on the measurement of permeability on site are described in some detail, both for carbonation- and chloride-induced corrosion. The models are presented chronologically.

9.3.1 Carbonation: Parrott's Model

Possibly the first model to predict service life on the basis of site permeability measurements was the one developed by Parrott (1994) for carbonation-induced corrosion.

The model is based on Hong-Parrott's method to measure the coefficient of air-permeability (see Section 4.3.2.2) and is expressed as

$$CD = a\,\frac{k^{0.4} \cdot t^{n}}{c^{0.5}} \tag{9.15}$$

where

CD = carbonation depth (mm) at age t (years)

k = coefficient of air-permeability of the *Covercrete* (10^{-16} m^2) at the prevailing relative humidity

n = exponent close to 0.5 for indoor exposure; obtained from a formula that accounts for environment relative humidity

c = CaO content in the hydrated cement matrix (in kg/m^3 of the cement matrix)

a = coefficient assumed = 64

A table is presented in Parrott (1994) which provides values of the parameters involved in the model as function of the RH. In the particular case of c, estimated values are given as function of the type of cement used.

In Eq. (9.15), when *CD* reaches the cover thickness *d*, *t* becomes the corrosion initiation time t_i.

The model also contemplates the calculation of the corrosion propagation time t_p, applying Eq. (9.9). The values of *CR* used to build the corresponding curve in Figure 9.2 were taken from a table in Parrott (1994).

This model was used to estimate the carbonation depth at 150 years of age of precast segments for the Miami Port Tunnel, see Section 11.3.1, where *k* was equalled to Torrent air-permeability *kT*. Compared with other estimates, Parrott's method yielded the most optimistic result (lowest values of *CD*).

9.3.2 Carbonation: South African *OPI* Model

This model is based on the determination of the South African Oxygen-Permeability Index (*OPI*), based on testing Ø70 × 30 mm cores drilled from the structure (see Section 4.3.1.3). It is worth reminding that *OPI*, which varies between 8 and 11, is the negative $\log_{10}$ of the coefficient of O_2-permeability, i.e. a higher *OPI* value means a less permeable concrete.

In Alexander et al. (2008), two approaches are proposed, namely, the "Deemed-to-satisfy" approach and the "Rigorous" approach, both described below. It has to be mentioned that a similar approach for chloride-induced steel corrosion was proposed, but based on the Chloride Conductivity Index (CCI), described in Annex A. Since this is not strictly a permeability test, it is beyond the scope of this book.

9.3.2.1 "Deemed-to-Satisfy" Approach

This approach follows the criterion of design codes by which the designer prescribes limiting values of the *OPI* and of the cover thickness which, when met, result in the structure "deemed-to-satisfy" the durability requirements. It applies only to exposure classes XC3 and XC4 of Table 9.1; for classes XC1 and XC2, it is assumed that if the minimum cover depth d_{min} is achieved, the risk of carbonation-induced corrosion is negligible. Table 9.2 shows some examples of this approach (Alexander et al., 2008).

An example of a real case application of the "deemed-to-satisfy" approach is presented in Table 9.3, where reduced-payment penalties apply if the

Table 9.2 *OPI* "deemed to satisfy" requirements of service life for exposure classes XC3 and XC4

Structure type	*Common*	*Monumental*	
Service life (years)	50	100	
d_{min} (mm)	30	30	40
OPI_{min} at 28 days	9.70	9.90	9.70

Table 9.3 Deemed to satisfy penalties for not meeting OPI_{min} and d_{min} requirements (Gauteng Freeway)

	Oxygen permeability index (OPI)		*Concrete cover*	
	OPI	*% Payment*	*Overall cover (mm)*	*% Payment*
Full acceptance	>9.70	100%	≥85% < (100% + 15 mm)	100%
Conditional acceptance[a]	>8.75 and ≤9.70	80%	<85% and ≥75%	85%
Conditional acceptance[b]	-	-	<75%	70%
Rejection	<8.75	Not applicable	<65%	Not applicable

[a] With reduced payment.
[b] With remedial measures as approved by engineer and reduced payment.

prescribed values of OPI_{min} and d_{min} are not met during the construction of the Gauteng Freeway Improvement Project (Alexander, 2018).

9.3.2.2 "Rigorous" Approach

In this approach, the designer is not limited by fixed values such as those presented in Table 9.2, but can establish its own limits by proper modelling that takes into consideration the specific conditions of the structure (exposure, design service life, cover thickness, materials, etc.).

9.3.2.3 Acceptance Criteria

A potential value OPI_{pot}, to be achieved on laboratory specimens, is defined as

$$OPI_{pot} = OPI_s + 0.10 \tag{9.16}$$

where OPI_s is the target value required for the site tests. The concrete producer has to demonstrate that the supplied concrete complies with the required OPI_{pot}.

The initial minimum sampling frequency suggested for site tests is one test per 50 m² of exposed concrete surface area or per structural element, to be extended to one test per 150 m².

The acceptance criterion involves the fulfilment of the following conformity rules:

- the average of three consecutive test results shall exceed the target *OPI* value 90% of the time
- no single test result is less than the target value OPIs minus 0.3

9.3.2.4 *Probabilistic Treatment*

A statistical analysis of a large number *N* of test results was reported by Nganga et al. (2013) for various projects having the same *OPI* specification (*OPI* > 9.70). The large number of samples collected in such projects allows a probabilistic assessment of the service life. In some of the reported cases, although the average *OPI* exceeded the specified limit, a relatively large percentage of "defectives" (individual values below the specified value) was detected, which calls for a statistical treatment of the results, see Alexander and Beushausen (2009).

9.3.3 "Seal" Method for Chloride-Induced Steel Corrosion

The "Seal" test method to measure the coefficient of air-permeability of concrete on site was described in Section 4.3.2.11.

A procedure to estimate the corrosion initiation time, based on the combined measurement of the air-permeability ("Seal" method) and the moisture content (by means of a conductive probe), was presented by S. Okazaki in pp. 291–294 of Beushausen and Fernández Luco (2016).

Following a calibration established at Ehime Univ., Japan, a value of the chloride-diffusion coefficient *D* (cm^2/y) can be obtained as function of the "Seal" air-permeability coefficient and of the moisture content, both measured on site, using a nomogram.

Once the value of *D* has been obtained, it can be used to estimate the corrosion initiation time applying the error function solution of Fick's second diffusion law (Eq. 3.5).

9.4 SERVICE LIFE ASSESSMENT OF NEW STRUCTURES APPLYING SITE *kT* TESTS

9.4.1 The "TransChlor" Model for Chloride-Induced Steel Corrosion

This model was developed at the EPFL (Lausanne, Switzerland) by David Conciatori, under the leadership of Prof. E. Brühwiler, on the basis of two PhD theses (Roelfstra, 2001; Conciatori, 2005). It is summarily described in Roelfstra et al. (1999, 2004) as a framework and, more recently, explicitly formulated as a numerical model in Conciatori et al. (2008, 2009b, 2011). Figure 9.3 presents the general approach followed by Roelfstra (2001). The model is applicable to steel corrosion induced by chlorides originated from the application of de-icing salts on highways and roads.

The two variables measured on site, indicated in bold characters in Figure 9.3, are the air-permeability *kT* and the cover thickness *d*. Other required inputs, shown in italics in Figure 9.3 are the type of exposure to

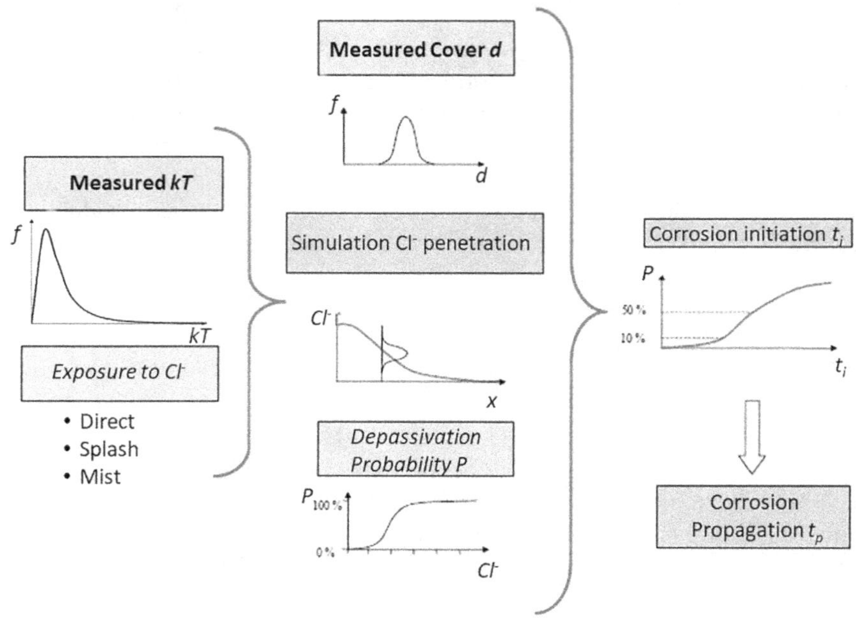

Figure 9.3 Probabilistic approach to model deterioration of highway bridges; adapted from Roelfstra (2001).

chlorides and the probability of steel depassivation as function of the total chlorides content. Figure 9.4 shows the probability of corrosion initiation of black steel as function of the chloride concentration at its level, adopted for the model. Similar probability distributions were proposed for different types of "corrosion-resistant" steel, with critical chloride thresholds that are 2, 3 and 5 times higher than that for carbon steel (Conciatori et al., 2018). A detailed analysis of the microclimatic conditions of structures exposed to de-icing salts solutions can be found in Conciatori et al. (2009a).

With the measured *kT*, the material concrete parameters to model the chloride ingress are defined, namely the sorptivity D_{cap} of the chloride solution, the coefficient of chloride-diffusion D_{Cl} and the coefficient of water diffusion D_w. The relation of *kT* with D_{Cl} and D_w, used in the TransChlor model, is presented in Table 9.4. The relation with D_{Cl} is represented by the dotted line in Figure 8.12, indicating that it is a reasonable assumption.

The "TransChlor" approach is quite original, as it is based on *in situ* measurements of *kT* and cover thickness *d* as input data (which have their own statistical distribution). The modelling is also innovative, as it is not based simply on the mathematical resolution of the second Fick's law to establish the chloride profile, but on a different approach. The mechanism of chloride penetration is conceived as, first, the rapid penetration of saline water by capillary suction, followed by the diffusion of chlorides through the water already penetrated in the concrete.

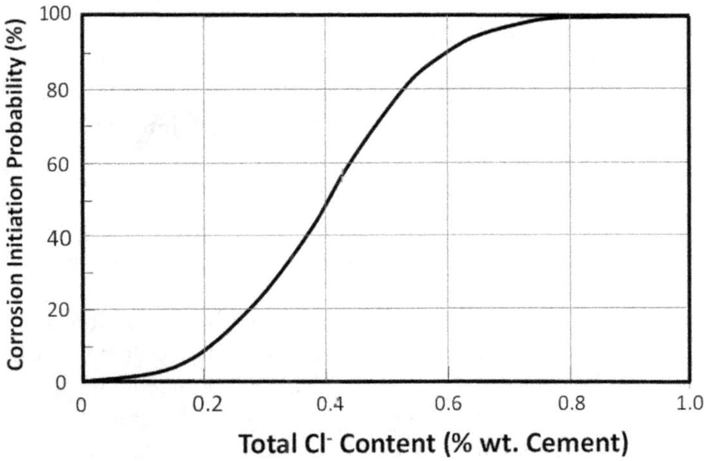

Figure 9.4 Probability of corrosion initiation as function of the total Cl^- content at "black" steel level (Roelfstra et al., 1999).

Table 9.4 Chloride and water diffusion coefficients as function of air-permeability kT (Roelfstra et al., 1999)

		D_{Cl} (10^{-12} m²/s)		D_W (10^{-12} m²/s)	
Cover concrete quality	*In situ kT* (10^{-16} m²)	*Mean*	*Std. dev.*	*Mean*	*Std. dev.*
Good	$kT < 0.2$	6	1.5	60	15
Average	$0.2 < kT < 2.0$	13	3.25	130	32.5
Bad	$kT > 2.0$	20	5.0	200	50

The transport of chloride ions is a function of the chloride ion diffusion through the pore water defined by the moisture content of concrete w, and the movement of entrained chloride ions dissolved in water through the concrete.

The main equations that model the penetration of chlorides into concrete are (Conciatori et al., 2008)

$$\frac{\partial C}{\partial t} = \text{div}\left(R_{\text{Cl}} \cdot c_f \cdot D_h \cdot \overrightarrow{\text{grad}(h_r)} + w(h_r,T) \cdot D_{\text{Cl}} \cdot \overrightarrow{\text{grad}(C)}\right) + R_{\text{Cl}} \cdot c_f \cdot \left(\overrightarrow{D_{\text{cap}}} \circ \overrightarrow{\text{grad}(h_r)}\right) \quad (9.17)$$

$$C = c_f \cdot w + c_f^{\beta} \cdot \gamma \quad (9.18)$$

where

C = total concentration of chlorides at depth x and time t

R_{Cl} = retardation of the chloride ion front with respect to the convection-induced water movement

c_f = concentration of free chlorides
D_h = water vapour diffusion coefficient (m²/s)
h_r = relative humidity of the concrete pores (-)
w = moisture content of the concrete (kg/m³)
T = temperature of the concrete (°C); taken into account through Arrehnius equation
D_{Cl} = coefficient of chloride-diffusion (m²/s)
D_{cap} = water sorptivity (mm/s)
β, γ = parameters of Freundlich isotherms

The complexity of the model is further increased by taking into account the carbonation process and its effect on freeing bound chlorides (Conciatori et al., 2010). The former is determined by the concentration of CO_2 in the environment, the molar concentration of $Ca(OH)_2$ and C–S–H in the concrete and the coefficient of diffusion of CO_2, as a function of air-permeability kT and moisture content w (Conciatori et al., 2008).

The boundary conditions can be realistically taken into account as time-dependent variables, involving temperature and RH of the air, precipitation and chloride concentration in the saline solution (function of the modality and application rate of de-icing salts).

Applying "TransChlor" requires a numerical resolution of the resulting complex system of differential equations, as described in Conciatori et al. (2008). Given its complexity, the probabilistic treatment of "TransChlor" model precludes the use of the Monte Carlo method; instead, the Rosenblueth method is used to analyse stochastic parameters in a numerical model to calculate the solution for independent variables (Conciatori et al., 2009b).

The resulting chloride profiles depart from the classical "error function law", showing a "hunch" relatively near the surface, which is often observed in actual chloride profiles of cores drilled from old structures.

A rather simple model for corrosion propagation was also included, defining corrosion rates as function of the exposure condition and of the *Covercrete* quality (defined by kT), see Table 9.5 (Roelfstra et al., 1999). It is recognized that the corrosion rate depends on many factors, such as O_2 availability, electrical resistivity of the concrete, the presence of cracks and the formation of macro-cells.

Table 9.5 Corrosion rates (mm/y) for different conditions (Roelfstra et al., 2009)

		Exposure zones		
Cover concrete quality	*In situ kT (10^{-16} m²)*	*Direct*	*Splash*	*Mist*
Good	$kT < 0.2$	0.004	0.02	0.02
Average	$0.2 < kT < 2.0$	0.004	0.02	0.02
Bad	$kT > 2.0$	0.02	0.08	0.08

9.4.2 Kurashige and Hironaga's Model for Carbonation-Induced Steel Corrosion

The model here discussed (Kurashige & Hironaga, 2015) is based on experimental results of kT and accelerated carbonation tests on a series of concrete mixes. Two series of three mixes each, with basically the same composition (w/c = 0.40; 0.50 and 0.60), one made with OPC (coded as N) and the other with BFSC (coded as B) were prepared in the laboratory. Specimens were prepared with the six mixes described above and subjected to four different curing conditions as detailed below, from better to worse:

- *W*: demoulded at 1 day and cured under water until the age of 28 days, then exposed
- *S*: demoulded at 5 or 7 days (for N and B concretes, respectively) and kept sealed until the age of 28 days, then exposed
- *5/7*: demoulded at 5 or 7 days (for N and B concretes, respectively), then exposed
- *1*: demoulded at 1 day, then exposed

After the indicated curing conditions, the specimens were exposed to 20°C/60% RH until the age of 91 days, when they were tested for kT and accelerated carbonation following Japanese standard (JIS, 2012).

The progress of accelerated carbonation could be expressed as

$$x_c = K_a \cdot t^{\beta} \tag{9.19}$$

where

x_c = carbonation depth (mm) at time t (years)
K_a = carbonation rate
β = accelerated carbonation index

Both K_a and β were related with kT (measured at 91 days), to the w/c ratio of the mix and to the cement type, through rather complex equations, obtained by test results data fitting.

Based on experimental results, an equivalent natural carbonation rate K'_c is calculated from K_a as follows:

$$K'_c = \frac{K_a}{21.2} \tag{9.20}$$

The design service life of the structure t_d (assumed equal to the corrosion initiation time) is computed as

$$t_d = 21.2^2 \cdot \left(\frac{x_{cr}}{K_a}\right)^{1/\beta} \tag{9.21}$$

$$x_{cr} = (0.9 \cdot d - 10)/(\gamma \cdot \beta_e) \tag{9.22}$$

where x_{cr} = critical carbonation depth (mm) and d = nominal cover thickness (mm)

γ is a combination of partial safety factors, usually adding to 1.32 and β_e an environmental factor, equal to 1.6 or 1.0 for dry and humid environments, respectively.

Using the equations relating K_a and β with kT and w/c ratio, from Eqs. (9.21) and (9.22) it is possible to build nomograms like the one shown in Figure 9.5, corresponding to the case OPC (N) and d = 40 mm. Similar charts are presented in Kurashige and Hironaga (2015), for other cases.

According to the model, a design service life of 100 years for a cover thickness of 40 mm can be achieved with a w/c ratio of 0.52 and $kT \approx 0.80 \times 10^{-16}$ m^2 for OPC concrete or with w/c = 0.46 and $kT \approx 1.5 \times 10^{-16}$ m^2. The charts for BFSC concrete, not shown here, reflect the higher sensitivity of this cement to carbonation. Also, if the OPC concrete is combined with a cover thickness of 60 mm, the 100-year service life can be attained with w/c = 0.58 and $kT \approx 1.0 \times 10^{-16}$ m^2.

For other cement types the system has to be recalibrated using experimental results of kT, K_a and β (Eq. 9.19), the last two parameters obtained through accelerated tests according to JIS (JIS, 2012).

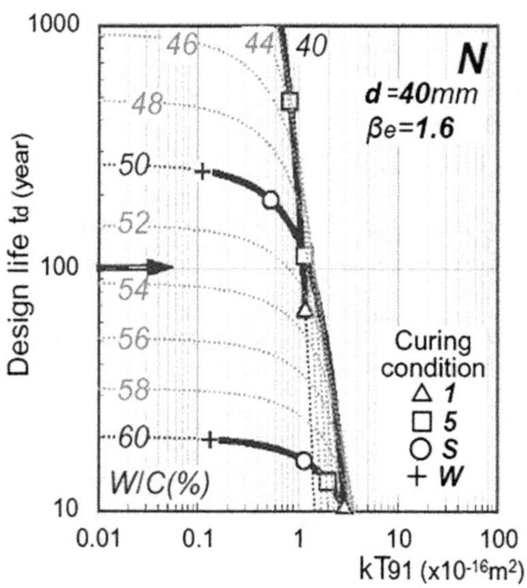

Figure 9.5 Nomogram for assessing the service life as function of *kT*, *w/c* ratio and cement type, example for OPC and 40 mm cover depth (Kurashige & Hironaga, 2015).

9.4.3 The "Exp-Ref" Method: Principles

The "Exp-Ref" method was originally developed by Torrent (2013, 2015) for chloride-induced steel corrosion, extended to carbonation-induced corrosion by Torrent and Fernández Luco (2014). It is applicable to relatively new structures, say not beyond 1 year of age. The general principles are the same for carbonation or chlorides and consist of three main elements:

- non-destructive experimental (hence "Exp") assessment of the permeability and thickness of the *Covercrete* through standard non-destructive site measurement of air-permeability kT (complemented by concrete surface moisture) (SIA 262/1, 2019) and cover thickness d (BS 1881-204, 1988; DBV, 2015)
- correlation between the measured coefficient of air-permeability and carbonation rate or chloride-diffusion
- definition of a reference condition (hence "Ref"), corresponding to a defined service life, that serves for calibration of the model

Figure 9.6 sketches the concept of the "Exp-Ref" method, taking as reference the requirements prescribed in standards (EN 206, 2013) and (EN 1992-1-1, 2004) (Eurocode 2).

For a given exposure severity of the structure (XC, XS or XD), the maximum water-cement ratio w/c_{max} and minimum cover depth d_{min} are prescribed by EN 206 (2013) and EN 1992-1-1 (2004), respectively (see Tables 9.6 and 9.7 for chloride- and carbonation-induced corrosion, respectively). These values are indicated in Figure 9.6 as white circles and

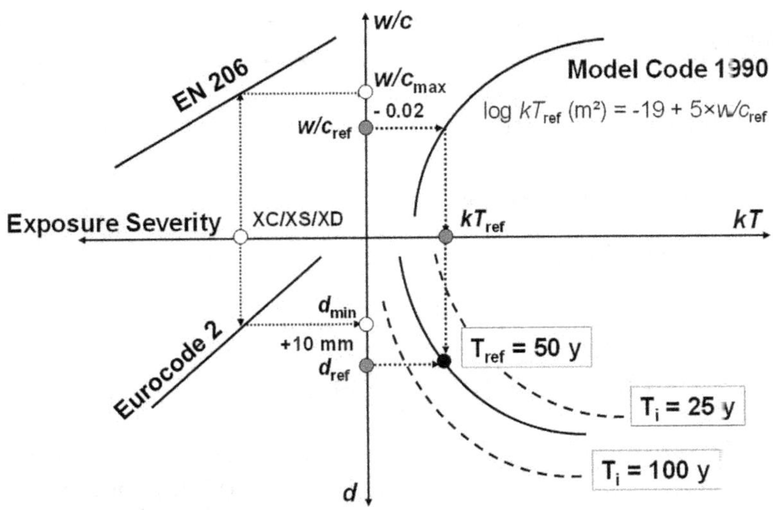

Figure 9.6 Sketch of the "Exp-Ref" method to estimate service life of new structures.

Table 9.6 Chloride exposure classes and parameters for service life estimate

<table>
<tr><th rowspan="3"></th><th colspan="6">Chloride exposure classes</th></tr>
<tr><th colspan="3">Sea water</th><th colspan="3">Other Cl[−] sources</th></tr>
<tr><th>XS1</th><th>XS2</th><th>XS3</th><th>XD1</th><th>XD2</th><th>XD3</th></tr>
<tr><td>w/c_{max}</td><td>0.50</td><td colspan="2">0.45</td><td colspan="2">0.55</td><td>0.45</td></tr>
<tr><td>d_{min} (mm)</td><td>35</td><td>40</td><td>45</td><td>35</td><td>40</td><td>45</td></tr>
<tr><td>w/c_{ref}</td><td>0.48</td><td colspan="2">0.43</td><td colspan="2">0.53</td><td>0.43</td></tr>
<tr><td>kT_{ref} (10^{-16} m^2)</td><td>0.25</td><td colspan="2">0.14</td><td colspan="2">0.45</td><td>0.14</td></tr>
<tr><td>d_{ref} (mm)</td><td>45</td><td>50</td><td>55</td><td>45</td><td>50</td><td>55</td></tr>
<tr><td>A</td><td>0.0156</td><td>0.0104</td><td>0.0086</td><td>0.0189</td><td>0.0153</td><td>0.0086</td></tr>
</table>

Table 9.7 Carbonation exposure classes and parameters for service life estimate

<table>
<tr><th rowspan="2"></th><th colspan="5">Carbonation exposure classes</th><th rowspan="2">Row</th></tr>
<tr><th>XC1a</th><th>XC1b</th><th>XC2</th><th>XC3</th><th>XC4</th></tr>
<tr><td>w/c_{max}</td><td colspan="2">0.65</td><td>0.60</td><td>0.55</td><td>0.50</td><td>1</td></tr>
<tr><td>d_{min} (mm)</td><td colspan="2">15</td><td colspan="2">25</td><td>30</td><td>2</td></tr>
<tr><td>w/c_{ref}</td><td colspan="2">0.63</td><td>0.58</td><td>0.53</td><td>0.48</td><td>3</td></tr>
<tr><td>kT_{ref} (10^{-16} m^2)</td><td colspan="2">1.41</td><td>0.79</td><td>0.45</td><td>0.25</td><td>4</td></tr>
<tr><td>d_{ref} (mm)</td><td colspan="2">25</td><td colspan="2">35</td><td>40</td><td>5</td></tr>
<tr><td>t_p (y)</td><td>45</td><td>10</td><td>10</td><td>25</td><td>2</td><td>6</td></tr>
<tr><td>t_{iref} (y)</td><td>5</td><td>40</td><td>40</td><td>25</td><td>48</td><td>7</td></tr>
<tr><td>β (y/mm^2)</td><td>0.24</td><td>1.91</td><td>0.78</td><td>0.38</td><td>0.42</td><td>8</td></tr>
</table>

correspond to an expected service life of 50 years. The reference values w/c_{ref} and d_{ref} are the target values that should be aimed at in order to satisfy those limiting values (grey circles in Figure 9.6):

$$w/c_{ref} = w/c_{max} - 0.02 \quad \text{Table 22 of EN 206 (2013)} \tag{9.23}$$

and

$$d_{ref} = d_{min} + 10\,\text{mm} \quad \text{Eq.(4.2) of EN 1992-1-1 (2004)} \tag{9.24}$$

The relation between gas-permeability and w/c ratio proposed in Eq. (2.1–107) of CEB-FIP (1991) Model Code 1990, applicable to air-permeability kT, as seen in Section 6.2.2.3, is used to convert the w/c_{ref} value into a reference air-permeability value kT_{ref} (grey circle in Fig. 9.6))

$$\log kT_{\mathrm{ref}}\left(10^{-16}\,\mathrm{m}^2\right) = -3 + 5\times w/c_{\mathrm{ref}} \tag{9.25}$$

A point of the concrete structure at which exactly the values kT_{ref} and d_{ref} are measured is expected to have t_{SL} = 50 years of service life (black circle in Figure 9.6). For 50 years' service life there are many compliant combinations of kT and d (full line in second quadrant). If the kT value measured is lower than kT_{ref} and/or the d value measured is higher than d_{ref}, the service life estimated at that point will be longer than 50 years, and vice versa (dotted lines).

In order to build the lines in the second quadrant, a relation between the rate of chloride or carbonation penetration as function of kT is needed, which is separately dealt with in Sections 9.4.3.1 and 9.4.3.2.

It shall be mentioned that the reference conditions could be those of codes or standards other than EN, provided that the prescriptive requirements are associated with a defined service life t_{SL}.

9.4.3.1 The "Exp-Ref" Method for Chloride-Induced Steel Corrosion

This approach is described in Torrent (2013, 2015). The corrosion initiation time t_i, due to the penetration of chlorides into concrete, by pure diffusion, is expressed by Eqs. (9.13) and (9.14), discussed in Section 9.2.2 and reproduced below:

$$t_i = \frac{d^2}{4\cdot D_0\cdot\left(t/t_0\right)^n}A^2 \quad \text{with } t = \mathrm{MIN}\left(t_i;t_d\right) \tag{9.13}$$

$$A = \frac{1}{erf^{-1}\left(1-\dfrac{C_{cr}}{C_s}\right)} \tag{9.14}$$

where

- d = cover thickness (mm)
- D_0 = coefficient of chloride-diffusion (mm²/y) measured at time t_0 (typically 28 days = 0.0767 year)
- n = exponent indicating the decay rate of D, with $0 < n < 1$
- t_d = duration of the diffusion coefficient decay period (y)
- C_s = surface chloride concentration (e.g. kg/m³)
- C_{cr} = critical chloride concentration in the concrete, capable of triggering the corrosion process (e.g. kg/m³)
- erf^{-1} = inverse error function

In Section 8.3.6.1, the correlation between the coefficients of chloride-diffusion D_0 and of air-permeability kT was discussed, resulting of the form indicated in Eq. (8.9) (for tests made typically at 28 days of age):

$$D_0 = 10\,\sqrt[3]{kT} \quad D_0\left(10^{-12}\,\frac{\text{m}^2}{\text{s}}\right);\,kT\,\left(10^{-16}\,\text{m}^2\right) \tag{8.9}$$

Introducing Eq. (8.9) into Eq. (9.13), we have

$$t_i = \frac{d^2}{40\cdot\sqrt[3]{kT}\cdot(t/t_0)^n}\,A^2 \quad \text{with}\, t = \text{MIN}(t_i;t_d) \tag{9.26}$$

and, applying Eq. (9.26) to the Reference case

$$t_{i\text{ref}} = \frac{d_{\text{ref}}^2}{40\cdot\sqrt[3]{kT_{\text{ref}}}\cdot(t/t_0)^n}\,A^2 \quad \text{with}\, t = \text{MIN}(t_i;t_d) \tag{9.27}$$

Dividing Eq. (9.26) by Eq. (9.27) and assuming that parameters C_s, C_{cr} (and therefore A), as well as the decay effect factor $(t/t_0)^n$, after decades of exposure, are the same for the reference case as for the investigated case, we get

$$t_i = t_{i\text{ref}}\,\frac{d^2}{d_{\text{ref}}^2}\,\frac{\sqrt[3]{kT_{\text{ref}}}}{\sqrt[3]{kT}} \tag{9.28}$$

or

$$t_i = \alpha\,\frac{d^2}{\sqrt[3]{kT}} \quad t_i\,(\text{years});\,d\,(\text{mm});\,kT\left(10^{-16}\,\text{m}^2\right) \tag{9.29}$$

$$\text{with} \quad \alpha = t_{i\text{ref}}\,\frac{\sqrt[3]{kT_{\text{ref}}}}{d_{\text{ref}}^2} \quad t_{i\text{ref}}\,(\text{years});\,d_{\text{ref}}\,(\text{mm});\,kT_{\text{ref}}\left(10^{-16}\,\text{m}^2\right) \tag{9.30}$$

The reference values w/c_{ref} and d_{ref} are calculated by applying Eqs. (9.23) and (9.24) to the limiting values w/c_{max} and d_{min}, specified in EN 206 (2013) and EN 1992-1-1 (2004), respectively, reported in Table 9.6. The reference air-permeability values kT_{ref} are computed with Eq. (9.25). Table 9.6 presents these parameters for the different chloride exposure classes and the coefficient α, computed with Eq. (9.30), for an expected corrosion initiation time $t_{i\text{ref}} = 50$ years.

In the case of chloride-induced corrosion, the corrosion propagation time is assumed as $t_p = 0$, therefore $t_{\text{SL}} = t_i$, hence

$$t_{\text{SL}} = \alpha\,\frac{d^2}{\sqrt[3]{kT}} \quad d\,(\text{mm});\,kT\left(10^{-16}\,\text{m}^2\right) \tag{9.31}$$

Equation (9.31) allows to estimate the expected service life at each point where parallel site measurements of d and kT are performed, considering the corresponding chloride exposure class of the element. Since the measurements of d and kT are non-destructive and fast, the evaluation can be made at many points in the structure, providing a probabilistic assessment of its service life (see real case of application in Section 11.3.2).

9.4.3.2 The "Exp-Ref" Method for Carbonation-Induced Steel Corrosion

This approach was first presented by Torrent and Fernández Luco (2014), which is reproduced here with some modifications. The starting point of the method is Hamada's Eq. (9.6), derived in Section 9.2.1, reproduced below.

$$CD = \sqrt{\frac{2 \cdot D \cdot C_s}{H}}\ \sqrt{t} \tag{9.6}$$

CD = penetration of carbonation front (mm) after t (years) of exposure
C_s = surface concentration of CO_2 (kg/m³)
D = coefficient of diffusion (mm²/y) of concrete to CO_2
H = unit $Ca(OH)_2$ content (kg/m³) of concrete

Taking into consideration Eqs. (9.1), (9.2) and (B.9) of DuraCrete (2000), Eq. (9.6) can be extended as follows:

$$CD = \sqrt{\frac{2 \cdot k_e \cdot k_c \cdot D_0 \cdot \left(\frac{t_0}{t}\right)^n \cdot C_s}{H}}\ \sqrt{t} \tag{9.32}$$

k_e = environment factor, function of cement type and exposure conditions (values in Table A.5 of DuraCrete (2000)). For laboratory conditions of exposure (65% RH) is $k_e = 1.0$
k_c = curing factor (values in Table A.12 of DuraCrete (2000))
D_0 = coefficient of CO_2 diffusion (mm²/y) measured at time t_0 (usually 28 days = 0.0767 year)
n = age factor, measuring the rate at which the coefficient of CO_2 diffusion decreases with time, function of cement type and exposure conditions (values in Table A.6 of DuraCrete (2000)). For laboratory conditions of exposure (65% RH) is $n = 0$.

In Section 8.3.7.1, the correlation between natural carbonation rate K_c and air-permeability kT was discussed. Based on laboratory results of natural carbonation at RH between 40% and 60%, supported by accelerated carbonation tests, a regression was established:

$$K_c = 1.67\ \ln\left(\frac{kT}{0.006}\right) \quad K_c\left(\text{mm}/\text{y}^{½}\right);\ kT\left(10^{-16}\ \text{m}^2\right);$$

$$\text{valid for } kT \geq 0.006 \times 10^{-16}\ \text{m}^2 \tag{8.10}$$

The carbonation rate can be derived from Eq. (9.32), remembering (DuraCrete, 2000) that, for laboratory tests with RH≤65%, is $n = 0$ and $k_e = 1$, whilst k_c (already included in kT) can be assumed as 1.0. Therefore, we have

$$K_c = CD/\sqrt{t} = \sqrt{\frac{2 \cdot C_s' \cdot D_0}{H}} \tag{9.33}$$

where C'_s is the natural CO_2 concentration in the laboratory where the specimens were stored.

Combining Eqs. (8.10) and (9.33), we get

$$2 \cdot \frac{D_0}{H} = \frac{2.79}{C_s'} \ln^2\left(\frac{kT}{0.006}\right) \tag{9.34}$$

Replacing Eq. (9.34) into (9.32)

$$CD = \sqrt{\frac{2.79 \cdot k_e \cdot \left(\frac{t_0}{t}\right)^n \cdot C_s \cdot \ln^2\left(\frac{kT}{0.006}\right)}{C_s'}}\ \sqrt{t} \tag{9.35}$$

The time of initiation of corrosion t_i is the time t at which the carbonation front CD has reached the cover thickness d. Therefore, from Eq. (9.35), t_i results

$$t_i = \frac{C_s'}{C_s} \frac{d^2}{2.79 \ln^2\left(\frac{kT}{0.006}\right) \cdot \left(\frac{t_0}{t_i}\right)^n \cdot k_e} \tag{9.36}$$

t_i = time of initiation of corrosion (years)
d = cover thickness (mm)

Equation (9.36) allows to estimate the initiation time t_i for carbonation-induced steel corrosion on the basis of site measurements of the cover thickness d and the air-permeability kT, provided that all other parameters in Eq. (9.36) are known, namely, C'_s, C_s, n and k_e. However, these parameters are difficult to know, especially the last two. Here enters the “Reference” aspect of the model, which is based on Table 9.7.

Table 9.1 showed the carbonation exposure classes and the limiting values w/c_{max} and d_{min} associated with them, according to EN 206 (2013) and EN 1992-1-1 (2004), respectively. They are transcribed in the first two rows of values in Table 9.7. From them, the reference values w/c_{ref} and d_{ref} are computed using Eqs. (9.23) and (9.24) and, from w/c_{ref}, the reference permeability values kT_{ref} are computed with Eq. (9.25). These values are indicated in rows 3–5 of Table 9.7.

As discussed in Section 9.2.1, for carbonation-induced corrosion, the propagation time cannot be neglected, as it may be the dominant factor of service life, particularly in dry environments. The corrosion propagation times t_p were calculated applying Eq. (9.9) to the corrosion rates shown in Figure 9.2 for the different exposure classes; they are indicated in row 6 of Table 9.7. The reference corrosion initiation times t_{iref}, which are the complement of t_p for a reference service life of 50 years, are indicated in row 7 of Table 9.7.

We can formulate the general case of Eq. (9.36) for the reference case:

$$t_{iref} = \frac{C_s^{'}}{C_s} \frac{d_{ref}^2}{2.79\ln^2\left(\frac{kT_{ref}}{0.006}\right)\cdot\left(\frac{t_0}{t_{iref}}\right)^n\cdot k_e} \tag{9.37}$$

If we divide Eq. (9.36) by Eq. (9.37) and assume that C_s and factor k_e are the same for the same exposure class, and that the decay factor $(t_0/t_i)^n$, after so many years of hydration, is also the same for the reference condition as for the investigated case, we can write

$$t_i = \frac{d^2}{d_{ref}^2}\cdot\frac{\ln^2\left(\frac{kT_{ref}}{0.006}\right)}{\ln^2\left(\frac{kT}{0.006}\right)}\cdot t_{iref} \tag{9.38}$$

Or, gathering all reference values in factor β:

$$t_i = \beta.\frac{d^2}{\ln^2\left(\frac{kT}{0.006}\right)} \tag{9.39}$$

with

$$\beta = \frac{\ln^2\left(\frac{kT_{ref}}{0.006}\right)}{d_{ref}^2}\cdot t_{iref} \tag{9.40}$$

The values of β, computed with Eq. (9.40), are indicated in the last row of Table 9.7, for the different carbonation exposure classes.

Indirectly, this approach is defining the rate of carbonation K_c as function of kT, for the different carbonation exposure classes, which becomes, from Eq. (9.39):

$$K_c = \frac{\ln\left(\frac{kT}{0.006}\right)}{\sqrt{\beta}} \qquad K_c = 0 \text{ for } kT < 0.006 \times 10^{-16}\,\text{m}^2 \tag{9.41}$$

Figure 9.7 shows graphically the relation K_c–kT of Eq. (9.41) for the different EN exposure classes.

It is interesting to see that, for the same kT, the carbonation rate K_c results highest for dry concrete (Class XC1a) and lowest for permanently wet concrete (Class XC1b), in agreement with the dotted line in Figure 9.2.

Finally, the service life of a structure can be calculated with Eq. (9.42) from values of d and kT measured on site, with parameters β and t_p obtained from Table 9.7 for the applicable carbonation exposure class.

$$t_{\text{SL}} = \beta \frac{d^2}{\ln^2\left(\frac{kT}{0.006}\right)} + t_p \quad \textit{valid for } kT \geq 0.006 \times 10^{-16}\,\text{m}^2 \tag{9.42}$$

Since the measurements of d and kT are non-destructive and fast, the evaluation can be made at many points in the structure, providing a probabilistic assessment of its service life.

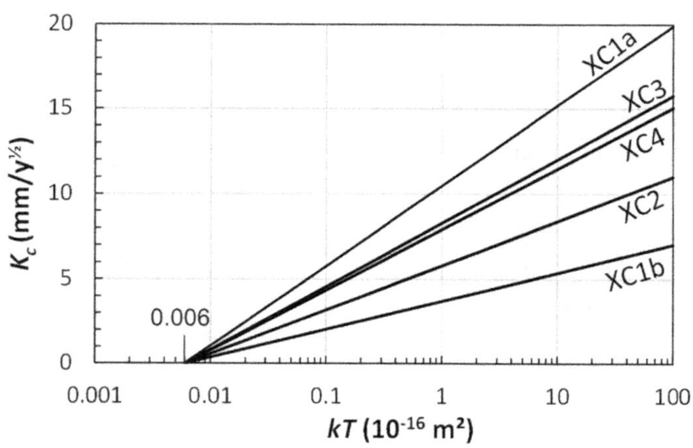

Figure 9.7 Carbonation rate K_c as function of air-permeability kT for EN carbonation exposure classes.

For some special cases, where CO_2 concentrations much higher than that for the reference case can be expected (e.g. tunnels, industrial areas), the initiation time can be modified by the factor C_{ref}/C where C_{ref} is the normal concentration of CO_2 in the air (typically 0.04%) and C is the concentration of CO_2 in the atmosphere to which the structural element will be exposed.

9.4.3.3 The CTK "Cycle" Approach

A more holistic approach was developed by Ebensperger (2020) as a part of a proprietary system called "CTK-ConcreLife®" (Construtechnik, 2019), registered in Chile.

This approach considers first a kind of a Cycle Approach including the different construction project phases, moving from *Theorecrete* → *Labcrete* → *Realcrete* (see Figure 9.8).

In this approach, in a first instance, the designer specifies design values of cover thickness d^d and air-permeability kT^d (both meant for the *Realcrete*), to achieve a design service life $t_{SL}{}^d$ under the expected exposure conditions (XC, XS, XD) linked to reinforcement steel corrosion. This can be done, using nomograms (built from Eqs. (9.31) and (9.42) for chloride- and carbonation-induced corrosion, respectively) like the one shown in Figure 9.9 for exposure class XS3. For instance, the designer can select, for a design service life $t_{SL}{}^d = 100$ years, a combination of $d^d = 50$ mm and $kT^d = 0.01 \times 10^{-16}$ m² or, alternatively, 70 mm and 0.07×10^{-16} m² (dotted lines in Figure 9.9).

Then, applying a transfer design factor δ^d (expected ratio between the jobsite kT^J and the laboratory kT^L), which is primarily function of the expected construction conditions, an equivalent laboratory kT^L is obtained (corresponding to well cured specimens tested at 28 days), value that has to be achieved by the concrete producer applying sound mix design practices (*Labcrete*).

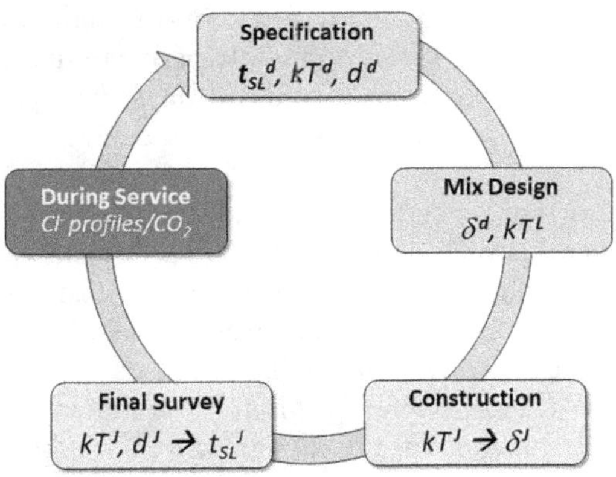

Figure 9.8 Sketch of the CTK "Cycle" approach.

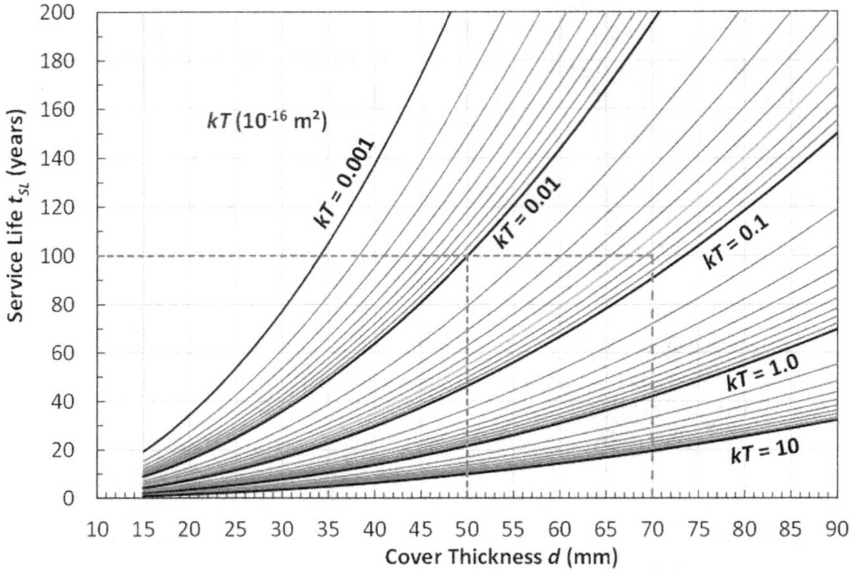

Figure 9.9 CTK design nomogram for exposure class XS3.

During construction, regular quality control samples are taken from the concrete delivered, to check that the required kT^L has been really achieved by the concrete producer. In parallel, jobsite measurements of kT^J are conducted on the finished elements built with the sampled concrete. This enables the calculation of δ^J (actual ratio between the jobsite kT^J and the laboratory kT^L), so as to verify that $\delta^J \leq \delta^d$. Same with cover thickness (using suitable covermeters) to check that the jobsite values are $d^J \geq d^d$. If not, the contractor has to improve the concreting techniques (placement, compaction, finishing, curing, positioning and fixing of the rebars) he is applying or, if not possible, order a concrete with a proportionally lower kT^L.

A survey of the cover thickness d^J and air-permeability kT^J of the structural elements (at ages between 28 and 180 days) will produce sufficient data for a probabilistic assessment of their service lives by the "Exp-Ref" method.

Finally, in important structures, it is advised to drill cores, from the structural elements themselves or from mock-up elements placed under the same exposure conditions to check the penetration of the carbonation front or the chloride profile, in order to take eventual remedial measures. They serve to calibrate the "Exp-Ref" model nomograms.

The core of the "*CTK-ConcreLife Model®*" contains a Design Module, allowing the user to specify the design cover thickness d^d and air-permeability kT^d, required to reach the design service life $t_{SL}{}^d$ and a Control Module where, entering the statistical results obtained at the jobsite for air

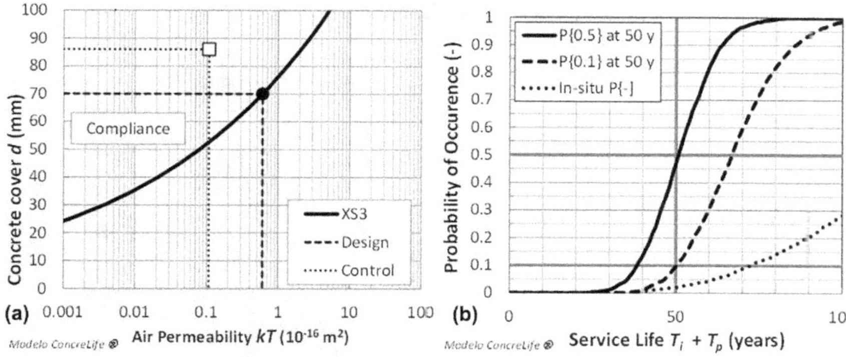

Figure 9.10 CTK-ConcreLife® model, example for exposure XS3: (a) design module chart; (b) control module chart.

permeability and cover thickness measurements, the probability of achieving the design service life is established.

An example of the model for exposure class XS3 is described with the help of Figure 9.10.

The Design Module shows that for a design cover thickness $d^d = 70$ mm, a design $kT^d = 0.60 \times 10^{-16}$ m² is needed for a 50% probability of reaching 50 years of $t_{SL}{}^d$ (black curves in Figure 9.10a and b. If the probability of occurrence of corrosion needs to be improved, e.g. to a 10%, a reduced design $kT^d = 0.26 \times 10^{-16}$ m² is required (broken line in Figure 9.10b).

The site testing produced mean values of $kT^J = 0.11 \times 10^{-16}$ m² and $d^J = 86$ mm (white square in Figure 9.10a), but with scatter much higher than specified, with a probability of less than 5% of corrosion at 50 years (dotted line in Figure 9.10b).

9.4.4 Belgacem et al.'s Model for Carbonation-Induced Steel Corrosion

This approach was developed by Belgacem, under the supervision of one of the authors and within the frame of a PhD internship (Belgacem et al., 2020). It follows the same theoretical principles of the approach proposed by Teruzzi (2009) for existing structures and further developed by Neves et al. (2018), presented in Section 9.5.2, but in this case, the principles were applied to air-permeability measurements carried out on young (non-carbonated concrete).

The approach assumes that the end of service life is defined by reinforcing steel depassivation due to concrete carbonation.

It is a full probabilistic service life design method, that allows the computation of failure probability for defined service lifetimes, based on concrete air-permeability coefficient and reinforcement cover of structures under

Table 9.8 Service life (years) of reinforced concrete according to Belgacem et al.'s model

	Exposure conditions					
	XC3			*XC4*		
	kT ($10^{-16}m^2$)			*kT* ($10^{-16}m^2$)		
Mean cover thickness (mm)	*0.01*	*0.1*	*1*	*0.01*	*0.1*	*1*
10	30	4.5	0.7	1.6	0.2	<0.1
20	125	19	2.8	37	5.4	0.8
30	300	44	6.5	125	18	2.7
40	540	80	12	250	37	5.5

exposure classes XC3 or XC4, as defined in EN 206 (2013), of concrete made with binder CEM II/A-L (EN 197-1, 2001).

The core of the model is a probabilistic relationship between concrete resistance to carbonation and *kT*, provided through a log-normal conditional probability density function. Besides, the model also encompasses an environmental factor to consider the difference between carbonation development under environmental conditions conforming to XC3 and to XC4. Belgacem et al. (2020) also recommend defining the allowed failure probability according to the consequences of attaining reinforcement depassivation in each of the considered exposure classes. They have adopted failure probabilities of 50% and 6.7%, for XC3 and XC4 exposure classes, respectively. Table 9.8 shows the model's results for the service life of reinforced concrete.

9.5 SERVICE LIFE ASSESSMENT OF EXISTING STRUCTURES APPLYING SITE *kT* TESTS

So far, the service life assessment of existing structures, on the basis of *kT* tests, has been developed only for carbonation-induced corrosion. Its applicability to structures exposed to chloride-rich environments is more complex, due to the strong effect of the type of exposure (in particular of the position of the structural element with respect to the chlorides sources) and to the lack of sufficient data.

Regarding carbonation, the "square root of time" rule, described by Eq. (9.7), is often used to estimate the service life of existing concrete structures. Indeed, measurements of the carbonation depth *CD* at time *t*, obtained destructively on drilled cores or fragments removed from the surface allow, by a simple application of Eq. (9.7), to know K_c and, therefore, to predict the corrosion initiation time t_i at which the carbonation front will reach the steel and depassivate it, i.e. when carbonation depth *CD* equals the cover thickness *d* (Eq. 9.8). *CD* is typically measured by spraying a pH indicator (usually phenolphthalein) on freshly broken concrete surfaces (CPC-18, 1988).

This simple approach faces two drawbacks: the destructive or damaging nature of the measurements and the high variability of *CD*, typically encountered in a single concrete structure, already discussed in Section 8.3.7.3. A wide range of *CD* was found in very old structures investigated in Chile (Rojas, 2006) (0–60 mm for a 98 years old bridge) and in not so old structures investigated in China (Liang et al., 2013) (0–140 mm in a 25 years old bridge); for more details, see Imamoto et al. (2016). As a result, in order to have a representative picture of the *CD* values in a structure, an unaffordable number of concrete samples is required.

Taking advantage of the good correlation observed between the carbonation rate K_c and the coefficient of air-permeability *kT* (Section 8.3.7), some service life models have been developed to minimize or even eliminate the need of testing *CD* on drilled cores.

9.5.1 Calibration with Drilled Cores

This approach was first used by Imamoto et al. (2013) for the condition assessment of Tokyo's National Museum of Western Art (NMWA), in order to assess the service life of the structure (this case is described in detail in Section 11.4.1). Here, the concept of the approach and some results are presented. Since the building belongs to the National Cultural Heritage of Japan, just three small cores were allowed to be drilled from the structure in order to measure the carbonation depth *CD*. Prior to proceeding with core drilling, the air-permeability *kT* was measured on the same spots. The values of *CD*, divided by the square root of 50 years (age at the investigation date), provided three pairs of data *kT*–K_c, which are plotted in Figure 9.11a (black dots).

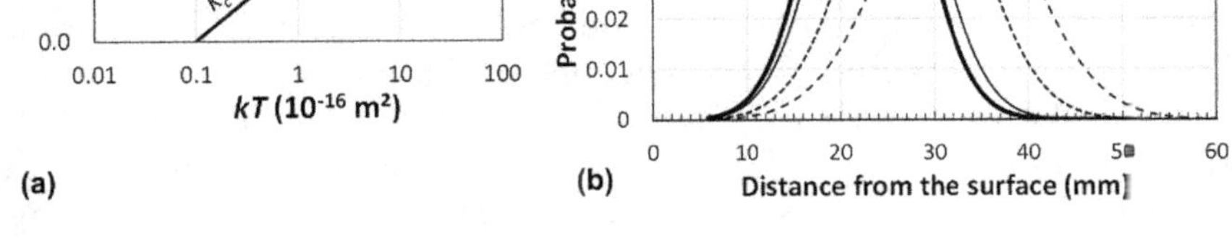

Figure 9.11 (a) Plot of measured *kT*–K_c values at Tokyio's NMWA and fitted regression line; (b) probability density functions of cover *d* and of estimated carbonation depth *CD* at 50, 75 and 100 years.

Imamoto et al. (2013) adopted the regression line that passed by the point (0.1; 0) and the point of the averages of K_c and kT of the three samples, line in Figure 9.11a, over which the regression equation is displayed.

At 111 different points of the structure's exposed exterior beams, parallel measurements of kT and the cover thickness d were made. The kT values were converted into estimated values of K_c through the regression equation shown in Figure 9.11a. Now, applying Eq. (9.7) the estimated carbonation depths CD can be calculated at different ages. Just for illustration purposes, the normal distributions corresponding to the cover thickness d (measured, thick full line) and the estimated carbonation depths CD at 50 (age of the investigation, thin full line), 75 and 100 years (broken lines) are presented in Figure 9.11b.

It can be seen in Figure 9.11b that, already at 50 years, the carbonation depth function is slightly to the right of the cover thickness d distribution, which indicates a small risk of failure, risk that increases as the distribution of CD moves to the right at longer ages. Imamoto et al. (2013) conducted a more interesting analysis of the risk of corrosion. At each of the 111 points, they calculated, at different ages, whether the corrosion initiation time had been reached:

If $CD\ (t) \geq d$ (failure), a value of 1 was attributed, otherwise, a value of 0 was given.

Then by adding the amount of failure cases and dividing them by 111, a probability of corrosion initiation was obtained as a function of age, as displayed in Figure 9.13 (white circles linked by full line). Interestingly, the model predicts 5.7% of corrosion initiation at 50 years of age (time of the investigation), which is consistent with the isolated corrosion cases observed.

The Architects' Institute of Japan recommends an intervention when the risk of corrosion reaches 20%. From the reported investigation at an age of 50 years, this should take place at an age of 75 years (see Imamoto et al.'s results in Figure 9.13), i.e. ≈25 years later.

9.5.2 Pure Non-destructive Approach

Another approach was developed to assess the probability of carbonation-induced steel corrosion in existing structures, in this case without the need of drilling cores for calibration. It was initially formulated by Teruzzi (2009), who applied it to a single building in Canton Ticino, Switzerland, later validated on a large number of structures by Neves et al. (2018).

The experimental basis of the method lies on the large number of parallel data of carbonation depth CD and kT, obtained from several old structures in Japan, Portugal and Switzerland (see Section 8.3.7.3). The ages of the structures ranged between 10 and 50 years; in order to combine the data, the carbonation rate K_c was calculated from the measured CD and the age of the measurement t, applying Eq. (9.7).

For a better understanding of the approach, the reader is referred to Figure 8.16. What can be observed in it is that, for air permeabilities in the low and very low range (see classes at the top of Figure 8.16), the carbonation rates measured are low (typically < 2 mm/y^{½} or 20 mm CD in 100 years). For higher values of kT, the values of K_c, in average, tend to increase, but the scatter grows too. This means that, if low values of kT are recorded, there is certainty that the carbonation rate will be low but, for high values of kT there is much uncertainty, because the K_c value may be low or high.

So, for a given measured value of kT, there is a probabilistic distribution of K_c values, that (Teruzzi, 2009) assumed as a Weibull distribution. That approach was applied by Neves et al. (2018) to the data in Figure 8.16, finding that the best fit of Weibull's function was

$$f(K_c \mid kT) = \frac{1.823}{K_c} \cdot \left(\frac{K_c}{0.848 \cdot \log_{10}(kT) + 2.636} \right)^{2.444} \cdot e^{-0.746 \left(\frac{K_c}{0.848 \cdot \log_{10}(kT) + 2.636} \right)^{2.444}} \tag{9.43}$$

where

$f(K_c|kT)$ is the conditional probability density function of K_c (mm/y^{½}) given a certain kT (10^{-16} m^2) value.

Figure 9.12a presents the plot of $f(K_c)$, following the Weibull distribution of Eq. (9.43), for different kT values.

Figure 9.12a reflects well the fact (shown in Figure 8.16) that both the expected (central) value of K_c and the scatter grow with increasing measured values of kT.

The importance of Eq. (9.43) lies in the fact that, for each kT value measured non-destructively, a probabilistic distribution of K_c can be inferred. If the cover thickness d is also measured non-destructively at the same or nearby location, then a corrosion initiation time can be estimated under optimistic, expected or pessimistic scenarios (Neves et al., 2018). These

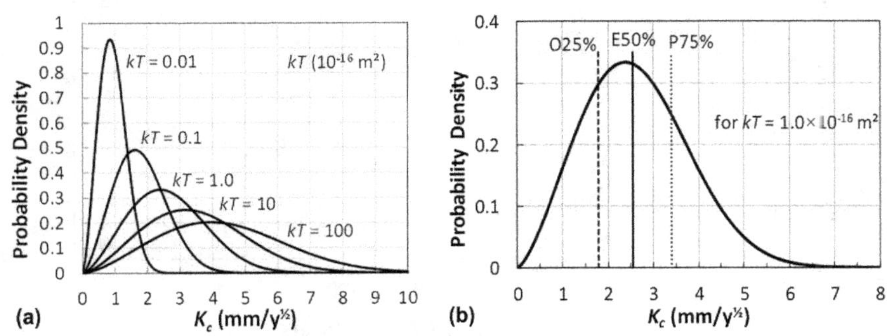

Figure 9.12 (a) Probability density functions of K_c, for several measured kT values; (b) K_c values corresponding to Optimistic, Expected and Pessimistic scenarios for $kT = 1.0 \times 10^{-16}$ m^2.

scenarios correspond to the K_c fractiles of 25%, 50% and 75% probability, see Figure 9.12b for the case in which the measured $kT = 1.0 \times 10^{-16}$ m².

The fractile of the Weibull distribution corresponding to a P (%) probability is

$$K_c\,(P) = f\left[-\ln\left(1 - \frac{P}{100}\right)\right]^{\frac{1}{g}} \tag{9.44}$$

With parameters:

$g = 2.444$

$$f = \frac{0.848\log_{10}(kT) + 2.636}{0.746^{\frac{1}{g}}} \tag{9.45}$$

For instance, for $kT = 1.0 \times 10^{-16}$ m² is $f = 2.972$ and K_c (25%) = 1.78 mm/y$^{1/2}$; K_c (50%) = 2.56 mm/y$^{1/2}$ and K_c (75%) = 3.40 mm/y$^{1/2}$, values represented by the vertical lines in Figure 9.12b.

If, for instance, a cover thickness d = 25 mm had been measured at the same point, the initiation time t_i can be calculated, applying Eq. (9.8), for the three scenarios: Optimistic, Expected and Pessimistic, giving: 197, 95 and 54 years, respectively.

This approach (Neves et al., 2018) has been applied to the 111 parallel data of kT and d, collected from Tokyo's National Museum of Western Art (NMWA), discussed in Sections 9.5.1 and 11.4.1, thus obtaining 111 t_i results, for each scenario, presented as probability curves in Figure 9.13.

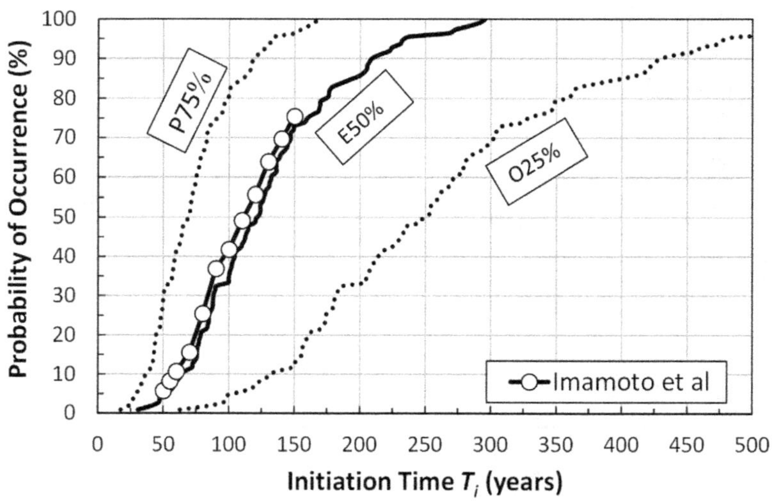

Figure 9.13 Probability of occurrence of corrosion initiation time t_i of NMWA, under different scenarios.

The three curves in Figure 9.13 represent the probability of occurrence of the corrosion initiation time t_i, under pessimistic, expected (full line) and optimistic scenarios, respectively, applying exclusively the (Neves et al., 2018) purely non-destructive approach (no use was made of the carbonation of the drilled cores). The dots linked by a line correspond to the assessment made by Imamoto et al. (2013), based on the carbonation measured on the drilled cores (Section 9.5.1). The coincidence with the expected non-destructive prediction is remarkable.

The validity of this approach has been confirmed by a recent experimental investigation, conducted on a bridge in Roding, Germany (Maack et al., 2021).

REFERENCES

AIJ (2016). "Recommendations for durability design and construction practice of reinforced concrete buildings". Architectural Institute of Japan (in Japanese).

Alexander, M.G. (2018). "Service life design and modelling of concrete structures – Background, developments, and implementation". *Revista ALCONPAT*, v8, n3, 224–245.

Alexander, M.G., Ballim, Y. and Stanish, K. (2008). "A framework for use of durability indexes in performance-based design and specifications for reinforced concrete structures". *Mater. & Struct.*, v41, 921–936.

Alexander, M.G., Bentur, A. and Mindess, S. (2017). *Durability of Concrete – Design and Construction*. CRC Press, Boca Raton, FL, 346 p.

Alexander, M.G. and Beushausen, H. (2009). "Performance-based durability testing, design and specification in South Africa: Latest developments". *Excellence in Concrete Construction through Innovation*. Limbachiya, M.C. and Kew, H.Y. (Eds.), Taylor & Francis, London, UK.

Andrade, C. (2006). "Multilevel (four) methodology for durability design". RILEM Proceedings PRO 47, 101–108.

Andrade, C. (2014). "Prediction of service life by considering the initiation and propagation periods". Keynote Lecture, RILEM Technical Day, São Paulo, Brazil, https://www.youtube.com/watch?v=NruQiViNU4c.

Angst, U., Elsener, B., Larsen, C.K. and Vennesland, Ø. (2009). "Critical chloride content in reinforced concrete – A review". *Cem. & Concr. Res.*, v39, n12, 1122–1138.

ASTM C1202 (2010). "Standard test method for electrical indication of concrete's ability to resist chloride ion penetration".

Belgacem, M., Neves, R. and Talah, A. (2020). "Service life design for carbonation-induced corrosion based on air-permeability requirements". *Constr. & Build. Mater.*, v261, 120507.

Bertolini, L., Elsener, B., Pedeferri, P. and Polder, R. (2004). *Corrosion of Steel in Concrete*. Wiley-VCH Verlag GmbH & Co., Weinheim, Germany, 394 p.

Beushausen, H. and Fernández Luco, L. (Eds.) (2016). "Performance-based specifications and control of concrete durability". RILEM State-of-the-Art Reports, v18, Chapters 10 & 11, 301–360.

Böhni, H. (2005). *Corrosion in Reinforced Concrete Structures*. CRC Press Ltd., Boca Raton, FL, 241 p.

BS 1881-204 (1988). "Testing concrete. Recommendations on the use of electromagnetic covermeters".

CEB-FIP (1991). "CEB-FIP model code 1990". Final Draft, CEB Bulletin d'Information N° 203, 204 and 205, Lausanne, Switzerland, July 1991.

Conciatori, D. (2005). "Effet du microclimat sur l'initiation de la corrosion des aciers d'armature dans les ouvrages en béton armé" (in French). Ph.D. Thesis, EPFL, Lausanne, 264 p.

Conciatori, D., Brühwiler, E. and Dumont, A-G. (2009a). "Actions microclimatique et environnementale des ouvrages d'art routiers". *Rev. Can. Génie Civ.*, v36, 628–638.

Conciatori, D., Brühwiler, E. and Gysler, R. (2011). "Brine absorption in concrete at low temperature: Experimental investigation and modeling". *J. Mater. Civil Eng.*, ASCE, June, 846–851.

Conciatori, D., Brühwiler, E. and Linden, C. (2018). "Numerical simulation of the probability of corrosion initiation of RC elements made of reinforcing steel with improved corrosion performance". *Struct. & Infrastruct. Eng.*, v14, n11, 1446–1454.

Conciatori, D., Brühwiler, E. and Morgenthaler, S. (2009b). "Calculation of reinforced concrete corrosion initiation probabilities using the Rosenblueth method". *Int. J. Reliability Safety*, v3, n4, 345–362.

Conciatori, D., Laferrière and Brühwiler, E. (2010). "Comprehensive modeling of chloride ion and water ingress into concrete considering thermal and carbonation state for real climate". *Cem. & Concr. Res.*, v40, 109–118.

Conciatori, D., Sadouki, H. and Brühwiler, E. (2008). "Capillary suction and diffusion model for chloride ingress into concrete". *Cem. & Concr. Res.*, v38, n12, 1401–1408.

Construtechnik (2019). "Concrete service life design & control. CTK-ConcreLife® model".

CONTECVET (2001). "A validated user manual for assessing the residual life of concrete structures". EC Innovation Programme IN 30902I, Madrid, Spain, 139 p.

CPC-18 (1988). "Measurement of hardened concrete carbonation depth". RILEM Recommendation, 3 p.

CSA A23.1 (2006). "Concrete materials and methods of concrete construction". Canadian Standards.

DBV (2015). "Betondeckung und Bewehrung nach Eurocode 2". DBV Merkblatt, (in German).

Di Pace, G. and Torrent, R. (2020). "Service life of reinforced concrete structures in marine environments using non-destructive site methods". fib *International Conference on Concrete* Sustainability, Prague, Czech Republic, September 16–18.

Duracrete (2000). "Probabilistic performance based durability design of concrete structures". The European Union–Brite EuRam III, BE95–1347/R17, CUR, Gouda, The Netherlands.

Ebensperger, L. (2020), "Modelo CTK-ConcreLife® Diseño y Control por Vida Útil", 2° *Simposio Nacional de Concreto y Construcción, 6° ExpoConVial*, Lima, Perú.

EN 197-1 (2001). "Cement–Part 1: Composition, specification and conformity criteria for common cements". European Standards.
EN 206 (2013). "Concrete – Specification, performance, production and conformity". European Standards.
EN 1992-1-1 (2004). "Eurocode 2: Design of concrete structures – Part 1: General rules and rules for buildings". European Standards.
EN 13670 (2009). "Execution of concrete structures". European Standards.
fib (2006). "Model code for service life design". fib Bulletin No. 34, Lausanne, February, 116 p.
Gulikers, J. (2006). "Considerations on the reliability of service life predictions using a probabilistic approach". *J. Phys. IV*, France, v. 136, 233–241.
Hamada, M. (1968). "Neutralization (carbonation) of concrete and corrosion of reinforcing steel". *Proceedings on* 5th *International Symposium on Chemistry of Cement*, Tokyo, Japan, 343–369.
Hunkeler, F., Kronenberg, P. and Chappex, Th. (2013). "Résistance à la carbonatation - une nouvelle exigence concernant les bétons". Bulletin TFB, n.4, Wildegg, Switzerland, November, 2 p.
Imamoto, K., Neves, R. and Torrent, R. (2016). "Carbonation rate in old structures assessed with air-permeability site NDT". *IABMAS* 2016, Paper 426, Foz do Iguaçú, Brazil, June 26–30.
Imamoto, K., Tanaka, A. and Kanematsu, M. (2013). "Non-destructive assessment of concrete durability of the National Museum of Western Art in Japan". *CONSEC'13*, Nanjing, China, September 23–25, 1335–1344.
JIS (2012). "Method of accelerated carbonation test for concrete". Japan Industrial Standard JIS A 1153.
Kurashige, I. and Hironaga, M. (2015). "Criteria for nondestructive inspections using Torrent air permeability test for lifespan of concrete structures". *Life-Cycle of Structural Systems*, Furuta, Frangopol and Akiyama (Eds.), Taylor & Francis, London, UK, 1221–1228.
Li, K. (2016). *Durability Design of Concrete Structures*. John Wiley & Sons, Fusionopolis Walk, 280 p.
Liang, M-T., Huang, R. and Fang, S-A. (2013). "Carbonation service life prediction of existing concrete viaduct/bridge using time-dependent reliability analysis". *J. Marine Sci. Technol.*, v21, n1, 94–104.
Life-365 (2018). "Service life prediction model and computer program for predicting the service life and life-cycle cost of reinforced concrete exposed to chlorides". Life 365 Consortium II, User Manual, 80 p.
Maack, S., Torrent, R., Ebell, G., Völker, T. and Küttenbaum, S. (2021). "Testing to reassess – Corrosion activity assessment based on NDT using a prestressed concrete bridge as case-study". Submitted to EUROSTRUCT2021, Padova, Italy, August 29–September 1.
Neves, R., Torrent, R. and Imamoto, K. (2018). "Residual service life of carbonated structures based on site non-destructive tests". *Cem. & Concr. Res.*, v109, 10–18.
Nganga, G., Alexander, M.G. and Beushausen, H. (2013). "Practical implementation of the durability index performance-based design approach". *Constr. & Build. Mater.*, v45, 251–261.
Parrott, L. (1994). "Design for avoiding damage due to carbonation-induced corrosion". ACI SP-145, 283–298.

Roelfstra, G. (2001). "Modèle d'évolution de l'état des ponts-routes en béton" (in French). Ph.D. Thesis, EPFL, Lausanne, 175 p.

Roelfstra, G., Adey, B., Hajdin, R. and Brühwiler, E. (1999). "The condition evolution of concrete bridges based on a segmental approach, non-destructive test methods and deterioration models". 78th *Annual Meeting Transportation Research Board*, Denver, April, 13 p.

Roelfstra, G., Hajdin, R., Adey, B. and Brühwiler, E. (2004). "Condition evolution in bridge management systems and corrosion-induced deterioration". *J. Bridge Eng.*, ASCE, v9, 268–277.

Rojas, K., L.A. (2006). "Estudio de la durabilidad de estructuras antiguas de hormigón armado, con énfasis en la corrosión de las armaduras". Civ. Eng. Dissert., Univ. de Chile, Santiago, December, 113 p.

SIA 262/1 (2019). "Construction en béton – Spécifications complémentaires". Swiss Standard.

SN EN 206 (2016). "Béton - Partie 1: Spécification, performance, production et conformité". Swiss Standard, 2013, updated in 2016.

Teruzzi, T. (2009). "Estimating the service-life of concrete structures subjected to carbonation on the basis of the air permeability of the concrete cover". EUROINFRA, Helsinki, Finland, October 15–16.

Torrent, R. (2013). "Service life prediction: Theorecrete, Labcrete and Realcrete approaches". Keynote Paper, *SCTM3 Conference*, Kyoto, Japan, August 18–21.

Torrent, R. (2015). "Exp-ref: A simple, realistic and robust method to assess service life of reinforced concrete structures". *Concrete* 2015, Melbourne, Australia, August 30–September 2.

Torrent, R. (2017). "Robust, Engineer's friendly service life design". Keynote Paper, *EASEC-15*, October 11–13, Xi'an, China.

Torrent, R. and Fernández Luco, L. (2014). "Service life assessment of concrete structures based on site testing". *DBMC* 2014 *Conference*, São Paulo, Brazil, September 1–5.

Tuutti, K. (1982). "Corrosion of steel in concrete". Research Report No.4.82, Swedish Cement and Concrete Research Institute (CBI), Stockholm, Sweden, 468 p.

Chapter 10

The role of permeability in explosive spalling under fire

10.1 EFFECT OF FIRE ON REINFORCED CONCRETE STRUCTURES

One of the most severe actions occasionally affecting reinforced concrete structures is their exposure to fire. Although concrete structures, when properly designed and built, offer a good resistance to fire, they undergo extensive damage that, under unfavourable circumstances, may cause their partial or total collapse. This results in loss of life and heavy financial losses, with the consequent negative repercussions in the media.

During a fire, the temperature may reach up to 1,100°C in buildings and even up to 1,350°C in tunnels, leading to severe damage in a concrete structure (*fib*, 2007). However, in some special cases, even much lower temperature may cause explosive destruction of concrete, thus endangering the bearing capacity of the structural element.

The focus of this chapter is placed on a phenomenon occurring when concrete structures are exposed to fire, called "explosive spalling", by which the concrete cover that protects the steel from the heat and flames is broken-off, leaving the rebars exposed to the fire (see Figure 10.1 (Bärtschi et al., 2018)). This phenomenon is more frequent in concretes of high gas-tightness, such

Figure 10.1 Spalling under fire of real structure (Bärtschi et al., 2018).

DOI: 10.1201/9780429505652-10

as Self Compacting and High-Strength Concrete (HSC). The fact that these concretes find application in structures where human lives are at most risk, e.g. high-rise buildings or tunnel linings, adds to the relevance of the topic.

Although concrete, as a material, suffers a loss of performance at high temperatures, steel is much more sensitive, becoming in this respect the weakest link in the reinforced concrete system. Due to its rather low coefficient of thermal conductivity, the conduction of heat through concrete is relatively slow which, if the cover is sufficiently thick, protects the steel from reaching too rapidly high temperatures that may put the structure into risk of collapse, thus giving opportunity to the occupants of the facility to escape.

This protective action of the concrete cover may vanish if explosive spalling takes place. Since spalling may occur already after 30 minutes of exposure to fire (Sertmehmetoglu, 1977), the time to collapse of the affected structural elements may become drastically reduced. Recently, severe cases of explosive spalling have occurred in tunnel fires (Channel Tunnel – 1996, Mont Blanc Tunnel – 1999, Gotthard Tunnel – 2001), causing casualties and important financial losses.

It has been found that HSC is more prone to experiencing explosive spalling than conventional concrete. As HSC is being increasingly used in tall buildings (for structural reasons) and in tunnels (for durability reasons), the failure of which under fire usually puts many people at risk, the phenomenon has been gaining interest among researchers. For instance, under the inspiring leadership of Prof. Mario Fontana, a series of investigations have been conducted at the ETHZ (Zürich, Switzerland) to better understand, model and predict this problem of high complexity (Klingsch, 2014; Lu, 2015; Van der Merwe, 2019).

10.2 EXPLOSIVE SPALLING OF CONCRETE COVER

From the scientific point of view, the problem involves different coupled phenomena: physical (heat transfer, change of state, mass transport), chemical (effect of heat on concrete constituents) and mechanical (state of stress and fracture), in a concrete material that is gradually increasing its temperature and, hence, changing its properties accordingly. For instance, it is generally accepted that the thermal conductivity of concrete decreases by a factor of ≈3 between 20°C and 900°C (EN 1992-1-2, 2004).

Good descriptions of the phenomena involved in concrete spalling can be found in Hager (2013), Jansson (2013), Lu (2015) and Van der Merwe (2019).

An important aspect of explosive spalling is the pore pressure build-up during heating, in which gas-permeability plays a prominent role. An early theory on the causes of explosive spalling, that still today finds acceptance, is based on the "moisture clog" concept (Shorter & Harmathy, 1961; Harmahty, 1965). As soon as concrete heats up to and above 100°C, evaporation of the free moisture in its pores takes place. The resulting pressure gradient drives part of the vapour towards the heated surface layers that escapes to the atmosphere whilst another part is driven towards the inner, cooler regions of the element. If the flow of vapour is not hindered, the pore pressure remains relatively low, see Figure 10.2a.

However, when vapour reaches the cooler zones it condenses, thus raising the moisture content in those regions. The zone near the surface regions dries out whilst the inner parts may become saturated. The low gas-permeability in the saturated zone makes further vapour movement nearly impossible, resulting in a "moisture or water clog", see Figure 10.2b. Depending on the heating rate and the gas-permeability of the concrete, two opposing forces result and govern the movement of the moisture clog. The pressure gradient between the high pore pressure in the saturated zone and atmospheric pressure at the heated concrete surface forces the moisture clog towards the heated surface. The temperature differential over the concrete cross section, on the other hand, results in a heat flux that forces the moisture clog deeper into the concrete section. When the moisture clog becomes trapped in this manner, the pore pressure in this zone increases until the tensile strength of the heated concrete is exceeded and explosive spalling occurs. A visual observation of the "moisture clog" was obtained by Jansson (2013) during a fire test of a slab.

Another explanation for spalling was offered by Bažant (2005). According to his theory, spalling results from brittle fracture and delamination buckling caused by compressive biaxial thermal stresses parallel to the heated surface. It has been claimed (Mindeguia, 2009) that the internal pore pressure build-up cannot be the only reason for spalling because when a crack is opened, the

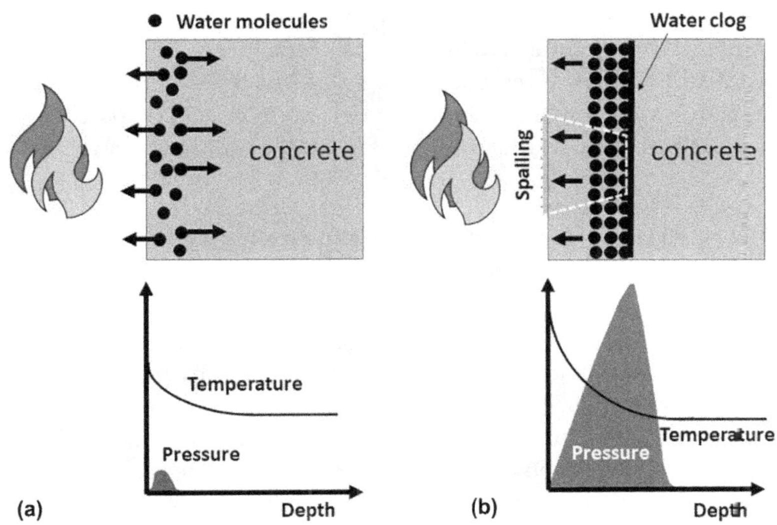

Figure 10.2 Scheme of pore pressure build up due to "moisture clog"; adapted from Hager (2013).

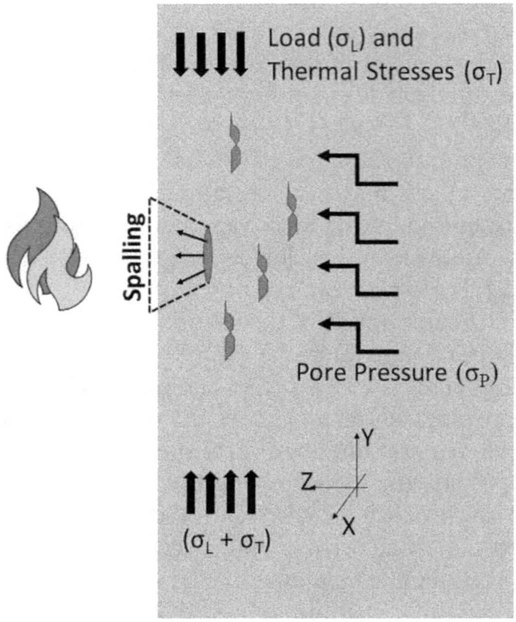

Figure 10.3 Sketch of Lu's model; adapted from Bärtschi et al. (2018).

internal pore pressure is immediately relieved. Thus, it has been concluded that the pore pressure can only act as a trigger for the spalling phenomenon. Once the pore pressure has triggered the crack, its growth and the resulting explosive spalling depend predominantly on the thermal stresses.

A more complex model, combining different theories, was proposed by Lu (2015), further developed by Van der Merwe (2019). These models contemplate three combined actions: pore pressure, thermal stresses and load-induced stresses, as illustrated in Figure 10.3 (Bärtschi et al., 2018).

10.3 THE ROLE OF CONCRETE PERMEABILITY IN EXPLOSIVE SPALLING

In the "moisture clog" theory, if the concrete is permeable enough, the water vapour will find open paths to move through, thus avoiding the build-up of high pore pressures and spalling.

Based on the "moisture clog" theory, Harmathy (1965) derived analytical expressions and proposed a limiting criterion for the pore saturation at which spalling can be expected. His analytical expressions were used to develop a curve relating the critical degree of saturation as function of the gas-permeability of the concrete, separating regions of risk and no-risk of spalling (see dotted line in Figure 10.4).

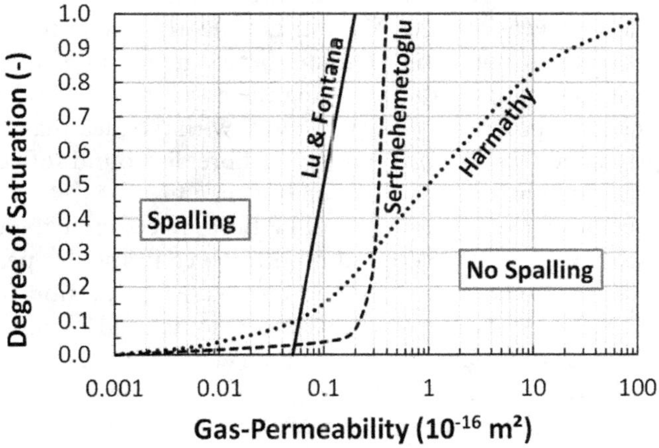

Figure 10.4 Spalling criteria based on intrinsic permeability and saturation degree of concrete; adapted from Lu and Fontana (2016), Van der Merwe (2019).

Sertmehmetoglu (1977) modified this model by accounting for the permeability and viscosity of vapour and the pressure difference between the moisture clog and the heated surface concrete to develop an alternative limit criterion. A comparison of his model (segmented line in Figure 10.4) with that of Harmathy (1965) indicates a much narrower permeability range associated with the boundary of spalling risk. Lu and Fontana (2016) conducted simulations on an unloaded slab, built with a rather poor concrete of 2.5 MPa tensile strength subjected to the ISO fire curve and developed the relation shown as a full line in Figure 10.4. Sertmehemetoglu (1977) and Lu (2015) indicate that the limit curves will shift to the left or right for concrete with greater or lower tensile strength, respectively.

Based on air-permeability *kT* measurements conducted by Lu (2015) and Lu and Fontana (2016) have proposed a limiting value of air-permeability $kT=0.2\times10^{-16}$ m^2, below which explosive spalling is prone to occur. This value is on the safe side, as it has been computed assuming a relatively weak concrete of 2.5 MPa tensile strength.

10.4 COPING WITH HSC

Despite their higher tensile strength, HSC is more prone to suffer explosive spalling than conventional concrete. This happens because its tighter microstructure, responsible for its higher strength, is usually accompanied by very low permeability, certainly lower than the limit proposed by Lu and Fontana (2016).

Intuitively, one way of making HSC more resistant to fire would be to reduce its moisture content (e.g. by drying in a precast plant). However, a series of tests brought (Diederichs et al., 2009) to the conclusion that moisture contents between 2% and 5% did not clearly influence the spalling behaviour of concretes of strength grades between 60 and 90 MPa.

The other, ingenious way of improving the resistance of HSC to fire was the introduction of polypropylene (PP) fibres, with the idea that they would melt at high temperature (melting point of PP fibres is around 160°C–170°C), thus relieving the pore pressure built-up during a fire.

A first attempt to assess the impact of PP fibres in concrete under fire was unsuccessful. Hannant (1978) found no difference in the fire performance of 0.9 m^2×50 mm panels tested at the Fire Research Station in the UK, made with concrete without and with 1.2% volume of PP fibres.

Later, numerous tests have shown (Nishida et al., 1995; Hoff et al., 2000; Kalifa et al., 2001; Niknezhad et al., 2019) that PP fibres improve the stability of high-performance concrete exposed to high temperature. In Section 6.8, it has been shown that the presence of PP fibres greatly increases the "Hot" and "Residual" air-permeability *kT* at temperatures above 150°C.

The effect of the type and geometry of fibres on the pore pressure of a HSC (*w/c*=0.30) at elevated temperatures was investigated by Bangi and Horiguchi (2012). They incorporated 0.1% vol. of PP monofilament fibres and of polyvinyl alcohol (PVA) fibres of different diameters and lengths, as well as hybrid systems, consisting of a combination of StF and PP or PVA fibres, totalling seven fibres systems (as well as control mixes without fibres). With each fibre system HSC, Ø175×100 mm cylinders were cast, containing pressure gauges capable of measuring the pore pressure at 10, 30 and 50 mm distance from the heated face; 3 months of curing under water was applied prior to heating. The pore pressure was monitored during heating up to 600°C. The main conclusions of the research by Bangi and Horiguchi (2012) were that all types of organic fibres significantly reduced the pore pressure, compared to the control mix without fibres, with PP fibres performing better than PVA fibres. In general, higher length/diameter ratio of the organic fibres improved the pore pressure reduction. The inclusion of StF, together with the organic fibres, slightly contributed to the pore pressure reduction.

In turn, Hager and Tracz (2010) found that the thickness and length of PP fibres affect their contribution to gas-permeability and, hence, to fire resistance. They made tests on a HSC of *w/c*=0.30 containing 0, 0.9 and 1.8 kg/m^3 of fibrillated PP fibres of three lengths: 6, 12 and 19 mm. Concrete discs of Ø150×50 mm were heated at different temperatures up to 200°C in a furnace, kept 2 hours at each temperature and cooled down in the furnace to ambient temperature. Then, the residual N_2-permeability was measured using the Cembureau method. Figure 10.5 shows the effect of the volume and length of the PP fibres on the gas-permeability of concrete (referred to the permeability measured at 105°C) at different temperatures.

Figure 10.5 illustrates that the length of the PP fibres plays a significant role on HSC gas-permeability at high temperatures. It shows that short fibres have almost no effect on the relative permeability, irrespective of its

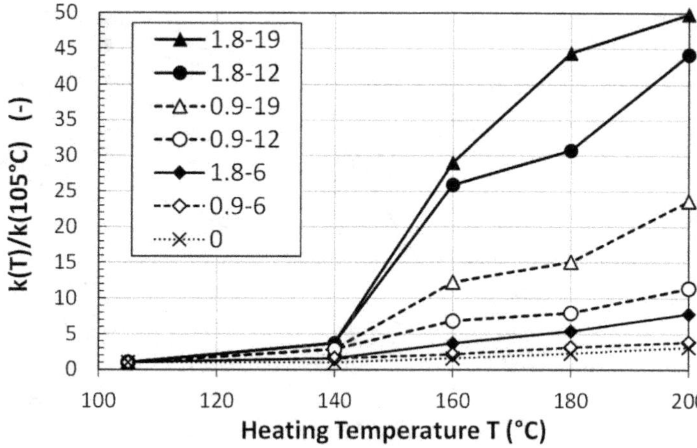

Figure 10.5 Effect of volume (kg/m³) – length (mm) of PP fibres on N_2 permeability at different temperatures (Hager & Tracz, 2010).

volume. On the other extreme, high volumes of long fibres (12 and 19 mm long) produce 45- to 50-fold increase in gas-permeability between 105°C and 200°C. The fibres' manufacturer declared a melting point of 163°C. It is interesting to observe that a sharp increase in permeability happened between 140°C and 160°C, i.e. below the fibres' melting point.

The exact mechanism by which PP fibres contribute to fire resistance of HSC may be different to the general belief that fibres' melting opens ways for gas flow. In a research by Kalifa et al. (2001), micro-cracking (MC) of heated concrete with PP fibres was observed, not found in concretes without fibres. This suggested that the presence of fibres in the cement matrix may be considered as a discontinuity that favours the initiation and development of MC during heating, contributing to an increase in concrete permeability.

Bošnjak et al. (2013) noted that the gas-permeability of concrete with PP fibres shows a sudden increase of approximately two orders of magnitude (from $\approx 0.03 \times 10^{-16}$ to $2.0 \times 10^{-16}\,m^2$) between 80°C and 130°C, i.e. well below the melting point of PP.

In a more recent research, Zhang et al. (2018) questioned the widespread belief that melting of PP fibres and the subsequent creation of pathways is the cause of increased spalling resistance. For this, Ø150×50 mm samples of UHPC (compressive strength ≈ 150 MPa), with and without PP fibres (designated UHPC/F and UHPC/O, respectively) were tested for permeability at high temperatures. Some samples were "Hot" tested at high temperature, applying an "ad-hoc" special testing technique (Li et al., 2016). Three discs were tested after cooling ("Residual permeability") applying the Torrent air-permeability test *kT*. Figure 10.6 shows the results obtained by Zhang et al. (2018).

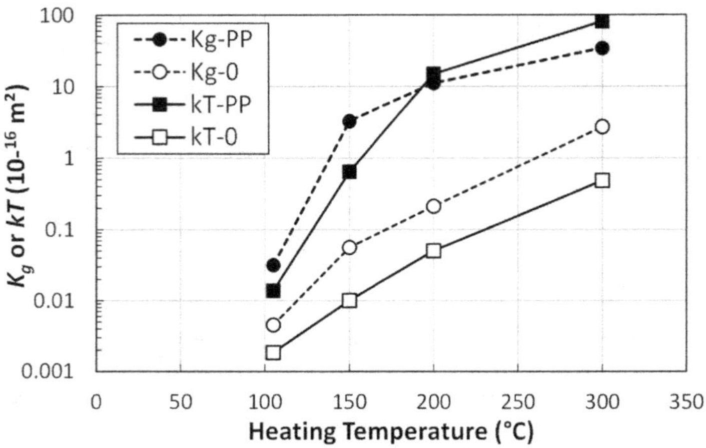

Figure 10.6 Effect of temperature on "Hot" K_g and "Residual" kT gas-permeability of UHPC without (0) and with (PP) fibres; data from Zhang et al. (2018).

Figure 10.6 clearly shows the higher "Hot" and "Residual" gas-permeability yielded by the samples containing Ø33 μm×12 mm PP fibres, compared with the samples without PP fibres (0). In addition, the samples without fibres showed an explosive behaviour, with the specimens broken into pieces, when exposed to the ISO 834 heating curve, whilst those with PP fibres kept their integrity.

Something interesting in Figure 10.6 is that the permeability of the PP samples rises more sharply between 105°C and 150°C, i.e. well below the melting point of the PP fibres (170°C), than between 150°C and 200°C (where the PP melting point lies). Trying to understand this behaviour, Zhang et al. (2018) investigated the microstructure around the PP fibres, by means of SEM, at different temperatures (27°C–300°C), as shown in Figure 10.7. It can be seen that already at 105°C microcracks (MCs) grow from the PP fibres locations, becoming wider and linking the fibres at 170°C when still many fibres have not yet started to melt. Beyond 170°C, not only the cracks get wider, but now the fibres are molten, opening new paths for the gas flow.

Based on their investigation, Zhang et al. (2018) claim that the beneficial effect of the addition of PP fibres to HSC in reducing explosive spalling, derives from their much higher (≈3 times higher) coefficient of thermal expansion. At high temperatures, this difference generates internal stresses that provoke the opening of cracks well below the melting point of PP, which results in a sharp increase of the gas-permeability, thus reducing the pore pressure and the risk of spalling. The further opening of channels resulting from the melting of the fibres only contributes marginally to the increase in permeability. One can also think that the much lower elasticity modulus of the PP fibres, compared to that of concrete, may create elastic discontinuities that give origin to crack formation.

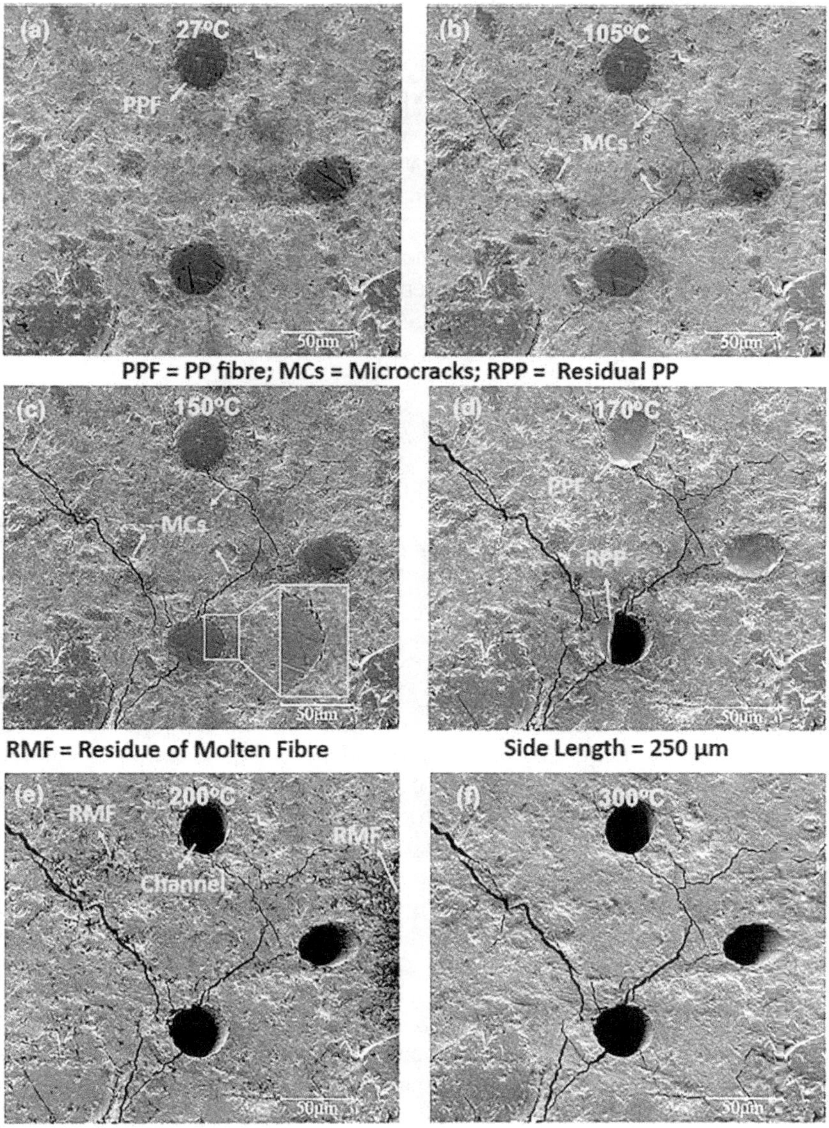

Figure 10.7 Evolution of microstructure and crack pattern in UHPC (PPF) with temperature (Zhang et al., 2018).

10.5 CONCLUDING REMARKS

There is enough evidence on that the gas-permeability of concrete plays an important role in the resistance against explosive spalling under fire and its serious consequences.

A good summary of the situation was presented by Bärtschi et al. (2018):

- permeability measurements allow quick assessment of the risk of explosive spalling, avoiding expensive fire tests which, in special cases, are nevertheless required
- predictions by modelling show, so far, 100% agreement with fire tests
- decision on protective treatment and its design can be optimized by computational design
- for existing buildings, the best solution is permeability measurements
- for new buildings, with too dense concretes, PP fibres should be a solution

REFERENCES

Bangi, M.R. and Horiguchi, T. (2012). "Effect of fibre type and geometry on maximum pore pressures in fibre-reinforced high strength concrete at elevated temperatures". *Cem. & Concr. Res.*, v42, 459–466.

Bärtschi, R., Lu, F. and Fontana, M. (2018). "Explosive spalling of concrete in fire. Assessment and protective methods". *Japanese-Swiss Workshop on* Durability Testing of Concrete, on Site and in the Lab, TFB, Wildegg, Switzerland, October 16, 18 slides.

Bažant, Z.P. (2005). "Concrete creep at high temperature and its interaction with fracture: Recent progress". Concreep 7 *International Conference on Creep, Shrinkage and Durability of Concrete and Concrete Structures 1*, 449–460.

Bošnjak, V., Ožbolt, J. and Hahn, V. (2013). "Permeability measurement on high strength concrete without and with polypropylene fibers at elevated temperatures using a new test setup". *Cem. & Concr. Res.*, v53, 104–111.

Diederichs, U., Alonso, M.C. and Jumpanen, U.M. (2009). "Concerning effects of moisture content and external loading on deterioration of high strength concrete exposed to high temperature". Workshop on Concrete Spalling under Fire, *Leipzig*, September 3–5, 269–278.

EN 1992-1-2 (2004). "Eurocode 2: Design of concrete structures – Part 1.2: General rules – Structural fire design". European Standards.

fib (2007). "Fire design of concrete structures – Materials, structures and modelling". *Bulletin 38*, Fédération Internationale du Béton, Lausanne, 97 p.

Hager, I. (2013). "Behaviour of cement concrete at high temperature". *Bull. Polish Academy Sci.,* Techn. Sci., v61, n1, 145–154.

Hager, I. and Tracz, T. (2010). "The impact of the amount and length of fibrillated polypropylene fibres on the properties of HPC exposed to high temperature". Archives of Civil Eng., LVI, 1, 57–68.

Hannant, D.J. (1978). *Fibre Cements and Fibre Concretes.* Wiley-Intersc., New York, 219 p.

Harmathy, T.Z. (1965). "Moisture in materials in relation to fire test". ASTM STP 385, 74–95.

Hoff, A., Bilodeau, A. and Malhotra, V.M. (2000). "Elevated temperature, effects on HSC, residual strength". Concr. Intern., 41–47.

Jansson, R. (2013). "Fire spalling of concrete – Theoretical and experimental results". PhD Thesis, KTH Royal Inst. of Technol., Stockholm, Sweden, 154 p.

Kalifa, P., Chéné, G. and Gallé, C. (2001). "High-temperature behaviour of HPC with polypropylene fibers: From spalling to microstructure". *Cem. & Concr. Res.*, v31, n10, October, 1487–1499.

Klingsch, E.W.H. (2014). "Explosive spalling of concrete in fire". Ph.D. Thesis, ETHZ, Zürich, Switzerland.

Li, Y., Tan, K.H., Garlock, M.E.M. and Kodur, V.K.R. (Eds.) (2016). "Effects of polypropylene and steel fibers on permeability of ultra-high performance concrete at hot state". Structures in Fire *(Proceedings of the Ninth International Conference)*, DEStech Publications, Princeton University, 145–152.

Lu, F. (2015). "On the prediction of concrete spalling under fire". PhD Diss. ETH No. 23092, ETHZ, Zürich, 139 p.

Lu, F. and Fontana, M. (2016). "Concrete permeability and explosive spalling in fire". *CONSEC* 2016, Lecco, Italy, September 12–14, Key Engng. Mater., v711, 541–548.

Mindeguia, J.C. (2009). "Contribution expérimentale à la compréhension du risque d'instabilité thermique des bétons". PhD Thesis, Univ. de Pau et des Pays de l'Adour, Pau, France.

Niknezhad, D., Bonnet, S., Leklou, N. and Amiri, O. (2019). "Effect of thermal damage on mechanical behavior and transport properties of self-compacting concrete incorporating polypropylene fibers". *J. Adhesion Sci. Technol.*, v33, n23, 2535–2566.

Nishida, A., Yamazaki, N., Inoue, H., Schneider, U. and Diederichs, U. (1995). "Study on the properties of high strength concrete with short polypropylene fibre for spalling resistance". Concrete Under Severe Conditions Environment and Loading, v2.

Sertmehmetoglu, Y. (1977). "On a mechanism of spalling of concrete under fire conditions". Ph.D. thesis, University of London, England.

Shorter, G.W. and Harmathy, T.Z. (1961). "Discussion: "The resistance of prestressed concrete beams" by L.A. Ashton and S.C.C. Bate". *Proc. Inst. Civil Engineers*, v20, n2, 313–315.

Van der Merwe, J. (2019). "Constitutive models towards the assessment of concrete spalling in fire". PhD Diss. ETH No. 26205, ETHZ, Zürich, 235 p.

Zhang, D., Dasari, A. and Tan, K.H. (2018). "On the mechanism of prevention of explosive spalling in ultra-high performance concrete with polymer fibers". *Cem. & Concr. Res.*, v113, 169–177.

Chapter 11

Real cases of *kT* test applications on site

11.1 INTRODUCTION

The main advantage of air-permeability *kT* test method over other test methods is its adequacy for the non-destructive assessment of the quality of the *Covercrete* on site, thus becoming a practice-oriented tool.

This chapter deals with practical *in situ* applications of the *kT* test on full-scale structural elements which, besides illustrating the potential usefulness of the test method, may stimulate the readers to make good use of it for site research, quality assessment or problem-solving purposes.

It also provides a valuable data bank on the values of *kT* (central values and scatter) that can be expected when testing structures made of conventional or high-performance concretes, be it when young or after several decades of service.

The chapter is subdivided into three parts, one dealing with planned investigations at a real scale, the second dealing with "new" structures, typically less than 1½ years old and the third part dealing with "old" structures, typically over 10 years old. A fourth section is added, dealing with applications not related to structural concrete or with applications to materials other than concrete.

11.2 FULL-SCALE INVESTIGATIONS

11.2.1 RILEM TC 230-PSC (Chlorides and Carbonation)

This case refers to the Application Test of RILEM TC 230-PSC (Performance-Based Specification and Control of Concrete Durability) (Beushausen & Fernández Luco, 2016). This endeavour was aimed at demonstrating the applicability of various performance-based approaches of durability assessment that rely on - site testing.

For that purpose, eight full-scale reinforced concrete panels were prepared at B|A|S headquarters in Venlo, The Netherlands, with four different

DOI: 10.1201/9780429505652-11

concrete mixes and following two different curing procedures, as described in Table 11.1 (more information can be found in Tables 11.1–11.4 of Beushausen and Fernández Luco (2016)). The front side of the panels had a nominal cover thickness of 30 mm and the rear side of 50 mm.

Several organizations participated in this sort of round-robin test; in particular, two rounds of air-permeability kT and cover thickness d tests were performed, one when the concrete panels had 14–21 days of age and the second when they were 101–108 days old. The data obtained during the first round corresponded to too young concretes (see data in Section 5.7.3.1), with surface moisture contents m that exceeded the limit of 5.5%, i.e. exceeding the limit prescribed by Swiss Standard SIA 262/1 (SIA 262/1, 2019). Therefore, only the second sets of data are considered here, which complied with that Standard. Between six and eight parallel tests of kT and

Table 11.1 Main characteristics of the eight panels investigated at Venlo (NL)

Mix	*Cement type*	*w/c*	$f'c_{28d}$ *(MPa)*	*Panel*	*Curing*
A	CEM I 52,5 N	0.45	78.0	1	Air curing
				2	Plastic sheet 7 days
B		0.54	46.0	3	Air curing
				4	Plastic sheet 7 days
C	CEM II/B-V (Fly-Ash)	0.41	63.0	5	Air curing
				6	Plastic sheet 7 days
D		0.59	34.0	7	Air curing
				8	Plastic sheet 7 days

Table 11.2 Statistical parameters of the test results of kT, ρ and d measured at Venlo and median t_{SL}

								t_{SL} *(years)*	
Panel	*Cement type – w/c*	kT_{gm} $(10^{-16}\ m^2)$	s_{LOG} (-)	ρ_m (kΩ.cm)	s_ρ (kΩ.cm)	d_m(mm)	s_d(mm)	*XD3/XS3*	*XC4*
1–30	O – 0.44	0.020	0.29	10.8	1.6	30.2	1.7	29	264
2–30		0.024	0.39	10.6	1.3	33.3	1.4	24	175
3–30	O – 0.54	0.35	0.66	4.9	0.7	32.9	0.9	13	27
4–30		0.45	0.75	4.1	0.7	30.3	1.8	10	21
5–30	F – 0.40	0.025	0.26	27.5	3.3	32.8	3.2	32	222
6–30		0.016	0.21	25.0	2.8	31.5	1.4	34	433
7–30	F – 0.59	1.76	0.52	27.7	11.9	34.3	1.0	8	15
8–30		0.86	0.58	16.0	3.9	29.0	3.6	8	14
1–50	O – 0.44	0.020	0.08	13.7	0.5	51.0	1.8	82	754
5–50	F – 0.40	0.021	0.36	33.7	3.0	50.3	1.4	79	677
7–50	F – 0.59	0.34	0.51	18.3	2.7	49.0	2.0	30	62

Note: O means OPC CEM I 52,5 N cement and F means PFA CEM II/B-V cement.

d (and surface moisture *m*) were performed by R. Torrent on each of 11 panel faces; the surface electrical resistivity ρ was also measured at the same locations, by the Wenner method. The data were presented in Table 11.13 of Beushausen and Fernández Luco (2016).

The central values and scatter of the measured variables *kT* and *d* are presented in Table 11.2, for the 11 panel faces investigated, where kT_{gm} = geometric mean of *kT*, s_{LOG} = standard deviation of $\log_{10}kT$, d_m = mean value of *d* and s_d = standard deviation of *d*. In the same Table 11.2, the estimated median service lives for chlorides severe exposure classes XD3/XS3 and for carbonation severe exposure class XC4 were computed, on the basis of the Exp-Ref model (see Section 9.4.3), applying Eqs. (9.29) and (9.42), respectively. The input variables were the central values kT_{gm} and d_m shown in Table 11.2. For exposure class XD3 or XS3 is $\alpha = 0.0086$ and for exposure class XC4 is $\beta = 0.42$ and $t_p = 2$ years, where α and β are parameters of Eqs. (9.29) and (9.42), respectively.

Figure 11.1 presents a more comprehensive view of the results. Since *kT* and *d* were measured at the same locations within the panels, Eqs. (9.29) and (9.42) can be applied to each individual pair of results and, thus, a distribution of service lives can be obtained for each panel. What is plotted in Figure 11.1 is the range (bar length) and the median of the predicted service life for each panel, differentiating the cases for XS3/XD3 (circles) and for XC4 (triangles). The designation of each investigated face panel is

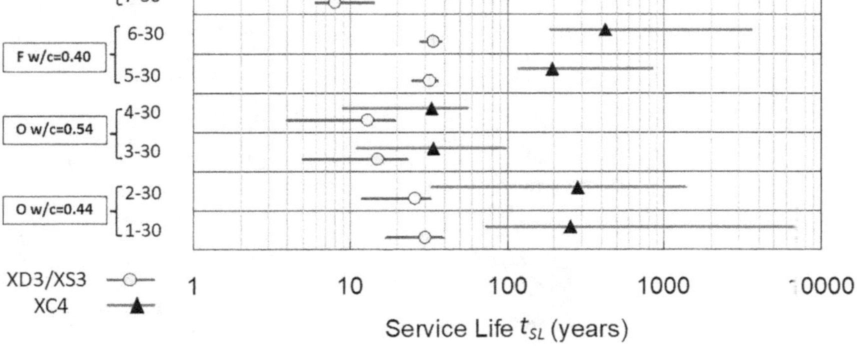

Figure 11.1 Range and median values of predicted service life t_{SL} for the 11 panel faces investigated in Venlo.

the panel number (see Table 11.1) followed by the nominal cover thickness (30 or 50 mm).

Both Table 11.2 and Figure 11.1 show the beneficial effect of low *w/c* (corresponding to low *kT* values) and thick cover *d* on the predicted service lives t_{SL} of the panels. They also show that the "Exp-Ref" model described by Eqs. (9.29) and (9.42) represents correctly the fact (supported by experience) that the same panel face, exposed to severe carbonation (XC4), shows a significantly longer t_{SL} than the same face exposed to a severe chloride environment (XD3/XS3).

From Table 11.2, it would be expected that panel 8–30, that has a large scatter in both *kT* and *d*, would also show a large scatter in t_{SL} which, looking at Figure 11.1, is not the case. A close observation of the individual pairs of tests showed that high values of *kT* corresponded, almost invariably, to low values of *d*, with a compensation effect that reduced the scatter in t_{SL}. Since both variables *kT* and *d* are independent, this happened by chance. A more realistic analysis would be to solve Eqs. (9.29) and (9.42) by the Monte Carlo method, assuming that *kT* and *d* are distributed according to log-normal distributions (to avoid negative values) with the parameters indicated in Table 11.2 for each panel face. This has been done for Panels 1, 5 and 7 (for both faces with nominal cover of 30 and 50 mm), with the results shown in Figure 11.2, for chloride exposure XD3/XS3.

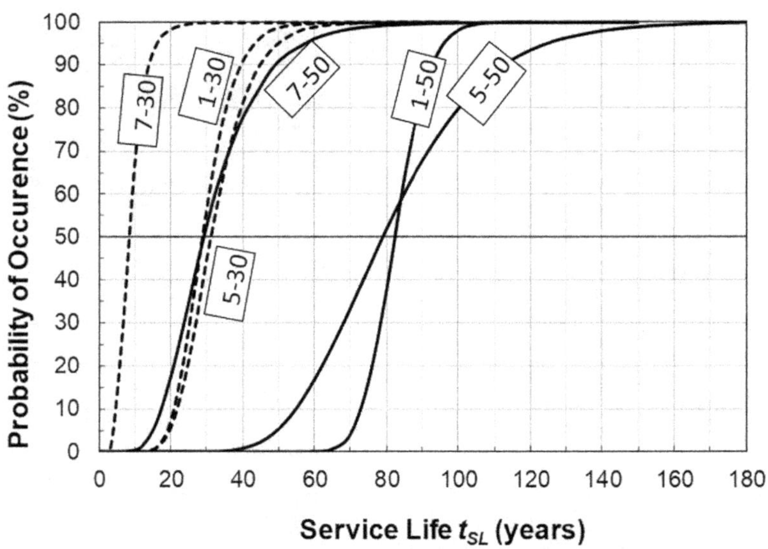

Figure 11.2 Probabilistic assessment of service life t_{SL} for six panel faces exposed to severe chlorides environment.

As validation of the probabilistic analysis, the predicted median service lives in Figure 11.2 (t_{SL} for 50% probability of occurrence) happened to be almost the same to those obtained from the application of Eqs. (9.29) to the central values of experimental results of kT and d (Table 11.2) and to those obtained computing the median of the six to eight t_{SL} values obtained from each individual pair of kT and d values measured (plotted as circles in Figure 11.2).

Figure 11.2 indicates that a conventional quality concrete (Panel 7 with $w/c = 0.59$ and $kT = 0.34 \times 10^{-16}$ m²) with a thicker nominal cover $d = 50$ mm has almost exactly the same service life characteristics as a good quality concrete (Panel 5 with $w/c = 0.40$ and $kT = 0.025 \times 10^{-15}$ m²) with a thinner nominal cover $d = 30$ mm. Both panels were made with same PFA cement. The probability lines in Figure 11.2 express very well the heterogeneity of concrete in terms of both its permeability kT and cover depth d, with several decades passing between the moment in which the first (weakest) point starts to corrode to when the last (strongest) point is expected to do it. This fact can be observed in reality looking at a corroding marine structure.

This case illustrates the potential of the "Exp-Ref" method, described in Section 9.4.3, to assess the service life of structural elements under risk of steel corrosion, induced by carbonation or chlorides, on the basis of site NDT measurements of kT and d. The "Exp-Ref" method has been calibrated on the prescriptive requirements of Eurocode (see Section 9.4.3); yet, as for any other model, more emphasis shall be placed on the relative performance of the panels than on the absolute service lives predicted.

11.2.2 Naxberg Tunnel (Chlorides and Carbonation)

11.2.2.1 Scope of the Investigation

This case refers to a full-scale investigation conducted in the 510 m long Naxberg Road Tunnel, located on A2 motorway, near the Gotthard Road Tunnel, Switzerland. The tunnel is located at an altitude of ca. 1,000 m AMSL. Due to its alpine location, the exposed surfaces of the tunnel are subjected, during winter time, to cycles of splash and spray with de-icing salt solutions rich in chloride ions; the minimum, mean and maximum temperatures measured inside the tunnel along 2½ years were 7.4°C, 11.6°C and 30.5°C, respectively, with relative moistures ranging typically between 60% and 80% (Ungricht, 2004).

As part of scheduled tunnel maintenance work at the end of 1990, the opportunity arose to place concrete test plates on its west-side wall, at a distance of 50 m from the south tunnel exit. For that purpose, a niche

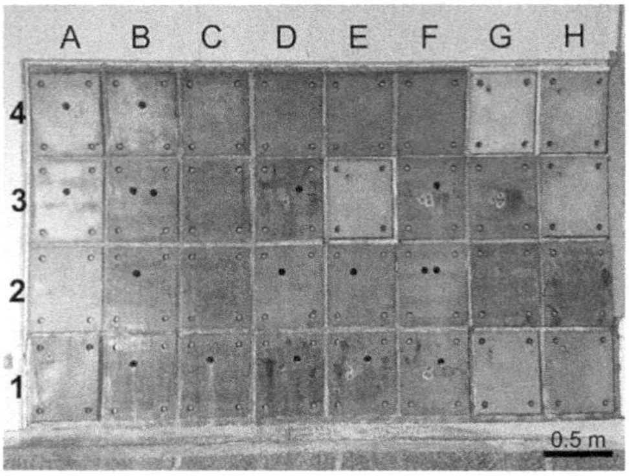

Figure 11.3 Location and identification of the test plates (Bisschop et al., 2016).

(4.0 × 2.5 × 0.18 m) was opened in the wall, where 32 plates (0.6 × 0.5 × 0.1 m each) were installed in four rows, as shown in Figure 11.3. The plates were fixed to the wall with stainless steel and the joints between plates were sealed. The plates were reinforced with carbon and corrosion-resistant steels with cover thicknesses between 10 and 45 mm.

The plates were instrumented with sensors to monitor concrete temperature and electrical resistivity as well as the corrosion potential and corrosion current of the steel bars.

As this was a long-term project, the information is fragmentarily reported, both regarding the original characteristics and performance of the plates (Hunkeler et al., 2002; Maître, 2002; Ungricht, 2004; Brühwiler et al., 2005; Jacobs, 2006) and after several years of exposure (Bisschop et al., 2016; Schiegg et al., 2017a, b; Donadio et al., 2019). The data presented here come from a thorough analysis of those sources.

11.2.2.2 Mixes Composition and Laboratory Test Results

Five concrete mixes were used to build the plates, with the characteristics described in Table 11.3, mix 9 being taken as "reference". For the preparation of the mixes, GGBFS of fineness ≈ 300 m^2/kg and "Hydrolent" PFA (both provided by Holcim Schweiz) and a SF 50/50 suspension (provided by MBT Schweiz) were added to mixes 11, 13 and 15, respectively (Ungricht, 2004).

The properties reported in the last five rows of Table 11.3 were measured on laboratory specimens, moist cured and tested at 28 and 360 days of age (Hunkeler et al., 2002). The "durability" test results, measured at 1 year of age, are comparatively shown in Figure 11.4.

Plates D1 to H1, made with the five different concrete mixes, have the same chloride exposure (bottom row in Figure 11.3); plates H1 to H4, made with "reference" mix 9, have different exposures to chlorides (extreme right

Table 11.3 Characteristics of the mixes used to build the 32 plates exposed in Naxberg Tunnel

Mix	*9*	*11*	*13*	*15*	*17*
Plates	A, B, C, H	D	E	F	G
Cement type	CEM I 42.5N	CEM I + GGBFS	CEM I + PFA	CEM I + SF	CEM I 42.5N
Cement (kg/m³)	300	249	269	277	367
SCM (kg/m³)	-	50	40	19	-
Spread (mm)	440	470	410	400	420
Air (% vol.)	0.9	1.6	2.1	1.9	1.9
w/b ratio	0.50	0.50	0.50	0.50	0.35
Density (kg/m³)	2,452	2,441	2,428	2,397	2,371
$f'c_{cube}$ (MPa)	49 → 57	43 → 54	43 → 55	53 → 59	66 → 74
kO (10^{-16} m²)	0.55 → 0.69	0.43 → 0.41	0.26 → 0.32	0.22 → 0.31	0.06 → 0.02
q_w (g/m²/h)	4.4 → 4.0	2.5 → 3.7	2.9 → 4.2	3.5 → 3.2	1.8 → 1.3
M (10^{-12} m²/s)	14.1 → 10.4	13.0 → 12.4	15.7 → 14.6	3.2 → 4.8	7.3 → 6.2
Q (Coulombs)	1,724	3,136	3,017	1,367	1,307

Spread, consistency according to EN 12350-5 test method; values at 28 days → 1 year of age; kO, coefficient of O_2-permeability (Cembureau method, Section 4.3.1.2) after 14 days drying at 50°C; q_w, *Wasserleitfähigkeit*, see Annex A of SIA 262/1 (2019), related to water sorptivity by Eq. (11.5); M, coefficient of chloride migration on Ø50 × 50 mm cores (Tang-Nilsson method, Annex A.2.1.2); Q, charge (ASTM C1202, 2012) (Annex A.2.1.1), measured at 360 days on Ø100 × 50 mm cores.

Table 11.4 Results of site *kT* and ρ measurements on Naxberg Tunnel plates

Plate	kT_{gm} (10^{-16} m²)	s_{LOG} (-)	ρ_{avg} (kΩ.cm)	s_ρ (kΩ.cm)
A1	0.365	0.20	999	0
A2	0.284	0.44	999	0
D1	0.568	0.24	62	12
D3	1.707	0.09	103	16
E1	0.526	0.89	96	29
F1	0.276	0.46	120	16
F3	0.698	0.30	114	17
G1	0.004	0.38	68	4
H1	0.092	0.80	41	10
H2	1.529	0.42	-	-
H3	0.337	0.54	58	15
H4	3.319	0.69	-	-

column in Figure 11.3). The above-mentioned eight plates have their natural surfaces exposed and were reinforced with normal black steel. The others had some surface treatments applied and/or contained re-bars made of special steels; to some mixes NaCl was intentionally added.

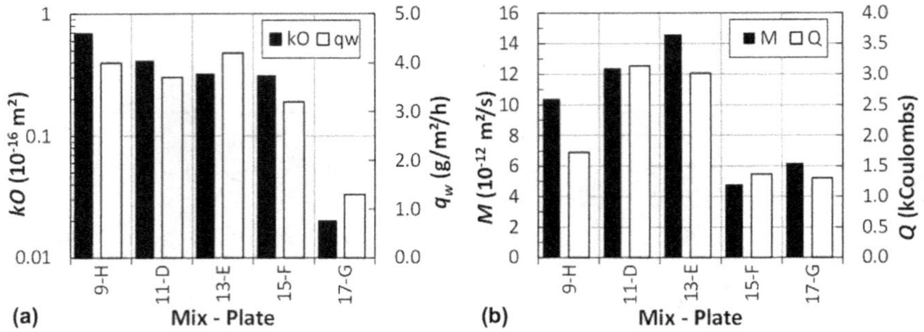

Figure 11.4 (a) O_2-permeability kO and water "conductivity" q_w, and (b) Cl^- migration coefficient M and charge Q passed in ASTM C1202 of the five mixes investigated.

In terms of permeability to gas and water, Figure 11.4a, mix 17, used for the G panels, shows a much better performance than the other four mixes, with mix 15-F coming in a far second rating. In terms of chloride migration, Figure 11.4b, mix 15-F and again mix 17-G clearly overperform the other three mixes.

11.2.2.3 Characteristics of the 32 Panels

Panels D1–D4, E1–E4, F1–F4, G1–G4 and H1–H4 contained normal carbon steel bars and were intended to assess the effect of the concrete mix design (see Table 11.3) and of their location with respect to the pavement on the carbonation rate, chloride ingress and corrosion behaviour.

Panels A-1 and A-2 received a hydrophobic surface treatment, apparently with triethoxy(2,4,4-trimethylpentyl)silane (Donadio et al., 2019). They were also reinforced with normal carbon steel.

To the mixes for Panels A-3, A-4, B-3, B-4, C-3 and C-4, 2 M% of NaCl was added.

Panels B and C were reinforced with two different types of corrosion-resistant steels.

11.2.2.4 On-Site Non-Destructive kT Measurements

The air-permeability kT and electrical resistivity ρ of several plates were measured on site, independently by EPFL (Table 1 of Maître (2002)) and by TFB (p. 124 of Jacobs (2006)), using different units of same instrument's brand (*TPT*, Proceq).

The tests were made when the concrete plates were about 2½ and 3½ years old, respectively. The results of kT obtained on same plates (not same spots) by both laboratories did not show a significant statistical difference and were merged; regarding ρ, only the data from Jacobs (2006) were

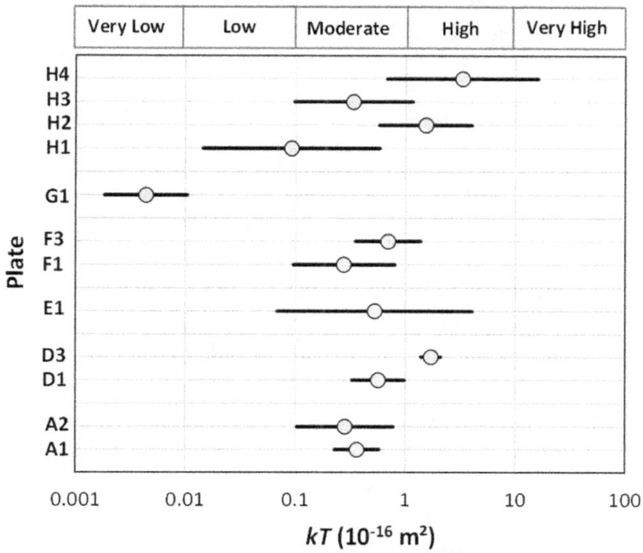

Figure 11.5 $kT_{gm} \pm s_{LOG}$ plot of the results obtained by both EPFL and TFB on 12 concrete plates.

considered, as the other set presents inaccurate results. Table 11.4 presents the results for each plate, showing the kT_{gm} and s_{LOG} values as well as the average (ρ_{avg}) and standard deviation (s_ρ) of the resistivity. Figure 11.5 presents the *kT* results in a graphical form.

As expected, the resistivity ρ of the plates made with mixes containing SCM (D, E and F) is higher than that of the OPC mixes (G and H). Interestingly, the plates that received the hydrophobic treatment showed a very high resistivity (beyond the sensitivity level of the instrument); the same effect was observed independently by Brühwiler et al. (2005), who suggest that the measurement of ρ has potential to detect whether a hydrophobic treatment has been applied or not.

Regarding the *kT* test results (Figure 11.5), the four plates H, cast with the "reference" mix 9, show highly variable results with, in one extreme, the bottom plate H1 showing a kT_{gm} corresponding to "Low" Permeability Class and, in the other, top plate H4 (same for plate H2) showing a kT_{gm} corresponding to "High" Permeability Class (see Table 5.2).

The plates D, E and F, cast with mixes 11, 13 and 15, i.e. with $w/b = 0.50$ (same as the "reference" mix), but containing SCMs, present mostly kT_{gm} values corresponding to "Moderate" Permeability Class, same as for the plates made with the "reference" mix 9, but subjected to the hydrophobic treatment of the surface (A1 and A2).

The plate that shows a clear distinct performance in terms of *kT* (kT_{gm} corresponding to "Very Low" Permeability Class) is G1, cast with mix 17, made with OPC and $w/c = 0.35$. This is in agreement with the results of laboratory tests, see Figure 11.4.

11.2.2.5 Core Drilling, Carbonation and Chloride Ingress

As reported in Bisschop et al. (2016), a program of cores' drilling and testing was executed, obtaining samples after exposure periods of 1.4, 2.5, 3.5, 4.4, 12 and 13 years; the initiation of the tunnel exposure started when the plates were about 6 months old. Due to the limited size of the plates (space further shrunk by the embedded steel and sensors), at each age just one Ø50 × 100 mm core was taken from each plate. Each core was cut into 10 mm thick slices and the total acid-soluble chloride content was measured after Swiss Standard (SN EN 14629, 2007).

The carbonation depth at an age of ≈13 years was measured on some plates and reported by Schiegg et al. (2017a). Figure 11.6 shows the carbonation at 13 years (in mm above each bar) and the corresponding value of the carbonation rate K_c in ordinates, for seven plates.

It can be seen that plates G1, G4, H1 and H3 present extremely low carbonation depths, plates A1 and E3 low carbonation depths, whilst plate H4 presents a moderate carbonation rate, according to AIJ (2003) rating.

The chloride profiles of some plates, particularly at later ages, do not show a "Fickean" behaviour or seem to have reached "saturation" across the 100 mm length. This was most evident in the case of plate H1, the results of which at different ages are shown in Figure 11.7a. Figure 11.7b presents the more "Fickean" chloride profiles obtained on six plates after 4.4 years of exposure (Bisschop et al., 2016).

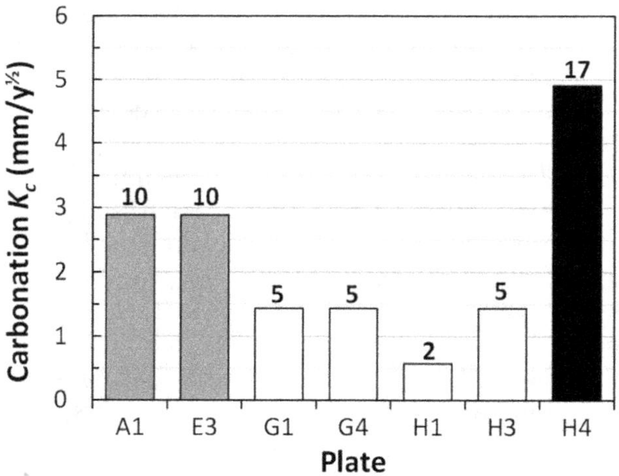

Figure 11.6 Carbonation of several plates at an age of 13 years; data from Schiegg et al. (2017a).

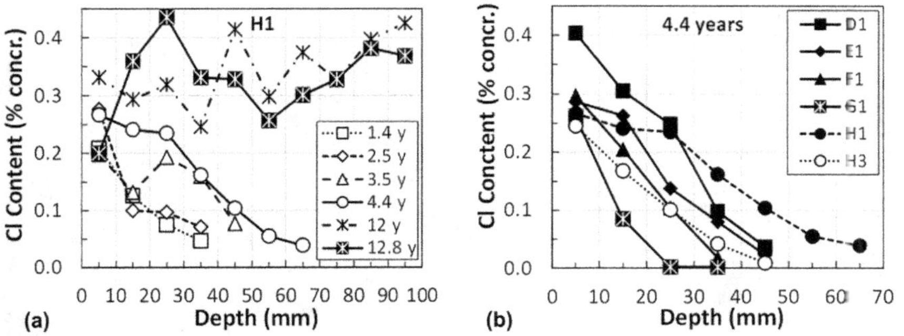

Figure 11.7 (a) Evolution of chloride profiles of plate H1 and (b) chloride profiles of six plates after 4.4 years exposure; data from Bisschop et al. (2016).

It is worth mentioning that plate A1 (hydrophobic treatment) showed a negligible penetration of chlorides after 4.4 and even after 13 years exposure.

The solution of Fick's second diffusion law, described by Eq. (3.5), was fitted to the chloride profiles of each panel, obtained at 4.4 years, Figure 11.7b, applying the tool available in Life 365 (2018), using all available points for the calculation (assuming $C_0 = 0$). The fitting to each chloride profile provides two computed properties: the surface concentration C_s (extrapolation to $x = 0$) and the apparent chloride-diffusion coefficient D^{app}. These characteristics are presented in Table 11.5.

It can be seen that most D^{app} values, rated after Table 4.2 of Gjørv (2014), correspond to chloride ingress resistance classes "very high" and "extremely high", with just plate H1 rated as "high".

Again, mix 17-plate G1 shows the best performance, followed closely by plates H3 and F1. Plate H1 shows a much higher D^{app} than its companion H3, the latter located at a higher level above the road.

Figure 11.8a presents the good correlation existing between *kT* and the carbonation rate K_c. The carbonation of the three H plates is worth discussing. According to Schiegg et al. (2017a), the higher carbonation of plate H4 "could be due to an increased CO_2 content of the air in the tunnel". It is hard to believe that the concentration of CO_2 inside the relatively short tunnel can vary significantly within 1.8 m difference in height. A more plausible explanation is the higher air-permeability *kT* measured on plate H4,

Table 11.5 Characteristics of the chloride profiles of the six plates analysed

Plate	*D1*	*E1*	*F1*	*G1*	*H3*	*H1*
C_s (% concr.)	0.482	0.360	0.377	0.391	0.325	0.319
D^{app} (10^{-12} m²/s)	3.34	3.11	1.71	0.50	1.28	8.14
Cl ingress resistance	Very high		Extremely high			High

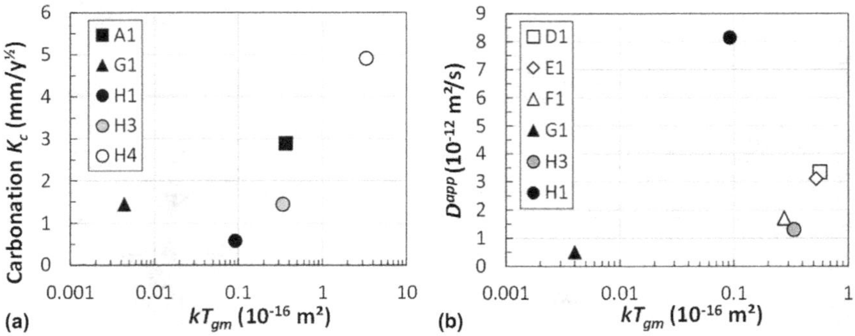

Figure 11.8 Relation between site *kT* and (a) site carbonation rate K_c and (b) site chloride diffusivity D^{app}.

perhaps helped by a lower moisture compared to plate H1, more exposed to splash (this also explains the intermediate result of plate H3).

Figure 11.8b shows the relation between the apparent chloride-diffusion coefficient D^{app}, measured after 4.4 years exposure, and *kT*. The correlation is not very good, although it would become better if the already mentioned rather high value for plate H1 was not considered.

At an age close to 13 years, some plates were crushed to recover the steel bars for inspection. Figure 11.9 shows the corroded areas and the maximum corrosion depth of the carbon steel bars, embedded with 10 mm cover thickness in several plates.

Plates A1 (mix 9 with hydrophobic treatment), G1 and G4 show little corrosion, both in terms of corroded area and of corrosion depth, whilst

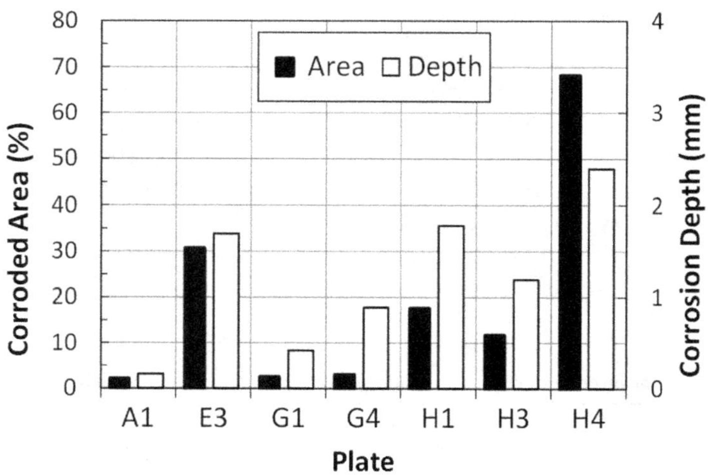

Figure 11.9 Corroded area and maximum corrosion depth of bars embedded with 10 mm cover in various plates; data from Schiegg et al. (2017a).

plate H4 shows an intensive corrosion (almost 70% of its area showing corrosion, with a maximum corrosion depth of almost 2.5 mm). The other three plates (E3, H1 and H3) show an intermediate behaviour.

11.2.2.6 Conclusions

The results presented above prove that plates G (made with mix 17, OPC with $w/c = 0.35$) are the best performing, presenting very low carbonation depth, chloride diffusivity and steel corrosion rate. This is in agreement with the good performance under permeability and chloride migration laboratory tests, confirmed by the very low air-permeability *kT* of plate G1.

Plates D, E, F and the "reference" H show, in general, and intermediate performance, with the exception of plate H4, which presents a high carbonation, accompanied by high steel corrosion rates. The main explanation for this rather poor performance can be found in its high air-permeability *kT*, the highest of all measured plates.

An interesting case is that of Plate A1, made with "reference" mix 9 (same as plates H), but receiving a hydrophobic treatment. Its carbonation rate is moderate, but it was almost "impermeable" to chlorides and showed negligible steel corrosion rate. The air-permeability *kT* of plate A1 is "moderate" but accompanied by such high electrical resistivity ρ that exceeded the range of the instrument. One explanation for this behaviour is that the hydrophobic treatment stops the saline solution but not the CO_2 nor the air during the *kT* test. The high resistivity may be attributed to some direct effect of the treatment on ρ and/or to an effect of lowering the moisture behind the barrier.

This research shows quite clearly the difficulties in performing site testing, particularly when involving costly core drilling and chloride profile analysis for diffusivity calculations. This real case shows the importance of conducting site tests of *kT* and ρ directly on the elements to be monitored, besides the laboratory characterization of the mixes, whenever such long-term performance tests under natural exposure sites are planned.

11.3 NEW STRUCTURES

11.3.1 Port of Miami Tunnel (Carbonation)

11.3.1.1 Description of the Tunnel

This investigation refers to the construction of a bored, under-seabed tunnel designed to avoid that the traffic of heavy vehicles to/from Miami Port (FL, USA) interferes with the urban traffic of the city. It was a PPP project with a total budget of 1,000 million USD.

The tunnel consists in the construction of a Ø13 m, 1.1 km long twin-tube tunnel between Watson and Dodge islands. The tunnel was bored using a

Figure 11.10 Aspect of Port of Miami Tunnel showing the precast curved elements.

tunnel boring machine which concurrently installs segmental precast concrete lining rings. Each ring panel is composed of eight precast curved segments that are typically about 4.8 m long, 1.7 m wide and 0.6 m thick (see Figure 11.10).

The project is a public-private partnership; the contract terms make it clear that durability and service life of the tunnel are of paramount importance, as the specified service life is 150 years.

The curved segments were cast in an "ad-hoc" precast concrete plant, compacted by heavy vibration in sturdy steel moulds. They contained a reinforcement cage, with a cover thickness of 76 mm in both faces. The concrete was poured through an opening of the upper side of the moulds, corresponding to the extrados face of the segment. More details on the whole construction process of the elements can be found in Torrent et al. (2013).

11.3.1.2 The Problem

At an age of 18 hours, the upper side of the mould was removed and the exposed surface of the extrados was cured by spraying an approved curing compound. Then, the elements were vacuum-lifted, turned upside down and taken to the storage yard. A relevant step, closely connected with the investigation, was the application of the curing compound on the sides of the element (Figure 11.11).

Figure 11.11 Spray-curing lateral sides of the element but not intrados.

The intrados surface of the segments should have also been sprayed, but, for some reason, it was not, which violated the specifications that prescribed three alternative curing methods for all the surfaces of the elements:

a. demoulding at 18 hours followed by spray-curing with approved compound (chosen);
b. leaving the elements 72 hours in the moulds;
c. demoulding at 18 hours followed by 2 days curing with water.

The chosen curing method was (a). When the inspection objected that the curing compound had not been sprayed on the intrados surface, the contractor made an analytical service life assessment based on the DuraCrete (2000) model. This analysis showed that the quality of the High-Performance Concrete mix used (w/b = 0.32; 188 kg/m^3 ASTM Type II OPC + 236 kg/m^3 GGBFS + 47 kg/m^3 Class F PFA), designed for resisting the aggressive chloride-rich environment in contact with the extrados for 150 years, was enough, even without curing, to withstand the risk of carbonation-induced corrosion, affecting the intrados.

The Inspection was not satisfied with just the analytical prediction; a decision was made to contact an USA expert, Dr. J. Armaghani, to investigate the matter. It is important to mention that, at that stage, 623 elements had already been produced, that were vital for the advancement of the boring machine. Therefore, besides the cost of the individual elements, the lost profit due to the delay in the perforation of the tunnel was huge.

11.3.1.3 Scope of the Investigation

Dr. Armaghani thought that measuring the air-permeability *kT* could be a good way of assessing the surface quality of the elements' intrados and contacted Dr. R. Torrent for support. Both conducted jointly an experimental field investigation of the quality of the intrados of the elements.

For that, five elements were identified that, cast during long week-ends, stayed 72 hours in the moulds, thus satisfying curing condition (b), accepted by the specifications. These elements were taken as reference, against which the elements that stayed only 18 hours in the moulds, without further curing were to be compared. Five elements cured just 18 hours in the moulds, of similar ages (≈4 months), were identified and selected for the comparative test. The selected ten elements were taken from the stockpiles and placed alongside the precast factory, covering them with tarpaulin some days before the tests, to protect them from the frequent rains happening in Miami.

Drs. Armaghani and Torrent conducted personally the measurements (Figure 11.12), completing 64 measurements of *kT* in 2 days of work, 33 and 31 on the segments with 72 and 18 hours of curing, respectively. Notice in Figure 11.12 the umbrella for protecting the instrument from strong Miami sunrays. The surface moisture content *m* (*Concrete Encounter*, Tramex) and the air-permeability *kT* (*PermeaTORR*, M-A-S) were measured on at least six points of each of the ten segments.

11.3.1.4 Site kT Test Results

The frequency charts of the 33 + 31 measurements are displayed in Figure 11.13, showing a typical asymmetrical distribution with positive skew.

Figure 11.12 Drs. Armaghani and Torrent site-testing the elements.

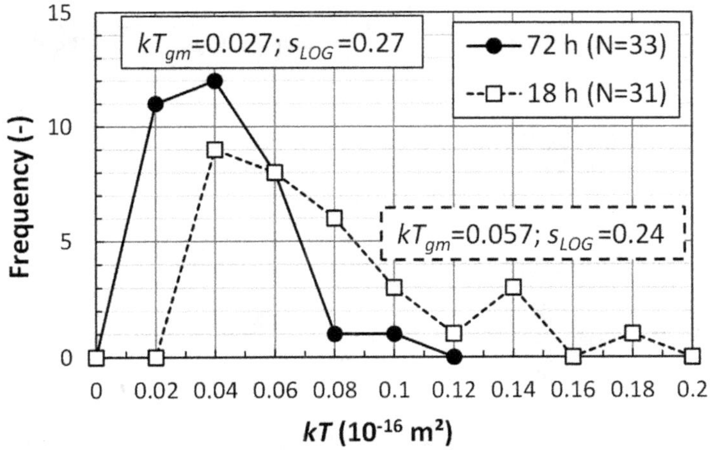

Figure 11.13 Frequency distribution of *kT* for both sets of elements.

The results indicate a high quality of manufacturing, expressed by the low scatter (see Figure 5.9) and also by the low geometric means ("Low" Permeability Class). A statistical analysis (F-test and t-test) of the logarithms of kT (assumed to be normally distributed) of the 33 and 31 readings on each set of elements revealed the following:

- the standard deviation s_{LOG} of the logarithms of kT is not significantly different between both sets
- the geometric means kT_{gm}, on the contrary, are significantly different, with the value for the elements cured 18 hours being statistically significantly higher than that of the elements cured 72 hours

Figure 11.14 presents the recorded kT results in another format. Here, each element is plotted separately with its individual central value kT_{gm} shown as squares (18-hours) and circles (72-hours) $\pm s_{LOG}$ as horizontal segments. The coarse vertical lines represent the overall kT_{gm} value for each set of five elements, those cured 18 hours at the top and those cured 72 hours at the bottom. The Permeability Rating, presented in Table 5.2, is indicated at the top of the chart.

What Figure 11.14 is telling is that, despite being significantly different statistically, both sets belong clearly to the same "Low" Permeability Class, indicating that from the practical point of view the difference is not relevant.

11.3.1.5 Modelling Carbonation at 150 Years

The Inspection was not satisfied with this conclusion, asking for an estimate of the carbonation depth after 150 years, certainly not a mean task. For

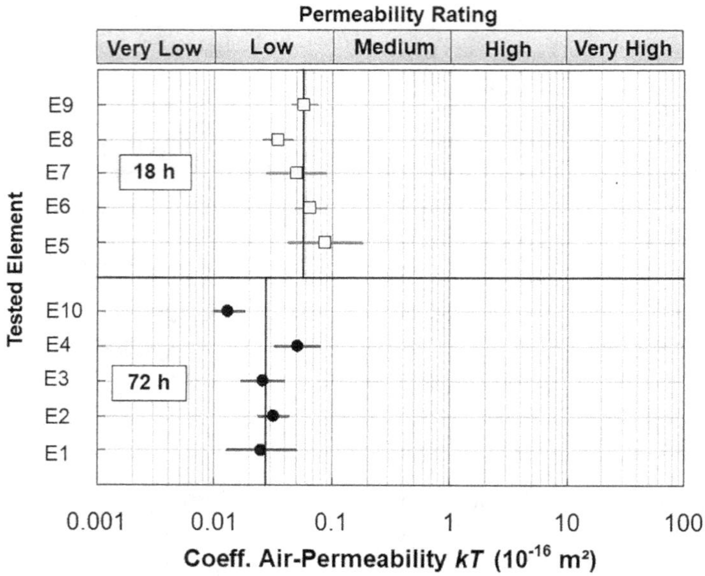

Figure 11.14 $kT_{gm} \pm s_{LOG}$ plot of test results of each individual element.

that, a couple of models were tried, summarized below. For more details, see Torrent et al. (2013):

- Parrott's Model (described in Section 9.3.1)
- extrapolation of old structures data

Parrott's Model Eq. (9.15) was applied under the following assumptions:

- the coefficient of air-permeability used in the equation, based on the Hong-Parrott test method (Section 4.3.2.2), is equal to *kT*. This assumption was supported, to some extent, by the data reported in Table C1 of Romer (2005);
- the most pessimistic value of parameter $c = 225$ kg/m³ was chosen (Parrott, 1994);
- the highest possible exponent n was chosen ($n = 0.52$).

The carbonation depth (mm) at 150 years was then computed as: $60\ kT^{0.4}$, with $kT = kT_{gm}$ in (10^{-16} m²).

Regarding the extrapolation of experimental results of old structures, the existing results of Figure 8.16 available at the time of the analysis were used (Figure 11.15). They consist of parallel measurements of the carbonation rate K_c and of air-permeability *kT* performed at ages of 12–60 years of age. As the *kT* data of the Tunnel's elements were obtained at 4 months of age,

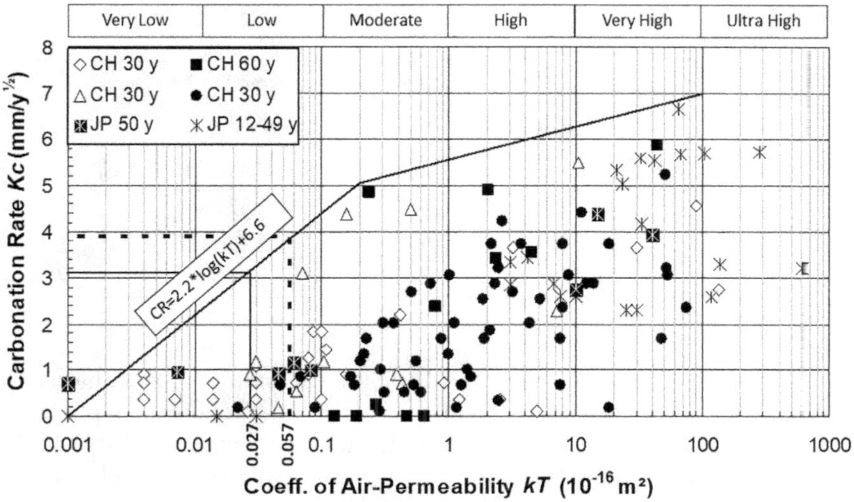

Figure 11.15 Estimation of K_c from kT, both measured on old structures.

and the data in Figure 11.15 at decades of age, a very pessimistic enveloping line was set to the data, of the form:

$$K_c = 2.2\log(kT) + 6.6 \quad \text{with } K_c\left(\text{mm/y}^{1/2}\right); kT = kT_{\text{gm}} \text{ in}\left(10^{-16}\,\text{m}^2\right) \quad (11.1)$$

As shown in Figure 11.15, the predicted carbonation rates for the elements cured 18 and 72 hours were 3.9 and 3.1 mm/y$^{1/2}$, respectively. Multiplying these values by 150$^{1/2}$ provides the predicted carbonation depth at 150 years.

The reported results are presented in columns 1 and 2 of Table 11.6 for both experimental models.

11.3.1.6 Conclusions

Column 4 of Table 11.6 shows the analytical prediction made by the contractor, applying the DuraCrete /DARTS model, that fall between the two experimental-based predictions (columns 1 and 2). Since all three predictions were far lower than the cover thickness d = 76 mm, the inspection finally accepted the 623 questioned elements but imposed the contractor, from then on, the obligation of spray-curing the intrados as established in the specifications.

At the time of the investigation, the "Exp-Ref" model had not yet been developed, but it is interesting now to see its predictions. The prediction of the carbonation rate K_c, by the Exp-Ref method (described in Section 9.4.3.2), consists in applying Eq. (9.41) to the measured kT_{gm} values. Parameter β was set equal to 0.42 for the most critical wet and dry cycles

Table 11.6 Predicted carbonation depths at 150 years according to experimental and analytical models

	Predicted carbonation depth (mm) at 150 years			
	Based on kT site measurements, applying model			
Curing time (h)	*Parrott*	*Old Structures*	*Exp-Ref (XC4)*	*Analytical Model*
72	14	38	28	24
18	19	48	43	35
Column	1	2	3	4

exposure XC4 (near the ends of the Tunnel). The predicted K_c values were 3.5 and 2.3 mm/y^{½} for the elements cured 18 and 72 hours, respectively. The resulting carbonation depths are indicated in column 3 of Table 11.6.

Observing the predictions of the four models included in Table 11.6, it can be concluded that "Exp-Ref" and "Old Structures". Models yield quite comparative data, not very different from the Analytical Model, which is somewhat more optimistic. Parrott's model predictions, at least applied with *kT* values for air-permeability, look too optimistic.

11.3.2 Hong Kong-Zhuhai-Macao Link (Chlorides)

The Hong Kong–Zhuhai–Macau (HZM) sea link project involves 28.8 km of sea bridges (four navigable spans), two offshore artificial islands and a submerged tunnel of 6.8 km, with a total investment of nearly 12 billion US dollars. The project is situated on the south-eastern coast of China and links three important cities (Hong Kong, Zhuhai and Macau) with six lanes in dual directions with a design speed of 100 km/h.

The reinforced concrete structures in the project include piers, bearing platforms and piles for sea bridges, the segments for the submerged tunnel and also retaining walls and breakwaters for the artificial islands. One of the technical challenges of the HZM project was to achieve a design service life of 120 years for the concrete structures in an aggressive marine environment.

An analytical service life design was performed applying a probabilistic model, based on the DuraCrete approach (DuraCrete, 2000), but defining the input parameters from a thorough analysis of 30 years of experimental data of an exposure station located in a similar environment to that of the project and from tests performed during construction. The model was applied to structures under the following marine environment exposures: Submerged, Tidal, Splash and Atmospheric Zones. For a detailed description of the applied analytical model, see Li et al. (2015, 2016, 2017).

This real case refers to non-destructive measurements of air-permeability *kT* and cover thickness *d* (also of electrical resistivity ρ, not discussed here) on

Figure 11.16 View and dimensions of tunnel precast segments of Hong Kong – Zhuhai – Macao link.

Table 11.7 Durability requirements for the construction of submerged precast tunnel elements

Requirements			*Intrados*	*Extrados*
Design service life		Years	120	
Durability limit state		-	Corrosion initiation	
Minimum cover thickness		mm	50	70
Concrete grade	@ 28 days	MPa	C45	
Chloride-diffusion coefficient	@ 28 days	10^{-12} m²/s	6.5	6.5
	@ 56 days		4.5	4.5
Allowable crack width		mm	0.2	0.2

the monumental precast sections of the submerged tunnel (see Figure 11.16). These segments were built on land and later floated to be towed and sunk at their precise positions. The durability requirements for the construction of the segments are described in Table 11.7.

Once the construction was started, samples of the placed concrete were taken, on which the coefficient of diffusion was measured at 28 and 56 of age, following the migration test (NT Build 492, 1999), described in Section A.2.1.2. At each age, 148 samples were tested with mean values of 4.68 and 2.95 (10^{-12} m^2/s) at 28 and 56 days, respectively, and coefficients of variation of 9.9% and 8.7% for same ages, respectively; i.e. in conformity with the requirements in Table 11.7.

An analytical assessment of the service life, regarded equal to the corrosion initiation time, was made assuming the probability distributions and parameters indicated in Table 11.8. A Monte Carlo simulation of Eqs. (9.11) and (9.12), making the variables adopt at random the probability functions indicated in Table 11.8, was run for different times t. For each t, the proportion of "failure" cases, i.e. cases where $C(x, t) > C_{cr}$ was computed, providing

Table 11.8 Variables and probability distributions used for the Analytical service life prediction

Variable	*Distribution*	*Parameters*	*Extrados values*
Initial chloride content C_0	Rectangle (uniform)	Lower limit (%)	0.02
		Upper limit (%)	0.04
Surface chloride content C_s	Log-normal	Average (%)	5.76
		Std. dev. (%)	0.86
Critical chloride content C_{cr}	Beta	Lower bound L (%)	0.45
		Upper bound U (%)	1.25
		Coefficient α (-)	0.22
		Coefficient β (-)	0.36
Chloride diffusion coefficient D_0 (at 28 days)	Log-normal	Average (10^{-12} m²)	4.71
		CoV (%)	9.8
Diffusion decay exponent *n*	Normal	Average (-)	0.47
		Std. dev. (-)	0.029
Cover thickness *x*	Normal	Average (mm)	73.4
		Std. dev. (mm)	3.9

the "Analytical" probability distribution of the initiation time illustrated in Figure 11.17 (plain full line). Consequent with its design criterion, it yields 5% probability of corrosion initiation at 120 years.

In addition, during construction, the cover thickness was measured non-destructively applying the common midpoint (CMP) technique, based on transmitting an electromagnetic wave pulse and receiving the reflected waves from the steel bars by two adjacent antennas (Halabe et al., 1993). The cover thickness was measured on 24 segments with a mean value of 73.4 mm and a standard deviation of 3.9 mm (used in the analytical service life prediction), showing that ≈80% of the values exceeded the minimum cover of 70 mm, stipulated in Table 11.7 (Li & Torrent, 2016).

The air-permeability *kT* (*TPT*, Proceq) of some segments (Case 1) was measured on 14 points at the age of 56 days, with a $kT_{gm} = 0.069 \times 10^{-16}$ m² and $s_{LOG} = 0.18$ (Li & Torrent, 2016). Another set of 15 *kT* results (Case 2) was reported in Table 2 of Wang et al. (2014) with $kT_{gm} = 0.022 \times 10^{-16}$ m² and $s_{LOG} = 0.42$. This set has a much lower central value kT_{gm} and larger scatter s_{LOG} than for Case 1.

The "Exp-Ref" model, expressed by Eq. (9.31), was applied to estimate the corrosion initiation time, using the Monte Carlo simulation method, assuming that *d* and *kT* have the distributions and parameters shown in Table 11.9. A value of 0.0104 was assigned to parameter α, corresponding to exposure class XS2 (submerged elements), see Table 9.6.

The Monte Carlo simulation was run through 8,000 instances, generating for each of them independent random values of *d* and *kT* according

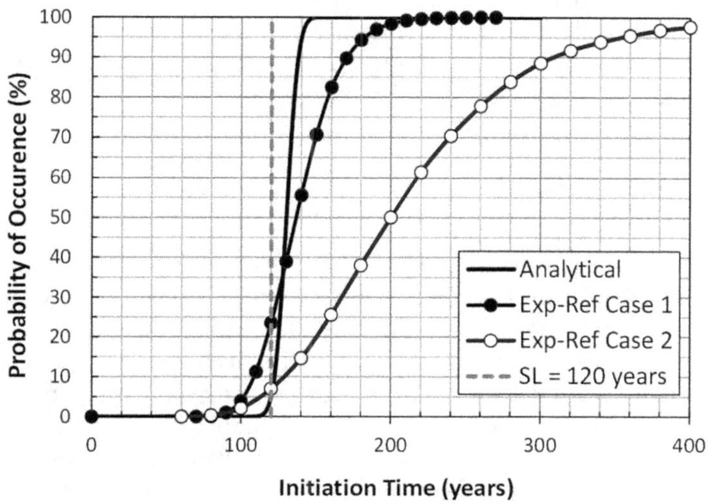

Figure 11.17 Probability of occurrence of corrosion initiation time (Analytical and Exp-Ref predictions).

Table 11.9 Variables and probability distributions used for the "Exp-Ref" service life prediction

Variable		*Distribution*	*Central value*	*Scatter*
Cover thickness *d*		Normal	Mean = 73.4 mm	Std. dev. = 3.9 mm
Air-permeability *kT*	Case 1	Log-normal	$kT_{gm} = 0.069 \times 10^{-16}$ m²	$s_{LOG} = 0.18$
	Case 2		$kT_{gm} = 0.022\ 10^{-16}$ m²	$s_{LOG} = 0.42$

to the distributions and parameters shown in Table 11.9 which, by applying Eq. (9.31), yielded 8,000 values of the corrosion initiation time. The resulting probability of occurrence for both cases is plotted in Figure 11.17 (black dots for Case 1, white dots for Case 2).

Figure 11.17 shows that the "Exp-Ref" prediction for Case 1 yields a median service life (50% probability) very similar to the Analytical prediction (≈135 years), but the probability of occurrence at 120 years is much higher for the "Exp-Ref" Case 1 (24%) than for the Analytical prediction (5%). The "Exp-Ref" Case 2 predicts at 120 years a probability of 7% (closer to the Analytical prediction) but a much longer median service life (≈200 years), reflecting the very low central value but high scatter of *kT* values, compared to Case 1.

It is difficult to compare the Analytical with the Exp-Ref predictions, because the principles of the models are quite different. Yet, since all three predictions are based on the same probability distribution of the cover thickness (squared), the differences can be attributed mostly to the "penetrability" (by diffusion or permeability) of the concrete cover. In this respect, the higher scatter of the Exp-Ref predictions probably reflects the higher variability of the quality (kT), measured on site, compared to that obtained on cast specimens (D_0).

11.3.3 Panama Canal Expansion (Chlorides)

The Panama Canal is a vital waterway to facilitate seaborne traffic between the Atlantic and Pacific oceans. Until 1914, year in which the Panama Canal was opened, ships travelling between both oceans have to sail round Cape Horn (South America), on waters that are particularly hazardous, owing to strong winds, huge waves, strong currents and icebergs.

The Panama Canal Expansion is a huge civil engineering project that consists in expanding the capacity of the Panama Canal to handle more and bigger cargo ships (from Panamax to New Panamax vessels, than can double the cargo capacity). Basically, Panama Canal operates by raising the ships entering it from one ocean, through a series of locks, to the level of internal Gatun Lake, which is navigated till finding the opposite set of locks that brings down the ships again to the opposite ocean.

The Panama Canal Expansion project consists in building several structures, in particular two new sets of locks, one on the Atlantic mouth and the other on the Pacific mouth of the canal, that should "be safe, structurally sound, economical, practical, and durable, with minimum maintenance costs and fit for a design life of 100 years" (PCE Specs, 2008a). The water in the locks that face the open seas have higher salinity than the others which have lower salinity as they take fresh water from the Gatun Lake.

Structural Marine Concrete structures are those directly exposed to saline water (lock walls, heads and floors, approach structures, water-saving basins and conduits, etc.) and are subjected to the specifications summarized in Table 11.10 for 100 years design service life (PCE Specs, 2008b).

In addition, the following was stated: "The use of fly ash or ground granulated blast-furnace slag in the mixture is encouraged. Consideration shall be given to heat dissipation, permeability, setting time, strength gain, curing time, especially for mass concrete placements". Many structures, in particular the locks, are massive reinforced concrete structures (see Figure 11.18, (Ferreira, 2014)) exposed to frequent wetting-drying cycles

Table 11.10 Specifications for structural marine concrete of Panama Canal expansion

Minimum cover	*f'c*	*w/c ratio*	*Permeability (ASTM C1202)*	*28-d shrinkage (ASTM C157)*	*Peak T*	*ΔT*	*Curing*
75 mm	Design	≤0.40	≤1,000 Coulombs	≤0.042%	≤70°C	≤20°C	≥7 days

Figure 11.18 Massive reinforced concrete blocks exposed to saline water (Ferreira, 2014).

of contact with the fluctuating salty water level. Hence, restrictions on the drying shrinkage, peak temperature and difference ΔT between concrete and air temperature were specified in Table 11.10.

Initially, the binder used consisted of cement Type II (ASTM C150, 2012), to which a "natural pozzolan" was added, the latter coming from grinding a basalt rock obtained from an external source. It has to be mentioned that the cement brand used for the Pacific locks was different than for the Atlantic locks, while the aggregates and "natural pozzolan" were the same. More details on the project and concrete characteristics and processing are provided in Andrade et al. (2016).

From the very beginning it was realized that, even lowering the *w/b* ratio well below the specified maximum 0.40, to limits that put the workability of the mixes at risk, the set of materials chosen would not be capable of complying with the 1,000 Coulombs specified as maximum allowed chloride permeability.

The contractor failed to produce mix designs with ASTM C1202 test results below the specified limit, seemingly due to the lack of activity of the so-called "natural pozzolan" which happened to be virtually inert. The contractor, on the other hand, suggested that the lack of compliance with ASTM C1202 was due to the basalt being electrically conductive. In order to investigate that claim, owner's consultants decided to incorporate another test to judge the quality of the mix designs being investigated that was not influenced by the conductivity of materials and/or pore solution of the concrete. Naturally, since it was not included in the original specifications and tender documents of the project, this test could not be used for compliance control, but to provide further information that could help decide whether the mixes proposed by the contractor were likely to ensure 100 years' service life of the project or not.

The chosen method was the air-permeability *kT*, with the Panama Canal Administration (ACP) acquiring two units of the *PermeaTORR* (M-A-S Ltd.), one for the Atlantic and the other for the Pacific side of the Canal. Fifteen members (engineers and technicians) of the quality control personnel were trained by Prof. Fernández Luco (Univ. of Buenos Aires) on theoretical and practical aspects of the measurement of air-permeability, later complemented by a visit of R. Torrent to confirm that the instruments were being used correctly, both in the laboratory and on site.

Just a few of the huge amount of quality control results obtained, made availability to the authors, will be discussed here, as much information is until now still classified. Figure 8.8 presented the correlation between the charge passed Q and kT for 92 pairs of data, among which are laboratory data from the Pacific side (Set 34) and the Atlantic side (Set 35) of the Canal. The results of the Canal merge reasonably well with those obtained from other sources, although it can be observed that most Atlantic results fall below the regression line whilst those of the Pacific side lie close to the regression line. The average of Q and the geometric mean of kT are as follows:

Atlantic side: $Q_m = 774$ Coulombs; $kT_{\mathrm{gm}} = 0.018 \times 10^{-16}$ m^2
Pacific side: $Q_m = 1{,}721$ Coulombs; $kT_{\mathrm{gm}} = 0.051 \times 10^{-16}$ m^2

The results confirm that the concrete produced in the Atlantic side is less permeable than that of the Pacific side, but discard the alleged effect of the conductivity of the basalt, as the results tend to follow the general trend of Figure 8.8.

During the visit and training by R. Torrent, nine measurements were conducted *in situ* on one Block ("A"). One aspect to remark is that the surface moisture *m*, measured with a *Concrete Encounter* instrument, showed values not exceeding 5.5% (on the sunny side of the block) despite two rainy days preceding the day of test, in a tropical climate. The shadow side, on the contrary, showed values above 5.5%. The air-permeability *kT* measured on eight out of the nine points showed uniform, quite acceptable low values. One point showed a high *kT*, measured in a zone of the block where some segregation was visible. Unfortunately, these optimistic results were not confirmed by tests performed on other blocks, as shown in Table 11.11 (Torrent, 2011). Indeed, blocks "B" and "C" showed high values of both kT_{gm} and s_{LOG}, opening some questions on their potential durability. As indicated in Torrent (2011), inappropriate consolidation with poke vibrators was observed during the visit, as well as serious

Table 11.11 On site air-permeability *kT* test results obtained on three blocks (Pacific side)

Block	*N*	kT_{gm} (10^{-16} m^2)	s_{LOG}
"A"	8	0.012	0.19
"A"	9	0.020	0.66
"B"	12	1.353	0.77
"C"	5	0.468	0.49

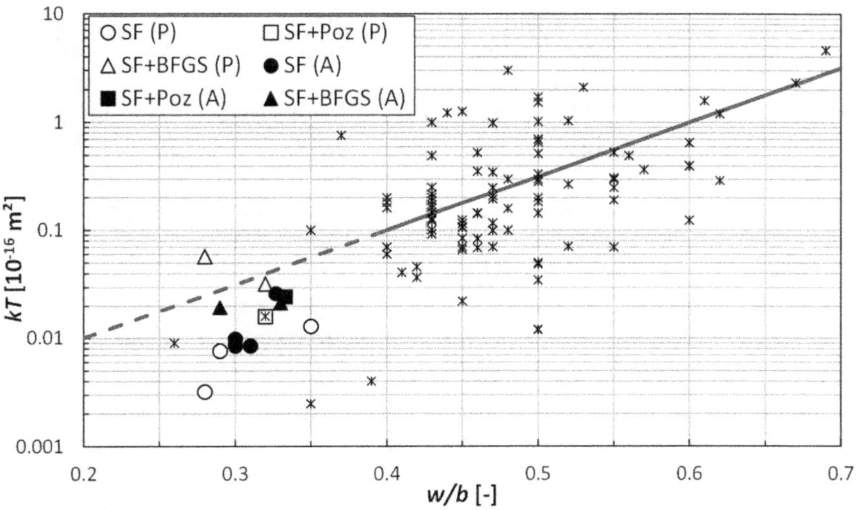

Figure 11.19 Effect of the addition of silica fume (SF) on the *kT* of low *w/b* ratio concretes.

honeycombing of a block that was being stripped while *kT* testing of block "A" was in progress.

Finally, the use of *kT* test to explore different alternatives to optimize the concrete binder is shown in Figure 11.19 where the impact of the addition of several SCMs on *kT* was investigated. Designations (A) and (P) indicate the use of materials from the Atlantic and Pacific sides, respectively.

The addition of silica fume (SF) alone and, to a lesser extent, in combination with the pozzolan in use (Poz) or with blast-furnace granulated slag (GBFS), looks very effective in reducing the permeability of concretes with very low *w/b* ratios. The chart in Figure 11.19 shows previous data (*) and the CEB-FIP relation (Eq. 6.2), already presented in Figure 6.5.

An expert, invited by the owner in May 2011, proposed the immediate use of SF along with a drastic reduction in ground basalt ("pozzolan") content, as the most effective solution. As a result, SF was incorporated into the mix designs for Marine Structural Concrete and ground basalt reduced to a minimum. In Andrade et al. (2014, 2016) a durability assessment approach is described, based on bulk diffusion and electrical resistivity tests (both described in Annex A) performed on cast cylinders moist cured during several months. As the results in Table 11.11 show, such assessment needs to be complemented by site testing to verify the durability quality of the end-product.

11.3.4 Precast Coastal Defence Elements (Sulphates)

This real case refers to a project aimed at building a floating regasification plant in the River Plate estuary (South America). The Liquified Natural Gas (LNG) terminal includes the construction of the breakwater and loading docks, capable of receiving ships up to 218 Mcm.

The construction of the facility required the production and installation of 20,000 coastal protection elements, made of plain precast concrete, out of which, at the time of the investigation some 4,000 had been produced. The elements were produced by a subcontractor in an "ad-hoc" precast plant, located inside the jobsite facilities. The elements were built under the Accropodes™ II technology, following the strict specifications of the system's owners (Accropode, 2011). Figure 11.20 shows the large stock of accumulated elements.

The subcontractor abandoned the project and, before accepting them for usage in the project, guarantees on the sulphate resistance of the 4,000 elements already produced were demanded by the owner. For this, a revision of the production control documentation available was conducted, as well as a visual inspection of the elements, including:

a. data on the aggressiveness of the water
b. data on the quality control of the cement used to build the elements

Figure 11.20 View of some of the ≈ 4,000 elements stored in the stockyard.

c. data on the quality control of the concrete used to build the elements
d. report on visual assessment of the elements

R. Torrent (M-A-S SRL) visited the jobsite, making a visual observation of several elements, agreeing on a sampling and testing plan and training the local personnel in charge of performing *kT* tests, discussed later.

The results of the investigation were reported in Torrent (2015) and later published in Aracil et al. (2016); here we provide a synthesis of the case.

11.3.4.1 Aggressiveness of the Water

The River Plate estuary is the widest river in the world, with a maximum width of 220 km, pouring into the South Atlantic Ocean the waters of two important South American rivers: Paraná and Uruguay. The water of the river becomes saltier the closer to the sea, reason why some investigation on the salinity of the water at the place of localization of the works was conducted.

A profile of the salinity of the water at different depths was obtained immersing a CTD M48 probe, which measures electrical Conductivity, Temperature and Depth, at five different stations along the project location. Figure 11.21 presents two out of the five profiles of salinity obtained, all similar, expressed in practical units of salinity [pus] (derived from the electrical conductivity).

Figure 11.21 shows a clear stratification in the salinity of water, with a low salinity at depths up to about 3 m and a high salinity at depths beyond 4 m. This suggests that sea water, of higher density, occupies the lower strata, whilst the river water occupies the upper strata.

Regarding the chemical characteristics of the water at shallow and deep depths, a series of chemical analyses were performed at the five stations, with the results shown in Table 11.12.

Although chlorides are the most aggressive ions for reinforced concrete elements in contact with salty water, as the coastal protection elements were made of plain concrete, the attention was focused on the sulphate content in the water.

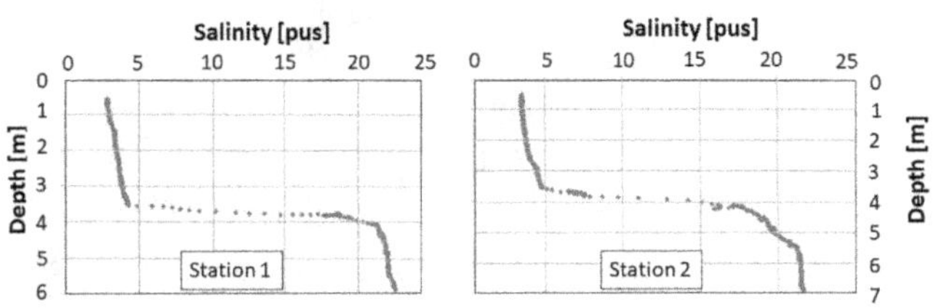

Figure 11.21 Salinity profiles of the water in contact with the coastal protection elements.

Table 11.12 Result of chemical analysis of the water from five stations at two different depths

	Parameters measured at five different sampling stations									
	Station 1		*Station 2*		*Station 3*		*Station 4*		*Station 5*	
Depth (m)	2.0	5.5	2.0	5.5	2.0	5.5	3.0	9.0	3.0	10.0
Salinity (pus)	6.0	21.0	4.3	21.9	4.4	21.9	5.1	22.2	6.0	22.4
SO_4^{2-} (ppm)	257	1,550	304	1,248	304	1,504	459	1,294	489	1,541
Cl^- (ppm)	1,825	12,340	2,325	10,337	2,202	11,338	3,650	13,187	3,696	13,110

11.3.4.2 Durability Requirements

The Accropode (2011) Specifications are based on European Standards and, regarding durability, wrongly classify the exposure class as XS3 from EN 206-1 (2004). Indeed, exposure class XS3 corresponds to tidal zone of marine structures under risk of chloride-induced steel corrosion. As the precast elements in question do not contain embedded steel, this classification is inapplicable to them.

Searching through EN 206-1 (2004), a gap (that subsists in today's version of EN 206) was found in that sulphate attack to plain concrete exposed to marine environment is not contemplated among the durability exposure classes.

On the contrary, ACI 318 (2011) considers explicitly that case, which is regarded as belonging to "Moderate Sulphate Attack" or class S1.

To resist the S1 exposure class, ACI 318 (2011) specifies the following requirements for the concrete mix:

- Moderate Sulphate-Resistant Cement (for OPC it means $C_3A \leq 8\%$)
- $w/c_{max} \leq 0.50$ and $f'c_{min} \geq 28$ MPa (measured on cylinders at 28 days)

11.3.4.3 Concrete Mix Quality Compliance

Of 58 quality certificates provided by the cement supplier, none of them showed a C_3A content above 8%, in compliance with ACI 318 (2011).

The mix design of the concrete used to build the precast elements was declared in a document issued by the ready-mixed concrete supplier, validated by the Quality Assurance and Control Authority.

According to that document, the concrete had the following declared main characteristics:

- cement content: 400 kg/m^3
- w/c ratio: 0.38
- $f'c$: 40 MPa, for 5% lower fractile
- maximum size of aggregate: 20 mm
- slump: 170 ± 30 mm

Therefore, according to the declared characteristics, the concrete mix complies with the requirements of $w/c_{max} = 0.50$ and $f'c_{min} = 28$ MPa of ACI 318 (2011) for exposure class S1.

The analysis of the available 1,129 records of 28-day compressive strength results, measured on Ø150 × 300 cylinders cast during the entire production yielded an overall mean strength $f'_{cm} = 52.8$ MPa, with a very high standard deviation $s = 7.7$ MPa which, applying Eq. (11.2) yielded a characteristic strength (5% fractile) $f'_{ck} = 40.1$ MPa.

$$f'_{ck} = f'_{cm} - 1.65\ s \tag{11.2}$$

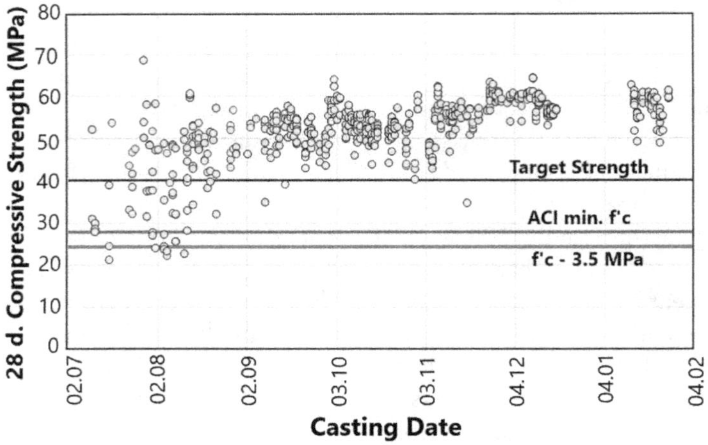

Figure 11.22 1,129 test results of 28-day compressive strength of concrete during elements' production.

This characteristic strength is just above the declared one (40 MPa) and largely above the required by ACI 318 (2011) (28 MPa for an even larger fractile of 9%). So, at first sight, the concrete supplied by the ready-mixed concrete producer was compliant.

Yet, a closer look at the 1,129 test results revealed some anomalies, see Figure 11.22. To start with, a huge variability during the first 2 months, in particular the first one, with extreme strengths of ≈20 and 70 MPa. Besides the 1,129 test results (circles), Figure 11.22 includes three horizontal lines, from top to bottom:

- mix design target characteristic strength of 40 MPa for 5% fractile
- minimum specified strength after ACI 318 (2011) of 28 MPa (to be complied with by the moving average of three consecutive samples)
- minimum strength after ACI 318 (2011) of 28–3.5 = 24.5 MPa (to be complied by each individual sample)

Figure 11.22 shows that, during this initial period, neither the mix design target characteristic strength nor even the minimum strength after (ACI 318, 2011) was complied with. After these initial 2 months, the concrete production tends to stabilize, with a gradual increase in the strength level and a reduction in variability.

11.3.4.4 Precast Elements' Compliance

Compliance of the concrete produced by the ready-mixed concrete supplier is not enough to ensure the performance of the precast elements. Indeed, inadequate compaction, insufficient moist curing, cracks and other defects

Figure 11.23 *kT* test in an area with bug-holes.

may impair the permeability of the concrete elements, which is widely regarded as the key factor for the durability of concrete exposed to seawater attack (Mehta & Monteiro, 2001; Richardson, 2002).

A visual examination of the precast elements was conducted and reported. The main surface defects detected were bug-holes (see Figure 11.23), some of them relatively large, resulting from the impossibility of air bubbles to escape during vibration, especially along top inclined surfaces of the metal forms (see the complex shape and large size of the elements in Figure 11.20). Some cracks and water streaks were also observed.

The visual inspection left some doubts on the extent to which the observed defects may affect the performance of the elements. In order to verify that, a program of tests was established to measure the coefficient of air-permeability *kT* directly on a sample of the precast elements produced. The results of the investigation are fully described in Aracil et al. (2016); here the main findings are summarized.

In a preliminary investigation, two elements (that were broken during handling) cast on the same day were intensively investigated, measuring *kT* in different parts of the elements and placing the central chamber of the vacuum cell (*PermeaTORR*, M-A-S Ltd.) exactly over a bug-hole as well as away from them.

Neither the location of the measurements nor the presence of bug-holes showed a clear effect on the measured values. In fact, the lowest *kT* values for both elements were obtained directly on bug-holes. This indicates that *per se*, the bug-holes do not have a detrimental effect on the permeability and potential durability of the elements (they would on the cover thickness in case of reinforced concrete). This topic was discussed in Section 5.7.5.4.

In addition, a sample of 28 precast elements was selected, one for each week of manufacturing. On each element, three measurements were performed: on top, on the nose and on the bottom, as cast. The elements were separated into lots of about 1,000 elements each (A, B, C and D in chronological order).

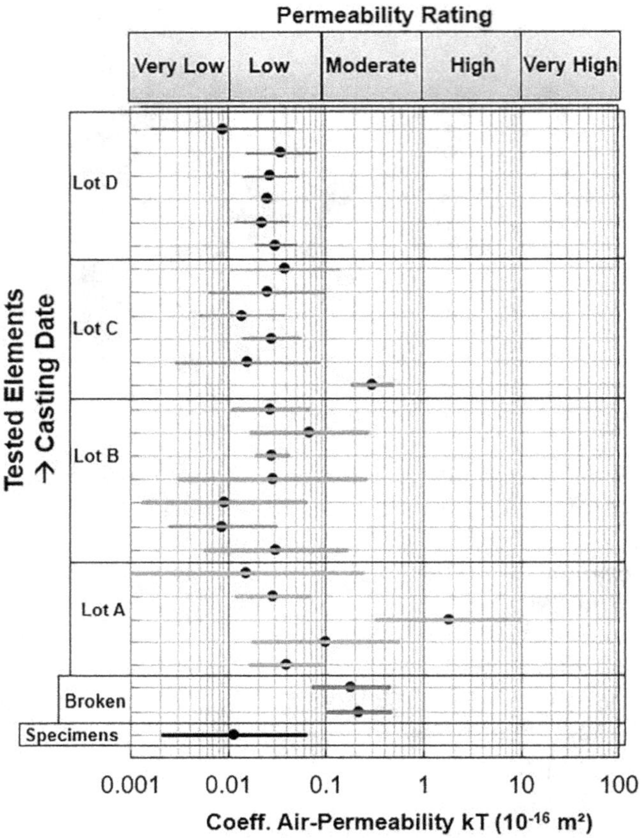

Figure 11.24 $kT_{gm} \pm s_{LOG}$ plot of test results obtained on the elements tested.

The results obtained on the precast elements tested are plotted in Figure 11.24, in a format that presents in logarithmic scale, for each element, the value of kT_{gm} (black dots) and a bar representing $kT_{gm} \pm s_{LOG}$. For comparison, the values of the preliminary trials on broken elements and on cylindric specimens are also plotted at the bottom.

From Figure 11.24, it can be seen that the kT_{gm} of most elements falls within the "Low" Permeability Class, with a few falling in the "Very Low", "Moderate" and "High" Permeability Classes.

The elements belonging to Lot A (those made at the initial stages) show higher and more scattered kT_{gm} values than the rest. Actually, the uniformity in kT_{gm} values increases with the date of casting (bottom to top). This is in line with the level and scatter of cylinder compressive strength results shown in Figure 11.22.

A more detailed analysis of compliance with equivalent requirements (also for $w/c_{max} = 0.50$) of the then available 2013 version of Swiss Standard

(SIA 262/1, 2019) can be found in Aracil et al. (2016), which showed that just 2 out of over 80 *kT* results exceeded the limit value of 2.0×10^{-16} m², applicable to this case, and both obtained on the same Lot A element.

11.3.4.5 Conclusions on the Durability of the Elements

Regarding the sulphate resistance of the elements investigated, the following conclusions were drawn:

1. the durability requirements for concrete exposed to sulphates from seawater are those specified for exposure class S1 (Moderate Severity) of ACI 318 (2011)
2. the cement used for preparation of the concrete corresponds to ASTM C150 Type II ("Moderate Sulphate Resistant"), in compliance with ACI 318 (2011) for class S1
3. the reported *w/c* ratio of the mix design of 0.38 is well below the maximum specified (0.50) in ACI 318 (2011) for class S1 (0.50)
4. the reported 28-day compressive strength results obtained from the third month of production onwards comply with the minimum requirements of ACI 318 (2011) for class S1 (28 MPa) and even with the mix design target of 40 MPa
5. the site air-permeability *kT* of the precast elements belonging to Lots B, C and D comfortably comply with the limiting value of SIA 262/1 (2019) of 2.0×10^{-16} m², corresponding to mixes with $w/c_{max} = 0.50$, applicable to sulphate attack from seawater
6. given that the results of the concrete specimens tested at the initial phases of the production show a high variability in strength that do not always comply with $f'c = 28$ MPa (let alone 40 MPa) and that the corresponding precast elements (belonging to Lot A) showed higher and more scattered *kT* values, the suitability of the precast elements produced during that period is questionable
7. therefore, elements produced from the third month of production onwards can be used with confidence that they will perform well under the planned exposure conditions
8. the elements produced during the first and second months of production may not have sufficient mechanical strength nor durability to perform adequately. Unless more detailed investigation on the actual mechanical strength and permeability of elements produced in that period demonstrate their suitability, they should not be used in the project

11.3.5 Buenos Aires Metro (Water-Tightness)

This real case refers to the construction of Line "H" of the Metro system of the city of Buenos Aires, Argentina (Di Pace & Calo, 2008). The first Metro Line "A" of Buenos Aires network opened in 1913, making it the 13th subway in the world, and first underground railway in Latin America, the Southern Hemisphere and the Spanish-speaking world, with the Madrid Metro opening 5 years later, in 1919. Buenos Aires Metro system experienced a fast growth until 1944 when it came to a halt until the construction was resumed only in the beginning of the 2000s. Today, its extension is far behind that of cities like Madrid and Santiago de Chile.

Construction of new Line "H" started in early 2001. The project involved the construction of eight train stations, three of them connecting to other existing subway lines and almost 8 km of tunnels. Water table level was kept low by means of pumping during construction. Most of the new line laid across clay soils rich in subsoil water, with one station standing 30 m of water head. The project involved the stabilization of excavation with shotcrete, followed by cast-in-place concrete for tunnel lining and for the train stations. Construction progress was good both in 2001 and 2002 but, during October 2002, concerns arose about the actual permeability of the concrete. Project specifications ruled that concrete for stations and some parts of tunnels had to be water-tight to prevent water infiltration and to ensure appropriate durability of structures, for at least 100 years.

Concrete specifications for the job established that cast-in-place concrete for stations and some of the concrete linings for tunnels had to comply with a maximum water penetration under pressure of 50 mm, when tested according to Argentine standard (IRAM 1554, 1983), similar to DIN 1048, see Section 4.1.1.2.

This method operates on cast specimens, not having the capability of measuring the actual water-tightness of the concrete effectively poured in the structures, taking into account not only concrete proportions but also its degree of compaction and curing conditions at the job site.

As construction progressed and the tunnel was excavated deeper and deeper into the ground, subsoil water pressure was found to be worse than expected and concerns arose about the effective water-tightness of concrete at the subway stations. At this point, it was decided to conduct, on site, measurements of air-permeability *kT* to assess the quality of the final product.

Therefore, a major survey was launched, strictly following the guidelines of the then available version 2003 of Swiss Standard (SIA 262/1, 2019) to assess the air-permeability *kT* of structural concrete elements of two train stations (Venezuela and Humberto 1°). It involved over 130 measurements of *kT* (*TPT*, Proceq) at randomly selected locations, corresponding to six different structures (see summary of results in Table 11.13, first stage).

The results obtained from the survey (Table 11.13, first stage) were very disappointing, showing a large variety in quality between elements but also within elements; the within element scatter, measured by s_{LOG}, was well above the average of 0.43 reported by Jacobs et al. (2009), see Figure 5.9.

As a consequence, the decision was made to improve the water-tightness of the concrete, stopping the concreting operations during 45 days, until a new mix was developed, tested and approved by the Engineer.

Table 11.13 Result of the survey of structural elements with *kT* tests (first and second stages)

	Structural element	*N*	kT_{gm} $(10^{-16}\ m^2)$	s_{LOG} (-)	kT_{gm} *Permeability Class*
First stage					
E1	Upper Beams, Venezuela Station	25	4.37	0.69	High
E2	Lower Beams, Venezuela Station	25	0.41	0.82	Moderate
E3	Water Canal (Upper), Venezuela Station	13	1.26	0.72	High
E4	Water Canal (Lower), Venezuela Station	14	0.079	0.78	Low
E5	Bottom slab, Humberto 1° Station	27	11.5	0.92	Very high
E6	Bottom slab, Venezuela Station	6	1.68	1.19	High
Second stage					
N1	Bottom Slab, Venezuela Station	27	0.076	0.41	Low
N2	Tunnel Lining, Humberto 1° Station	17	0.057	0.45	Low

The new mix was carefully proportioned, incorporating a waterproofing agent to reduce permeability without altering 28-day strength and elasticity modulus. In addition, stricter controls were imposed on the sand moisture compensation of the ready-mixed concrete producer and on observing a minimum 7-day curing by the contractor.

Before authorizing the use of the new mix, both its maximum water penetration (IRAM 1554, 1983) and the air-permeability *kT* (on *labcrete* Ø150 × 100 mm specimens, oven-dried at 50°C) were measured, yielding mean values of 47 mm and $0.20 \times 10^{-16}\ m^2$, respectively. The values fit very well to the correlation expressed by Eq. (8.5), see Set 23 point in Figure 8.7.

The new concrete mix was implemented under a restrained approval for the completion of Venezuela station slab (2/3 of total slab), provided satisfactory air-permeability test results were obtained on site. The slab was poured and cured for at least 7 days; the air-permeability *kT* was measured at five locations with a $kT_{gm} = 0.34 \times 10^{-16}\ m^2$ and $s_{LOG} = 0.25$. The results, both in terms of central value and scatter, are much better than those shown in Table 11.13 for the first stage.

The new mix was approved and full re-start of the works was granted, establishing a *kT* target value of $0.20 \times 10^{-16}\ m^2$ for site measurements at an age of at least 28 days.

After about one and a half month, a second stage of measurements was launched. In this stage, two new zones were investigated: remaining concrete poured at the slab of Venezuela Station and randomly selected zones of tunnel concrete linings in the vicinity of Humberto 1° Station. Table 11.13 (second stage) summarizes the test results obtained from this second stage.

It is remarkable that only 4 tests out of the 27 recorded at Venezuela Station Slab exceeded $0.20 \times 10^{-16}\ m^2$, whilst none of the 17 Tunnel Lining results exceeded that target value.

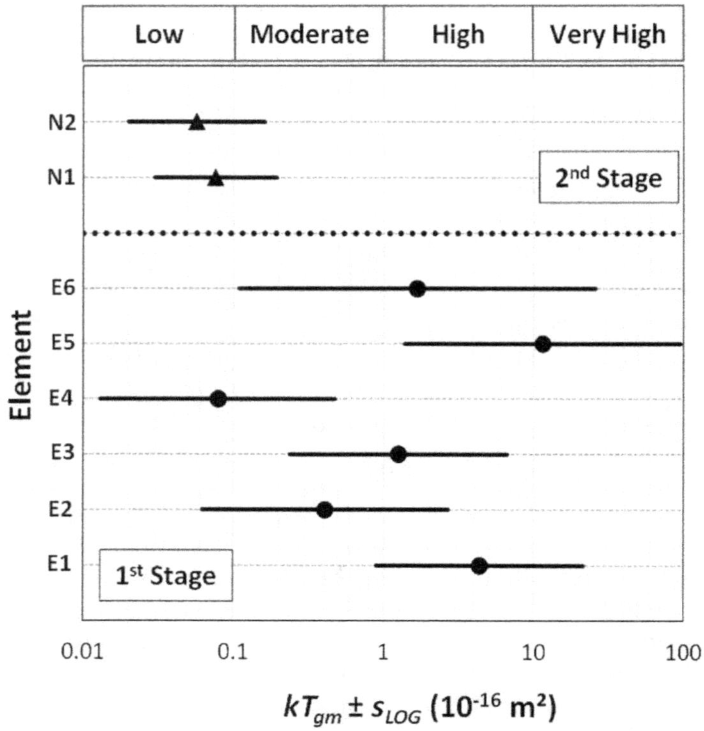

Figure 11.25 $kT_{gm} \pm s_{LOG}$ plot of test results obtained in the first and second stages of construction.

The much better performance obtained in this second stage can be better appreciated in Figure 11.25, where the values of $kT_{gm} \pm s_{LOG}$ are plotted in log-scale for both stages. The results of the second stage (upper plots) not only have a lower kT_{gm} (indicated by black triangles in the chart), but also the uniformity was greatly improved, testimony of a careful processing of the new concrete mix by both the contractor and the ready-mixed concrete supplier.

A lesson to be learnt from this real case is the importance of on-site measurement of *kT* to check the performance (water-tightness in this case) of the end-product as a tool to improve the quality of the work of all players involved in the concrete construction process.

11.3.6 HPSFRC in Italy (Water-Tightness)

11.3.6.1 Description of the Case

This real case refers to an industrial building meant for a company devoted to manufacturing equipment for the pharmaceutical industry; it is described in detail in Torrent et al. (2018). The two-storey building, illustrated in

Figure 11.26 View of the investigated building at an advanced construction stage.

Figure 11.26, is situated in the city of Como, Italy, located at the southwestern tip of Lake Como, bordering with Canton Ticino, Switzerland.

Stringent requirements were set for the building; in particular a very low permeability was required against the relatively high water-table. The design of the building, as well as the quality of the concrete, was concerned with providing an impermeable foundation barrier, besides structural safety, durability and functionality.

Different concrete qualities were used for different parts of the building. The most critical areas were built with steel fibre reinforced concrete (SFRC), cast *in situ*. Some slabs were cast with self-compacting SFRC (SCC-SFRC).

To verify the degree of water-tightness reached by the end-product, on site air-permeability *kT* measurements were carried out on representative elements of the structure, as detailed in Table 11.14.

Table 11.14 Characteristics of the elements tested for site air-permeability *kT*

Code	*Function*	*Concrete type*	*Mix design*	*Cast*	*Age at test (d)*
FB	Foundation beam	SCC-SFRC	E	On site	163
FS1	Foundation slab	SCC-SFRC	G	On site	86
FS2	Foundation slab	SCC-SFRC	G	On site	82
SS	Suspended slab 1st floor	RC	S	On site	144
EW	External wall	RC	P	On site	175
IW1,2	Internal walls	RC	P	On site	176
PP1-3	Internal precast columns	RC	Columns	Plant	n.a.

SCC, self-consolidating concrete; SFRC, steel-fibre reinforced concrete; RC, reinforced concrete.

11.3.6.2 Characteristics of the Concretes Used for the Different Elements

Table 11.15 shows the main characteristics of the mixes used to build the different elements, more details and meaning of fracture properties of Mixes E and G in Torrent et al. (2018); Mix S is proprietary and was not disclosed.

11.3.6.3 Air-Permeability kT Tests Performed

Air-permeability *kT* tests were performed on site with two instruments (*PermeaTORR* and *PermeaTORR AC*, M-A-S Ltd.), on the as-cast concrete surface condition, following the prescriptions of version 2013 of SIA 262/1 (2019); prior to testing the air-permeability, the surface moisture *m* of the concrete was measured (*CMEXpert II*, Tramex), to check that the indication was not above 5.5%, which was the case for all tests.

In addition, *kT* tests were performed on cast specimens used for material characterization and quality control. The results of the site tests are shown on the left-hand side of Table 11.16, whilst those obtained on laboratory specimens are shown on the right-hand side of the Table.

Table 11.15 Main characteristics of the concrete mixes used in the investigated elements

Component or property	*Mix E*	*Mix G*	*Mix P*	*Columns*
Cement (kg/m^3)	470	380	360	480
Limestone filler (kg/m^3)	-	100	-	-
Superplasticizer (% cem. wt.)	1.62	1.58	1.20	3.00
Shrinkage reducer (% cem. wt.)	0.85	1.44	-	-
Steel fibres (kg/m^3)	35	35	-	-
w/c ratio	0.40	0.43	0.38	0.46
$f'_{c,\text{cube}}$ @ 28 days (MPa)	-	59.3	35.0	67.0

Table 11.16 Statistical parameters of air-permeability test results obtained on site and in the laboratory

Site tests				*Laboratory tests*			
Element	*No. of tests*	kT_{gm} (10^{-16} m^2)	s_{LOG} (-)	*Specimen*	*No. of tests*	kT_{gm} (10^{-16} m^2)	s_{LOG} (-)
FB	7	0.058	0.71	Cubes 150 mm	11	0.330	0.57
FS1	6	0.023	0.22	Notched beams*	13	0.103	0.34
FS2	6	0.0045	0.27	*L: 1,500; H: 500; W: 300 (mm)			
SS	12	0.123	0.64				
EW	31	0.210	0.82				

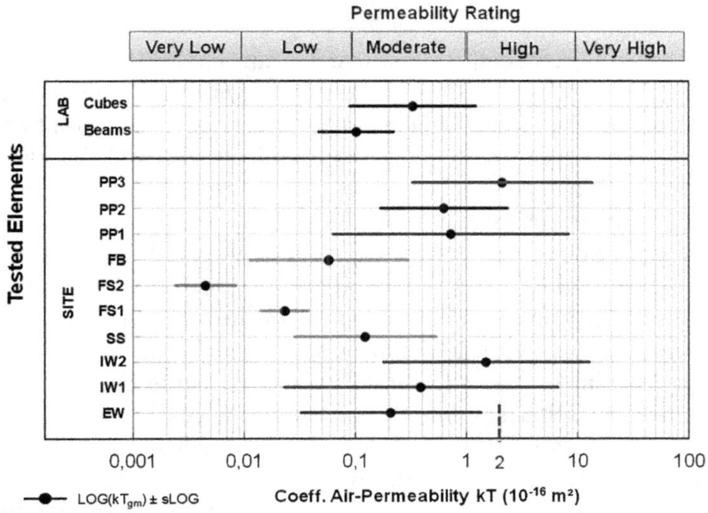

Figure 11.27 $kT_{gm} \pm s_{LOG}$ plot of test results obtained on site (b) and in the laboratory (a).

The results presented in Table 11.16 are plotted in graphical form in Figure 11.27 in the usual way ($kT_{gm} \pm s_{LOG}$). The short vertical segment at $kT = 2.0 \times 10^{-16}$ m² indicates the upper limit of kT prescribed by neighbouring Switzerland for surfaces exposed to severe carbonation (XC4).

What is immediately obvious from Figure 11.27 is the wide range of kT values obtained on the different concretes tested. Indeed, the values of kT_{gm} recorded *in situ* on the different elements span four orders of magnitude, from "Very Low" to "High" Permeability Classes (see classification at the top of Figure 11.27).

11.3.6.4 Performance of SCC-SFRC Elements

The better permeability performance of the SCC-SFRC foundation elements (FB, FS1, FS2), compared with the RC elements (including the cast-on-site walls and the precast columns), is evident in Figure 11.27. This can be attributed to a mix design with low *w/c* ratio, to the inclusion of a shrinkage-reducing admixture and, furthermore, to the positive contribution of steel fibres in controlling shrinkage cracks.

Looking at the kT values of SCC-SFRC in more detail, it results that the foundation slabs on the ground (FS1 and FS2) present lower permeability than the foundation beams and even than the laboratory specimens (flexural beams). The foundation beam was made with mix E, while the two foundation slabs were made of the same nominally identical mix G which is improved by the addition of filler. Moreover, this better performance may lie on the fact that the ground slabs were kept ponded with water for several weeks after casting, thus receiving a better curing than the other elements and even than the specimens.

The kT_{gm} value recorded for foundation slab FS1, 0.0045×10^{-16} m^2, is among the lowest ever recorded by the authors on site. Compare it with the values, recorded on precast elements made with concretes meant to last over 100 years in a marine environment, reported in Sections 11.3.2 and 11.3.3. This indicates that the SFRC concrete cast on site in this building can be classified as a High-Performance HP-SFRC.

It is worth noticing that the kT_{gm} values measured on the foundation slabs on the ground (0.023×10^{-16} and 0.0045×10^{-16} m^2) are much lower than that obtained on the laboratory flexural beams, made with the same concrete mix (0.103×10^{-16} m^2). This is a rather unusual case as, in general, laboratory specimens show lower kT values than the corresponding site concrete, due to the better compaction and curing conditions applied to the specimens (see Section 7.1.5). The inversion of this general rule can be due to the exceptional conditions under which the foundation slabs were built: self-compacting workability and several weeks of curing by ponding; this confirms the importance of good concreting practices on the performance of the end products.

11.3.6.5 Performance of Walls

The case of the External Wall merits some analysis. The external face of the building walls will be directly exposed to the environment. The city of Como has a Mediterranean climate, quite rainy (around 1,300 mm/y) and with temperatures occasionally dropping below 0°C in winter. It can be classified as XC4 (severe carbonation risk) and XF1 (mild frost risk), according to EN 206. For climates XC4 and XF1, Swiss Standard (SIA 262/1, 2019) specifies a "statistical upper limit" of the air-permeability $kT_s = 2.0 \times 10^{-16}$ m^2. It is worth mentioning that the building in question lies less than 10 km of the Italian/Swiss border, where those requirements apply. A statistical analysis (Torrent et al., 2018) showed that the intensively tested External Wall (EW) comfortably complies with the (SIA 262/1, 2013) specification, something that can be visualized in Figure 11.27.

Just as a speculation, if internal walls IW1 and IW2 had been external, i.e. exposed to the same environment as wall EW, they would have not complied with (SIA 262/1, 2019) requirements, especially IW2.

11.3.6.6 Performance of Precast Columns

The values obtained on the precast columns are rather disappointing. They show a high kT_{gm} and a high scatter. A closer look to the surface of the elements showed a crazed skin, which may have affected the air-permeability measurements. However, on one spot, the removal of the skin revealed a visible crack on the concrete underneath; given that these internal elements play no role in terms of water-tightness or durability of the structure, the matter was not pursued further.

11.3.6.7 Conclusions

Based on the results of this experimental investigation, the following conclusions can be drawn:

- the tests conducted on site showed a wide range of air-permeability *kT* values for the different elements investigated, reflecting the different mix designs and concreting processes involved
- the self-compacting, steel-fibre reinforced foundation concretes showed permeabilities in the range of "Low" and "Very Low" classes, suitable for their function as water-tight barrier against the high water-table level underneath. Based on their high performance, they can be classified as HP-SFRC
- foundation slabs FS1 and FS2 even yielded lower site *kT* values than laboratory samples made with the same mix, a rare example to be attributed to their self-compacting workability and long-term curing by water ponding
- the intensively tested external wall, exposed to an environment that can be classified as XC4, XF1 (EN 206), showed results of *kT* in compliance with the upper limit $kT_s = 2.0 \times 10^{-16}$ m^2, specified in neighbouring Switzerland
- the internal cast-on-site walls and precast columns showed "Moderate" to "High" air-permeability values, that are rather disappointing, but have no consequences on the serviceability and durability of the structure, given their indoors location

11.3.7 UHPFRC in Switzerland (Chlorides)

A successful line of research was pursued at the EPFL (Lausanne, Switzerland) under the leadership of Prof. E. Brühwiler, to develop a durable solution for the repair of deteriorating bridges, which was extended internationally under the SAMARIS project (Sustainable and Advanced MAterials for Road InfraStructure). See also Section 7.2.2.

Several documents describe the approach in detail (Denarié, 2005; Denarié & Brühwiler, 2006; Denariè et al., 2006; Brühwiler, 2007a, b; Schmidt et al., 2007; Oesterlee et al., 2009), on which this section is based.

The concept is founded on the observed fact that, often, conventional reinforced concrete does not resist chloride-induced corrosion for sufficiently long periods. This is of special relevance for important structures, like bridges, meant to last for at least 100 years.

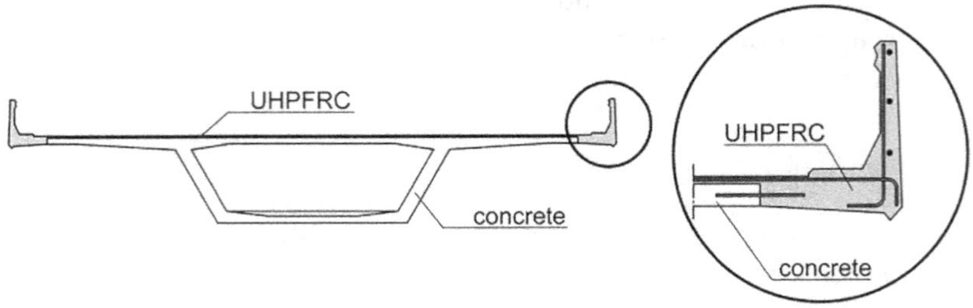

Figure 11.28 Example of selective use of UHPFRC in critical elements of concrete bridges (Brühwiler, 2007b).

The bridge elements most vulnerable to de-icing salts chloride-induced corrosion are those where Ultra High-Performance Fibre Reinforced Concrete (UHPFRC) can be selectively used to improve durability (see example in Figure 11.28).

UHPFRC is a fine cementitious composite of following composition range (Brühwiler, 2007b):

- 1,000–1,400 kg/m^3 of CEM I 52.5 or CEM II 32.5
- silica fume (7%–26% of cement content)
- superplasticizer (1% of dry extract mass of the cement content)
- quartz sand of 0.5 mm maximum size
- w/b ratio = 0.12–0.16
- 235–705 kg/m^3 of steel micro- and macro-fibres (3%–9% by volume)

UHPFRC can be produced in a wet-process ready mix concrete plant and transported to the jobsite by truck; preliminary trials are recommended to "train the team". It has a self-compacting workability, with slump flow > 600 mm and works at ambient temperatures between 7°C and 35°C.

The mechanical performance of UHPFRC is far superior to that of conventional concrete (CC), as shown in Table 11.17. Figure 11.29 illustrates the σ–ε diagram of UHPFRC tested in direct tension, where points *1* (first crack) and *2* (maximum load) are indicated.

Table 11.17 Mechanical properties of UHPFRC compared with conventional concrete (CC) (Brühwiler, 2007b)

Property	*Compressive strength*	*E-modulus*	*Tensile strength*	*First crack strength*	*Strain-hardening*
Material	*MPa*	*GPa*	*MPa*	*MPa*	*%*
CC	≈ 40	≈ 35	≈ 3	≈ 3	0
UHPFRC	160–250	48–60	9–20	7–16	0.05–0.2

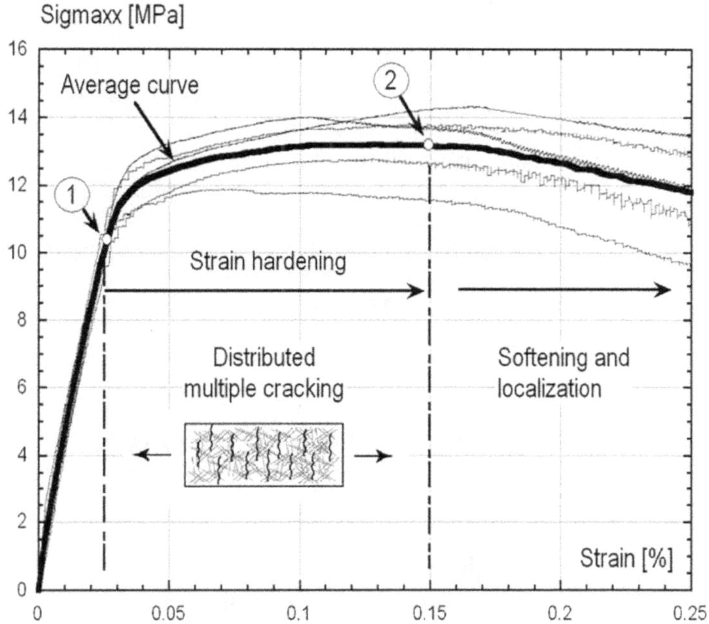

Figure 11.29 σ-ε diagram of UHPFRC tested in direct tension (Brühwiler, 2007b).

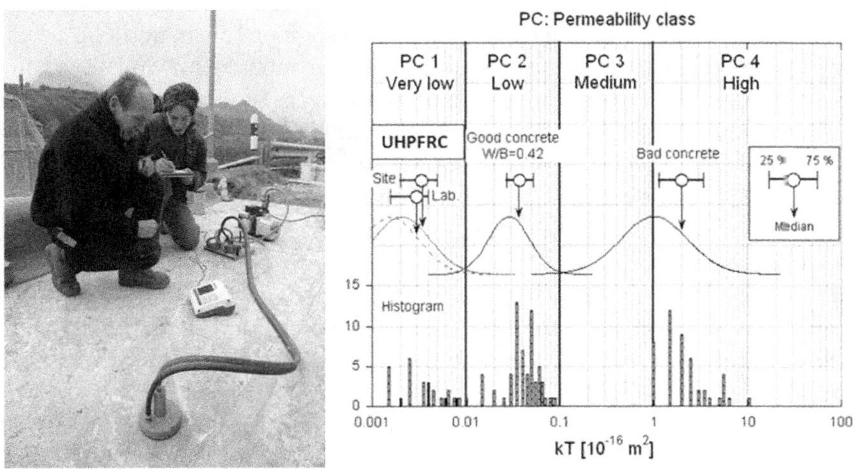

Figure 11.30 Site *kT* test of UHPFRC and reported results (Brühwiler, 2007b).

The air-permeability *kT* (*TPT*, Proceq) was measured on UHPFRC specimens and on site, with the results summarized in Figure 11.30, confirming its very low permeability: $kT = 0.004 \times 10^{-16}$ m^2 on average as reported by Denarié & Brühwiler (2006). Interesting to remark is that the same average value was obtained, also on site, 2 years after the rehabilitation with UHPFRC of a 36-year old bridge in Slovenia (Sajna et al., 2012).

In terms of durability, the UHPFRC showed much better performance than conventional concrete, not only in terms of *kT* but also of N_2-permeability, Cl^- migration and frost-thaw-salts resistance (Schmidt et al., 2007).

Among the specifications of UHPFRC for these applications, an extremely low *kT* value of 0.005×10^{-16} m² for the 75% fractile is required for outstanding protective function (Denarié et al., 2006).

11.3.8 Field Tests on Swiss New Structures

This topic is discussed jointly with field tests of old structures in Section 11.4.2.

11.3.9 Field Tests on Portuguese New Structures

11.3.9.1 Bridge at the North of Lisbon (Quality Control/Carbonation)

This real case concerns the production control of precast elements to be applied in the construction of a bridge 40 km to the North of Lisbon, Portugal. The bridge has a total length of nearly 12 km, the second longest in Europe at the time of construction, with a total budget of 218 M€. The concrete consumption was 400,000 m³ and it was constructed in less than 2 years. The bridge spans two rivers and comprises three major structures: the main bridge, 972 m long with a maximum span of 133 m, and two access viaducts. One of the viaducts is quite long (around 9 km) and to ensure timely construction, the girder was built with precast concrete elements: four U-beams with a precast slab that, when assembled, would work as multiple box girder (Figure 11.31). It was decided to carry out air-permeability *kT* tests on the precast beams and slabs, with the aim of checking the concrete uniformity concerning carbonation resistance (Ribeiro & Gonçalves, 2009).

Air-permeability tests, using the *TPT* instrument (Proceq), were performed on 51 beams (35 m long) and 244 slab panels (2.1 × 4.2 m). Each beam was tested on six different locations (two each on the left web, right web and flange) always on the outer side, whereas each panel was tested once on the bottom surface.

The concrete mix design for beams and slabs had a blend of CEM I 52.5R and CEM IV 32.5R (EN 206-1) totalling 465 kg/m³, a *w/c* ratio of 0.35 and was set for a strength class C40/50 according to EN 206-1. All elements were cured for 24 hours and then left exposed to the outside air (Ribeiro & Gonçalves, 2009).

The results of the 550 *kT* tests, obtained at ages ranging between 3 and 155 days, are presented in Figure 11.32, adapted from Figures 3 and 4 of Ribeiro and Gonçalves (2009), who investigated but found no correlation between air-permeability *kT* and testing age.

Figure 11.31 Ground view of Lisbon viaduct's multiple box girder.

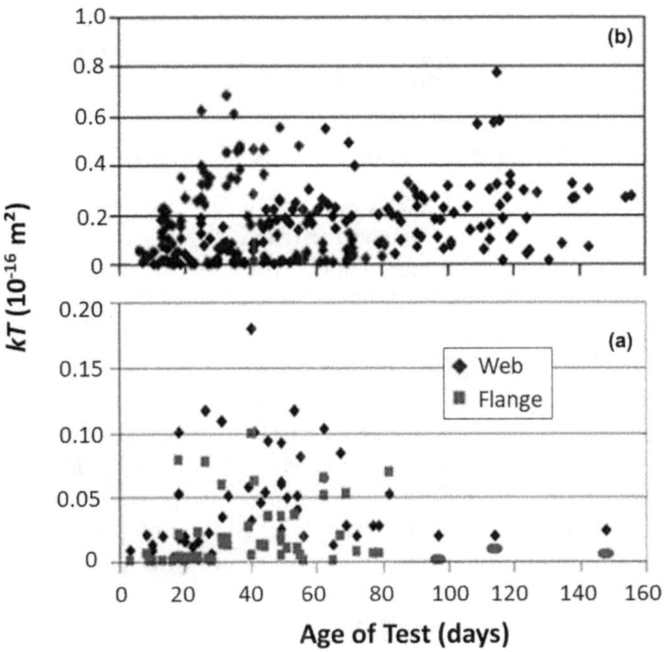

Figure 11.32 *kT* results at different ages of Lisbon's viaduct: (a) on the beams; (b) on the slabs (Ribeiro & Gonçalves, 2009).

It is interesting to note the typical distribution associated with air-permeability results, positively skewed. The permeability in beams (Figure 11.32a) is rated as "very low" (only 2% of the results are greater than $0.01 \times 10^{-16} m^2$), see classification in Table 5.2. On the other hand, the permeability of the slabs (Figure 11.32b) can be rated as "low" to "moderate", as the largest share of the results is equally divided between "Low" and "Moderate" Permeability Classes. The differences in permeability between beams and slabs (built with the same concrete mix design) were attributed to microcracking in slab panels originated by the handling of these slender elements after demoulding (Ribeiro & Gonçalves, 2009). These authors also stressed the difference between web and flange permeabilities, which they attributed to the sheltering of flanges (bottom surface of the beam) that allowed further cement hydration and thus led to a more refined porous network. Arithmetic means of 0.020, 0.046 and $0.167 \times 10^{-16} m^2$ for *kT* on beam flanges, beam webs and slab panels, respectively, were reported by Ribeiro and Gonçalves (2009).

Production uniformity was evaluated through results variability. It was considered that the scatter of results was within the usual range for gas-permeability assessments and no considerable fluctuations were noticed along production time. To the best of the authors' knowledge, this was the first real case application of Torrent test method in Portugal, and it was considered by its users as a promising method for *in situ* concrete evaluation (Ribeiro & Gonçalves, 2009). Furthermore, they also acknowledged the differentiation capability of the method, concerning the presence of micro-cracks, endorsing its use within the frame of quality control.

11.3.9.2 Urban Viaduct in Lisbon (Quality Control)

During the construction of a viaduct in Lisbon, Portugal, built to close a gap in an urban highway, it was agreed with the Owner to carry out air-permeability *kT* tests. The viaduct has a total length of ca. 800 m and the design had several constraints, due to the vicinity to the airport and to nearby roads and buildings. A solution comprising a single box-girder in pre-stressed concrete, 32.4 m wide and with a maximum span of 105 m, was adopted. The main objective of testing was to assess the influence of placing, compaction and curing procedures on concrete performance. Air-permeability as absolute value was not relevant, since the viaduct was to be coated with a protective paint.

Air-permeability tests, using the *TPT* instrument (Proceq), were performed on the box-girder (webs, upper slab and internal struts), piers and abutments and in companion laboratory specimens. The laboratory specimens were small slabs (360×250×120 mm), cast when ready-mix concrete was delivered to the construction site, carefully vibrated, cured in water for 7 days and then kept in a room at 22°C/65% RH until testing at 28 days. The concrete in structural elements had no special protection procedures beside the formwork, that was removed 0.5 (precast struts), 1 (piers and abutments) and 3+ (girder webs and flanges) days after casting. The site air-permeability tests were performed at 28 days. In some situations, the tests had to be performed a few days after, to ensure that the testing surfaces had not been in contact with water for at least 15 days (Neves & Santos, 2008).

Differences between site and laboratory *kT* air-permeability were observed, as illustrated in Figure 11.33. The values on the top of the bars indicate the ratio between site *kT* and lab *kT*, ranging from situations where *kT* measured on site was lower than in laboratory companion specimens, to site *kT* resulting four times higher than in the laboratory. The major differences were observed in girder concrete. Neves and Santos (2008), attributed this fact to the large volume castings, with high rates of pouring concrete, to the existence of several work fronts and to the large number of accumulated labour hours, that led to a lower quality in placing and compacting concrete when compared with smaller castings.

The first three groups of elements in Figure 11.33 correspond to a concrete mix with a *w/b* ratio of 0.42, strength class C40/50, whilst the abutment was cast with a mix of *w/b* ratio 0.51, strength class C30/37, which explains its higher *kT*. Following the Permeability Classes, regardless of being site or laboratory results and the different curing/protection times, it is clear that the *w/b* = 0.42 concrete belongs to a "Low" Permeability Class, whereas the *w/b* = 0.51 concrete belongs to a "Moderate" Permeability Class. Despite undergoing the lowest protection/curing time, among the different situations for *w/b* = 0.42 mix, the precast struts show the smallest difference between site and laboratory concrete tests. On the contrary, the box-girder, where the site protection/curing was the longest, recorded the highest difference between site and laboratory results. This confirms the

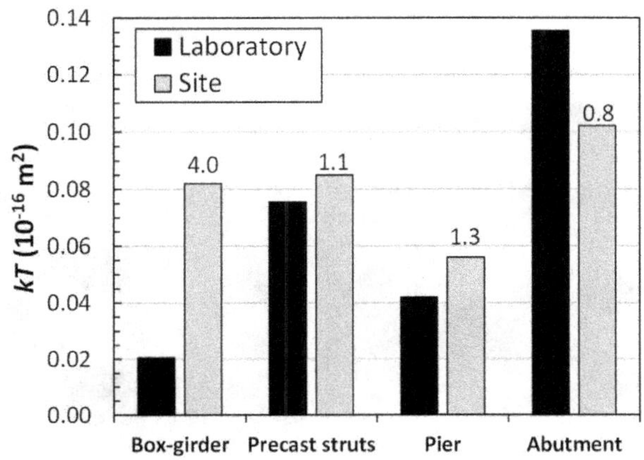

Figure 11.33 Comparative *kT* results obtained on site and in the laboratory for different structural elements.

belief that the main causes of the differences between site and laboratory concrete air-permeability were placing and compaction.

11.3.9.3 Sewage Treatment Plant (Chemical Attack)

In the construction of a sewage treatment plant, in Portugal, precast slabs were used. The plant was designed for a biological treatment capacity of 3.3 m^3/s peak flow, and twice that flow through physical-chemical treatment, serving a population of nearly 750,000 people. A curious feature of this plant is its landscape architectural plan, comprising the covering of the tanks with vegetation to make the plant resemble an urban garden park.

During the production of the precast slabs (3.3 × 1.9 m), there were doubts about their durability, due to the appearance, in some of them, of entrapped air bubbles at the top-as-cast surface (Figure 11.34). Therefore, it was decided to test two slabs, about 3 months old, for air-permeability. A total of 43 air-permeability *kT* tests was performed on opposite surfaces of the two slabs, using the *TPT* instrument (Proceq), to assess whether the permeability was affected by the presence of air bubbles (on top-as-cast surface) or not.

The analysis of the results led to conclude that between the four sets of results (2 slabs × 2 surfaces) only one was statistically different. The permeability results obtained on the two top-as-cast surfaces were comparable with the results from one of the bottom-as-cast surfaces. Unexpectedly, the different set (that corresponded to the other bottom-as-cast surface) presented higher *kT* than the rest, especially than those whose aspect had raised concerns about concrete durability. The concrete mix had a *w/c* ratio of 0.28 and the compressive strength at 28 days was 58 MPa (Neves et al., 2012). The *kT* of the three "comparable" sets was $0.026 \times 10^{-16} m^2$, whist the *kT* of the "different" set was $0.074 \times 10^{-16} m^2$.

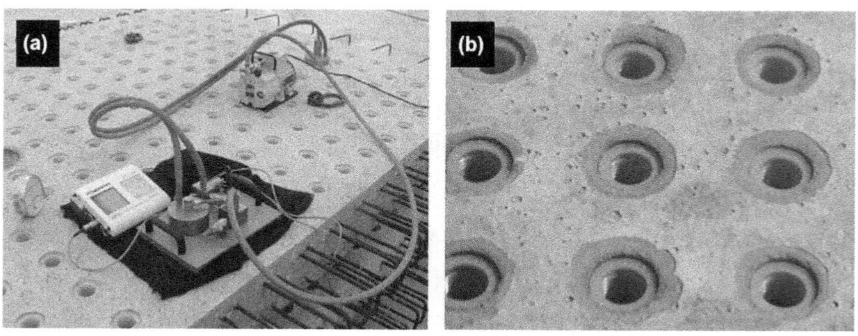

Figure 11.34 (a) Air-permeability *kT* testing of precast slab; (b) air-bubbles on top-as-cast surface.

11.3.10 Delamination of Industrial Floors in Argentina ("Defects" Detection)

The concept of Shrinkage-Compensating Concrete (ShCC) has been described in Section 7.2.4, acknowledging it as a way of achieving very low concrete permeability on site. This technique allows the construction of large industrial floors' slabs (typically 1,200 m^2 in surface area) without joints which, after a first successful project (Fernández Luco et al., 2003) gained acceptance in Argentina.

High performance industrial floors, usually subjected to heavy-duty traffic of lifting vehicles, benefit from the lack of joints which, besides being a weak point of the floor, also damage the wheels of those vehicles. In order to improve the abrasion resistance of the floor, often a wear-resistant, top 20 mm thick layer is built, made with quartz sand and cement, which receives a power floating finishing (see Figure 11.35).

Figure 11.35 illustrates, in the foreground, the placement and spreading of the quartz-cement mortar, its compaction and, in the background, the power-floating with "helicopters". If these operations are not properly and timely conducted, delamination of the wear layer may happen, with subsequent damage to the floor during service. This kind of damage arose in a floor, the causes of which were investigated. For that, Ø150 mm cores were drilled, slicing the top 50 mm for observation under the stereomicroscope, whilst the rest was used for compressive strength testing.

A first assessment of the damage was made by the contractor, by knocking the concrete surface with a light hammer. The Argentine Portland Cement

Figure 11.35 Construction of a ShCC industrial floor. (Courtesy of Bautec S.A., Argentina.)

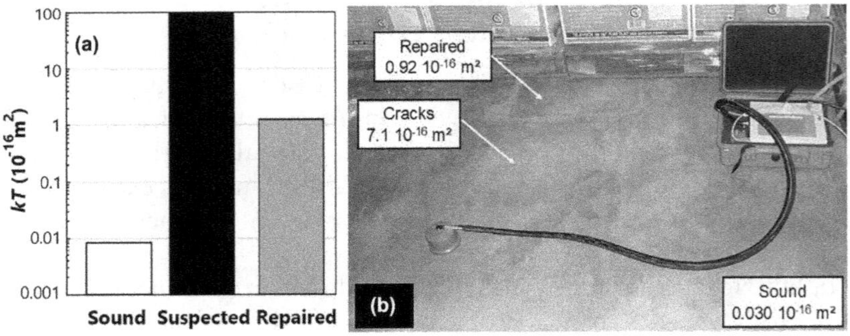

Figure 11.36 *kT* measured on different zones; results of two damaged ShCC floor.

Institute (ICPA) was called to confirm the extent of failure, by applying a more sensitive non-destructive technique. ICPA measured the air-permeability *kT* (*TPT*, Proceq) on different points of the floor, covering sound, suspected and repaired areas, with the results shown in Figure 11.36a, (Fernandez Luco & Pombo, 2005).

It can be seen that the sound areas showed very low permeability, whilst the suspected areas showed extremely high *kT* values, possibly due to air flowing along the gap between the unbonded wear layer and the core concrete. This helped identifying the exact extent of the area to be repaired. The repair was done by removing the unsound wear layer and replacing it with a similar quartz-cement mortar with an acrylic admixture to improve bond with the core concrete and avoid cracking. As shown in Figure 11.36a, the *kT* of the repaired areas, although much lower than before the repair, is still considerably higher than that of the sound concrete.

A similar investigation related to the appearance of cracks was reported by Torrent (2008) on another, similar floor, applying a *PermeaTORR* (M-A-S Ltd.). In this case, the sound concrete was of "low" permeability, see Figure 11.36b and the repaired part of just "moderate" permeability. Measurements near the repaired zone revealed tiny microcracks which raised the *kT* to "high" permeability level.

11.4 OLD STRUCTURES

11.4.1 Old Structures in Japan

Two real cases, involving emblematic structures investigated in Japan, are described in this section, namely:

- Tokyo Museum of Western Art, recognized by UNESCO as World Heritage Site
- Jyugou Bridge, Japan's First Post-Tensioning PC Bridge

In addition, other relevant cases are reported at the end of this section.

11.4.1.1 Tokyo's National Museum of Western Art (Carbonation)

Tokyo's National Museum of Western Art (NMWA) houses a valuable collection of paintings and sculptures from renowned western artists. The collection belonged to the rich Japanese businessman Kojiro Matsukata, president of the Kawasaki Dockyard Co. Ltd. He was an enthusiastic and wealthy collector, having his artwork pieces stored in museums in the UK and Japan (both mostly destroyed during World War II) and in France, confiscated at the end of the war. France agreed to return the masterpieces to Japan, stipulating that the museum to house the collection be designed by a French architect. In the end, the task was assigned to the Swiss-French architect Charles-Édouard Jeanneret-Gris (better known as Le Corbusier) who, together with three Japanese architects (Junzo Sakakura, Kunio Maekawa and Takamasa Yoshizaka), designed the building in 1955. The construction was completed in 1959; the building has been recognized by UNESCO as a World Heritage Site.

In the garden outside, famous August Rodin's sculptures are displayed; Figure 11.37 shows the sculpture "Burghers of Calais" in the foreground with the museum in the background. Inside the building, paintings of famous western artists are displayed.

Figure 11.37 Rodin's "Burghers of Calais" sculpture with NMWA building in the background.

Figure 11.38 Aspect of the visual damage detected in Tokyo's NMWA.

Le Corbusier had a great affinity to concrete as building material, which he used profusely in the interior and exterior of NMWA building (NMWA, 2016).

In 2009, when the building had reached 50 years of age, a visual inspection revealed incipient signs of corrosion in some parts of the concrete elements (see Figure 11.38), predominantly at the edges of the exterior beams. Carbonation-induced steel corrosion was identified as the main cause of the damage.

A detailed investigation was commissioned to Tokyo University of Science, under the direction of Prof. K. Imamoto, described in detail in Imamoto et al. (2012, 2013) and Imamoto (2012). Details of the mix design can be found in Imamoto et al. (2012); summarizing it had *w/c* = 0.62 and cylinder *f'c* at 28 days = 18 MPa, not of high quality for today's standards.

An important restriction was posed by the Museum authorities: the investigation should not be intrusive, i.e. should be based entirely on NDT, with no damage to the building, given its architectural and historic value. Finally, the authorities tolerated the extraction of just three small cores (Ø20 ×≈50 mm) to measure the carbonation rate by the phenolphthalein method. An elaborate method to reinsert the cores in the drilled holes, without leaving traces on the surface, is described in Imamoto et al. (2012).

Prior to the extraction of the three cores, the coefficient of air-permeability kT was measured on the same spots, which allowed to calibrate a relation between kT and the carbonation rate K_c of the form (see Figure 9.11a):

$$K_c = 1.4 + 0.63\ln(kT) \quad K_c\left(\text{mm/y}^{1/2}\right);\, kT\left(10^{-16}\,\text{m}^2\right) \tag{11.3}$$

for $kT \leq 0.1 \times 10^{-16}$ m^2 is $K_c = 0$

K_c was computed as the measured carbonation depth divided by the square root of the age (50 years).

Besides measurements in other elements, 111 paired measurements of *kT* and the cover thickness *d* were performed along the more exposed exterior beams. The cover thickness was measured by means of a GPR (ground-penetrating radar) instrument. At each measurement point, the corrosion initiation time t_i (years) was calculated applying Eq. (9.8) = Eq. (11.4):

$$t_i = \left(\frac{d}{K_c}\right)^2 \quad t_i\,(\text{years});\; K_c\left(\text{mm/y}^{1/2}\right); d\,(\text{mm}) \tag{11.4}$$

where K_c was obtained from the measured *kT* value through the application of Eq. (11.3).

Thus, 111 values of t_i were obtained, which allowed the probabilistic assessment described in Section 9.5.1 and illustrated in Figure 9.13 (white circles linked with full line). The criterion of the Architects' Institute of Japan is to recommend an intervention when the risk of corrosion reaches 20%, which happens at an age of 75 years, i.e. 25 years after the investigation.

This was one of the most elaborate and successful application of the *kT* test to assess service life of existing structures. A different approach with a similar result (Neves et al., 2018), described in Section 9.5.2, was later applied to this case.

11.4.1.2 Jyugou Bridge (Condition Assessment)

Jyugou Bridge is Japan's first post-tensioned concrete (PC) road bridge, with the characteristics shown in Table 11.18, depicted in Figure 11.39a at the time of construction and in Figure 11.39b nowadays.

According to construction records of the Jyugou Bridge, the concrete was cast with shovel and stick, and cured with flooding water and moisture matting for slab and handrail, respectively.

Table 11.18 Characteristics of Jyugou bridge

Type of structure	**Post-tensioned PC**
Bridge length	7.85 m
Girder length	7.80 m
Bridge width	7.20 m
Type of PC cable	12 Ø5 Freyssinet
Design strength	
Main girder	36.8 MPa
Between girder	24.5 MPa

Figure 11.39 Past (a) and present (b) of Jyugou Bridge.

This historical bridge marked its 60th birthday in 2013 and still remains in service. A detailed inspection of the Jyugou Bridge was carried out on that date to evaluate its soundness (Akiyama et al., 2014).

The air-permeability *kT* was measured at different points of five girders (three measurements per girder), with kT_{gm} values between 0.001 and 0.008×10^{-16} m^2 (three measurements on each girder are reported). All values correspond to the "Very Low" Permeability Class.

A couple of cores were drilled from the structure to measure the compressive strength and carbonation depth at the age of 60 years. The recorded strength was 78.3 MPa whilst the measured carbonations (average of five readings around the core) were 0.0 and 1.6 mm (K_c = 0 and 0.2 mm/y^{½}, respectively). These results, not included in Figure 8.16, confirm the fact that concretes of "very low" permeability ($kT < 0.01 \times 10^{-16}$ m^2) display a negligible carbonation rate.

11.4.1.3 Other Japanese Structures (Condition Assessment)

Some real cases, involving *kT* measurements on eight existing structures located in different parts of Japan, with ages ranging between 3 weeks and 53 years, are described in Kishi et al. (2008). It is interesting to remark that, among the old structures, the lowest *kT* value (0.005×10^{-16} m^2) and the highest (12.7×10^{-16} m^2) were obtained on railway bridges of 53 and 28 years of age, respectively. Prestressed girders of three bridges, built with same nominal strength grade concrete (40 MPa), showed quite different *kT* values, ranging from 0.005×10^{-16} m^2 ("Very Low" Permeability Class after Table 5.2) to 0.173×10^{-16} m^2 ("Moderate" Permeability Class), confirming that durability cannot be judged only on the basis of compressive strength. Measurements of a reinforced concrete culvert at 3 months showed a 3-fold increase of *kT* with respect to that measured at 3 weeks of age.

A field investigation of the condition of a 23-year old bridge in Japan (Lidoi et al., 2021) revealed an extremely wide scatter of *kT* values within the range 0.001–100 ($\times 10^{-16}$ m^2), accompanied by low Wenner electrical resistivity (< 10 kΩ.cm). The amount of chlorides around corroded steel bars was not very high (0.31 kg/m³), the corrosion being attributed to damage of the resin coating applied to the steel bars.

An extremely interesting case is the survey, conducted by Sato (2017), of the condition of several bridges located in the Tohoku Region, at the N end of the main Japanese island of Honshū. He found that 40% of the bridges needed prompt repairs, some being just about 25 years old. The main deterioration mechanisms detected were frost damage, ASR and steel corrosion, enhanced by the extensive use of de-icing salts on concretes affected by construction-related defects. In the survey, extensive use of non-destructive tests was made, namely air-permeability *kT* and water sorptivity SWAT (see Section 4.2.2.5). He attributed at least part of the failures to the lack of provisions regarding cover concrete quality in road bridges specifications, promoting efforts to disseminate information and create awareness of the importance of the issue, both among the staff and the construction industry as a whole. In order to prove the importance of such provisions, draft specifications were prepared for reconstruction works of a bridge deck, establishing upper limits or the permeability measured on site at 28 days of age, namely $kT \leq 1 \times 10^{-16}$ m² and SWAT $p_{600} \leq 0.5$ ml/(m².s). The use of additions such as PFA and GGBFS was specified, as well as an intensive moist curing system. After the floor slab was constructed, non-destructive tests such as air-permeability coefficient *kT* and electrical resistivity were conducted on the actual floor slab, which indicated a high durability. In addition, it was confirmed by laboratory tests that it had high resistance to salt damage, frost damage, and ASR. From these results, it was confirmed that the design and construction of the trial work was appropriate. Based on the data acquired in this pilot construction, a guide for reinforced concrete floor slabs was prepared, that was applied to three more bridges of the Minami-Sanriku National Highway Office. As a result, by utilizing the guide, highly durable floor slabs were built without causing any major problems. Generalization of the guide, which is described in Sato (2017), is judged as useful in disseminating new technologies.

11.4.2 Old (and New) Swiss Structures (Chlorides + Carbonation)

11.4.2.1 *Investigated Structures and Tests Performed*

The data reported and discussed here were collected from a relatively large number of new and old structures, more thoroughly described and discussed in Frenzer et al. (2020). The sample of investigated structures comprises 17 new structures (with ages at the time of testing between 18 and 400 days) and 11 old structures (with ages at the time of testing between 17 and 60 years).

The 28 concrete structures (new + old) investigated, involved the testing of (38 + 46) structural elements with a total of (333 + 223) measurement areas. For the sake of comparison of the performance of concrete in the structures with that obtained in the laboratory, test results obtained on laboratory cast specimens kept 21 days in a dry room (20°C/50% RH) after 0, 7 or 28 days moist curing, reported in Tables 3.1-III and 3.2-IV of Torrent and Ebensperger (1993) and in Tables 1.2.1.1, 1.2.1.2 and 3.2.1.1 of Torrent and Frenzer (1995), are often included in the analysis.

The following NDT methods were applied directly on the surface of the structural elements, typically without any previous preparation:

- coefficient of air-permeability *kT* [m²]
- surface electrical resistivity ρ [kΩ.cm], Wenner test method, Section A.2.2.2. *kT* and ρ were measured using the *TPT* instrument (Proceq). It is worth mentioning that ρ was measured with the intention of assessing the moisture conditions of the surface at the moment of measuring *kT*; hence, both properties were measured at the prevailing temperature and moisture conditions at the moment of test, without any effort to artificially wet or dry the surface. This situation, combined with the fact that different cements were used to build the structures, known to have a strong effect on the electrical resistivity ρ, led to poor or simply no correlation between ρ and the rest of durability and strength characteristics of the concretes. For that reason, the ρ results are not discussed in this Section. The interested reader can consult (Frenzer et al., 2020).
- surface moisture content *m* [%], electrical impedance method, using a *Concrete Encounter* device (Tramex)

The following tests were applied in the laboratory, on drilled cores diamond-cut to size:

- O_2-permeability *kO* [m²], (Cembureau test method), measured on Ø100 × 50 mm discs, conditioned by 6 days oven drying at 50°C, followed by 1-day cooling to 20°C in a desiccator
- water sorptivity a_{24} [g/m²/s^{½}], HMB procedure, was obtained by placing the same discs used for *kO* in contact with 3 mm of water and monitoring the mass increase due to capillary suction (along the lines of Annex A of Standard (SIA 262/1, 2019)). The mass of water absorbed by unit surface area of the specimen [g/m²] after 24 hours of contact, divided by the square root of 24 hours [s^{½}] is the water sorptivity a_{24}
- water sorptivity a_{24} [g/m²/s^{½}], TFB procedure, was obtained on Ø50 × 50 mm discs, conditioned by 2 days oven drying at 50°C, followed by 1-day cooling to 20°C in a desiccator. The values originally reported were of the so-called "water conductivity" q_w [g/m²/h]

(Annex A of Standard (SIA 262/1, 2019)). For comparison with HMB results, the q_w values were converted into a_{24} [g/m²/s½], through Eq. (11.5), developed at TFB from regression analysis of many test data:

$$a_{24} = 3.4 + 1.142\,(q_w - 2.26) \tag{11.5}$$

with q_w in [g/m²/h] and a_{24} in [g/m²/s½]

- pore characteristics, from MIP analysis of diamond-cut small specimens, about 10 × 20 × 40 mm each, that could fit into the (Carlo Erba Series 2000 WS with macro-pore unit 120) MIP instrument, capable of measuring pore radii between 3.7 and 300,000 nm. The reported values are the total porosity V_t [%] and the mean pore radius r_m [nm].
- carbonation rate K_c [mm/y½], based on the measurement of the carbonation depth x_c [mm] (phenolphthalein method), on freshly exposed surfaces of cores of old structures, divided by the square root of the age of the structure [years]
- chloride content Cl [% of cement weight] obtained by titration analysis of a 10 mm thick slice cut at 10–20 mm from the surface of a drilled core (old structures)
- compressive strength, measured after 28 days of moist curing on 150 mm cubes (f'_{c28}) for new structures or on Ø100 × 100 mm cores drilled from the structures (f'_c) for old structures; both in [MPa]

11.4.2.2 Combined Analysis of Results

Figure 11.40 presents the relation of the air-permeability *kT* vs. *kO* and a_{24}, whilst Figure 11.41 presents the relation of the air-permeability *kT* vs. r_m and V_t, including new and old structures. *kT* was measured on site and the other four properties were measured in the laboratory on saw-cut samples from cores drilled from the structures. As reference, the values obtained in

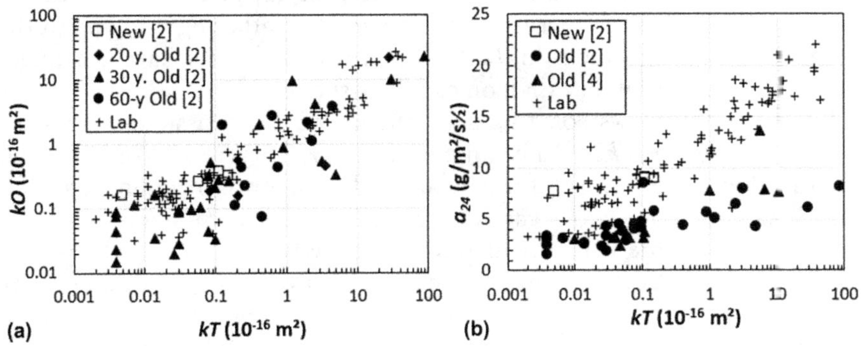

Figure 11.40 Relation *kT* vs. (a) O_2-permeability *kO* and (b) water sorptivity a_{24}

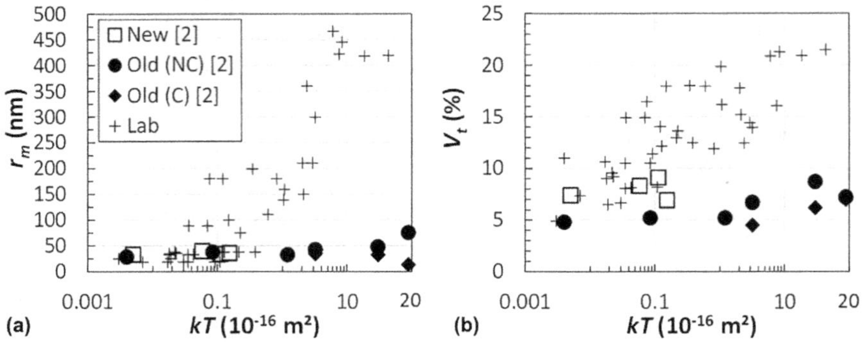

Figure 11.41 Relation kT vs. MIP parameters (a) mean pore radius r_m and (b) pore volume V_t.

the laboratory, under the conditions described in Section 11.4.2.1, are also plotted as + symbols.

Figure 11.40a shows that the coefficient of air-permeability kT, measured *in situ* on new and old structures, correlates well with the coefficient of O_2-permeability kO, measured in the lab on preconditioned cores drilled (at the same place) from the same structures. Most interesting, the relation obtained on new and old structures fits quite well to that obtained on specimens, cast, cured, preconditioned and tested in the laboratory (+ symbols). It is worth mentioning that the kT values obtained on the 30-years-old structure (black triangles in Figure 11.40a) span five Permeability Classes, see Table 5.2.

Figure 11.40b shows a rather different picture. The water sorptivity a_{24} of old structures is significantly lower, for the same site kT, than that obtained on new structures, the latter fitting quite well the relation obtained on laboratory specimens (+ symbols). The water sorptivity a_{24} of old structures is less than half the value obtained on laboratory specimens for the same kT, especially for kT values above 0.1×10^{-16} m². Two explanations can be advanced for this phenomenon: (1) the carbonation of old structures has a stronger effect on the capillary suction than on gas-permeability, and (2) high kT values in old structures, after years of weathering, are due to the appearance of microcracks in the ITZ and/or matrix, that have a stronger effect on gas-permeability than on capillary suction.

Figure 11.41 throws some light on the previous discussion. It presents the relation between kT and mean pore radius r_m (Figure 11.41a) and total porosity V_t (Figure 11.41b), for new and old structures, including also values obtained on laboratory cast specimens as reference (+ symbols). It can be seen that the pore structure (determined by MIP) of old concretes of permeability above 0.1×10^{-16} m² is much tighter (lower r_m and total porosity V_t) than what would be expected from the tests on laboratory specimens. Figure 11.41 includes data of old structures obtained from both non-carbonated (NC) and carbonated (C) parts of the samples. As expected, the pore structure of the carbonated zone is even tighter.

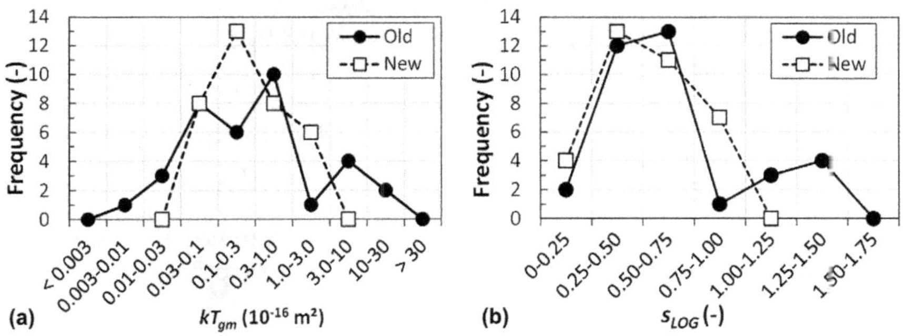

Figure 11.42 Frequency distribution of kT_{gm} (a) and s_{LOG} (b) for new and old structures.

An analysis of the data presented in Figures 11.40 and 11.41 indicates that the microstructure of the concrete in the old structures remains quite tight (low r_m and V_t values) after decades of exposure to the mildly severe Swiss climate (that yet includes frost-thaw cycles and, for bridges and tunnels, de-icing salts too). The very high air- and O_2-permeability (kT and kO) of the old concretes can be attributed to defects (bond or matrix microcracks) that allow the flow of gas under pressure, but that do not influence as much the capillary suction (relatively low a_{24} values).

Figure 11.42a shows the frequency distribution of the central value of kT (kT_{gm}) obtained on elements of new and old structures, whilst Figure 11.42b shows the frequency distributions of the scatter of kT values (s_{LOG}). The number of cases analysed for new and old structures amounts to 35 each.

Although the results were not obtained on the same structures at young and later ages, some patterns can be observed. First, the kT_{gm} values of new structures span 5 class intervals, whilst those of old structures span 9 class intervals. This may be attributed, a bit speculatively, to a phenomenon by which concrete that is originally of "good" quality (low kT_{gm}) and durable improves with the passage of time, whilst concrete that is originally of "insufficient" quality (high kT_{gm}) impairs after decades of service and weathering exposure.

The same phenomenon can explain why the scatter of kT (s_{LOG}) within each element tested becomes larger for old structures compared with that of new structures, as shown in Figure 11.42b.

11.4.2.3 Conclusions of the Investigations

The analysis of the results of these comprehensive investigations yielded the following conclusions:

- the *in situ*, non-destructive measurement of the coefficient of air-permeability kT, of new (some months) and old (several decades) structures, provides a good indication of the gas-permeability of the *Covercrete* (kO), measured in the laboratory

- measuring *in situ* the coefficient of air-permeability kT of new structures, provides a good indication of the water sorptivity of the material (a_{24}), measured in the laboratory
- some old structures present high values of kT, whilst the intrinsic pore structure of the concrete (determined by MIP) is quite tight, which can be attributed to damage (microcracks) after decades of service and weathering
- as a result, high values of kT in old structures are not always accompanied by proportionally high values of water sorptivity (a_{24}), which is believed to be less affected by microcracks
- possibly for a similar reason, high values of kT in old structures are not always accompanied by deep carbonation, although the microclimate (especially moisture of the concrete) may also play a role
- the range of (geometric) mean values of kT in old structures is much wider than in new structures, suggesting that the "durable" concrete improves with time, whilst the "non-durable" concrete impairs after decades of weathering
- similarly, higher values of the scatter of kT (s_{LOG}), within a single structural element, can be found in old structures, compared with new structures
- the high variability in the concrete properties measured in old structures constitutes a challenge for modellers, even when applying probabilistic approaches

11.4.3 Permeability and Condition of Concrete Structures in the Antarctic

11.4.3.1 The "Carlini" Base

This real case corresponds to a condition assessment of buildings in the "Carlini" Base in the Antarctic, described in detail in Benítez and Polzinetti (2016, 2019). The "Carlini" Base is a permanent scientific Antarctic Base of the Argentine Republic, located in the Potter Peninsula of the 25 de Mayo Island (62°14′18″S 58°40′00″W). It was originally established in 1953 and named after the late Argentine scientist Dr. Alejandro Carlini. It is located in a region of great biodiversity, which justifies its intensive involvement in glacier, atmospheric, oceanographic and marine research activities (Argentine and International). It houses the first cinema (52 seats) ever built in the Antarctic and *Ataque*, possibly the southernmost disco dancing in the world, providing entertainment to the staff (up to 80 people).

The Base houses a laboratory to study the green-house effect, within the frame of the Global Atmospheric Watch, which is located in two buildings: Cabildo and Catedral. This laboratory was installed in cooperation with the Italian *Programma Nazionale di Ricerche in Antartide*. In 2001, a seismological station was installed, in cooperation with Italian Istituto Nazionale di Oceanografia e Geofísica Sperimentale.

Finally, the Base houses the Dallmann Laboratory and Aquarium, installed in 1994 and 2001 in cooperation with the German Alfred Wegener Institute for marine research. Since 2004 it is equipped with facilities for diving activities.

11.4.3.2 The Climate

The "Carlini" Base is subjected to a rigorous (albeit slightly milder) typical Antarctic climate, with average monthly minimum temperatures in winter reaching – 8°C. Winds with velocities exceeding 150 km/h, predominantly from NE, hit the Base, dropping the thermal comfort to –50°C. Precipitation is in the form of snow in winter with some drizzles in summer time. Figure 11.43 presents a view of "Carlini" Base in winter; all photos shown in this section are courtesy of A. Benítez.

11.4.3.3 Buildings Construction and Exposure

The Base includes some 30 buildings, gradually built, containing residential, sanitary and laboratory facilities, plus meteorological, IT and communications stations.

The investigation focused on four buildings, known as Main House (MH), Cabildo Building (CB), Argentine Laboratory (AL) and Dallmann Laboratory

Figure 11.43 Aspect of Antarctic "Carlini" Base in summer.

(DL). The age at the time of the survey was approximately 30 years for the MH and CB, 20 years for DL and 5 years for AL.

The buildings are displayed along a predominant E-W axis, with the main façade facing N (sun). Most buildings are located at short distance (typically 5–10 m) from the seaside, with the prevailing winds blowing from the sea.

The CB building is located 10 m from the seashore, somewhat farther away than the other three buildings investigated which are located along the front row, very close to the sea (3–5 m).

Predominantly, constructions in the Base consist of reinforced concrete framed buildings, resting on isolated foundations, not deeply seated below the ground level. The superstructure is composed of timber or laminated steel structures with timber panels or timber-corrugated steel mixed panels, combined with thermal insulating materials, such as expanded polystyrene or polyurethane foam. The construction was carried out by military personnel devoted to the maintenance of the Base, without knowledge and experience in construction techniques. The concrete was prepared on site, with the limitations associated to the local prevailing conditions, namely, unqualified manpower, inadequate constituents and equipment for site concreting, lack of technical specifications or documentation and cold weather.

The aggregates, both coarse and fine, were quarried locally, although some evidence of the use of expanded clay lightweight aggregates was observed (CB). According to verbal declaration of the Base personnel, mixing water was taken from melting snow, sometimes helped by ice heating.

The situation of DL building is entirely different as the superstructure was built and assembled in Argentina's mainland, disassembled and transported by ship to the Base, where it was reassembled in 1994. So, it showed a better construction planning.

11.4.3.4 Scope of the Investigation

In 2013, the Argentine Antarctic Administration asked the National Institute of Industrial Technology (INTI) to conduct a comprehensive survey of the conditions of the buildings in "Carlini" Base, reporting the identified pathologies and recommending remedial measures.

The survey included the following aspects: structures' condition, electrical installations, drinking water supply, liquid and solid waste management and fire protection facilities. The survey was conducted *in situ* by a multidisciplinary team of specialized professional and technical personnel.

Here, just the results of the survey corresponding to the condition of the concrete structures are presented and discussed, with special emphasis on the site measurement of the coefficient of air-permeability *kT*.

11.4.3.5 Identified Pathologies

The main aggressive actions affecting the concrete structures were

- damage due to freezing and thawing
- steel corrosion induced by marine chlorides, acting in the form of spray and splash
- wetting and drying cycles, leading to leaching of hydration products and efflorescences

The poor concreting techniques applied were revealed by the following pathologies:

- exposed coarse aggregate particles consisting of smooth rounded gravel, with maximum size above 30 mm, incompatible with the geometry of the structural elements, with the distance between steel bars and with the cover thickness
- insufficient or null cover thickness over steel bars
- segregation, due to incorrect aggregate grading and/or mix design
- inadequate compaction
- insufficient or inexistent curing
- scaling and cracks due to freeze-thaw cycles, aggravated by wetting and drying cycles, due to the accumulation and melting of snow and ice, leading to high saturation degrees of not air-entrained concretes

11.4.3.6 On-Site Measurements of Air-Permeability *kT*

In order to check the effect of the poor concreting techniques applied and of the tough Antarctic environment on concrete quality, the coefficient of air-permeability *kT* was measured on site (*PermeaTORR* instrument, M-A-S Ltd.).

Between four and eight points were chosen for *kT* measurements on the exposed concrete surfaces of the four buildings investigated, avoiding areas with visible honeycombing or exposed rebars. The tests were performed under natural outdoors conditions (no heating); only when necessary the cell and instrument were shaded from direct sunlight.

Figure 11.44 shows the conditions under which the measurements were made on vertical surfaces. Some of the concrete surfaces were covered with a coating, asphaltic in the case of MH and DL buildings and latex in the case of CB.

The air temperature was measured and recorded, ranging between 2°C and 9°C, which was sometimes below the lower limit of 5°C for *kT* measurements, established by Swiss Standard (SIA 262/1, 2019).

The test results obtained on the four investigated buildings are summarized in Figure 11.45, with each point representing the geometric mean of

Figure 11.44 Testing one of "Carlini" Base buildings.

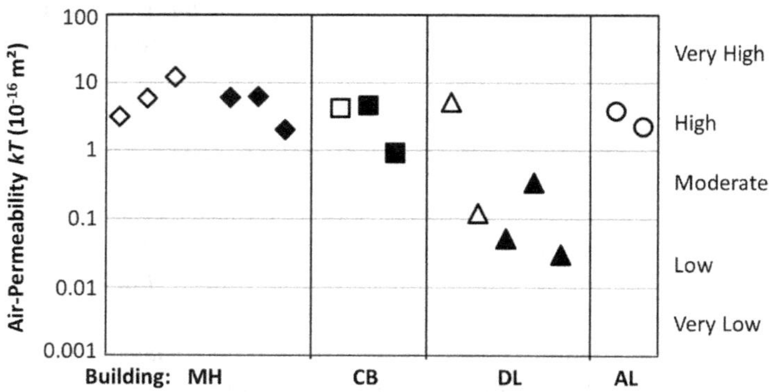

Figure 11.45 Results of *kT* obtained on four buildings of "Carlini" Base.

the data obtained on each element. The black symbols refer to tests made on coated surfaces, whilst the white ones on uncoated surfaces. The qualitative scale, shown at the right-hand side of the chart, indicates the classical *kT* Permeability Classes.

The first conclusion that can be drawn from the results in Figure 11.45 is that the coatings of the concrete surfaces seem to have little or no effect on the measured *kT* values. The other important conclusion is that buildings MH, CB and AL present air-permeabilities within the "High Permeability" range, whilst DL building results fall mainly within the "Moderate/Low Permeability" ranges (i.e. one to two orders of magnitude lower *kT* values).

The significantly better concrete quality displayed by the DL building, revealed by its lower permeability, can be attributed to better mix design and concrete practices, corresponding to a better planning of the building's construction, as described earlier. The *kT* results obtained showed a concrete of poor quality for buildings MH, CB and AL, which certainly contributed to the pathologies described in Section 11.4.3.5. On the contrary, DL building, resulting from a more careful planning (and possibly construction as well), showed lower *kT* permeability results. It can be estimated that, other conditions constant, the expected service life of this building is about 4–5 times longer than for the rest.

The following recommendations can be formulated for future durable constructions in the Antarctic:

- define an expected service life, to be used in the design
- structural design, specifications and construction should involve well-trained personnel
- concrete mixes should comply with the requirements and include air-entrainment
- equipment should be adequate for cold weather concreting conditions.
- fixing of steel must leave sufficient room between bars and ensure a generous cover thickness
- the feasibility of using precast construction elements, manufactured in mainland and shipped and assembled on site, should be seriously considered

11.4.4 Permeability of a Concrete Structure in the Chilean Atacama Desert

The Chilean National copper mining company CODELCO runs several mining operations along the central and northern parts of the Chilean territory. The "El Salvador" operation is located in the southern part of the Atacama Desert. The nearest human settlements are the city of "El Salvador" and "ghost town" Potrerillos, at an altitude of ≈2,900 m AMSL, where the copper foundry is located. The remote interior of the Atacama Desert is regarded as the driest desert in the world. The climatic conditions are characterized by warm temperatures, intense solar radiation, almost total absence of rain and low relative humidity.

The mining operations include the exploitation of open-pit and underground mines, with nearly 1,500 people working at the operations that produced over 50,000 tons of fine copper in 2019.

The real case described corresponds to a condition survey of the concrete foundations of a steel tank containing sulphuric acid in "Potrerillos" foundry (CTK, 2019). The foundation's external diameter is 11 m and the wall thickness 0.75 m. Figure 11.46 shows a view of the tank, in which the concrete foundation is clearly visible.

Figure 11.46 View of the investigated tank foundation.

As part of the survey, six measurements of the air-permeability *kT* (*PermeaTORR*, M-A-S Ltd.) were performed in each of three defined lots described below, where the air temperature/relative humidity and CO_2 concentration, as well as the mean surface moisture of the concrete *m*, are indicated:

- *Lot 1*: external side of the foundation (19.7°C/25%; 327 ppm); $m = 3.3\%$
- *Lot 2*: along the thickness of the external wall (19.7°C/25%; 327 ppm); $m = 2.3\%$
- *Lot 3*: internal side of the foundation (25°C/10%; 406 ppm); $m = 1.9\%$

The tested areas contained some minor incrustations that were removed prior to the measurements. Due to the wetness of the surface, it was not possible to make measurements on the acid-attacked parts of the foundation. The age of the concrete at the moment of test was 21 years.

The recorded values of *kT* were extremely high, typically between 70 and 700×10^{-16} m^2, which would indicate concretes of "very high" to "ultra-high" permeability, according to the classification in Table 5.2. However, due to the climate conditions of the place, the surface moisture *m* of the tested concrete was extremely low (between 1.8% and 3.6%). Therefore, in order to get a representative picture of the true quality of the concrete,

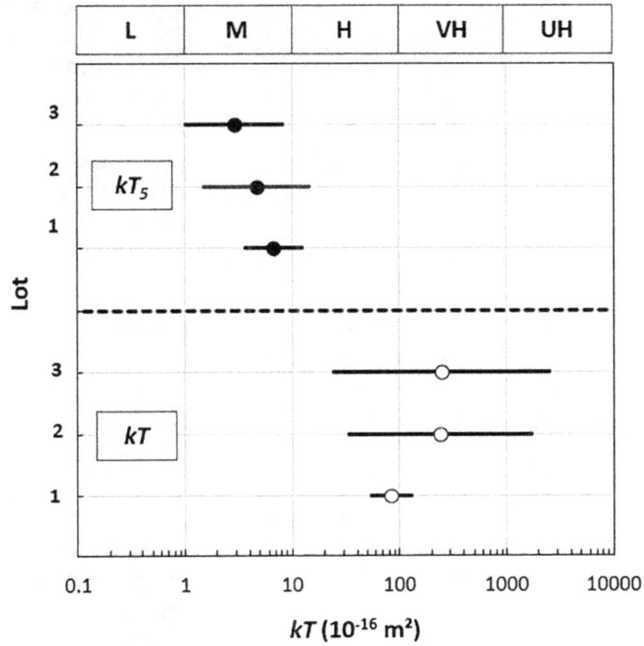

Figure 11.47 Original *kT* measured on the tank and moisture-compensated kT_5 values.

the moisture compensation, described in Section 5.7.2.2, was applied to the original kT results so as to get the equivalent values kT_5 at the reference surface moisture $m = 5.0\%$, using Eqs. (5.41) and (5.42).

Figure 11.47 presents the results obtained on the three lots, on the bottom part (○) the original measured kT values are plotted, whilst on the upper part (●) the values compensated for surface moisture kT_5 are plotted. The impact of the compensation is dramatic, reducing the air-permeability one order of magnitude for Lot 1 where the moisture was not so low, and two orders of magnitude for the other two lots where the surface moisture was much lower.

This real case shows the importance of the moisture compensation when dealing with very dry concretes, so as to get a representative quality of the tested concrete.

11.5 UNCONVENTIONAL APPLICATIONS

11.5.1 Concrete Wine Vessels

Ovoid vessels for wine conservation and transport have been used for many centuries. Indeed, the "amphora", a ceramic two-handled jar with a narrow neck was used by the ancient Greeks and Romans to carry wine or oil.

Its point-ended base allowed it to be driven into sand or to stow them in boats. The handles facilitated the use of ropes and hooks for its handling (Río, 2016; Twede, 2002). The amphoras apparently originated in today's Lebanon and Syria around XV century b.c. and spread around the ancient world, especially into Greece and Rome.

Later, the ceramic vessels were replaced by wooden barrels, as referred by historians Homer (900 b.c.) and Herodotus (~450 b.c.). The use of wooden barrels to store wine became consolidated around the XV century a.d. (Muñoz, 2006). The wooden barrel became a synonym of good wine, becoming limited to wine maturation, whilst glass bottles are used for transportation of the finished product. Wood from chestnut, cherry, acacia, elm, etc. was tried to build the barrels, the preferred choice being finally oak, due to its convenient physical-mechanical properties (Río, 2016).

In the second half of the XIX century, the use of cementitious vessels was introduced in France, with a patent to build them assigned to Joseph Mornier (Girini, 2014).

The use of ovoid vessels, without internal coatings, is a relatively new technology that is gaining acceptance among wine producers of France, Argentina, Chile, etc., which has not been fully investigated. It is claimed that the intrinsic permeability of concrete to air favours the micro-oxygenation of the stored most and wine, playing an important positive role in the fermentation and maturation of the wine (Blouin & Peynaud, 1977; Sánchez-Iglesias, 2007). It has been proved that atmospheric oxygen penetrates a wood barrel full of wine at rates measured by different researchers (Del Alamo-Sanza & Nevares, 2018) and a similar phenomenon should happen also for a porous material like concrete.

The focus of Río's dissertation work (Río, 2016) was to investigate the permeability to air *kT* of concrete wine vessels produced in France and in Argentina, of about the same size. Figure 11.48a shows the aspect of an Argentine ovoid vessel; notice that it is made in two pieces, glued together with epoxy resin, whilst the French vessels are made in one piece.

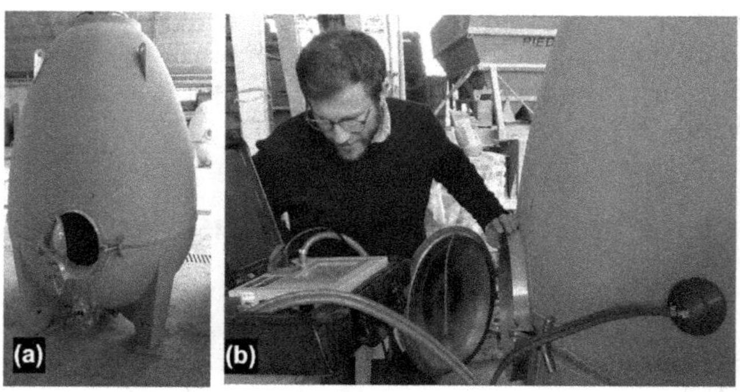

Figure 11.48 (a) Argentine concrete ovoid wine vessel and (b) P. Río testing extrados' *kT*.

Table 11.19 Description of the elements tested

Code	*Element*	*Origin*	*Condition*	*Number*
FU	Concrete vessel	France	Used for ~3 years	2
AN		Argentina	New	2
AU			Used for ~2 years	4
OD	Oak dowels	France	Used	11

With that aim, supported by R. Torrent, P. Río conducted several tests on wine vessels of both origins, including new and used vessels, measuring the permeability, moisture and surface hardness of their intrados and extrados surfaces (see extrados' test in Figure 11.48b).

The investigated elements are described in Table 11.19. The French concrete vessels were manufactured by Nomblot (France) and were tested at El Buho winery, whilst the Argentine vessels were produced by Obras Premoldeados S.A. The Argentine new vessels were tested at the factory whilst the used vessels at El Zorzal winery. All testing sites were located in Mendoza Prov., Argentina. Few data, obtained on oak barrels at Bodega Cerrón, Albacete, Spain and on a piece of oak wood sampled in Switzerland, not included in the dissertation (Rio, 2016), are also presented.

The surface moisture content *m* (*CMEXpert II*, Tramex) and the coefficient of air-permeability *kT* (*PermeaTORR*, M-A-S Ltd.) were measured according to version 2003 of Swiss Standard (SIA 262/1, 2019). In addition, surface hardness was measured by the rebound number *R* (*Original Schmidt N/NR*, Proceq) according to Argentine Standard (IRAM 1694, 1984).

The *kT* test results are summarized in Figure 11.49 where the circles and triangles represent the geometric mean kT_{gm} of the values obtained in the intrados and extrados of the vessels, respectively, and the segments' length represents the scatter s_{LOG} (in log scale).

A statistical analysis of the data was performed (Río, 2016), which yielded the following conclusions (5% significance level):

- the air-permeability *kT* of the extrados is significantly higher than that of the intrados for all elements, except FU2 and AU2. Two possible reasons for this difference were:
  a. the extrados is drier than the intrados;
  b. the pores of the intrados have been blocked by small particles in suspension in the wine.

  Hypothesis (a) has to be rejected, because there is not statistically significant difference between the surface moisture *m* of both faces of the concrete vessels.

  Hypothesis (b) has to be rejected too, because the difference in permeability was observed also on vessels AN1 and AN2 that were never in contact with wine.

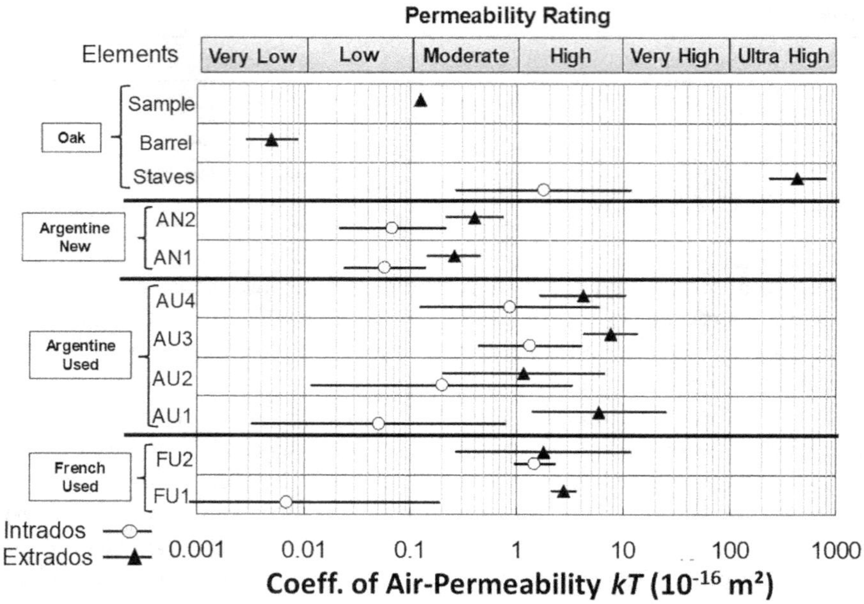

Figure 11.49 $kT_{gm} \pm s_{LOG}$ plot of test results obtained on the concrete vessels and oak samples.

It is worth mentioning that the surface hardness R also showed a significant difference between the intrados (lower) and the extrados (higher), except for (again) elements FU2 and AU2 plus FU1. This is somewhat paradoxical, because the surfaces that were harder were, in most cases, more permeable. The reasons for the difference, at least for the Argentine vessels, may lie on the construction technique that revealed some surface defects (bug-holes) on the extrados face (Figure 11.50a). These surface defects were not visible in the final vessels due to a cosmetic treatment (Figure 11.50b).

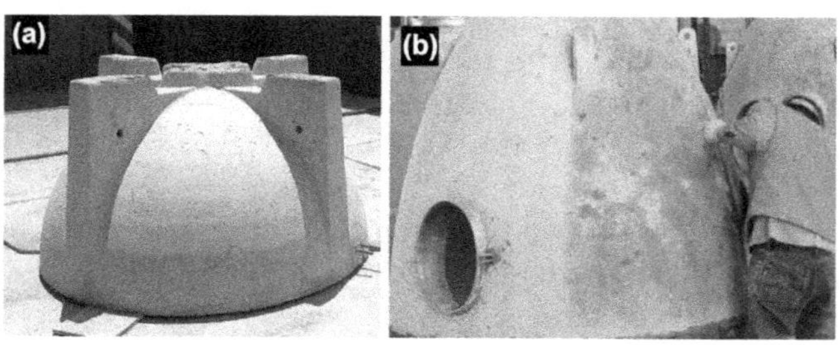

Figure 11.50 Original aspect (a) and surface treatment (b) of Argentine vessels.

- the air-permeability *kT* of the Argentine used vessels was significantly higher than that of the new vessels, both for the intrados and the extrados surfaces
- the air-permeability *kT* of the intrados surface of the Argentine used vessels was significantly higher than the same surface of the used French vessels, whilst it was not significantly different regarding the extrados surfaces

Regarding the data obtained on oak (barrels, staves and samples), the results are somewhat erratic, ranging between very low and ultra high permeability (see scale at the top of Figure 11.49). As discussed in Section 11.5.3 the permeability of wood is extremely variable between and also within species and, even for the same wood trunk, being such an anisotropic material, most properties depend strongly on the direction of flow. Typically, the flow in the longitudinal direction is 4–8 orders of magnitude higher than in the other two directions, seemingly the radial flow being the lowest of the three directions (Illston et al., 1979; Del Alamo-Sanza & Nevares, 2018). In addition, in the case of the barrel (results shown in Figure 11.49), the oak was treated with a sealer and varnish. Definitely, a more focused investigation is required to clarify this issue.

The measurement of the air-permeability *kT* of the concrete vessels allows quantifying the coefficient of O_2 diffusion of the walls, through the relation described in Section 8.3.2. The O_2 diffusivity is a key parameter for modelling the already mentioned micro-oxygenation effect on the quality of the stored wine.

11.5.2 Rocks and Stones

Two major functions of rocks and stones concerning mankind are, besides mineral extraction, their application as building material and their role in the exploitation of natural oil and gas reservoirs. Both construction and oil and gas industries have established methods to characterize the permeability of stone and rocks.

11.5.2.1 Permeability of Stones as Building Material

In construction, stone has been used for centuries for masonry, cladding, flooring and ornamental purposes. Its presence is quite common in most nations' heritage. In Belém's area of Lisbon, Portugal, several examples can be found (Figure 11.51). In general, construction building stones are subject to weathering that causes their decay, deterioration that often can be observed in the form of dark stains (Figure 11.51).

The stone behaviour, when subjected to water-induced weathering processes, can be indirectly foreseen through the permeability of the stone,

Figure 11.51 Jeronimos' Monastery, Lisbon: UNESCO's World Heritage, with masonry and ornamental stone.

same as the efficiency of stone consolidation can be assessed by means of permeability tests (Mertz et al., 2016).

A research on the suitability of air-permeability *kT* test method for testing building stone has been carried out using a *PermeaTORR* instrument (M-A-S Ltd.). The test method was applied on 15 varieties of stones (Sena da Fonseca et al., 2015). Their investigation comprised the repeatability of the test, the between-specimen variability, the test sensitivity to surface finishing and the relationship with other established methods for transport properties of dimension stones.

The *kT* test method exhibited an excellent repeatability, as the mean relative difference between consecutive measurements on the same spot was 5.6%. The mean relative between-specimen variation was 21.6%. This value must be addressed not only from the perspective of the method itself, but also from the macro-scale heterogeneity of the tested stone variety. The relative between-specimen variation ranged from 5.6% to 68.6%. The smallest difference was found for "Moleanos" stone (limestone rock), whereas the largest difference was observed for samples of "Gascogne blue" stone (also limestone rock).

An interesting result was found when investigating the test sensitivity to surface finishing. The test was applied on a polished surface and on a honed surface. The results were quite comparable, providing strength to the usefulness of the outer chamber, absorbing the spurious air incoming through

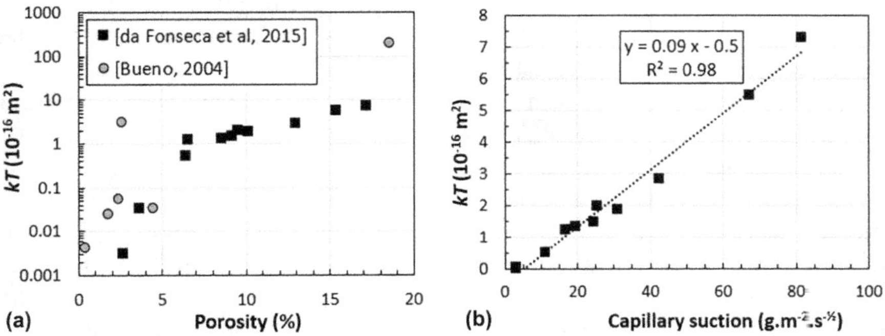

Figure 11.52 *kT* vs. porosity (a) and vs capillary suction (b); data from Sena da Fonseca et al. (2015), Bueno (2004).

surface irregularities. The air-permeability *kT* results were compared with those from porosity tests, performed according to EN 1936 (2001), water absorption at atmospheric pressure tests, carried out following (EN 13755, 2005), and capillary water-absorption test, performed according to EN 15801 (2009).

The correlation of porosity with the logarithm of *kT* can be seen (black squares) in Figure 11.52a, where values obtained by Bueno (2004) by a different technique, discussed in next Section, are also included (grey circles). A correlation coefficient $R = 0.84$ was obtained. However, the correlation improves if only values of $kT > 0.1 \times 10^{-16} m^2$ are considered. It shall be noted that for porosities ≤2.2% (and water absorptions ≤0.83%), the *kT* values were below the sensitivity limit of the instrument and, therefore, were not considered in the analyses.

An important mechanism of building stone degradation is the penetration of (acid) rain through capillary suction. In this respect, even more relevant than the relation of *kT* with porosity or absorption is its relation with the rate of capillary suction, which is excellent as shown by the linear correlation ($R^2 = 0.98$) presented in Figure 11.52b. Following the experimental observations, the air-permeability *kT* test method is only capable to estimate capillary suctions higher than 5.5 $g/(m^2.\sqrt{s})$ in dimension stones.

11.5.2.2 Permeability of Rocks for Oil and Gas Exploitation

Within the frame of oil and gas industry, the permeability of the reservoir rocks is a fundamental property for well's productivity (Fanchi & Christiansen, 2017; Yancheng et al., 2017). Simplifying the matter, the porosity of the rock determines the storage capacity of a reservoir, whilst its permeability determines the easiness with which oil and gas can be extracted. For instance, the increasing scarcity of conventional wells has led to the exploitation of those involving very impermeable rock strata, requiring the so called "fracking" technology, by which the "impermeable" rock is fractured by liquids under pressure, so as to open ways for the extraction of oil and gas. A test method for measuring the air-permeability of small samples of rock, similar in conception to the Cembureau method for concrete, is standardized (ASTM D4525, 2013).

Recognizing the relevance of rocks permeability for the oil and gas industry, Bueno (2004) investigated the suitability of air-permeability *kT* method to assess the permeability of rocks. Possibly the first application of the *kT* test on rocks was made within this context (Bueno, 2004), however, on rocks not necessarily related to the oil and gas industry. Her report includes a survey of test methods used by the industry to measure porosity and permeability of rocks and preliminary *kT* test results (*TPT*, Proceq) on rock samples supplied by EMPA and SUPSI (Switzerland). As cultural remark, Figure 11.53b shows the façade of San Lorenzo Cathedral, Lugano, Switzerland, which was built in 1517 using the same Saltrio Stone tested by Dr. Bueno (Figure 11.53a).

Afterwards, from the same locations where *kT* was measured, cores were drilled from six stones. Later, "plug" samples≈Ø25 × 65 mm were obtained from those cores at Schlumberger Laboratory in Maracaibo, Venezuela. These small "plugs" were tested for porosity with Boyle's porosimeter using Helium (PHI-220 instrument) and for air-permeability using two instruments: Frank Jones and KA-210. It shall be noted that these permeameters are designed for testing rocks in conventional natural reservoirs, having permeability typically within the range $10–10{,}000 \times 10^{-16}\,m^2$ (Fanchi & Christiansen, 2017), i.e. much higher than the permeability of the tested rocks.

The test results obtained are reported in Table 11.20 (Bueno, 2004); in five out of six samples, the KA-210 instrument failed to display a result (samples too "impermeable" for the sensitivity of the apparatus), reason why its data are not included in the table.

It should be mentioned that the *TPT* instrument was unable to display the permeability of the cracked Tertär sandstone and of the Fribourg Sandstone

Figure 11.53 (a) Test of Saltrio Stone sample rock used in (b) San Lorenzo Cathedral, Lugano, Switzerland.

Table 11.20 Permeability and porosity of the tested rock samples

Instrument	*TPT*	*Frank Jones*	*PHI-220*
Rock sample	*kT* (10^{-16} m²)	K_{FJ} (10^{-16} m²)	*He porosity (%)*
Calcarenite (Saltrio Stone)	0.024	0.039	1.80
Onsernone Gneiss	0.004	0.019	0.45
Fribourg Sandstone	188	-	18.54
Jurassic Limestone (sound)	0.031	0.041	4.47
From Nüremberg (hairline crack)	0.035	-	-
Germany (porous)	3.08	0.021	2.63
Tertär Sandstone (sound)	0.052	0.021	2.42
From Zürich (cracked)	333	-	-

(too permeable, the latter not measurable by Frank Jones instrument either); so, the *kT* values had to be manually computed from the recorded pressure rise, applying Eq. (5.13).

It is interesting to see from Table 11.20 that the range of *kT* values obtained on the different rocks spans more or less the same range than for concretes of "very low" to "ultra high" permeability (see Table 5.2). The *kT* values match very well those recorded by the Frank Jones instrument, except for the porous Jurassic Limestone. The difference may be attributed to the fact that the small sample used by that instrument (actually, it showed lower He porosity than the sound sample) was not truly representative of the porous zone.

The relation found between *kT* and He porosity, shown (grey circles) in Figure 11.52a, is quite consistent with that obtained by Sena da Fonseca et al. (2015), who measured porosity by the classical gravimetric method.

The results of the reviewed researches show that *kT* air-permeability method is suitable to assess the permeability of rocks.

11.5.2.3 Permeability of Rocks for Nuclear Waste Disposal

The permeability of the surrounding rocks is very important when planning nuclear waste disposal sites, as it determines, in general, the transport of radionuclides. A very elaborate research was carried out to investigate the permeability characteristics of rock outcrop in the vicinity of Rokko fault in Japan (Manaka et al., 2020). The investigation involved field testing of air-permeability by Figg method (see Section 4.3.2.1) using the *Porosiscope* instrument and also by measuring *kT* using a *PermeaTORR* instrument, complemented by laboratory tests on rock samples from the same sites. The *kT* values obtained in the field (in some cases an adaptor had to be interposed to cope with rock surface irregularities) were rated, according to the concrete quality scale of Table 5.2, as of 'Very High' and 'Ultra High' permeability. The results obtained were related to the distance from the fault, as well as to the weathering grade of the tested rock. Regarding the suitability of the *kT* test, Manaka et al. (2020) concluded that "Based on comparisons of intrinsic permeabilities at the same site, the on-site surface suction test was found to be effective for measuring a wide range of intrinsic permeabilities".

11.5.3 Timber

A tree trunk has three functions: support the crown, transport mineral substances and store food. The entire trunk fulfils the support function, whereas transport and storage are restricted to the outer zone, known as *sapwood*. The zone no longer fulfilling these two functions is known as *heartwood* (Illston et al., 1979). Figure 11.54 shows the main structural features of a hardwood tree, with its strong anisotropy, since 90%–95% of the cells are elongated and vertically oriented, with 5% of cells arranged in radial directions and none in tangential direction (DoITPoMS, 2008).

The term "flow" in timber is associated mainly with impregnation and durability and is synonymous with passage of liquids, gases and thermal energy.

The longitudinal cell arrangement, designed to allow the transport of nutrients from the roots to the crown are preferential paths for the flow of matter and the permeability of timber along the longitudinal direction is expected to be the highest. Both longitudinal and tangential flow paths are predominantly by way of the bordered pits while the horizontally aligned ray cells constitute the principal pathway for radial flow, though it has been suggested that very fine capillaries within the cell wall may contribute slightly to radial flow (Illston et al., 1979).

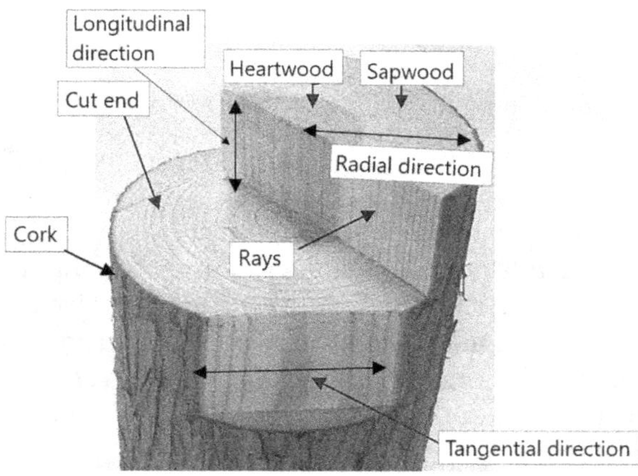

Figure 11.54 Diagrammatic illustration of the macrostructure of wood.

Regarding permeation, the flow theory of liquids, Darcy's law (see Section 3.5.1) will be applied as follows:

$$Q = \frac{K_p A \Delta P}{\eta L} \tag{11.6}$$

where Q is volume rate of flow, A is cross-sectional area of capillary and K_p is liquid viscous permeability coefficient; ΔP is pressure drop across capillary length L and η is the viscosity of the liquid.

Regarding permeation flow theory of gases (see Sections 3.5.1 and 3.6), two types of flow can occur: laminar flow and slip or Knudsen flow

Laminar flow occurs in capillaries where the rate of flow is relatively low. In laminar flow, Darcy's flow is slightly modified considering compressibility of gases along. The equation is

$$Q = \frac{\pi r^4 \Delta P}{8\eta L} \cdot \frac{P_{av}}{P_0} \tag{11.7}$$

where Q is volume rate of flow, r is capillary radius, ΔP is pressure drop across capillary length L, η is viscosity, P_{av} is mean gas pressure within capillary and P_0 is the gas pressure where Q was measured.

Slip flow along a capillary of circular section is given by

$$Q_s = \frac{2\pi r^3 \delta_1 V_{av}}{3L} \tag{11.8}$$

where δ_1 is a factor on fraction of molecules undergoing diffusion and V_{av} is molecular mean thermal velocity. The gas flow with slip in timber will be given by adding Eqs. (11.7) and (11.8):

$$Q \frac{P_0}{P_{av}} = \frac{\pi r^4 \Delta P}{8\eta l} + \frac{2\pi r^3 \delta_1 V_{av}}{3l} \tag{11.9}$$

Results of permeability of *sugi* (cryptomeria japonica) wood were reported by Tanaka et al. (2015), along the three directions, for sapwood and heartwood, respectively. Regarding difference of permeability with the flow direction, longitudinal permeability is highest and that of radial permeability is lowest, corresponding to the macrostructure of wood discussed above. The permeability of *heartwood* is about an order of magnitude less than the permeability of *sapwood* in the same direction due to deposition of encrusting materials over the torus and margo strands and also within the ray cells.

Permeability is not only directionally dependent, but it also varies for earlywood and latewood. Since earlywood cells possess larger and more frequent bordered pits, the flow through the earlywood is considerably greater than that through the latewood.

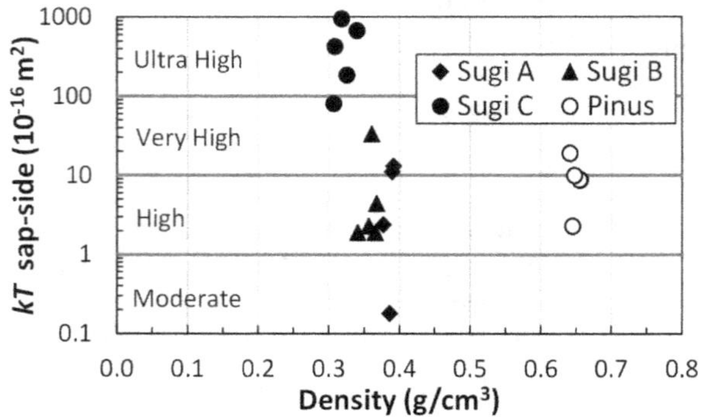

Figure 11.55 *kT* test results on different samples of *Sugi* and *Pinus* wood.

Figure 11.55 presents test results, obtained by K. Imamoto, of air-permeability *kT* (*PermeaTORR*, M-A-S Ltd.) along the radial direction, on *Sugi* wood samples of different densities and of a *Pinus* wood sample. Figure 11.55 indicates that permeability of *Sugi* increases drastically with a decrease in density. However, density by itself is not a good general indicator of the permeability of wood, as the *Pinus* sample (density ca. 0.65 g/cm^3) shows similar permeability to that of *Sugi* of density ca. 0.37 g/cm^3. It is interesting to note that, depending on the sample, *Sugi* wood spans four orders of magnitude of *kT*, corresponding to Permeability Classes (for concrete) from "Moderate" to "Ultra High".

Although more research is needed, it seems that *kT* test method has potential to measure the gas-permeability of timber.

11.5.4 Ceramics

Under the general designation of "ceramics", a large variety of materials can be included. According to Jastrzebski (1959), "Ceramics are inorganic, non-metallic materials that are processed and/or used at high temperatures", which can be grouped into three broad divisions: clay products, refractories and glasses.

Typical clay products used in the construction industry are bricks and tiles. Permeability plays a role in the performance of roof tiles. A water-permeability test is included in ASTM C1167 (2011) and an air-permeability test in ASTM C1570 (2003). Both are simulation tests, the former is intended to ensure that the tiles are watertight to fulfil their protection function and the latter is intended for the tiles to be sufficiently permeable to air, to avoid being airlifted by the depression caused by wind blowing over them. So, both requirements are a bit in contradiction. The National Technological University of Córdoba (Argentina) has performed routine air-permeability *kT* tests on roof tiles, both ceramic and concrete, without finding significant differences between them (Positieri et al., 2013).

In consulting projects, Materials Advanced Services Ltd. has conducted *kT* tests on samples of ceramic materials of widely different types, intended for different purposes, supplied by their manufacturers.

In one of them, *kT* was measured on Ø150 × 35 mm ceramic discs that were pre-conditioned by oven drying at 105°C (electrical-impedance surface moisture ranged between 0% and 1.6%). It is interesting to remark that two samples, with density ≈ 1,000 kg/m^3, showed *kT* values of 14 and $28 \times 10^{-16} m^2$, whilst the remaining six samples, with density ≈ 2,800 kg/m^3, showed *kT* values between 6.0 and $28 \times 10^{-16} m^2$.

In another project for another customer, high density (≈4,500 kg/m^3) ceramic specimens were tested. Two out of the three specimens tested yielded a null *kT* (no effective pressure rise detected during the test), whilst the third one yielded a value $kT = 0.0002 \times 10^{-16} m^2$, which is below the sensitivity limit of the *PermeaTORR AC* instrument used, requiring calculation of the value from the recorded pressure rise applying Eq. (5.13).

It is clear that the *kT* test method can be applied to ceramic materials which, incidentally, show a large variety of permeabilities.

REFERENCES

Accropode (2011). "Abstract of Accropode™ II technical specifications". CLI, Tours, France, 20 p.

ACI 318 (2011). "Building code requirements for structural concrete". ACI, 510 p.

AIJ (2003). "Japanese architectural standards specification (JASS 5: reinforced concrete work)". Architectural Inst. of Japan.

Akiyama, K., Hara, M., Yoshida, M. and Abe, T. (2014). "Soundness inspection of the 'Jyugou Bridge', Japan's first post-tensioning PC bridge". *Concr. J.*, v52, 1067–1074 (in Japanese).

Andrade, C., Rebolledo, N., Castillo, A., Tavares, F., Pérez, R. and Baz, M (2014). "Use of resistivity and chloride resistance measurements to assess concrete durability of new Panama Canal". *International Workshop on Performance-based Specification and Control of Concrete* Durability, Zagreb, Croatia, June 11–13, 411–418.

Andrade, C., Rebolledo, N., Tavares, F., Pérez, R. and Baz, M. (2016). "Concrete durability of the new Panama Canal; background and aspects of testing". *Marine Concrete Structures: Design, Durability and Performance*. Alexander, M.G. (Ed.), Woodhead Publishing, Duxford, UK, Ch. 15, 429–458.

Aracil, F., Dufour, G., Wauters, P., de Schutter, G. and Torrent, R. (2016). "Real case: Assessing sulfate resistance of coastal protection precast elements". *Key Eng. Mater.*, v711, 327–334.

ASTM C150 (2012). "Standard specification for portland cement".

ASTM C1167 (2011). "Standard specification for clay roof tiles"

ASTM C1202 (2012). "Standard test method for electrical indication of concrete's ability to resist chloride ion penetration".

ASTM C1570 (2003). "Standard test method for wind resistance of concrete and clay roof tiles (air permeability method)".

ASTM D4525 (2013). "Standard test method for permeability of rocks by flowing air".

Benítez, A. and Polzinetti, M. (2016). "Relevamiento de edificaciones de hormigón en la base Carlini, Antártida Argentina". *VI Congreso Intern. y XXI Reunión Técnica AATH*, Salta, Argentina, September 28–30, 8 p.

Benítez, A. and Polzinetti, M. (2019). "Permeability and condition of concrete in the Argentine Antarctic 'Carlini' base". *CONSEC 2019*, Porto Alegre, Brazil, June 5–7, 10 p.

Beushausen, H. and Fernández Luco, L. (Eds.) (2016). "Performance-based specifications and control of concrete durability". *RILEM State-of-the-Art Reports*, v18, Chapters 10 & 11, 301–360.

Bisschop, J., Schiegg, Y. and Hunkeler, F. (2016). "Modelling the corrosion initiation of reinforced concrete exposed to deicing salts". Bundesamt für Strassen, Report No. 676, February, 91 p.

Blouin, J. and Peynaud, E. (1977). "Enología práctica, conocimiento y elaboración del vino". Ediciones Paraninfo, Madrid, Spain, 360 p.

Brühwiler, E. (2007a). "Lifetime oriented composite concrete structures combining reinforced concrete with Ultra-High Performance Fibre Reinforced Concrete". 3rd *International Conference on Lifetime-Oriented Design Concepts*, November 12–14, Ruhr-Universität Bochum, Germany, 10 p.

Brühwiler, E. (2007b). "A structural engineer's view: More attractive value-added solutions for concrete structures". *Holcim Workshop*, Zürich, Decmber 12, 44 slides.

Brühwiler, E., Denarié, E., Wälchli, Th., Maître, M. and Conciatori, D. (2005). "Applicabilité de la mesure de perméabilité selon Torrent pour le contrôle de qualité du béton d'enrobage" (in French), Office Fédéral Suisse des Routes, Rapport n. 587, Avril, 120 p.

Bueno, V. (2004). "Estudio de factibilidad de un nuevo ensayo de permeabilidad en rocas". Univ. del Zulia, Macaraibo.

CTK (2019). "Permeabilidad al Aire del Hormigón – Informe de Medición in Situ – Potrerillos". Construtechnik Report, Santiago de Chile, November 28, 7 p.

Del Alamo-Sanza, M. and Nevares, I. (2018). "Oak wine barrel as an active vessel: A critical review of past and current knowledge". *Crit. Rev. Food Sci. Nutr.*, v58, n16, 2711–2726.

Denarié, E. (2005). "Full scale application of UHPFRC for the rehabilitation of bridges – from the lab to the field". SAMARIS Report AM_GE_DE22v03_01, Dec. 12, 77 p.

Denarié, E. and Brühwiler, E. (2006). "Structural rehabilitations with ultra-high performance fibre reinforced concretes (UHPFRC)". *Restoration of Buildings and Monuments, Bauinstandsetzen und Baudenkmalpflege*, v12, n5–6, 453–465.

Denarié, E., Brühwiler, E., Lestuzzi, P and Tilly, G. (2006). "Guidance for the use of UHPFRC for rehabilitation of concrete highway structures". SAMARIS Report SAM_GE_DE25v04_01, May 23, 64 p.

Di Pace, G. and Calo, D. (2008). "Assessment of concrete permeability in tunnels". *SACoMaTIS*, Varenna, Italy, September 1–2, v1, 327–336.

DoITPoMS (2008). "The structure and mechanical behaviour of wood". Dissemination of IT for the Promotion of Materials Science (DoITPoMS), Univ. of Cambridge. www.doitpoms.ac.uk/tlplib/wood/index.php.

Donadio, M., Bakalli, M. and Dan, Z. (2019). "Reinforced concrete corrosion prevention/reduction by hydrophobic impregnation". MATEC Web of Conferences, v289, 05002, 7 p.

DuraCrete (2000). "Probabilistic performance based durability design of concrete structures". The European Union–Brite EuRam III, BE95–1347/R17 CUR, Gouda, The Netherlands.

EN 13755 (2005). "Natural stone test methods: Determination of water absorption at atmospheric pressure". European Standard.

EN 15801 (2009). "Conservation of cultural property". Test methods. Determination of water absorption by capillarity. European Standard

EN 1936 (2001). "Natural stone test methods: determination of real density and apparent density, and of total and open porosity". European Standard.

EN 206-1 (2004). "Concrete – Part 1: Specification, performance, production and conformity". European Standard.

Fanchi, J.R. and Christiansen, R.L. (2017). *Introduction to Petroleum Engineering.* John Wiley & Sons, Inc., Hoboken, New Jersey, USA, 352 p.

Fernández Luco, L. and Pombo, R. (2005). "Innovative non-destructive assessment of adherence failure of the top layer in an industrial floor". *ICCRRR*, Cape Town, South Africa, November 21–23, 5 p.

Fernández Luco, L., Pombo, R. and Torrent, R. (2003). "Shrinkage-compensating concrete in Argentina". *ACI Concr. Intern.*, May, 49–53.

Ferreira, L. (2014). "Avances del programa de ampliación del Canal de Panamá". *XXII CONEIC & X COLEIC*, Arequipa, Perú, August 4–8, 177 slides.

Frenzer, G., Jacobs, F. and Torrent, R. (2020). "*In situ* durability characteristics of new and old concrete structures". submitted for publication in *JACT*, Japan Concr. Institute.

Girini, L. (2014). "La Revolución Vitivinícola en Mendoza. 1885–1910. Las transformaciones en el territorio, el paisaje y la arquitectura". Editorial Idearium, 70–80.

Gjørv, O.E. (2014). *Durability Design of Concrete Structures in Severe Environments.* 2nd Ed., Taylor & Francis, UK, 254 p.

Halabe, U.B., Sotoodehnia, A., Maser, K.R. and Kausel, E.A. (1993). "Modeling the electromagnetic properties of concrete". *ACI Mater. J.*, v90, n6, 552–563.

Hunkeler, F., Ungricht, H. and Merz, C. (2002). "Vergleichende Untersuchungen zum Chloridwiderstand von Betonen" (in German). Office Fédéral Suisse des Routes, Rapport n. 568, November, 203 p.

Iidoi, T., Maeshima, T., Koda, Y., Miyamura, M., Ueda, H., Ishida, T. and Iwaki, I. (2021). "A proposal on maintenance methodology of existing road bridge PC superstructure focusing on water behaviors". *Proc. Struct. Engng.*, v67A, 659–672 (in Japanese).

Illston, J.M., Dinwoodie, J.M. and Smith, A.A. (1979). "Concrete, timber and metals: The nature and behaviour of structural materials". Van Nostrand Reinhold, New York, 663 p.

Imamoto, K. (2012). Presentation of Paper 180 Slides, *Microdurability* 2012, Amsterdam, April 11–13.

Imamoto, K., Tanaka, A. and Kanematsu, M. (2012). "Non-destructive assessment of concrete durability of the National Museum of Western Art in Japan". Paper 180, *Microdurability* 2012, Amsterdam, April 11–13.

Imamoto, K., Tanaka, A. and Kanematsu, M. (2013). "Non-destructive assessment of concrete durability of the National Museum of Western Art in Japan". *CONSEC'13*, Nanjing, China, September 23–25, 1335–1344.

IRAM 1554 (1983). "Hormigón de cemento pórtland. Método de determinación de la penetración de agua a presión en el hormigón endurecido".

IRAM 1694 (1984). "Hormigón de cemento pórtland: Método de ensayo de la dureza superficial del hormigón endurecido mediante la determinación del número de rebote empleando el esclerómetro de resorte".

Jacobs, F. (2006). "Luftpermeabilität als Kenngrösse für die Qualität des Überdeckungsbetons von Betonbauwerken" (in German), Office Fédéral Suisse des Routes, Rapport n. 604, September, 100 p.

Jacobs, F., Denarié, E., Leemann, A. and Teruzzi, T. (2009). "Empfehlungen zur Qualitätskontrolle von Beton mit Luftpermeabilitätsmessungen". Office Fédéral des Routes, VSS Report 641, December, Bern, Switzerland, 53 p.

Jastrzebski, D. (1959). *Nature and Properties of Engineering Materials.* Wiley & Sons, Japan, 571 p.

Kishi, T., Akiyama, H., Inoue, S. and Yoshida, R. (2008). "Quantitative comparison on surface quality of historic concrete structures by field investigation - quality evaluation on cover concrete of existing structures by air permeability test". *Seisan Kenkyu*, v60, n5, 118–121 (in Japanese)

Li, Q., Li, K.F., Zhou, X., Zhang, Q. and Fan, Z. (2015). "Model-based durability design of concrete structures in Hong Kong–Zhuhai–Macau sea link project". *Struct. Safety*, v53, March, 1–12.

Li, K. and Torrent, R. (2016). "Analytical and experimental service life assessment of Hong Kong-Zhuhai-Macau Link". *IABMAS 2016*, Paper 427, Foz do Iguaçú, Brazil, June 26–30.

Li, K., Li, Q. and Fan, Z. (2016). "Hong Kong-Zhuhai-Macau sea link project, China". *Marine Concrete Structures – Design, Durability and Performance*, Alexander, M.G. (Ed.), Woodhead Publishing, Duxford, Ch. 13, 339–370.

Li, K., Zhang, D., Li, Q. and Fan, Z. (2019). "Durability for concrete structures in marine environments of HZM project: Design, assessment and beyond". *Cem. & Concr. Res.*, v115, 545–558.

Life 365 (2008). "Life-365™ service life prediction Model™". Version 2.2.3, September.

Maître, M. (2002). "Tunnel de Naxberg - Perméabilité à l'air du béton d'enrobage (méthode Torrent)". EPFL, Rapport d'essais n° MCS 02.09-01, Lausanne, November, 9 p.

Manaka, M., Shimizu, T. and Takeda, M. (2020). "Comparison of rock outcrop permeability with increasing distance from the Rokko fault zone". *Engng. Geology*, v271, 105591.

Mehta, P.K. and Monteiro, P. (2001). *Concrete: Microstructure, Properties and Materials.* 3rd Ed., McGraw-Hill, New York, Chicago, San Francisco, Athens, London, Madrid, Mexico City, Milan, New Delhi, Singapore, Sydney, Toronto, 660 p.

Mertz, J.-D., Colas, E., Ben Yahmed, A. and Lenormand, R. (2016). "Assessment of a non-destructive and portable mini permeameter based on a pulse decay flow applied to historical surfaces of porous materials". *13th International Congress on the Deterioration and Conservation of Stone*, Glasgow, Scotland, September 6–10, 415–422.

Muñoz, P. (2006). “Alternativas a la crianza en barrica”. *Revista de Enología*, n72, ACE (Associació Catalana d’Enolegs), Cataluña.

Neves, R., Branco, F. and de Brito, J. (2012). “About the statistical interpretation of air permeability assessment results”. *Mater. & Struct.*, v45, April, 529–539.

Neves, R. and Santos, J. (2008). “Air permeability assessment in a reinforced concrete viaduct”. SACoMaTIS, Varenna, Italy, September 1–2, v1, 299–308.

Neves, R., Torrent, R. and Imamoto, K. (2018). “Residual service life of carbonated structures based on site non-destructive tests”. *Cem. & Concr. Res* , v109, 10–18.

NMWA (2016). “The National Museum of Western Art 1959”. Tokyo, March 31, 37 p.

NT Build 492 (1999). “Concrete, Mortar and cement-based repair materials: Chloride migration coefficient from non-steady migration experiments”. Nordtest Method, Nordtest, Espoo, Finland.

Oesterlee, C., Denarié, E. and Brühwiler, E. (2009). “Strength and deformability distribution in UHPFRC panels”. *ConMat'09*, Nagoya, Japan, August 24–26, 390–397.

Parrott, L. (1994). “Design for avoiding damage due to carbonation-induced corrosion”. *ACI SP-145*, 283–298.

PCE Specs (2008a). “Specifications for lock structures”. Section 01 81 16, Locks Performance and Design Criteria, November.

PCE Specs (2008b). “Specifications for lock structures”. Section 03 30 00, Locks Performance and Design Criteria, October.

Positieri, M., Gaggino, R., Oshiro, A., Baronetto, C. and Irico, P. (2013). “Una opción para las tejas: diseño clásico y material sustentable”. *1° Congreso ALCONPAT*, Mendoza, Argentina, May 9–11, 14 p.

Ribeiro, A. and Gonçalves, A. (2009). “Use of durability related tests for quality control: Variability obtained in a real case”. *Concrete in Aggressive Aqueous Environments – Performance, Testing, and Modeling*, Toulouse, France, June 3, 515–522.

Richardson, M.G. (2002). *Fundamentals of Durable Reinforced Concrete*. Spon Press, London, UK and New York, USA 260 p.

Río, P. (2016). “Estudio de la permeabilidad al aire de las paredes de vasijas ovoides de hormigón”. Graduate Dissertation, Facultad de Agronomía y Cs Agroalimentarias, Universidad de Morón, Argentina, 82 p.

Romer, M. (2005). “Comparative test – Part I – Comparative test of penetrability methods”. *Mater. & Struct.*, v38, December, 895–906.

Sajna, A., Denarié, E. and Bras, V. (2012). “Assessment of a UHPFRC bridge rehabilitation in Slovenia, two years after application”. *3rd International Symposium on Ultra-High Performance Concrete – HiPerMat*, Kassel, Germany, March 7–12, 937–944.

Sánchez-Iglesias, M. (2007). “Incidencia del tratamiento de microoxigenación sobre la composición fenólica y el color en vinos tintos jóvenes y de crianza de Castilla y León”. Tesis doctoral. Universidad de Burgos, España.

Sato, K. (2017). “Practical research on ensuring the quality and durability of concrete structures in reconstruction road projects based on the deterioration of the Tohoku region”. Dissertation, Univ. of Tokyo, January, 194 p. (in Japanese).

Schiegg, Y., Bisschop, J. and Von Greve-Dienfeld, S. (2017a). "Monitoring rebar corrosion propagation in concrete – Results of the Naxberg field experiment after 12 years". *EUROCORR* 2017.

Schiegg, Y., Hunkeler, F. and Keller, D. (2017b). "Massnahmen zur Erhöhung der Dauerhaftigkeit - Fortsetzung des Feldversuchs Naxbergtunnel". ASTRA Bericht Nr. 683, Bern, Schweiz, 159 p.

Schmidt, M., Brühwiler, E., Fehling, E., Denarié, E., Leutbecher, T. and Teichmann, T. (2007). "Mix design and properties of Ultra-High Performance Fibre Reinforced Concrete for the construction of a composite UHPFRC - concrete bridge". *Ultra High Performance Concrete (UHPC)*, Univ. Kassel, Heft 7, 44–54.

Sena da Fonseca, B., Castela, A.S., Duarte, R.G., Neves, R. and Montemor, M.F. (2015). "Non-destructive and on site method to assess the air-permeability in dimension stones and its relationship with other transport-related properties". *Mater. & Struct.*, v48, 3795–3809.

SIA 262/1 (2019). Swiss Standard SIA 262/1:2019, "Construction en béton. Spécifications complémentaires". Annex E: 'Perméabilité à l'air dans les structures. Norme Suisse, March 1, 60 p. (in French and German).

SN EN 14629 (2007). "Produkte und Systeme für den Schutz und die Instandsetzung von Betontragwerken - Prüfverfahren - Bestimmung des Chloridgehaltes von Festbeton". Swiss Standard, 16 p.

Tanaka, T., Kawai, Y., Sadanari, M., Shadi, S. and Tsuchimoto, T. (2015). "Air permeability of sugi (cryptomeria japonica) wood in the three directions". *Maderas: Ciencia y Tecnología*, v17, n1, Concepción, Chile, 17–28.

Torrent, R. (2008). "Permeability tests on sound and damaged areas of a shrinkage-compensating concrete industrial floor". M-A-S Report No. 09-002, Materials Advanced Services Ltd., Buenos Aires, November.

Torrent, R. (2011). "Visit of R. Torrent to Panama Canal Project, 1–2 November 2011". M-A-S Report 11-001, November 5, Buenos Aires, Argentina, 11 p.

Torrent, R. (2015). Report MAS 15/01. Materials Advanced Services, Buenos Aires, Argentina, 24 p.

Torrent, R., Armaghani, J. and Taibi, Y. (2013). "Evaluation of Port of Miami Tunnel Segments: Carbonation and service life assessment made using on-site air permeability tests". *ACI Concr. Intern.*, May, 39–46.

Torrent, R., di Prisco, M., Bueno, V. and Sibaud, F. (2018). "Site air-permeability of HPSFR and conventional concretes". *ACI SP-326*, Paper 84.

Torrent, R. and Ebensperger, L. (1993). "Methoden zur Messung und Beurteilung der Kennwerte des Ueberdeckungsbetons auf der Baustelle -Teil I". Office Fédéral des Routes, Rapport No. 506, Bern, Suisse, January, 119 p.

Torrent, R. and Frenzer, G. (1995). "Methoden zur Messung und Beurteilung der Kennwerte des Ueberdeckungsbetons auf der Baustelle -Teil II". Office Fédéral des Routes, Rapport No. 516, Bern, Suisse, October, 106 p.

Twede, D. (2002). "Commercial amphoras: The earliest consumer packages?" *J. Macromarket.*, v22, n1, 98–108.

Ungricht, H. (2004). "Wasserhaushalt und Chlorideintrag in Beton – Einfluss der Exposition und der Betonzusammensetzung". PhD Thesis, ETH Zürich, 203 p.

Wang, Y.F., Dong, G.H., Deng, F. and Fan, Z.H. (2014). "Application research of the efficient detection for permeability of the large marine concrete structures". *Appl. Mechan. Mater.*, v525, 512–517.

Yancheng, L., Xianbo, L., Kai, K., Tingli, L., Jiang, S., Zhang, J., Zhang, Z. and Yunting, L. (2017). "Permeability characterization and directional wells initial productivity prediction in the continental multilayer sandstone reservoirs: A case from Penglai 19-3 oil field, Bohai Bay Basin". *Petrol. Explor. Develop.*, v44, n1, 97–104.

Chapter 12

Epilogue: the future

At the risk of sounding presumptuous, this book might be considered a state-of-the-art document on the property permeability of concrete, its measurement, the variables that influence it, its relation with durability and its practical applications. With that intention it has been written by the authors.

The book describes the *status quo* of the knowledge on the subject, although it may be incomplete and imprinted by the personal experiences of the authors.

In this closing chapter, a vision on the possible incoming developments on the subject is given, assuming the enormous risks involved in trying to predict the future.

12.1 CHAPTER 1: DURABILITY

This chapter dealt with the basics of durability and how the main deterioration mechanisms are invariably related to the permeability ("penetrability" to be more general) of the concrete. The recognition that the prescriptive approach is inadequate to ensure the expected service life is generating a "magmatic" movement towards the performance approach. Several national codes and standards (Argentina, Canada, Japan, Portugal, South Africa, Spain, Switzerland) have included some sort of permeability tests, like water penetration under pressure, capillary suction and/or gas-permeability (and/or chloride-migration tests) as durability indicators, specifying limiting values as function of the exposure conditions.

The main international codes (ACI, EN), however, have stuck so far to the classical prescriptive approach, reflecting a conservative attitude or the lack of agreement whenever many countries are involved in the decision process. It seems that the chauvinist syndrome "not invented here" plays a role in the delay for adopting a performance approach.

A breakthrough may happen through the coming version 2020 of *fib* Model Code, decidedly performance oriented.

But, possibly, the advent of systems that promote sustainable construction (LEED, BREEAM, DGNB, MINERGIE, etc.) (NNBS, 2018) will

DOI: 10.1201/9780429505652-12

accelerate the P2P path (P2P: Prescriptive to Performance). Indeed, it has been recognized that prescriptive requirements (maximum *w/c*, minimum cement content) conspire against the sustainability of concrete constructions, placing them at considerable competitive risk against other solutions.

In North America, the Strategic Development Council (SDC) was created in 2007, stating:

> "The vision of the North American concrete industry is to transform the built environment by improving the way concrete is designed, specified, produced, transported, installed, maintained, and recycled to ensure an optimum balance between environmental, social, and economic conditions for the industry and the world."
>
> *SDC (2008)*

Two initial objectives were declared:

"*By 2010*: Create and adopt action plans with all relevant parties for achieving significant improvements in the sustainability of the built-environment, to include measurable process objectives".

"*By 2020*: Improve the sustainable characteristics of the built-environment through the efficient and effective use of concrete in green building, improving design to take full advantage of concrete's attributes, and <u>adopting specifications that facilitate innovation in product design</u>". The underlined text (by this book's authors) refers to the obstacle prescriptive standards pose to innovation. Fixing or restricting the concrete composition obstructs differentiation, as all concrete producers will be delivering essentially the same concrete, hindering innovative players to introduce solutions that can achieve the desired performance with a better environmental impact. One of the Goals of SDC (2008) reflects the current situation: "Work to remove technical constraints in building codes, standards, and specifications that may prevent concrete and construction from reaching its full sustainability potential".

Given the situation in 2020, it has to be admitted that the SDC initiative has not been entirely successful, as little has changed, at least in the area of specifications.

There is a certain negative perception in society against concrete, with coined terms such as "concrete or cement jungle" meaning "an area in a city with large modern buildings that is perceived as dangerous and unpleasant" (TFD, 2020) or "concrete: the most destructive material on Earth" (The Guardian, 2019). The industry needs to take actions to change this negative image and making concrete more sustainable is one of them.

In the authors' opinion, the tide towards performance specifications is slow but unstoppable, which certainly will involve concrete permeability as a key durability indicator.

12.2 CHAPTER 2: PERMEABILITY

This chapter was devoted to "Permeability as Key Concrete Property", describing many applications for which a knowledge of the coefficient of permeability is critical for the design of the structures or buildings. Many of these applications are well established, but others explore incipient fields of application (evacuated tunnels for high-speed trains, underground gas reservoirs for energy storage or even CO_2 sequestration, etc.).

The XX century was the century of space exploration; the XXI century may become the one of space colonization, which means that building activities shall be conducted, at least probably in the Moon (Cemento, 2001).

Several materials are good candidates for use in lunar construction, as reviewed by Ruess et al. (2006), concrete among them. Indeed, in reports gathered in a special ACI publication (ACI, 1991), it has been established that the raw materials required to produce Portland cement and concrete exist in the moon, except water (needed to sustain life anyway) which, in fact, may even exist in frozen state in the Moon (Ruess et al., 2006). Very recently (NASA, 2020; Página 12, 2020), the existence of water, albeit in extremely tiny quantities (≈1/100 of that existing in the Sahara Desert), has been detected on the sunlit surface of the Moon. This discovery indicates that water may be distributed across the lunar surface, and not limited to cold, shadowed places. Paul Hertz, Director of the Astrophysics Division in the Science Mission Directorate at NASA Headquarters in Washington stated: "Now we know it is there. This discovery challenges our understanding of the lunar surface and raises intriguing questions about resources relevant for deep space exploration".

Regolith, by far the most abundant material in the moon (Ruess et al., 2006), the characteristics of which are also described in Meyer (2003), is a suitable candidate as lunar concrete aggregate. It shows good potential as building material (GCR, 2018).

The lunar structures, besides protecting the dwellers from meteorites, radiations, extreme temperatures and resisting loads (gravity is only 1/6 that of the Earth), will be subjected to a pressure gradient of near 1 bar; in fact, a smaller value of 0.69 bar is proposed by Ruess et al. (2006), stating "The enclosure structure must contain this pressure, and must be designed to be failsafe against catastrophic and other decompression". This means that, among other important properties of lunar concrete, air-permeability will be relevant, particularly assuming that, for economic reasons and low gravity, the lunar structures will have to be thin (Osio-Norgaard & Ferraro, 2016). Possibly, some adapted sort of UHPFRC systems might be appropriate (Young, 1985).

12.3 CHAPTER 3: MICROSTRUCTURE AND TRANSPORT THEORIES

Here, with the advent of increasingly sophisticated experimental techniques to explore the micro- and nano-structure of hydrated cement and concrete, coupled with more elaborated models and software, major advances can be expected in the knowledge and interpretation of the inner intricacies of the material. Same as MIP is a useful tool to predict gas-permeability, as discussed in Section 3.8, it may well be that new techniques of experimental analysis of the microstructure of rather small samples may provide an accurate assessment of different properties of the bulk concrete, permeability among them.

Regarding the theory, although the main laws of transport of matter through concrete are well established (Hagen-Poiseuille, Darcy, Fick, Nernst-Planck), further refinement cannot be discarded. An important field, often neglected, is the complex flow of gases under pressure which, particularly under vacuum, involves a mixed transport mode (viscous permeation flow and molecular diffusion). This topic has recently been addressed by Sakai (2020) and the new RILEM TC GDP (Test Methods for Gas Diffusion in Porous Media) may also contribute in this respect.

12.4 CHAPTER 4: PERMEABILITY TEST METHODS

The interest in testing concrete permeability is near 100 years old. Although in a first stage this interest was focused on water-tightness, later, in the 70s of the XX century, it has shifted to a durability viewpoint and started to also include permeability to gases. Figure 12.1a provides a chronology of permeability test methods development that is based on those presented in Chapter 4.

The awareness of concrete durability issues and their dependence on concrete transport properties, namely permeability, has triggered a boom of new test methods in the 80s. Nevertheless, still new methods continue to be developed, including considerable efforts to the improvement of existing methods. This has been accompanied by a trend to include the best performing permeability methods in standards.

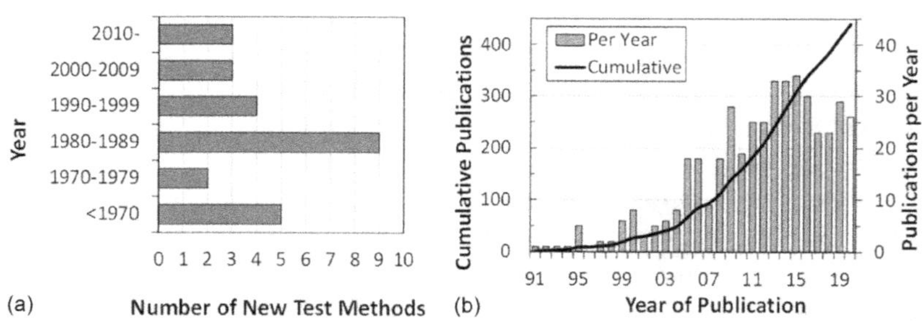

Figure 12.1 (a) Evolution of permeability test methods' development and (b) publications on *kT* test method since its origin.

Possibly, the most widely accepted permeability test method for concrete is the "crude" test of water penetration under pressure, described in Section 4.1.1.2. With variants, it is standardized in Europe, China, Argentina and Chile and reportedly is used also in the USA (Pacheco, 2018).

It is likely that more and better permeability tests are developed in the future, for laboratory as well as for site applications (preferably non-destructive); same for other transport properties (dealt with in Annex A). The possibility of having available test methods that can "scan" the microstructure of concrete from its surface should not be ruled out (same as what is done today for the macrostructure with georadar or ultrasonic scanning techniques).

The theoretical background of a new test to measure the effective porosity and air-permeability (with some resemblance to the *kT* test) of cylindrical specimens of rocks and concrete-like materials has recently been presented by Torrent and Zino (2021a, 2021b).

12.5 CHAPTER 5: *kT* AIR-PERMEABILITY TEST METHOD

This test method, developed by one of the authors, has been gaining recognition and acceptance in the concrete community. One symptom is the increasing number of publications dealing with the test method and its applications. Figure 12.1b shows the annual (bars) and cumulative (line) number of publications available to the authors (there may be more), totalling over 430 and growing. The general rate since 2005 is one to two related papers per month.

The basic principles of the method are simple and easily understandable for the users, as summarily described in Section 5.3.1. The derivation of the formula to compute the air-permeability coefficient and its assumptions is also presented (see Section 5.3.4). The practical implementation of the test method can be carried out through different means. In fact, there are several suitable instruments on the market from different manufacturers and versions. The interest in this test method has led to several improvements from the original prototype (Torrent & Szychowski, 2016), as well as concerning the interpretation of results (Neves, 2015). Further improvements are still foreseen.

It is expected that this test method, currently standardized in Switzerland, Japan, Jiangsu Province (China) and Argentina, will be included in more standards worldwide; see Section 8.5 and Annex B.

12.6 CHAPTER 6: FACTORS INFLUENCING CONCRETE PERMEABILITY

Despite the extensive coverage in Chapter 6 (one of the book's longest) on the effects key technological parameters of concrete exert on the permeability of concrete, the topic is far from being exhausted. More testing efforts on increasing number of concretes (especially those resulting from innovative constituents and solutions) will add to the current knowledge.

12.7 CHAPTER 7: THEORECRETE, LABCRETE, REALCRETE AND COVERCRETE

When looking at cement and concrete-related journals it become obvious that the majority of the research efforts are focused on rather small specimens, cast with the concretes, mortars or pastes under investigation, thoroughly moist-cured at constant temperature and analysed or tested in the laboratory. This type of research is essential for understanding mechanisms and concrete behaviour in general, for developing new materials or improving the existent ones, etc. There are many reasons for that, some discussed in the controversial Chapter 8: "Research" of Neville (2003).

Without fuelling the debate, it is clear that performing laboratory tests, under controlled conditions, is more comfortable, less risky, less costly and more likely of getting fast publishable results than performing experimental work on site.

What Chapter 7 addresses in some detail are the enormous differences existing between *Labcrete* and *Realcrete* (and its more specific component, the *Covercrete*) and the lack of representativity of the former to describe the quality and performance of the latter.

Concrete construction is in many cases handled by poorly trained personnel, sometimes even without a minimum knowledge of the effect their misdoings may have on the performance of the structure they are building; just two examples: improper positioning and fixing of reinforcing bars or the lack of curing, both magisterially described in Neville (2003). Let us quote from him in the section of Chapter 8 "Features of macro-scale research":

> "Good concrete is not a commodity found in abundance. After all, concrete is considered by many to be a simple material – nothing but a mixture of cement, aggregate and water – which any fool can make; the trouble is that he or she does...."

It is interesting to remark that in Japan, a country distinguished for its concern with quality, several construction companies have acquired instruments to measure air-permeability *kT* on site in order to demonstrate their competence and allegiance to good concrete construction practices. Unfortunately, in many other countries, the same sector is reluctant to be subjected to any kind of control of the end-product.

It is the authors' view that more and more research efforts should be focused on site investigations to assert at a real scale the true benefits of improving mix designs, concreting practices and innovative products and solutions, such as self-curing, self-healing (including bacteria), sealers, curing compounds and the long list that may follow. Often, solutions that appear feasible at the laboratory scale may prove inadequate when tested under real conditions (one of the authors had a bad experience when developing a self-compacting repair material, that worked well in the laboratory but failed on site). A good example of a comprehensive development is that of a UHPFRC as a highly durable material, carried out by E. Brühwiler, E.

Denarié and co-workers at the EPFL (Lausanne, Switzerland), that went all the way from the laboratory formulation up to the performance evaluation in real constructions, see Section 11.3.7.

Another important field for which site testing is relevant is service life modelling, the challenging object of which is, basically, to predict the durability performance of concrete structures subjected to an aggressive environment, typically one that promotes steel corrosion. Some of the existing models recognizing the aleatory nature of the problem, affected by countless factors many of them out of control of the engineers, attempt the so-called probabilistic approaches. The results of these predictions are in the form of probability of "occurrence" of certain phenomena (e.g. that the carbonation front reaches the steel or the appearance of cracks). The enormous spatial variability of the carbonation and chloride penetration rates observed in real structures (Torrent et al., 2020) poses a huge challenge to model-developers. The only way to assess the accuracy of the models is by site testing of real structures and comparing the estimated central values and variabilities with real hard data. Section 11.4.2 indicates that making tests at early ages gives a useful indication of the overall quality of the building which, as times go by, may change due to ageing, weathering and the action of service loads.

As a wish rather than a forecast: it is of extreme importance that more research efforts are focused on new and existing structures and buildings if a positive answer to the question "Is Our Research Likely to Improve Concrete?", in Section 8.1 of Neville (2003), is sought, at least for what matters in terms of durability.

12.8 CHAPTER 8: *kT* AIR-PERMEABILITY AS DURABILITY INDICATOR

Chapter 8 provides abundant evidence on the merits of the air-permeability *kT* test method to serve as durability indicator. More research and new applications may extend the arguments in favour; in parallel, it would be interesting also to report possible negative examples, trying to understand why the test method may have provided a wrong indication and, thus, better define the scope of application and limitations of the test method and stimulate possible improvements.

This chapter is closely linked to Chapter 5, already discussed.

12.9 CHAPTER 9: MODELLING BASED ON SITE PERMEABILITY TESTS

Service life modelling is an area where huge developments are to be expected, as more and more structures will have durability as one of the key design criteria.

Some of the available models for steel corrosion, like Life-365 (chloride-induced corrosion) rely on the coefficient of chloride-diffusion as key input material parameter of concrete, be it calculated from the mix composition (Theorecerete) or from results of ASTM C1556 bulk diffusion test (Labcrete). Other well-known models for chloride-induced corrosion rely on the NT Build 492 chloride migration test as key material input parameter (Duracon, Duracrete, *fib*) or, the last two, on accelerated carbonation tests for carbonation-induced corrosion. In all cases, the key input material parameters of concrete are obtained on tests made on laboratory specimens (*Labcrete*).

Despite their acceptance (perhaps for lack of better alternatives), these models have shown limitations, chiefly derived from the need to define parameters, the values of which are quite elusive. One of the developers of DuraCrete model claims that many of the parameters tabulated in DuraCrete (2000) are "invented" and only of academic value (Andrade, 2014). Yet, the values attributed to these parameters have a capital influence on the computed service life and may even lead to absurd results (Gulikers, 2006; Gulikers & Groeneweg, 2017; Torrent, 2017, 2018).

As this situation is far from satisfactory, these models need to be improved (some steps in this direction are being taken for the incoming *fib* Model Code 2020) and new ones developed. Moreover, so far, the available models deal with the very important – but by no means unique – problem of deterioration due to steel corrosion. It is also expected to count in the future with solutions that model other deterioration processes, like freeze-thaw damage, chemical attack (especially sulphate attack), etc.

In Chapter 9, various models have been described in which the key input material parameter is the coefficient of gas permeability, measured directly on site or in the laboratory on cores drilled from the structure (*Realcrete/Covercrete*). Since these models refer also to steel-corrosion induced by carbonation or chlorides, the other key input parameter is the cover thickness which, for the *kT*-based models, is assessed non-destructively on site in parallel to the air-permeability determination. These approaches have the advantage that a more realistic picture of the thickness and quality of the vital surface layers of the site concrete is contemplated, especially of their significant space variability (to a large extent reflecting the quality of the concreting practices displayed by the contractor). These variabilities are converted into probabilistic assessment of the service life. Then, even if the actual estimation of the expected years of service life may be inaccurate (the examples shown in Sections 11.3.1, 11.3.2 and 11.4.1.1 indicate the contrary), the approach will help identifying potential "weak areas" (those with high gas-permeability and/or low cover thickness), where the deterioration is more likely to start, thus allowing a more focused monitoring for taking early remedial actions, avoiding later costly intervention (de Sitter's "Law of Fives" described in Section 1.4). Measuring on site the permeability and thickness of the concrete cover is essential for producing a realistic "Birth Certificate" (Bartholomew, 2020).

It is also to be expected that methods, based on site testing, will be improved in the future, adjusting its algorithms and parameters, as new laboratory, but especially site data become available.

12.10 CHAPTER 10: GAS PERMEABILITY AND FIRE PROTECTION

The role of gas permeability in the risk of explosive spalling under fire is a relatively new field of research, with a huge potential for development as well as for inclusion in the codes.

It is likely that, in the future, codes may introduce *minimum* air-permeability limits to the concrete exposed to potential fires, to be assessed on site. These limits may enter in conflict with the *maximum* air-permeability limits specified for durability. One successful solution seems to be the inclusion of synthetic fibres in low-permeability concretes that, at the high temperatures reached during a fire, crack the concrete and melt, increasing the permeability and offering escape paths to the pressurized vapour in the concrete pores.

We can expect important developments in this field.

12.11 CHAPTER 11: APPLICATIONS OF AIR-PERMEABILITY *kT* TESTS

Chapter 11 presents over 20 cases of application of *kT* measurements on site, covering full-scale investigations and new and old real structures. It constitutes a token of the versatility of the test method that was applied, under extreme conditions like in the Antarctic or high-altitude deserts, on a wide variety of concrete structures or elements.

It has been used to verify the quality of concrete that needed to be water-tight, to assess the effect of curing on the carbonation at 150 years!! and the risk of steel corrosion of submerged marine structures at 120 years!!., Also to assess the service life of new and old structures, as a tool for condition assessment surveys, etc.

It can be asserted, without fear of being mistaken, that the list will grow in the future as more real cases will be reported in the literature, in particular if the wishful prediction of research more focused on the Realcrete is realized.

But there are simpler applications for which *kT* testing on site may be of help. Let us consider companies promoting a new concrete (e.g. with innovative constituents, HP or UHP, etc.), new curing systems (improved curing compounds or self-curing solutions), the advantages of permeable formwork membranes or of solutions that reduce the gas-permeability of the *Covercrete* (sealers, pore/crack-fillers, bacterial self-healing, vacuum-treatment). What more convincing than offering the reluctant client to apply the solution on a part of the structure or on a mock-up element and compare the results of air-permeability in front of his/her eyes with those obtained on the current situation, taken as reference?

Another new field where measuring *kT* on site will probably find application is 3D concrete printing (additive manufacturing), since the lack of compaction may create in-between-layers preferential paths for the penetration of moisture or aggressive species. This is particularly relevant in the case of 3D-printed reinforced concrete elements (Kloft et al., 2020).

The example in Section 11.2.2 indicates the importance of characterizing, not just the concrete mixes involved in long-term research projects that monitor real structures or samples in ad-hoc exposure stations, but the exposed elements themselves, by measuring their *kT* air-permeability, for a correct evaluation of the results.

It is expected that Chapter 11 acts as catalyst of the creativity of scientists, engineers and architects in finding new applications for *kT* testing, in the laboratory but especially on site, where its potential is fully exploited.

REFERENCES

ACI (1991). “Lunar concrete”. *ACI SP-125*, 20 papers, 300 p.

Andrade, C. (2014). “Prediction of service life by considering the initiation and propagation periods”. Keynote lecture, *XIII DBCM*, São Paulo, Brazil, September 3–5. Retrieved from: www.youtube.com/watch?v=NruQiViNU4c.

Bartholomew, M. (2020). “Birth certificate and through-life management documentation”. Technical Report, *fib Bulletin 93*, June, 90 p.

Cemento (2001). “Pasado, Presente y Futuro del Hormigón”. Torrent, R. (Ed.). *Cemento*, n1, Buenos Aires, Argentina.

DuraCrete (2000). “Probabilistic performance based durability design of concrete structures”. The European Union–Brite EuRam III, BE95–1347/R17, CUR, Gouda, The Netherlands.

GCR (2018). “Europe’s space scientists plan to turn moon dust into lunar concrete”. *Global Constr. Review News*, August 22.

Gulikers, J. (2006). “Considerations on the reliability of service life predictions using a probabilistic approach”. *Journal de Physique IV (Proceedings)*, v136, 233–241.

Gulikers, J.J.W. and Groeneweg, T.W. (2017). “Residual service life of existing concrete structures – Is it useful in practice?”. *fib Symposium*, Maastricht, The Netherlands, paper 211.

Kloft, H., Empelmann, M., Hack, N., Herrmann, E. and Lowke, D. (2020). “Reinforcement strategies for 3D-concrete-printing”, Report, *Civil Eng. Des.*, v2, 131–139.

Meyer, C. (2003). “Lunar regolith”. *NASA Lunar Petrographic Educational Thin* Section *Set*, 46–48.

NASA (2020). “NASA’s SOFIA discovers water on sunlit surface of moon”. Release 20–105, October 26.

Neves, R. (2015). "Enhancing the interpretation of Torrent air permeability method results". Concrete Repair, Rehabilitation and Retrofitting IV, Leipzig, Germany, October 5–7, 281–286.

Neville, A. (2003). "Neville on concrete". ACI, Farmington Hills (MI).

NNBS (2018). "Landkarte Standards und Labels Nachhaltiges Bauen Schweiz" (in German). Sustainable Construction Network Switzerland, Zürich, October 24, 48 p.

Osio-Norgaard, J.M. and Ferraro, C.C. (2016). "Permeability of sulfur based lunar concrete". *Earth and Space 2016 – Engineering for Extreme Environments.*

Pacheco, J. (2018). Personal Communication, December 5.

Página 12 (2020). "La NASA confirmó que hay agua en la Luna". Buenos Aires, Argentina, October 26.

Ruess, F., Schaenzlin, J. and Benaroya, H. (2006). "Structural design of a lunar habitat". *J. Aerospace Eng.*, v19, n3, July, 133–157.

Sakai, Y. (2020). "Relationship between air diffusivity and permeability coefficients of cementitious materials". *RILEM Technical Letter* 5, 26–32.

SDC (2008). "Concrete sustainability – A vision for sustainable construction with concrete in North America". *Strategic Development Council*, Draft 7.3, November 14, 8 p. Retrieved from https://www.acifoundation.org/Portals/12/Files/PDFs/Sustainability-Vision.pdf.

TFD (2020). The free dictionary. Retrieved from https://www.thefreedictionary.com/Cement+jungle.

The Guardian (2019). "Concrete: the most destructive material on Earth". 25 February.

Torrent, R. (2017). "Robust, Engineers' friendly service life design". Keynote Lecture, *EASEC-15*, Xi'An, China, 11–13 Oct., 14 p.

Torrent, R.J. (2018). "Bridge durability design after EN standards: PRESENT and future". *Struct. & Infrastruct. Eng.*, January 12, 1–14.

Torrent, R., Imamoto, K. and Neves, R. (2020). "Resistance of concrete to carbonation and chloride penetration assessed on site through nondestructive tests". fib Struct. Concr., September 8, 1–14.

Torrent, R. and Szychowski, J. (2016). "The active cell principle for ND testing of air-permeability of concrete". *Microdurability 2016*, Nanjing, China, October 24–26.

Torrent, R.J. and Zino, G.O. (2021a). "Analytical modelling of a novel test for determination of porosity and permeability of porous materials". *10th Intern. Conf. on Mathem. Modeling in Phys. Sci.*, September 6–9.

Torrent, R.J. and Zino, G. (2021b). "Hacia un ensayo simultáneo de porosidad y permeabilidad al aire de rocas y materiales cementicios". CADI/CLADI/CAEDI, Buenos Aires, Argentina, October 5–7.

Young, J.F. (1985). "Concrete and other cement-based composites for lunar base construction". *Lunar Bases and Space Activities of the 21st* Century. Mendell, W.W. (Ed.), Lunar and Planetary Institute, Houston, TX, 391–396.

Annex A: Transport test methods other than permeability

In Chapter 3, the main mechanisms and laws governing mass transport through concrete were discussed, including their links to the pore structure of the material. Chapter 4 was devoted to describing test methods to measure permeability of concrete to gases and liquids (including capillary suction).

In this Annex, relevant test methods aimed at measuring concrete properties associated with the other two transport mechanisms, namely diffusion and migration, are described. The list is not exhaustive, preference being given to test methods referred to in the book.

A.1 DIFFUSION TESTS

Gaseous diffusion of water vapour, CO_2 and O_2, as well as Cl^- ion diffusion in aqueous solution through concrete are relevant transport mechanisms regarding steel rebars corrosion. Standard and relevant methods to measure the diffusivity of concrete to gases and ions are described in the following subsections.

A.1.1 Gas-Diffusion

A.1.1.1 Gas-Diffusion Tests

The Swiss Federal Laboratory EMPA has implemented an oxygen-diffusion test method for concrete, based on the design developed by Wong (2006), sketched in Figure A.1; more information in Leemann et al. (2017) and Vilani et al. (2014). The tested concrete specimen is a Ø100 × 50 mm disc, which prior to the test is dried 7 days at 35% RH, followed by 7 days oven-drying at 50°C. The specimen is sealed laterally in the diffusion cell (similar to the Cembureau oxygen-permeability cell, Section 4.3.1.2), picture shown in Figure A.1. Oxygen is circulated on the upper face of the specimen (double arrowed broken lines), whilst nitrogen is circulated on the opposite face

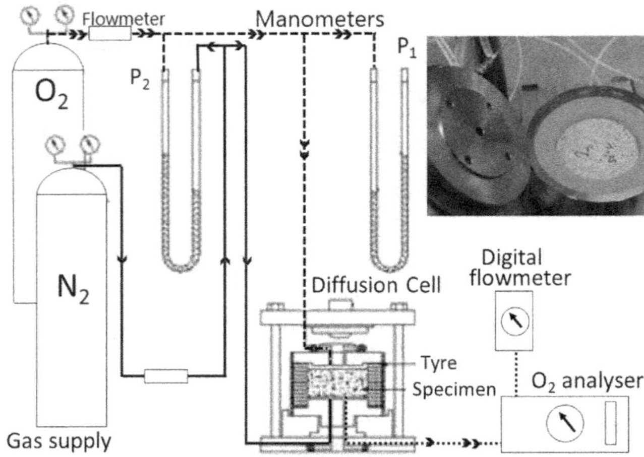

Figure A.1 Sketch and details of testing cell of the method to measure O_2-diffusivity of concrete (Wong, 2006).

(single arrowed full lines), both gases kept at the same pressure (to avoid permeation phenomena). The oxygen content in the nitrogen flow (dotted line) is measured using a zirconium oxygen analyser until steady-state is reached. The oxygen-diffusion coefficient is then calculated applying the first Fick's law (Section 3.4.1). It is claimed that the O_2-diffusion coefficient is very similar to that obtained for other gases (e.g. CO_2) (Leemann et al., 2017).

Values of the coefficient of oxygen-diffusion within the range from 10^{-9} to 10^{-7} m^2/s, obtained with EMPA method, were reported by Moro and Torrent (2016) and Leemann and Moro (2017), on the 18 mixes discussed in Section 8.3.2.

Similar methods have been reported by Schwiete et al. (1969) and Kobayashi and Shuttoh (1991). RILEM TC GDP "Test Methods for Gas Diffusion in Porous Media" has recently been created, which certainly will contribute to this topic.

A.1.1.2 Carbonation: Natural and Accelerated Tests

Although, in principle, it is possible to carry out CO_2-diffusion tests in concrete, this particular gas reacts with hydration products, changing the concrete porosity (Ngala & Page, 1997); therefore, it is almost impossible to obtain an intrinsic CO_2-diffusion coefficient. Instead, the concrete resistance to carbonation is usually assessed, as it is more interesting from an engineering viewpoint.

Carbonation under natural conditions is a slow process requiring, for ordinary concretes, at least 1½ year to have a reasonably accurate evaluation of concrete resistance to carbonation. Natural carbonation tests are conducted exposing concrete samples to the ambient air, be it inside a laboratory room with controlled temperature and relative humidity (typically 20°C and 50%–65%) or outdoors, protected from or exposed to rain. At different exposure times t, the carbonation depth x_c is measured by one of the methods described below (phenolphthalein being the more popular, due to its simplicity) (CPC-18, 1988). The slope of the linear relation, usually obtained between x_c and $\sqrt{t}$, provides the carbonation rate or carbonation rate K_c, typically expressed in $mm/\sqrt{y}$. The results obtained depend strongly on the environmental exposure conditions.

To have a timelier information, accelerated conditions are often applied. After a period of preconditioning, the tests are carried out exposing the specimens to a strictly controlled environment (constant T and RH) and a CO_2 concentration well above that of natural air (3%–5% are commonly adopted, against around 0.03%–0.04% for natural air). Thus, concrete "accelerated" carbonation is assessed. Some of such methods have been standardized (ISO 1920-12, 2015; SIA 262/1, 2019; EN 12390-12, 2020).

There are several techniques to measure the carbonation depth, based on concrete pH, amount of calcium carbonate (the main carbonation product) or CO_2. In general, the available techniques are scanning electron microscopy (SEM), quantitative X-ray diffraction analysis (QXRDA), Fourier-transform infrared spectroscopy (FTIR), thermogravimetric analysis (TGA), gamma densitometry, thin-sections optical microscopy and pH measurements.

In most cases, concrete carbonation depth from the exposed surface is evaluated by means of a pH indicator (typically phenolphthalein solution), sprayed on a freshly broken surface. This popular indicator (there are others, like healthier thymolphthalein (Moro et al., 2014)) shows a bright pink colour when in contact with a highly alkaline medium and stays colourless when the pH falls below ≈8; thus, a clear boundary between the carbonated and non-carbonated parts can be easily seen and measured.

Results of natural and accelerated carbonation tests were discussed in Section 8.3.7.

A.1.1.3 Accelerated Carbonation Test after Swiss Standard SIA 262/1-Annex I

Annex I of Swiss Standard (SIA 262/1, 2019) defines a method to estimate the natural carbonation rate, based on the measured accelerated carbonation rate. The layout of the test method is similar to that of the recently approved (EN 12390-12, 2020).

The specimens used are horizontally cast beams 120 × 120 × 360 mm or cores (Ø at least 50 mm; length at least 100 mm) drilled from larger specimens. The prisms follow a complex curing regime starting with 3 days moist curing and ending 18 days in a climatic chamber (20°C/57% RH, ≤0.15% CO_2) whilst the cores are stored 21 days in the climatic chamber.

After 0, 7, 28 and 63 days of exposure to the accelerated carbonation environment (20°C/57% RH, 4.0% CO_2), the specimens are split mechanically, removing slices ≈50 mm thick each time, on which the mean carbonation depth is measured by the phenolphthalein method.

With the four test results of accelerated carbonation, a linear regression is fit between carbonation depth and square root of time, the slope of which yields an accelerated carbonation rate K_a, which is later converted into an equivalent natural carbonation rate K'_c, applying Eq. (8.11) (Section 8.3.7.2). Maximum values of K'_c are specified for XC3 and XC4 exposure classes (typically 5.0 mm/y$^{1/2}$ for 50 years' service life). The test method has proved being sensitive, repeatable and reproducible (Moro et al., 2014).

Another standard method (ISO 1920–12, 2015) for accelerated carbonation is based on measuring the carbonation depth after exposing the specimen for 70 days to an environment of 22°C/55% RH and 3% CO_2 concentration.

A.1.2 Diffusion of Chloride Ions

There are several methods to assess chloride ion diffusivity in concrete. The conditions for a steady-state regime of chloride ion diffusion through concrete are seldom found in practice. In addition, reaching steady-state conditions in experiments requires long times and thin specimens, more appropriate for paste than concrete testing (Page et al., 1981), see also Andrade (2002) and Tang et al. (2012). For this reason, the preferred methods for testing the diffusivity of chlorides in concrete involve non steady-state conditions.

The standard test methods to measure the coefficient of chloride-diffusion *D*, under non steady-state conditions, are based on placing a preconditioned concrete specimen in contact with a concentrated chloride solution (typically 3% NaCl solution). This contact can be achieved by ponding the upper side of the specimen with the solution (AASHTO T259, 2002) or by submerging the specimen into the solution, with all faces but one sealed (NT Build 443, 1995; ASTM C1556, 2016).

After a relatively long period of exposure to the chloride solution (the longer the better the quality of the concrete tested), but typically of 2–3 months, cores are drilled from the exposed surface. The cores are diamond-cut or profile ground so as to get samples of the material at different small depth intervals, measuring the chloride ions' concentration (usually acid-soluble but sometimes water-soluble) at each depth. In this manner, a chloride profile is obtained. Now, the error function solution of Fick's second law (Eq. 3.5) is applied to the data, getting the best fit, from which the chloride-diffusion coefficient D is obtained. A routine is available to obtain both D and the surface chloride concentration C_s in a free software (Life-365, 2018). The chloride-diffusion coefficient, obtained in this manner, is usually designated as D^{ns} (non steady-state diffusion coefficient).

The peculiarities of some standardized test methods are outlined below.

AASHTO T259 ponding test is possibly the first non steady-state method established to determine the chloride-diffusion coefficient, also known as "ponding test" (a version existed already in 1980). The specimens are moist cured for 14 days and then stored in a dry room (50% RH) for 28 days. The sides of the specimens are sealed but the bottom and top face are not. A concentrated (3% NaCl) solution is permanently placed in a pond over the investigated concrete surface during 90 days. The specimen is kept in a dry room (50% RH) during the whole duration of the test.

Some criticism of the test lies in that it does not measure a "pure" diffusion transport, but a mixed one, as capillary suction plays a role – at least initially – when placing the relatively dry specimen in contact with the NaCl solution. The fact that the bottom face of the slab is in contact with dry air creates a so-called "wick effect" that enhances the penetration of chlorides (Stanish et al., 1997). Another problem, experienced by Torrent and Ebensperger (1993), is that the thickness of the slices (15–20 mm) is too coarse for an accurate chloride profile determination, at least for the 1980 version. This shortcoming can be overcome by taking thinner slices (e.g. 5 mm). On the other hand, the test enhances the penetration of chlorides (capillary suction is a more efficient driver than diffusion) and it is not far from real conditions, typically encountered in the field.

The ASTM C1556 test method, based on the original NT Build 443, consists in immersing a concrete specimen, all surfaces but one being sealed, in a chloride solution (usually 165 g of 2.8 M NaCl per litre) for at least 35 days (90 days for high-quality concretes). The specimen (at least Ø75 × 75 mm) is fully saturated in lime water prior to initiating the test. Afterwards, the chloride profile (chloride content vs. depth) is measured by analysis of the total chloride content in powder samples obtained at different depths of the specimen by profile grinding. The thickness of the powder samples varies between 1 mm (for $w/cm \leq 0.35$) to 5 mm (for $w/cm = 0.70$). The total chloride content, in the case of ASTM C1556, is determined according to the analytical method specified in ASTM C1152 (2004).

The chemical analysis of chloride content may be carried out applying different techniques like potentiometric titration, spectrophotometry, X-ray fluorescence (XRF) or direct potentiometry by means of a chloride selective electrode. It has been proved that XRF is a good practical alternative to potentiometric titration (Kanada et al., 2006a, b) and is regularly used in Japan.

The duration and technical complexity (and associated costs) of both non steady-state chloride-diffusion tests make them unappealing for current routine quality control testing.

A.2 MIGRATION TESTS

Migration is a transport mechanism by which ions move along the pore network of concrete under the influence of an external electric field (Section 3.4.2). Migration plays a role in the steel corrosion process, since OH^- ions should move through an electrolyte (pore solution) to close the electrochemical cell circuit. The facility with which OH^- ions move is one factor determining the rate at which corrosion will proceed.

A.2.1 Chloride-Migration Tests

Although migration is not a transport mechanism with relevant implications on the penetration of aggressive ions into real structures, it allows a fast assessment of the concrete resistance to the movement (and penetration) of chloride ions. Migration tests are similar to diffusion tests where, instead of a concentration gradient, an electric field is applied across the specimen by means of electrodes and using a DC power supply. Equation (3.6) shows that the same parameter, diffusion coefficient, governs both diffusion and migration transport processes. Therefore, diffusion coefficients obtained from migration tests should not be very different from those obtained in diffusion tests, with the advantage that by applying a reasonable voltage, an acceleration effect can be obtained.

The acceleration factor depends on the applied potential, which should not be less than 10 V, to keep negligible the relative amount of chloride transport by diffusion (Geiker et al., 1995). On the other hand, when applying voltages near 60 V or higher, considerable heat may develop in low-quality concretes and artificially influence (increase) the ion flow (Andrade, 1993).

In the following, the procedures and techniques defined in existing migration testing standards are addressed.

A.2.1.1 "RCPT" ASTM C1202

The first migration-based test method was developed by Whiting (1981), which led rapidly to an AASHTO Standard test (AASHTO T277, 1983); the same method was later adopted as ASTM C1202 (2019).

It is interesting to note the different names adopted by AASHTO ("Standard Method of Test for Rapid Determination of the Chloride Permeability of Concrete") and, more cautiously, by ASTM ("Standard Test Method for Electrical Indication of Concrete's Ability to Resist Chloride

Ion Penetration"). The former led to the now common identification of the test as RCPT (Rapid Chloride Permeability Test).

The method consists in applying a 60 V potential across two sides of a Ø100 × 50 mm specimen for 6 hours, while recording the current intensity. One side of the specimen is in contact with a 3% NaCl solution (where the cathode is placed), whilst the opposite side is in contact with a Na(OH) solution (where the anode is placed), see Figure A.2a. The test relies on the fact that negatively charged ions (in particular Cl^-) will migrate from the cathode to the anode, establishing an electric current, the intensity of which is monitored during the test. After 6 hours of migration, the intensity vs time data records are integrated, providing the total amount of electric charge passed, measured in Coulombs (hence the also common designation of "Coulomb" test)

This test has become extremely popular for specification and quality control, often forgetting its limitations. Indeed, the test is not meant to give a definitive diffusion property, ASTM C1202 stating

> "This test method covers the determination of the electrical conductance of concrete to provide a rapid indication of its resistance to the penetration of chloride ions. This test method is applicable to types of concrete where correlations have been established between this test procedure and long-term chloride ponding procedures such as those described in AASHTO T259."

Therefore, ASTM C1202 recognizes that it is truly an electrical conductivity test.

Among the possible interferences, ASTM C1202 mentions the addition of calcium nitrite and the presence of embedded steel in the specimen.

The test result is the total charge (Coulombs) passed during the 6 hours' duration; ASTM C1202 provides a qualitative indication of the chloride ion penetrability of concrete based on the measured values, as indicated in second column of Table A.1.

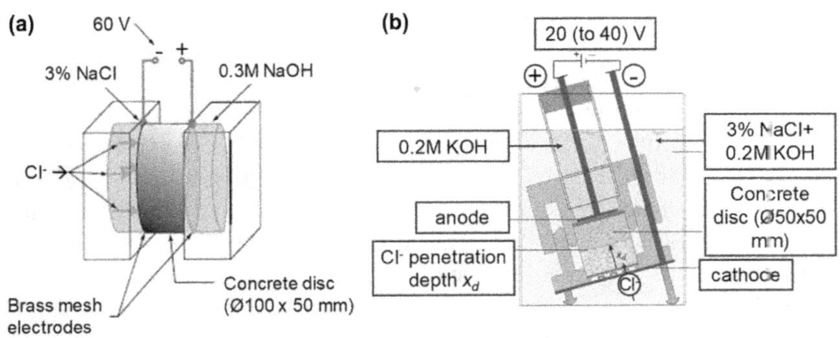

Figure A.2 Sketch of cell assemblage for (a) ASTM C1202 test, adapted from CI (2006) and (b) Tang-Nilsson test set-up, adapted from SIA 262/1 (2019).

Table A.1 Chloride ion penetrability based on charge passed in ASTM C1202 and on surface electrical resistivity test

Chloride ion penetration	Charge passed (Coulomb)	Surface resistivity (kΩ.cm) for a = 38 mm	
		Ø100 × 200 mm cylinders	Ø150 × 300 mm cylinders
High	>4,000	<12	<9.5
Moderate	2,000–4,000	12–21	9.5–16.5
Low	1,000–2,000	21–37	16.5–29
Very low	100–1,000	37–254	29–199
Negligible	<100	>254	>199

When using Table A.1, one shall bear in mind that the rating was established for testing specimens with 100 mm diameter and 50 mm thickness. If different dimensions are used, the proposed values should be corrected by a ratio between 50 mm and the actual thickness and/or by the square of the ratio between the actual diameter and 100 mm.

Therefore, this method provides a durability indicator that, however, is not directly suited for service life modelling. To help that, a correlation between the chloride-diffusion coefficient D and the charge Q passed in the ASTM C1202 test was established (Eq. 8.8), confirmed by independent test results (see 8.3.6.1).

Although it is a fast method, especially when compared with natural diffusion tests, and it is widely used, its nature as well as the interpretation of its results have raised some questions. Stanish et al. (1997) sum up the main criticisms to this method: the electric current crossing the specimen results not only from the chloride ions but also from other ions present in concrete; the current monitoring is carried out when a steady state is not yet reached; the high potential causes a temperature variation in concrete, particularly those with lower quality, that causes itself an increase in charge passed, biasing their rating. Similar objections were brought up by other researchers (Andrade, 1993; Shi, 2003).

Yet, the test has gained worldwide acceptance and limiting values are specified in the Canadian Standard (CSA, 2006): ≤1,500 Coulomb for structurally reinforced concrete exposed to chlorides with or without freezing and thawing conditions or structurally reinforced concrete exposed to severe manure and/or silage gases, with or without freeze-thaw exposure or concrete exposed to the vapour above municipal sewage or industrial effluent, where hydrogen sulphide gas may be generated. The above limit is reduced to 1,000 Coulomb for structurally reinforced concrete exposed to chlorides or other severe environments with or without freezing and thawing conditions, with higher durability performance expectations. The test age should not exceed 56 days (it is known that that the result of the test decreases significantly with the age of the specimens). In the New Panama Canal, the limit of 1,000 Coulomb was specified, without indication of the testing age which led to serious contractual problems (see Section 11.3.3).

A.2.1.2 Tang-Nilsson Chloride-Migration Test

The rapid migration test was proposed by Tang and Nilsson (1993), becoming standardized in Scandinavia as a Nordic Standard (NT Build 492, 1999). It has later been adopted also by Swiss Standards, Annex B ("Chloride resistance") of SIA 262/1 (2019).

Comparing it with the ASTM C1202 test, both tests operate under non-steady-state conditions, the main differences lying on the applied potential and on the evaluation of the test result. In the NT Build 492 test, the applied potential is 30 V, being adjustable as function of the developed current intensity. The test duration can vary between 6 hours and 4 days, the normal duration being 24 hours, depending on the observed current after potential adjustment. The test set-up used in the SIA 262/1 (2019) Standard is depicted in Figure A.2b; the original NT Build 492 specifies a Ø100 × 50 mm disc as specimen.

At the end of the test, the chloride penetration depth x_d is determined by spraying silver nitrate solution into a broken surface after longitudinally splitting the specimen (see Figure A.2b).

Based on test conditions and measurements, it is possible to compute a non steady-state chloride (migration) diffusion coefficient through Eq. (A.1), which is based on Eq. (3.6)

$$D_m = \frac{RTL}{zF\,(U-2)\,t} \times \left(x_d - 2\sqrt{\frac{RTLx_d}{zF\,(U-2)}} \right) \times erf^{-1}\left(1 - \frac{2c_d}{c_0}\right) \tag{A.1}$$

where

D_m = coefficient of chloride ion diffusion in concrete, obtained through migration test (m^2/s)
z = absolute value of ion valence; $z = 1$ for chloride ion
F = Faraday constant; $F = 9.6485.10^4$ J/(V mol)
U = absolute value of applied potential (V)
R = gas constant; $R = 8.314$ J/(K mol)
T = mean temperature in the anolyte solution (K)
L = thickness of the specimen (m)
x_d = chloride penetration depth (m)
t = test duration (s)
erf^{-1} = inverse error function
c_d = chloride concentration at which silver nitrate solution colour changes ($c_d = 0.07$ N for OPC)
c_0 = chloride concentration in the catholyte solution ($c_0 = 2$ N)

Classification criteria of concrete quality, based on the result of this test, have been proposed by Tang (1996) and Nilsson et al. (1998), see Table A.2.

Table A.2 Rating of concrete quality based on Tang-Nilsson test

Criterion of Tang (1996)		Criterion of Nilsson et al. (1998)	
D_m ($10^{-12}m^2/s$)	Concrete quality	D_m ($10^{-12}m^2/s$)	Resistance to chloride ingress
<2	Very good	>15	Low
2–8	Good	10–15	Moderate
8–16	Normal	5–10	High
>16	Poor	2.5–5	Very high
		<2.5	Extremely high

In Switzerland, a maximum limit of $10 \times 10^{-12}m^2/s$ is specified for Labcrete and of $12 \times 10^{-12}m^2/s$ for Realcrete, see Section 7.1.5.4, of structures exposed to chlorides (Torrent & Jacobs, 2014).

Several service life models, based on chloride penetration according to the second Fick's law (DuraCrete, 2000; Ferreira, 2004; Ferreira et al., 2004; *fib*, 2006), use D_m as input for the chloride-diffusion coefficient. Comparative data (Tang & Sørensen, 2001) indicate that the D_m value is slightly higher than the diffusion coefficient obtained by immersion tests. Li (2016) claims that D_m is about twice the value of the immersion diffusion coefficient, claim that has been confirmed experimentally by Ren et al. (2021).

A.2.1.3 On-Site Chloride Migration Tests

Whiting (1984) attempted to extend his invention of the Rapid Chloride Permeability Test, meant for the laboratory, to site testing, attempt that was not very successful.

More recently, a site chloride migration test has been developed (Basheer et al., 2005) under the brand name PERMIT.

A.2.2 Electrical Resistivity Tests

The application of a potential and the measurement of the generated current intensity, or the opposite, allows the determination of concrete's electrical resistivity (or its reciprocal, electrical conductivity). The electrical resistivity is a material property that characterizes its resistance to the flow of electric current, typically expressed in Ω .m or in kΩ .cm (1 kΩ .cm = 10 Ω .m). It is used as input for assessing the rate of corrosion in DuraCrete (2000) model and, in the model by Andrade and d'Andrea (2008), also for assessing the time for corrosion initiation.

Einstein's law establishes a direct relationship between ions' diffusivity and electrical resistivity (Andrade & d'Andrea, 2008):

$$D_e = \frac{K_{Cl}}{\rho_{es}} = K_{Cl} \cdot \sigma \tag{A.2}$$

where

D_e = effective diffusion coefficient
K_{Cl} = factor depending on the external ionic concentration
ρ_{es} = electrical resistivity of concrete saturated with water
σ = electrical conductivity of concrete saturated with water

There are different ways of measuring electrical resistivity/conductivity of concrete (TC-154-EMC, 2000), from which three have been standardized and will be discussed here.

A.2.2.1 Bulk Resistivity Test

This is a laboratory test, covered by two ASTM Standards (ASTM C1760, 2012; ASTM C1876, 2019), applied typically on Ø100 × 200 mm specimens or cores. The former requires the use of the ASTM C1202 cell. Here we will focus exclusively on the more practical (ASTM C1876, 2019) test method, known also as "direct" method. Figure A.3 shows two commercial instruments of the kind.

The age of the test specimen has a significant effect on the test results, depending on the type of concrete and the curing procedure. Most concretes, if properly cured, become progressively and significantly less conductive with time. To obtain meaningful results the specimens must be fully saturated.

The direct method is applicable to laboratory cast specimens or drilled cores and consists in applying electrodes on opposite ends of a specimen, through which an electric current of intensity I is applied, and in measuring the generated potential U. The bulk electrical resistivity can be calculated through

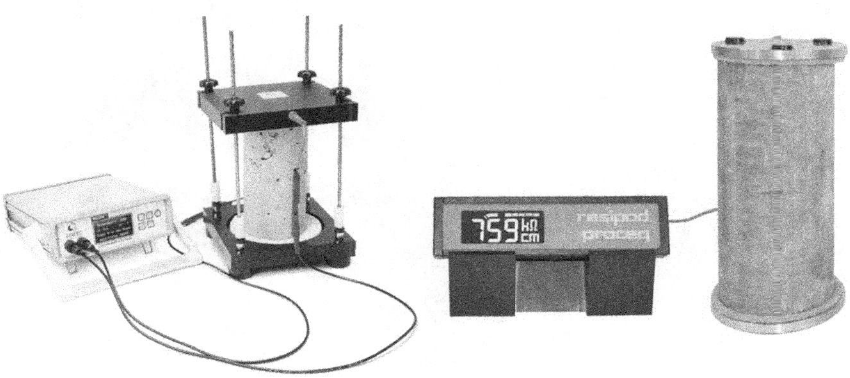

Figure A.3 View of two commercial bulk electrical resistivity instruments.

$$\rho = \frac{U}{I} \times \frac{A}{L} \tag{A.3}$$

where

ρ = electrical resistivity (Ω.m)
I = intensity of applied current (A)
U = measured potential (V)
A = area of specimen cross section (m^2)
L = thickness of the specimen (m)

A.2.2.2 Surface Resisitivity Test (Wenner Method)

This test method consists in placing, on the tested concrete surface, a linear arrangement of four electrodes, the tips of which are moistened to ensure electrical contact (see the sketch in Figure A.4a) and two commercial instruments in Figure A.4b and c.

A current of intensity I is applied between the two external electrodes and the resulting potential U, established between the two internal electrodes, is measured. The electrical resistivity ρ is calculated as

$$\rho = \frac{2 \cdot \pi \cdot a}{K} \cdot \frac{U}{I} \tag{A.4}$$

where

ρ = electrical resistivity (Ω.m)
I = intensity of applied current (A)
U = measured potential (V)
a = separation of the electrodes (m)
K = geometric factor (-); $K = 1$ for application on a semi-infinite surface

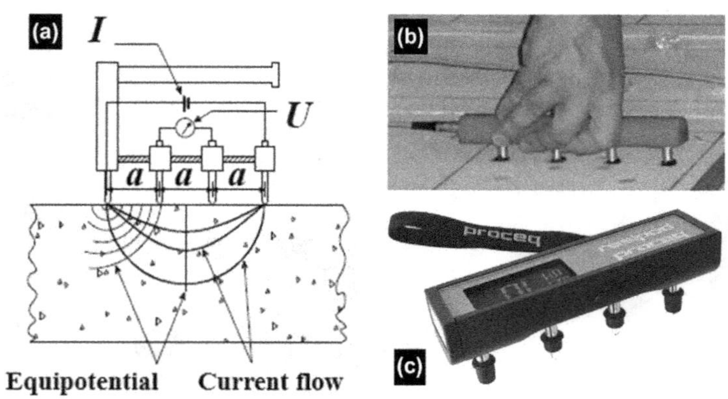

Figure A.4 (a) Sketch (Rupnow & Icenogle, 2011) and (b) and (c) view of commercial instruments for surface concrete resistivity measurement.

Values of *K* for different geometric arrangements can be found in Morris et al. (1996).

The test method is covered by Standards (AASHTO T358, 2017) and, in Spain, by UNE 83988-2 (2014), both intended for laboratory applications.

AASHTO T358 covers the determination of the electrical resistivity of concrete in the laboratory. The specimens/cores used measure Ø100 × 200 or Ø150 × 300 mm that are moist cured or immersed in a lime-saturated water bath (the latter yields 10% lower resistivity results). Four readings are performed along four lateral lines at 90° from each other. Same as for the direct method, the surface resistivity method produces meaningful results only when applied on fully saturated concrete.

As reflected by its name ("Standard Method of Test for Surface Resistivity Indication of Concrete's Ability to Resist Chloride Ion Penetration"), AASHTO T358 is intended as a possible substitute for ASTM C1202 test. Hence, it proposes a rating of the chloride ion penetration, based on the measured surface resistivity, see two last columns in Table A.1.

The values shown in Table A.1 already have the geometric factor *K* included, so they correspond to the reading of the instrument for $K = 1$.

Incidentally, the name "surface" electrical resistivity indicates that ρ is measured within the *Covercrete*, to a depth approximately equal to the separation of the electrodes *a*.

This is a simple and fast technique that could be easily applied on site. However, the results are quite sensitive to several external factors (Spragg et al., 2013; Azarsa & Gupta, 2017), such as humidity, temperature, cracks, steel fibres and nearby steel reinforcing bars, that are not easy to control on site, making the method more appropriate for laboratory testing. In particular, achieving a high degree of saturation of the site concrete is rather difficult (Presuel-Moreno et al., 2010; Torrent et al., 2020). The air-permeability *TPT* instrument has the option of including, as accessory, a Wenner probe for moisture compensation purposes (see Section 5.7.2).

A.2.2.3 Chloride Conductivity Test

In South Africa, a durability concept, based on durability indices, was developed at the end of the 90s (Alexander et al., 1999a, b). It consists in drilling Ø68–70 mm cores from the finished structures and cutting them to a thickness of 25–30 mm, after slicing off the outermost 5 mm of the core. After specific preconditioning processes, the discs are tested under laboratory conditions to measure the Oxygen-Permeability Index (OPI), see Section 4.3.1.3, and the Chloride Conductivity Index (CCI). A third index was initially adopted, that seems to having been abandoned (water sorptivity index). The indexes obtained in this manner are used to check the expected service life of structures exposed to carbonation (OPI), see Section 9.3.2, and to chlorides (CCI). Both tests, OPI and CCI, are covered by South African standards (CO3-2, 2015; CO3-3, 2015).

In this section the Chloride Conductivity Test will be described in some detail (Streicher & Alexander, 1995; Alexander et al., 1999c; Mackechnie & Alexander, 2001; Otieno & Alexander, 2015).

The test method consists in applying a 10 V DC potential across the above-mentioned disc, vacuum-saturated with 5 M NaCl solution, placed in between two compartments, both filled with a 5 M NaCl – calcium hydroxide-saturated – solution, and measuring the generated current, see sketch (a) and picture (b) of the instrument in Figure A.5.

The Chloride Conductivity Index is calculated as

$$CCI = \frac{I}{U}\frac{e}{A} \tag{A.5}$$

where

CCI = chloride conductivity index (mS/cm); mS/cm = $(\text{k}\Omega.\text{cm})^{-1}$
I = intensity of applied current (mA)
U = measured potential (V)
e = thickness of the specimen (cm)
A = area of specimen cross-section (cm^2)

The test is claimed to provide a good assessment of concrete quality, especially concerning resistance to chloride penetration; Table A.3 presents a concrete quality rating based on the measured *CCI*.

The *CCI* is used in connection with performance-based specification of concrete in South Africa (Nganga et al., 2013; Alexander, 2004).

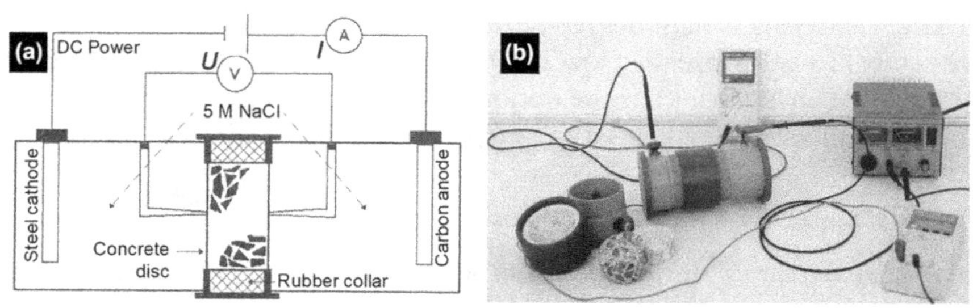

Figure A.5 (a) Sketch and (b) view of Chloride Conductivity Test (Alexander et al., 1999c).

Table A.3 Classification of concrete quality based on CCI

Concrete quality	*CCI (mS/cm)*
Very good	<0.75
Good	0.75–1.50
Poor	1.50–2.50
Very poor	>2.50

REFERENCES

AASHTO T259 (2002). "Standard method of test for resistance of concrete to chloride ion penetration".

AASHTO T277 (1983). "Standard method of test for rapid determination of the chloride permeability of concrete".

AASHTO T358 (2017). "Standard method of test for surface resistivity indication of concrete's ability to resist chloride ion penetration", 10 p.

Alexander, M.G. (2004). "Durability indexes and their use in concrete engineering". *International RILEM Symposium on Concrete Science and Engineering: A Tribute to Arnon Bentur*, 9–22.

Alexander, M.G., Ballim, Y. and Mackechnie, J.R. (1999a). "Guide to the use of durability indexes for achieving durability in concrete structures". Research Monograph n2. Univs. of Cape Town and of the Witwatersrand South Africa, 35 p.

Alexander, M.G., Ballim, Y. and Mackechnie, J.R. (1999b). "Concrete durability index testing manual". Research Monograph n4. Univs. of Cape Town and of the Witwatersrand. South Africa, 33 p.

Alexander, M.G., Streicher, P.E. and Mackechnie, J.R. (1999c). "Rapid chloride conductivity testing of concrete". Research Monograph n3. Univs. of Cape Town and of the Witwatersrand. South Africa, 35 p.

Andrade, C. (1993). "Calculation of chloride diffusion coefficients in concrete from ionic migration measurements". *Cem. & Concr. Res.*, v23, n3, 724–742.

Andrade, C. (2002). "Concepts on the chloride diffusion coefficient". *Third RILEM Workshop on Testing and Modelling the Chloride Ingress into Concrete*, Madrid, Spain, September 9–10, 15 p.

Andrade, C. and d'Andrea, R. (2008). "Electrical resistivity as microstructural parameter for the calculation of reinforcement service life". *1st International Conference on Microstructural Durability of Cementitious Composite*, Nanjing, China, October 13–15, 1483–1490.

ASTM C1152 (2004). "Standard test method for acid-soluble chloride in mortar and concrete".

ASTM C1202 (2019). "Standard test method for electrical indication of concrete's ability to resist chloride ion penetration".

ASTM C1556 (2016). "Standard test method for determining the apparent chloride diffusion coefficient of cementitious mixtures by bulk diffusion".

ASTM C1760 (2012). "Standard test method for bulk electrical conductivity of hardened concrete".

ASTM C1876 (2019). "Standard test method for bulk electrical resistivity or bulk conductivity of concrete".

Azarsa, P. and Gupta, R. (2017). "Electrical resistivity of concrete for durability evaluation: A review". *Adv. Mater. Sci. Eng.*, v2017, 30 p.

Basheer, P.A.M., Andrews, R.J., Robinson, D.J. and Long, Aww.E. (2005). "'PERMIT' ion migration test for measuring the chloride ion transport of concrete on site". *NDT & E Intern.*, v38, n3, 219–229.

CI (2006). "Concrete Q&A: Rapid chloride permeability tests". *Concr. Intern.*, August, 98–100.

CO3-2 (2015). "Civil engineering test methods. Part CO3-2: Concrete durability index testing – Oxygen permeability test". South African Standard SANS 3001-CO3-2:2015.

CO3-3 (2015). "Civil engineering test methods. Part CO3-3: Concrete durability index testing – Chloride conductivity test". South African Standard SANS 3001-CO3-3:2015.

CPC-18 (1988). "Measurement of hardened concrete carbonation depth". *RILEM Recommendation*, 3 p.

CSA (2006). "Concrete materials and methods of concrete construction/methods of test and standard practices for concrete". CAN/CSA-A23.1–04/CAN/CSA-A23.2–04, August, 526 p.

DuraCrete (2000). "Probabilistic performance based durability design of concrete structures". *The European Union–Brite EuRam III*, BE95–1347/R17, CUR, Gouda, The Netherlands.

EN 12390-12 (2020). "Testing hardened concrete – Part 12: Determination of the carbonation resistance of concrete – Accelerated carbonation method".

Ferreira, M. (2004). "Probability based durability design of concrete structures in marine environment". Doctoral Dissert., Dept. Civil Engng., Univ. of Minho, Guimarães, Portugal.

Ferreira, M., Årskog, V., Jalil, S. and Gjørv, O.E. (2004). "Software for probability-based durability analysis of concrete structures. *Proceedings on 4th CONSEC*, v1, Seoul, Korea, 1015–1024.

fib (2006). "Model code for service life design". *fib Bulletin 34*, February 2006, 112 p.

Geiker, M., Grube, H., Luping, T., Nilsson, L.-O. and Andrade, C. (1995). "Laboratory test methods". *RILEM Report 12*, Hilsdorf, H.K. and Kropp, J. (Eds.), 135–179.

ISO 1920-12 (2015). "Testing of concrete – Part 12: Determination of the carbonation resistance of concrete – Accelerated carbonation method". ISO, 12 p.

Kanada, H., Ahn, T.H. and Uomoto, T. (2006a). "Non-destructive inspection methods for componential analysis of concrete". www.kci.or.kr/wonmun/KCI_4_2006_11_933(C).pdf.

Kanada, H., Ishikawa, Y. and Uomoto, T. (2006b). "On-site elemental analysis of concrete by portable energy dispersive X-ray fluorescence analyzer". *Concr. J.*, v44, n6, June, 16–23 (in Japanese).

Kobayashi, K. and Shuttoh, K. (1991). "Oxygen diffusivity of various cementitious materials". *Cem. & Concr. Res.*, v21, n2–3, March–May, 273–284.

Leemann, A., Loser, R., Münch, B. and Lura, P. (2017). "Steady-state O_2 and CO_2 diffusion in carbonated mortars produced with blended cements". *Mater. & Struct.*, v50, 247.

Leemann, A. and Moro, F. (2017). "Carbonation of concrete: The role of CO_2 concentration, relative humidity and CO_2 buffer capacity". *Mater. & Struct.*, v50, n30, 14 p.

Li, K. (2016). *Durability Design of Concrete Structures*. 1st Ed., John Wiley & Sons, Singapore, 280 p.

Life-365 (2018). "Service life prediction model for reinforced concrete exposed to chlorides". v2.2.3, September.

Mackechnie, J.R. and Alexander, M.G. (2001). "Practical considerations for rapid chloride conductivity testing". *RILEM Proceeding PRO 19*, 451–459.

Moro, F., Hunkeler, F., Nygaard, P. and Cuchet, S. (2014). "Carbonation resistance: Accelerated determination and limiting values based on Swiss Standard method". *International Workshop on Performance-based Specification and Control of Concrete Durability*, Zagreb, Croatia, June 11–13.

Moro, F. and Torrent, R. (2016). "Testing fib prediction of durability-related properties". *fib Symposium 2016*, Cape Town, South Africa, November 21–23.

Morris, W., Moreno, E.I. and Sagüés, A.A. (1996). "Practical evaluation of resistivity of concrete in test cylinders using a Wenner array probe". *Cem. & Concr. Res.*, v26, n12, December, 1779–1787.

Ngala, V.T. and Page, C.L. (1997). "Effects of carbonation on pore structure and diffusional properties of hydrated cement pastes". *Cem. & Concr. Res.*, v27, n7, July, 995–1007.

Nganga, G., Alexander, M. and Beushausen, H. (2013). "Practical implementation of the durability index performance-based design approach". *Constr. & Build. Mater.*, v45, August, 251–261.

Nilsson, L.-O., Ngo, M.H. and Gjørv, O.E. (1998). "High-performance repair materials for concrete structures in the Port of Gothenburg". *2nd CONSEC*, v2, 1193–1198.

NT Build 443 (1995). "Accelerated chloride penetration into hardened concrete". Nordtest, Espoo, Finland.

NT Build 492 (1999). "Concrete, mortar and cement-based repair materials: Chloride migration coefficient from non-steady state migration experiments". Nordtest, Espoo, Finland, 8 p.

Otieno, M. and Alexander, M. (2015). "Chloride conductivity testing of concrete – Past and recent developments". *J. South African Inst. Civ. Eng.*, v57, n4, 55–64.

Page, C.L., Short, N.R. and El Tarras, A. (1981). "Diffusion of chloride ions in hardened cement pastes". *Cem. & Concr. Res.*, v11, n3, May, 395–406.

Presuel-Moreno, F.J., Suares, A. and Liu, Y. (2010). "Characterization of new and old concrete structures using surface resistivity measurements". FDOT, Final Report, August 1, 279 p.

Ren, F., Li, L., Wang, W. and Zhou, C. (2021). "Transport properties of surface layers from full-scale reinforced concrete member". Struct. Concr., v22 (Suppl. 1), E1062–E1073.

Rupnow, T. and Icenogle, P. (2011). "Evaluation of surface resistivity measurements as an alternative to the rapid chloride permeability test for quality assurance and acceptance". FHWA/LA.11/479 Report, July, 68 p.

Schwiete, H.E., Bohme, H.J. and Ludwig, U. (1969). "Measuring gas diffusion for the valuation of open porosity on mortars and concretes". *Matériaux et Constr.*, v2, n1, January, 43–48.

Shi, C. (2003). "Another look at the Rapid Chloride Permeability Test (ASTM C1202 or AASHTO T277)". FHWA Resource center, Baltimore, MD, 15 p.

SIA 262/1 (2019). "Concrete structures – Supplementary specifications" Swiss Standard (in French and German).

Spragg, R., Bu, Y., Snyder, K., Bentz, D. and Weiss, J. (2013). "Electrical testing of cement-based materials: Role of testing techniques, sample conditioning, and accelerated curing". FHWA/IN/JTRP-2013/28 Report, Joint Transportation Research Program, Purdue Univ., November, 28 p.

Stanish, K.D., Hooton, R.D. and Thomas, M.D.A. (1997). "Testing the chloride penetration resistance of concrete: A literature review". Univ. of Toronto, FHWA Contract DTFH61–97-R-00022, 33 p.

Streicher, P.E. and Alexander, M.G. (1995). "A chloride conduction test for concrete". *Cem. & Concr. Res.*, v25, n6, August, 1284–1294.

Tang, L. (1996). "Chloride transport in concrete – Measurement and prediction". PhD thesis, Chalmers Univ. of Technology, Sweden, October, 104 p.

Tang, L. and Nilsson, L.-O. (1993). "Rapid determination of the chloride diffusivity in concrete by applying an electric field". *ACI Mater. J.*, v89, n1, 40–53.

Tang, L., Nilsson, L.-O. and Basheer, M. (2012). *Resistance of Concrete to Chloride Ingress. Testing and Modelling.* Spon Press, Abingdon, VA, 246 p.

Tang, L. and Sørensen, H.E. (2001). "Precision of the Nordic test methods for measuring the chloride diffusion/migration coefficients of concrete". *Mater. & Struct.*, v34, n8, 479–485.

TC-154-EMC (2000). "Electrochemical techniques for measuring metallic corrosion". RILEM Recommendation, *Mater. & Struct.*, v33, December, 603–611.

Torrent, R. and Ebensperger, L. (1993). "Studie über Methoden zur Messung und Beurteilung der Kennwerte des Überdeckungsbetons auf der Baustelle". Office Fédéral des Routes, Rapport No. 506, Bern, Suisse, Januar 1993, 119 p.

Torrent, R., Imamoto, K. and Neves, R. (2020). "Resistance of concrete to carbonation and chloride penetration assessed on site through nondestructive test". *Struct. Concrete*, DOI: 10.1002/suco.201900474, September.

Torrent, R. and Jacobs, F. (2014). "Swiss Standards 2013: World's most advanced durability performance specifications". *3rd Russian International Conference on Concrete and Ferrocement*, Moscow, May 12–15, 8 p.

UNE 83988-2 (2014). "Durabilidad del hormigón – Métodos de Ensayo - Determinación de la resistividad eléctrica – Parte 2: Método de las cuatro puntas o de Wenner".

Vilani, C., Loser, R., West, M.J., Di Bella, C., Lura, P. and Weiss, W.J. (2014). "An inter lab comparison of gas transport testing procedures: Oxygen permeability and oxygen diffusivity". *Cem. & Concr. Composites*, v53, 357–366.

Whiting, D. (1981). "Rapid determination of the chloride permeability of concrete". Report No. FHWA/RD-81/119, August, NTIS DB No. 82140724.

Whiting, D. (1984). "In situ measurement of the permeability of concrete to chloride ions". *ACI SP 82*, 501–524.

Wong, H.S. (2006). "Quantifying the pore structure of cement-based materials using backscattered electron microscopy". PhD Thesis, Univ. of London and Imperial College London, 272 p.

Annex B: Model standard for measuring the coefficient of air-permeability *kT* of hardened concrete

This document is prepared in a format suitable for presentation before National or International Standards' Organizations. The operational aspects of the test, common to laboratory and field applications, are based on those corresponding to Annex E of Swiss Standard SIA 262/1:2019 and Argentine Standard IRAM 1892:2021; the sampling procedures for site testing as well as the conformity assessment with specified values are also taken from those standards. It also serves as guidance for the correct application of the test method so as to get representative and meaningful results.

B.1 OBJECT AND SCOPE

The object of this model standard is to describe a non-destructive test method to measure the coefficient of air-permeability *kT* of hardened concrete, applying the double-chamber vacuum cell technique.

The method is applicable:

- in the laboratory on concrete specimens
- *in situ* on finished concrete elements of new structures
- *in situ* on finished concrete elements of existing structures

The degree of water-saturation of the concrete pores affects the air-flow and, hence, the result of the test, reason why the surface moisture content of the concrete to be tested shall be measured to check that it is sufficiently low to obtain *kT* values that are representative of the true quality of the concrete.

B.2 REFERENCED DOCUMENTS

Standards:

- *ASTM F2659*: "Standard Guide for Preliminary Evaluation of Comparative Moisture Condition of Concrete, Gypsum Cement and Other Floor Slabs and Screeds Using a Non-Destructive Electronic Moisture Meter", 6 p.
- *IRAM 1892*: "Hormigón – Método de ensayo para la determinación del coeficiente de permeabilidad al aire (*kT*) del hormigón endurecido". Argentine Standard, 2021
- *ISO 1920–3*: "Testing of concrete – Part 3: Making and curing test specimens"
- *SIA 262/1*: "Construction en béton. Spécifications complémentaires". Annex E: 'Perméabilité à l'air dans les structures'. Swiss Standard, 1 March 2019, 60 p. (also in German) Partial English translation at: http://www.m-a-s.com.ar/pdf/Translation%20of%20SIA%20262-1-2013%20referred%20to%20kT%20v1.1.pdf

Other:

- Jacobs, F., Denariè, E., Leemann, A. und Teruzzi T., "Empfehlungen zur Qualitätskontrolle von Beton mit Luftpermeabilitätsmessungen". Bundesamt für Strassenbau, VSS Report 641, December 2009, Bern, Switzerland
- Torrent, R., Neves, R. and Imamoto, K. (2022). *Concrete Permeability and Durability Performance – From Theory to Field Applications.* CRC Press, Boca Raton, FL, USA and Abingdon, Oxon, UK, 570 p.
- User Manuals of: *Torrent Permeability Tester*, *PermeaTORR*, *PermeaTORR AC*, *PermeaTORR AC+*

B.3 SIGNIFICANCE AND USE

The durability performance of concrete structures subjected to various aggressive environments depends, to a large extent, on the "penetrability" of the concrete pore system.

The present knowledge on suitable test methods and correlations between the individual transport parameters justifies the extensive use of gas-permeability measurements to characterize concrete transport properties. Although not directly linked to degradation, this parameter offers close correlations with the diffusion coefficient of gases, the diffusion of aggressive ions in the liquid phase, the rate of water absorption as well as with the permeability to water or diluted solutions in concrete. Therefore, this single parameter may characterize the "penetrability" of concrete in a variety of different cases, thereby covering various corrosion and deterioration mechanisms.

The coefficient of air-permeability of a concrete surface depends on many factors including:

a. concrete mixture proportions
b. chemical composition and physical characteristics of the binder
c. characteristics of the aggregate and of the aggregate-paste inter-facial transition zone
d. placement method including consolidation and finishing
e. the type and duration of curing
f. the degree of hydration or age
g. the presence of microcracks
h. the presence of surface treatments

The air-permeability is also strongly affected by the moisture condition of the concrete at the time of testing.

The coefficient of air-permeability *kT* is, therefore, a relevant durability indicator of the concrete tested (Torrent et al., 2012; Torrent & Ebensperger, 2013). The background and relevance of the test method are described in detail in Chapters 5 and 8 of Torrent et al. (2022).

Moreover, in combination with the thickness of the cover over the reinforcement, it can provide an estimate of the initiation time of steel corrosion (closely linked to the service life) by applying several models (see Chapter 9 of Torrent et al. (2022)):

- the "Exp-Ref" method for steel corrosion of new structures, induced by chlorides (Torrent, 2013, 2015) or by carbonation (Torrent & Fernández Luco, 2014)
- two methods for carbonation-induced steel corrosion of new structures (Kurashige & Hironaga, 2015; Belgacem et al., 2020)
- two methods for carbonation-induced steel corrosion of old structures (Imamoto et al., 2013; Neves et al., 2018)

B.4 PRINCIPLES OF THE TEST METHOD

The principles of the test method are:

a. A vacuum cell, composed of two concentric chambers, is applied on the concrete surface to be tested (Figure 5.2). A vacuum in the two chambers is created by a vacuum pump, the external atmospheric pressure pressing the cell onto the concrete surface and sealing the two chambers by means of two soft concentric rubber rings (Figure 5.3)

b. After 60 seconds, the central, test chamber (then at a pressure typically of 0–50 mbar) is isolated from the pump, moment at which its pressure starts to rise due to the air (at atmospheric pressure ≈ 1,000 mbar) in the concrete pores flowing into the chamber through the concrete (Figure 5.2)
c. The pump continues to run, operating exclusively on the external chamber, extracting just the necessary amount of air for its pressure P_e to balance the pressure of the central chamber P_i (this is achieved by a high-precision pressure regulator); so that, at any time, $P_e = P_i$
d. The rate of increase in pressure in the central chamber during the test, that is higher the more permeable the concrete tested, is recorded, allowing the calculation of the coefficient of air-permeability *kT*, by applying Eq. (B.1)
e. After storing the results in the instrument's memory, a venting valve is opened to restore the whole system to atmospheric pressure, moment in which the cell can be detached from the concrete surface, ready for a new test
f. The characteristic feature of the test method is the presence of the external chamber which, keeping the same pressure as the central chamber, acts as a guard-ring. The resulting controlled flow of air into the central chamber, that can be assumed to be unidirectional (see Figure 5.2), allows the derivation of Eq. (B.1)

B.5 APPARATUS

B.5.1 Permeability Tester

The equipment for testing air-permeability basically consists of:

- a two-chamber vacuum cell composed of two concentric chambers: internal (test chamber) and external (guard-ring) and soft elastomeric rings that divide the chambers and seal the cell onto the concrete surface when under vacuum
- a control system, consisting of valves, pressure sensors and a pressure regulator capable of keeping the pressure of both vacuum chambers permanently balanced within the tolerance $|P_e - P_i| \leq 5$ mbar (P_e = pressure in the external chamber; P_i = pressure in the internal chamber);
- a vacuum pump capable of reaching a pressure of 15 mbar or less when applied on an impermeable surface

The vacuum pump may be embedded in the instrument or may consist of a separate unit connected to the instrument.

The instrument shall carry an impermeable plate (made of metal or polycarbonate) for conditioning and calibration purposes.

B.5.2 Surface Moisture Meter

A device, based on the electrical impedance principle, capable of measuring non-destructively the surface moisture of the concrete within the range of 0.0%–6.9% with an accuracy of ±0.1%, according to ASTM F2659 Standard.

B.5.3 Abrasive Stone or Angle Grinder

In most cases, the test can be done by applying the vacuum cell directly on the natural concrete surface (previously cleaned of dust). In some cases, it might be necessary to polish the concrete manually, using an abrasive stone. For extreme cases of irregular surfaces (e.g. shotcrete), it may be necessary to apply an angle grinder machine (Bosch GWS 6–115 Professional or similar).

B.5.4 Brush, Dry Sponge or Vacuum Cleaner

To remove the dust that may have accumulated on the concrete surface. In certain occasions, especially when testing dirty horizontal surfaces like floors, it may be convenient to use a domestic vacuum cleaner.

B.5.5 Cover Meter (for Reinforced Elements)

To identify the position of steel bars and estimate their cover thickness. The cover thickness provides an essential input to models to assess the service life of the structures in risk of steel corrosion induced by carbonation or chlorides.

The cover meter may operate by creating a magnetic field and monitoring the change of magnetic reluctance or on the generation of "eddy currents" due to the vicinity and size of the steel bars. Alternatively, it can be based on the electromagnetic detection of reflecting microwaves generated by the instrument, better known as Ground Penetrating Radar.

B.5.6 Laboratory Ventilated Oven (for Laboratory Testing)

A sufficiently large oven allowing for air circulation and capable of maintaining a temperature of 50°C±2°C, used to precondition the laboratory specimens' moisture prior to the air-permeability test.

B.5.7 Thermometer

To measure the air and concrete temperature within the range 0°C–50°C with an accuracy of 0.1°C. Some instruments measure the temperature of the air inside the central chamber.

B.5.8 Visual Aids

Accessories to help identifying defects (e.g. microcracks) on the concrete surface that may cause anomalous results, such as magnifying lenses, lamps, isopropyl alcohol for spraying, crack-width gauges, etc.

B.6 TEST OBJECTS AND TESTING CONDITIONS

B.6.1 Type of Samples

Both for laboratory and site testing, the maximum size of the aggregate in the concrete shall not exceed 75% of the diameter of the central chamber (i.e. 38 mm or 1½ in for the standard cell).

Concretes made intentionally with very high porosity, such as cellular or pervious concrete, are not suitable for testing with this method.

B.6.1.1 Laboratory Tests

The surface of the specimen to be tested shall have a minimum dimension of 150 mm with a minimum thickness of 50 mm.

The test can be performed on concrete cylinders, cubes or prisms. The preparation of the specimens shall be made according to ISO 1920–3 or equivalent.

> *Note 1*: Cubes or prisms (150 mm or larger) are ideal specimens, as they present four nominally identical surfaces for testing *kT*.

B.6.1.2 Site Tests

The vacuum cell shall be placed at a distance of at least 50 mm from the nearest edge of the element and of at least 200 mm from the nearest location of other tests (distance measured from the edge of the vacuum cell).

B.6.1.3 Special Cases

Testing specimens of 100 mm minimum size or elements of pronounced curvature require the use of special cells or adaptors, see Section 5.7.5.2 and Figure 5.25 of Torrent et al (2022). Consideration shall be given to eventual changes in the V_c/A ratio of Eq. (B.1) when interposing an adaptor.

B.6.2 Curing and Age of Testing

B.6.2.1 Laboratory Tests

The specimens shall be moist cured according to ISO 1920–3 or equivalent for 28 days and tested immediately after completing the preconditioning process described in Section B.8.4.

Note 2: Shorter periods of moist curing can be established if the effect of lack of curing wants to be investigated; in that case, it is recommended to store the specimens, after the moist curing period, in a room with RH=75% or less. Experience indicates that little is gained in air-permeability by extending the moist curing beyond 28 days.

B.6.2.2 Site Tests on New Structures

For new structures, an age of testing between 28 and 180 days is recommended, with at least 7 days after the end of moist curing.

B.6.2.3 Site Tests on Structures in Service

The method is applicable for condition assessment and potential residual service life of structures in service, for which there are no limits on their age.

Note 3: The effect of ageing and service loads may have an impact on the test results. For this application, neither the sampling (B.7.2) nor the evaluation of results (B.10.2) is mandatory.

B.6.3 Temperature and Moisture Conditions

B.6.3.1 Laboratory Tests

After curing and before testing, the specimens shall be dried in the ventilated oven at a temperature of 50°C±2°C, leaving a free distance of at least 20 mm between specimens and with the walls of the oven. The drying shall continue until the moisture meter indicates a surface moisture of the specimens within the range of 4.0%–5.5% (normally this is achieved within 4±2 days of drying for specimens originally at or near saturation).

Note 4: Experience indicates that the moisture can be measured when the specimens are still at 50°C temperature, without significant difference to room temperature.The room and concrete temperature during the test shall be within the range of 5°C–50°C.

B.6.3.2 Site Tests

The air and concrete temperature at the moment of test shall be within the range of 5°C–50°C. In case of intense solar radiation, both the tested surface and the instruments shall be protected with umbrellas or canopies.

The surface moisture of the concrete at the moment of test shall not exceed 5.5% (see B.8.3).

Note 5: The required moisture conditions are usually reached after 2–3 consecutive days with RH<80% of the last contact of the element with water (rain, splash, etc.).

B.7 SAMPLING

B.7.1 Laboratory Tests

To characterize a given concrete, at least four air-permeability measurements shall be performed on at least two different companion specimens. This can be achieved, for instance, by testing the opposite lateral faces of two cubes or one plane face of four cylinders (testing always the same face, top or bottom surface as cast, the latter being preferable).

Note 6: The test surface (finished, lateral or bottom of the mould) may have an influence on the permeability (effects of compaction, settlement, bleeding). Therefore, it is advisable to test always the same surface of the specimens.

B.7.2 Site Tests

The following procedure shall be followed:

a. organize the structural elements in Groups, representing:
   - same concrete type, i.e. concrete mixes that belong to the same consistency class, same strength class, subjected to the same exposure class and produced with the same constituents
   - same nominal concreting practices (placing, consolidation, finishing, curing, etc.)
   - if specified, have the same limiting value kT_s (see B.10.2)

b. within each Group, arrange the elements chronologically

c. define Lots within each Group. Each Lot will be subjected to an individual conformity decision and involves the smallest of the following surface areas:
   - 500 m^2 of exposed surface
   - three concreting days

d. within each resulting lot, select randomly six measurement points, respecting what was stipulated in Section B.6.1.2

B.8 TESTING PROCEDURE

B.8.1 Inspection of the Concrete Surface

Prior to initiating a test, the surface to be tested shall be visually inspected to check for the absence of sharp irregularities, grooves, micro- or macro-cracks, bug-holes or other defects that may influence the measurements and invalidate the test result. Use can be made of the elements described in Section B.5.8.

B.8.2 Condition of the Concrete Surface

If the irregularities are such that prevent attaining sufficiently low initial vacuum (P_i < 100 mbar), a manual or mechanical polishing of the area to be tested is required, using the elements indicated in Section B.5.3.

The surface shall be free of dust, oil, grease, paint or any other substance that may affect the instrument or the test result. If the test has to be performed on a coated area, the situation shall be described in the test report.

Remove loose dust from the surface to be tested as described in Section B.5.4 before applying the vacuum cell.

B.8.3 Determination of Surface Moisture

At each measurement point, two readings of the surface moisture meter (see B.5.2) shall be performed, along approximately perpendicular directions and at about 45° of the determined or expected alignment of the steel bars. Make sure that all the instrument electrodes are firmly pressed onto the concrete surface during the readings.

The result is the arithmetic mean of both readings and shall be recorded as part of the testing protocol.

B.8.4 Conditioning and Calibration of the Instrument

Before starting the measurements, the instrument – with the vacuum cell placed on the supplied calibration plate – must be conditioned (20 minutes evacuation of both chambers' pneumatic systems). This is done in order to extract moisture, volatiles, etc., that may have accumulated inside the pneumatic system.

Immediately after completing the conditioning, the instrument – with the vacuum cell always placed on the supplied calibration plate – shall undergo two successive calibrations. The data of pressure rise with time, from the last calibration, are stored in the instrument's memory. For the calculation of the coefficient of air-permeability (Eq. B.1), the effective pressure rise is used, which is the difference between the pressure rise measured during the test and that recorded during the calibration at the same time.

The calibration is valid if the maximum pressure rise recorded after the second calibration does not exceed 5.0 mbar and does not differ by more than ±0.5 mbar from that of the previous calibration. If these conditions are not met, further calibration(s) shall be conducted until their achievement for two successive calibrations (Note 7).

> *Note 7*: Usually two calibrations are sufficient to meet the requirements, eventually three. For laboratory testing, conducted under reasonably stable temperature conditions, just one complete conditioning + calibration process is recommended before starting the tests. In the field, it

is recommended to apply this procedure twice a day, once just before starting the measurements and, again, around noon.

Note 8: An accurate calibration is especially important for concretes of low-permeability, where a small change in the calibration pressure has a strong effect on the relatively small pressure rise due to the air coming from the sample. The calibration changes with temperature, hence the need to calibrate twice a day.

The maximum pressure rise during the calibrations shall be recorded as part of the testing protocol.

B.8.5 Air-Permeability Test

Just one measurement shall be performed at each measurement point, defined in B.6.1.1 and B.6.1.2. If the central chamber lies in coincidence with the location of steel bars, make sure that the cover thickness is of at least 20 mm.

The formula to calculate the coefficient of air-permeability *kT* (Eq. B.1) assumes that, initially, all the pores in the concrete contain air at atmospheric pressure. Therefore, it is required to observe a waiting period of at least 20 minutes between successive measurements at the same point.

Note 9: It is recommended to draw a circle around the vacuum cell with pencil, chalk or marker, in case a repetition of the test is required at the same location (see B.6.1.2 and B.10.2).

B.9 CALCULATIONS

The coefficient of permeability to air *kT* of hardened concrete is expressed in m² and is calculated with Eq. (B.1).

$$kT = \left(\frac{V_c}{A}\right)^2 \cdot \frac{\mu}{2 \cdot \varepsilon \cdot P_a} \cdot \left(\frac{\ln\left(\frac{P_a + \Delta P}{P_a - \Delta P}\right)}{\sqrt{t_f} - \sqrt{t_0}}\right)^2 \tag{B.1}$$

where

V_c = volume of the inner test chamber pneumatic system (m³)
A = area of the inner test chamber (m²)
μ = dynamic viscosity of air (N.s/m²)
ε = open porosity of the concrete (-) which by default is taken as 0.15
P_a = atmospheric pressure (N/m²)

ΔP = increase of effective pressure in the inner chamber between time t_0 and t_f (N/m^2)
t_0 = time from which the increase in pressure of the central chamber is measured (60 seconds)
t_f = time at which the test is finished (s)

If the calculated value of kT is below 0.001×10^{-16} m^2, it should be reported as "$kT < 0.001 \times 10^{-16}$ m^2" as it is out of the accuracy limit of the instrument.

The maximum penetration depth L (m) of the atmospheric pressure front is calculated with Eq. (B.2).

$$L = \sqrt{\frac{2 \cdot kT \cdot P_a \cdot t_f}{\varepsilon \cdot \mu}} \tag{B.2}$$

Both kT and L are indicated by the instrument at the end of the test, together with other relevant data.

If the penetration of the test L exceeds the thickness of the element e, a correction is required applying Eqs. (5.29) and (5.30) of Torrent et al (2022). Some instruments make the correction automatically after entering the value of e.

> *Note 10*: The coefficient of air-permeability (kT) is usually expressed in 10^{-16} m^2 units and ranges between 0.001×10^{-16} and 100×10^{-16} m^2.
>
> *Note 11*: For the case of application of the test method on mortar or cement paste, it is recommended to use in Eqs. (B.1) and (B.2) the porosity ε measured experimentally. If not available, it is recommended to adopt default values of 0.25 for mortar and of 0.55 for cement paste.
>
> *Note 12* : Depending on the kT of the tested concrete, t_f ranges between 75 seconds and a maximum of 720 seconds, maximum that some instruments reduce optionally to 360 seconds.
>
> *Note 13*: At the time of publication, all available commercial instruments compute kT and L according to Eqs. (B.1) and (B.2) and report their values.

B.10 EVALUATION OF TEST RESULTS

B.10.1 Laboratory Tests

Based on the assumption that the distribution of test results is well represented by a log-normal distribution (see Section 5.8 of Torrent et al (2022)), the central value of a series of measurements conducted on companion specimens of the "same" concrete, as described in Section B.7.1, is the geometric mean kT_{gm} of the four or more individual results, which is the n-root of the product of the n individual results kT_i obtained:

$$kT_{\text{gm}} = \left(\prod_{i=1}^{n} kT_i \right)^{\frac{1}{n}} \tag{B.3}$$

The scatter of the n individual kT_i results is represented by the s_{LOG} value, which is the standard deviation of the $\log_{10}$ of the individual kT_i results:

$$s_{\text{LOG}} = \sqrt{\frac{\sum_i^N \left(kT_i - kT_{\text{gm}}\right)^2}{n-1}} \tag{B.4}$$

Using the properties of the log-normal distribution, it is possible to compute a characteristic value of the air-permeability kT_k as

$$kT_k = 10^{\left(\text{loglog}\ kT_{\text{gm}}\ +\ z.\ s_{\text{LOG}}\ \right)} = kT_{\text{gm}} \times 10^{\left(z.\ s_{\text{LOG}}\right)} \tag{B.5}$$

where z is the argument of the normal distribution for a given probability p of finding kT values lower than kT_k in the population.

When the test is used for quality control purposes, it shall be indicated whether kT_{gm} or kT_k shall comply with the specified maximum requirement; in the latter case, a value of $z = 1$ is recommended.

Table 5.2 of Torrent et al (2022) provides a classification of air-permeability from "Negligible" to "Ultra High", as function of the kT values measured.

B.10.2 Site Tests

In the case of site tests, the following conformity criterion should be applied to the six individual kT_i results obtained as described in Section B.7.2, with respect to the specified kT_s limiting value:

a. Not more than one kT_i result out of the six tests shall exceed kT_s
b. If just two out of the six test results exceed kT_s, a new set of six tests is performed at different random places within the same lot. From this new set of six results, not more than one kT_i result shall exceed kT_s

If either condition (a) or (b) is complied with, the lot is considered in conformity with the specified kT_s value. If none of conditions (a) or (b) is complied with, the lot is considered not in conformity with the specified kT_s value, requiring correcting, remedial or even compensation actions.

Table 8.4 of Torrent et al (2022) presents the limiting values kT_s recommended by Swiss Standard SIA 262/1 as function of the exposure class.

The operating characteristic (O-C) curve of the above-mentioned conformity criterion is presented in Figure 8.20 of Torrent et al (2022), which provides a clear meaning of the kT_s value. The O-C curve carries in abscissas the proportion of the concrete surface in a lot with kT higher than the specified value kT_s, i.e. the proportion of "defective" concrete. In ordinates, the chart presents the probability of accepting a lot (applying conformity rules 1 and 2) containing a given proportion of "defectives".

The O-C curve indicates that a lot containing 10% of "defectives" will have a 97% probability of being accepted, whilst for one with 50% "defectives", the probability drops to just 13%.

Eq. (B.3 and B.4) can be used to evaluate a series of kT measurement on site, both for new and for existing structures.

B.11 REPORT

The report of a series of measurements should include the following information:

- description of jobsite/project and ordering person/body
- dates of concreting and of testing → age at testing
- mix design code or details (binder content, *w/c* ratio, strength class, etc.)
- exposure condition of the elements (e.g. EN XC4 or ACI C1)
- specified air-permeability kT_s limiting value, if any
- ambient conditions (temperature and relative humidity). For site tests, weather conditions on the day of test and on preceding 2 days
- maximum pressure rise during first and second calibration (ΔPc_1 and ΔPc_2); for site tests, in the morning and around noon
- data on the operator/s and instruments used
- for each test:
  instrument's sequence test (for easy identification of data in its memory)
  lot and specimens or elements investigated
  location of measurement point
  if measured, cover thickness at the measurement point
  temperature of the concrete
  surface moisture of concrete
  reported anomalies (cracks, bug-holes, coatings, need of polishing, etc.)
  individual test results kT_i and penetration depths L (indicating eventual thickness correction)
- Compliance of the lots with specified *kTs* limits
- geometric mean kT_{gm} and s_{LOG} of the individual test results. Optionally, graphical presentation of results as described in Section 5.8.2

ACKNOWLEDGEMENTS

The authors are indebted to the members of the IRAM Standards Committee for their contribution to the development of Standard IRAM 1892, on which this Annex is based, namely: Alejandra Benítez, Eduardo Alberto Castelli, María Carolina Domizio, María Emilia Ferreras, Sebastián Laprida, Vanesa Tamara Lupori, Noemí Graciela Maldonado, Gerardo Andrés Martínez, Carlos Alberto Milanesi, Ariel Muñoz Baltar, Matías Daniel Polzinetti, María Josefina Positieri, Bárbara Belén Raggiotti, Matías Agustín Rodríguez, Verónica Roncoroni and Mónica Inés Suárez.

REFERENCES

Belgacem, M.E., Neves, R. and Talah, A. (2020). "Service life design for carbonation-induced corrosion based on air-permeability requirements". *Constr. & Build. Mater.*, v261, 120507, 10 p.

Imamoto, K., Tanaka, A. and Kanematsu, M. (2013). "Non-destructive assessment of concrete durability of the National Museum of Western Art in Japan". *CONSEC'13*, Nanjing, China, September 23–25, 1335–1344.

Kurashige, I. and Hironaga, M. (2015). "Criteria for nondestructive inspections using Torrent air permeability test for lifespan of concrete structures". *Life-Cycle of Structural Systems.* Furuta, Frangopol and Akiyama (Eds.). Taylor & Francis, London, 1221–1228.

Neves, R., Torrent, R. and Imamoto, K. (2018). "Residual service life of carbonated structures based on site non-destructive tests". *Cem. & Concr. Res.*, v109, 10–18.

Torrent, R. (2013). "Service life prediction: Theorecrete, Labcrete and Realcrete approaches". Keynote Paper, *SCTM3 Conference*, Kyoto, Japan, August 18–21.

Torrent, R. (2015). "Exp-Ref: A simple, realistic and robust method to assess service life of reinforced concrete structures". *Concrete* 2015, August 30–September 2, Melbourne, Australia.

Torrent, R., Denarié, E., Jacobs, F., Leemann, A. and Teruzzi, T. (2012). "Specification and site control of the permeability of the cover concrete: The Swiss approach". *Mater. & Corrosion*, v63, n12, December, 1127–1133.

Torrent, R.J. and Ebensperger, L. (2013). "Site air-permeability test: Credentials as 'durability meter'". *CONSEC'13*, Nanjing, China, September 23–25, 1345–1360.

Torrent, R. and Fernández Luco, L. (2014). "Service life assessment of concrete structures based on site testing". *DBMC* 2014 *Conference*, São Paulo, Brazil, September 1–5.

Torrent, R., Neves, R. and Imamoto, K. (2022). *Concrete Permeability and Durability Performance – From Microstructure to Field Applications.* CRC Press, Boca Raton, FL, USA and Abingdon, Oxon, UK, 570 p.

Index

Note: **Bold** page numbers refer to significant discussions.

CPSIA information can be obtained
at www.ICGtesting.com
Printed in the USA
LVHW082053250422
717165LV00004B/148

9 781138 584884